Transport Planning and Traffic Engineering

Edited by

CA O'Flaherty

Contributing authors

MGH Bell, BA (*Cantab*), MSc, PhD (*Leeds*), FIHT
PW Bonsall, BA (*Oxon.*), DipTRP, MCIT
GR Leake, BSc, MSc (*Birmingham*), DipTE, CEng, MICE, MIHT
AD May, MA (*Cantab*), FEng, FICE, MIHT, FCIT
CA Nash, BA (Econ), PhD (*Leeds*), MCIT
CA O'Flaherty, BE (*NUI*), MS PhD (*Iowa State*), Hon LLD (*Tasmania*),
CEng, FICE, FICEI, FIEAust, FCIT, FIHT

A member of the Hodder Headline Group
LONDON • SYDNEY • AUCKLAND
Copublished in North, Central and South America by
John Wiley & Sons, Inc., New York • Toronto

First published in Great Britain in 1997 by
Arnold a member of the Hodder Headline Group,
338 Euston Road, London, NW1 3BH

Copublished in North, Central and South America by
John Wiley & Sons, Inc., 605 Third Avenue,
New York, NY 10158-0012

British Library Cataloguing in Publication Data
A catalogue record for this book is available from the British Library

Library of Congress Cataloging-in-Publication Data
A catalog record for this book is available from the Library of Congress

ISBN: 0 340 66279 4
ISBN: 0 470 23619 1 (Wiley)

Typeset by GreenGate Publishing Services, Tonbridge, Kent
Printed by J.W. Arrowsmith Ltd., Bristol

Contents

About the contributors   **xiii**

Preface   **xv**

Acknowledgements   **xvi**

Part I: Planning for transport

Chapter 1 Evolution of the transport task   **2**
C.A. O'Flaherty

1.1   The road in history   2
1.2   Railways, bicycles and motor vehicles   5
1.3   Some changes associated with the motor vehicle   10
1.4   Britain's road network   18
1.5   A final comment   19
1.6   References   20

Chapter 2 Transport administration and planning   **21**
C.A. O'Flaherty

2.1   Transport administration in Great Britain   21
2.2   The statutory land use planning process   26
2.3   Finance   27
2.4   Some transport planning considerations   28
2.5   References   41

Chapter 3 Transport policy   **42**
A.D. May

3.1   Introduction   42
3.2   A logical approach to transport policy formulation   42
3.3   Problem-oriented planning and the objectives-led approach 44
3.4   Types of objective   45
3.5   A possible set of objectives   46
3.6   Quantified objectives and targets   51
3.7   Problem identification   55
3.8   The instruments of transport policy   56
3.9   Infrastructure measures   57
3.10   Management measures   61
3.11   Information provision   67
3.12   Pricing measures   69

3.13   Land use measures   72
3.14   Integration of policy measures   74
3.15   References   76

Chapter 4 Economic and environmental appraisal of transport improvement projects   80
         C.A. Nash

4.1   Economic efficiency   81
4.2   Economic efficiency and markets   82
4.3   Valuing costs and benefits   84
4.4   Valuing environmental effects   89
4.5   Equity considerations   93
4.6   Economic regeneration considerations   94
4.7   Budget constraints   95
4.8   Appraisal criteria   95
4.9   Appraisal of pricing policies   98
4.10   Public transport appraisal   99
4.11   Final comment   100
4.12   References   101

Chapter 5 Principles of transport analysis and forecasting   103
         P.W. Bonsall

5.1   The role of models in the planning process   103
5.2   Desirable features of a model   104
5.3   Specification, calibration and validation   105
5.4   Fundamental concepts   106
5.5   Selecting a model   108
5.6   Classes of model available to the transport analyst   110
5.7   Transport modelling in practice   128
5.8   References   131

Chapter 6 Transport planning strategies   132
         C.A. O'Flaherty

6.1   Do-minimum approach   133
6.2   The land use planning approach   138
6.3   The car-oriented approach   140
6.4   The public transport-oriented approach   141
6.5   The demand management approach   147
6.6   Transport packaging   151
6.7   References   153

Chapter 7 Developing the parking plan   154
         C.A. O'Flaherty

7.1   Parking policy – a brief overview   154
7.2   Planning for town centre parking – the map approach   155

7.3   Park-and-ride   166
7.4   References   169

Chapter 8 Planning for pedestrians, cyclists and disabled people   170

   G.R. Leake

8.1   Introduction   170
8.2   Identifying the needs of pedestrians, cyclists and disabled people   171
8.3   Identifying priorities of need   173
8.4   Pedestrian and cyclist characteristics and requirements influencing design   174
8.5   Special needs of elderly and disabled people   177
8.6   References   179

Chapter 9 Technologies for urban, inter-urban and rural passenger transport systems   181

   G.R. Leake

9.1   Introduction   181
9.2   Role of passenger transport systems in urban and non-urban areas   182
9.3   Desired characteristics of public transport systems   186
9.4   Urban, inter-urban and rural technologies   187
9.5   Final comment   199
9.6   References   199

Chapter 10 Planning for public transport   201

   C.A. Nash

10.1   Appropriate public transport modes   201
10.2   Commercial services   202
10.3   Subsidised services   205
10.4   Socially optimal pricing and service levels in public transport   207
10.5   Public transport provision in practice   209
10.6   Ownership and regulation   210
10.7   Conclusions   212
10.8   References   212
10.9   Appendix: Alternative objectives for public transport   213

Chapter 11 Freight transport planning – an introduction   214

   C.A. Nash

11.1   Trends in freight transport   214
11.2   Roads and economic growth   215
11.3   Policy issues   216
11.4   Potential for rail and water   218
11.5   Conclusions   219
11.6   References   220

Part II: Traffic surveys and accident investigations

Chapter 12 Issues in survey planning and design **222**
P.W. Bonsall

12.1	Defining the data requirements	223
12.2	Secondary sources	223
12.3	Choice of survey instrument	225
12.4	Design of sampling strategy	225
12.5	The survey plan	228
12.6	Cross-sectional and time series surveys	230
12.7	Training and motivation of staff	230
12.8	Administration	231
12.9	References	231

Chapter 13 Observational traffic surveys **232**
P.W. Bonsall and C.A. O'Flaherty

13.1	Inventory and condition surveys	232
13.2	Vehicle flow surveys	234
13.3	Vehicle weight surveys	238
13.4	Spot speed surveys	239
13.5	Journey speed, travel time and delay surveys	241
13.6	Origin–destination cordon and screenline surveys	244
13.7	Parking use surveys	245
13.8	Surveys of pedestrians, cyclists and public transport use	247
13.9	Environmental impact surveys	248
13.10	References	250

Chapter 14 Participatory transport surveys **252**
P.W. Bonsall and C.A. O'Flaherty

14.1	Group discussion	252
14.2	Household interview surveys	252
14.3	Trip end surveys	255
14.4	En-route surveys	255
14.5	Public transport user surveys	257
14.6	Attitudinal surveys	257
14.7	References	259

Chapter 15 Accident prevention, investigation and reduction **261**
C.A. O'Flaherty

15.1	Traffic accident terminology	261
15.2	Accident prevention	261
15.3	Accident investigation and reduction	262
15.4	References	269

Part III: Design for capacity and safety

Chapter 16 Introduction to traffic flow theory   272

A.D. May

16.1   Introduction   272
16.2   The principal parameters   272
16.3   The fundamental relationship   275
16.4   References   280

Chapter 17 Road capacity and design-standard approaches
to road design   281

C.A. O'Flaherty

17.1   Capacity definitions   281
17.2   The *Highway Capacity Manual* approach   282
17.3   The British design-standard approach   290
17.4   References   298

Chapter 18 Road accidents   299

C.A. O'Flaherty

18.1   International comparisons   299
18.2   Accident trends in Great Britain   301
18.3   Accident costs   307
18.4   Reducing the accident toll   308
18.5   References   318

Chapter 19 Geometric design of streets and highways   320

C.A. O'Flaherty

19.1   Design speed   320
19.2   Sight distance requirements   324
19.3   Horizontal alignment design   327
19.4   Vertical alignment design   333
19.5   Cross-section elements   339
19.6   Safety audits   353
19.7   References   354

Chapter 20 Intersection design and capacity   356

C.A. O'Flaherty

20.1   Types of intersection   356
20.2   Overview of the design process   357
20.3   Priority intersections   364
20.4   Roundabout intersections   369
20.5   Traffic signal-controlled intersections   377
20.6   Intersections with grade-separations   381
20.7   References   399

Chapter 21 Introduction to computer-aided design of junctions and highways

400

G.R. Leake

21.1	Role of computer-aided design	400
21.2	What is CAD?	401
21.3	Data input requirements	402
21.4	Outputs from CAD programs	404
21.5	References	408

Chapter 22 Design of off-street parking facilities

409

C.A. O'Flaherty

22.1	Car parking standards	409
22.2	Locating off-street car parking facilities	412
22.3	The design-car concept	413
22.4	Surface car parks	415
22.5	Off-street commercial vehicle parking	423
22.6	Types of multi-storey car park	425
22.7	Self-parking multi-storey car parks: some design considerations	430
22.8	Fee-collection control and audit	432
22.9	References	434

Chapter 23 Road lighting

435

C.A. O'Flaherty

23.1	Objectives	435
23.2	Lighting terminology	436
23.3	Basic means of discernment	440
23.4	Glare	443
23.5	Lamps	444
23.6	Luminaires	444
23.7	Mounting height	445
23.8	Luminaire arrangements	445
23.9	Overhang, bracket projection and setback	446
23,10	Spacing and siting	447
23.11	References	447

Part IV: Traffic management

Chapter 24 Regulatory measures for traffic management

450

C.A. O'Flaherty

24.1	Speed limits	450
24.2	Restriction of turning movements	451
24.3	One-way streets	452
24.4	Tidal-flow operation	454
24.5	Priority for high-occupancy vehicles	456

24.6   Waiting restrictions and parking control   458
24.7   References   464

Chapter 25   Physical methods of traffic control   **465**

   C.A. O'Flaherty

25.1   Traffic calming   465
25.2   Pedestrian priority   473
25.3   Cyclist priority   480
25.4   References   482

Chapter 26   Signal control at intersections   **484**

   M.G.H. Bell

26.1   Hardware   484
26.2   Intersection design   486
26.3   Safety and fairness   490
26.4   Control variables   493
26.5   Capacity   494
26.6   Performance   498
26.7   Off-line signal plan generation   500
26.8   On-line microcontrol   501
26.9   On-line proprietary systems   504
26.10   References   505

Chapter 27   Signal control in networks   **506**

   M.G.H. Bell

27.1   Off-line control   506
27.2   On-line control   512
27.3   References   515

Chapter 28   Driver information systems   **517**

   M.G.H. Bell, P.W. Bonsall and C.A. O'Flaherty

28.1   Conventional traffic signs   517
28.2   Variable message signs   521
28.3   Road markings   522
28.4   Guide posts   525
28.5   In-vehicle information systems   526
28.6   Issues in the provision of in-vehicle information and guidance   529
28.7   References   531

Index   **532**

About the contributors

Michael G.H. Bell

Michael G. H. Bell has a BA in Economics from the University of Cambridge, and MSc and PhD degrees from the University of Leeds. He joined the University of Newcastle upon Tyne in 1984 where he is currently Professor of Transport Operations and Director of Transport Operations Research Group. He is author of over one hundred papers in fields related to transportation engineering, and is an Associate Editor of *Transportation Research B*.

Peter W. Bonsall

Peter W. Bonsall has a Geography BA with Honours from Oxford University and a Diploma in Town and Regional Planning. His first employment was as a Systems Analyst with Software Sciences Ltd. He joined the staff of Leeds University as a Research Assistant in 1974 and was seconded to work with West Yorkshire Metropolitan County Council from 1979 to 1984. He has headed the Institute for Transport Studies' MSc (Eng.) Programme in Transport Planning and Engineering since 1992 and became Professor of Transport Planning in 1996. He has been author of four books, over one hundred other publications and several items of commercially available software as well as being a contributor to two television documentaries.

Gerald R. Leake

Gerald R. Leake graduated from the University of Birmingham with degrees in civil engineering and traffic engineering. He worked with Freeman Fox and Partners (consulting civil engineers) and Liverpool Planning Department before joining the Department of Civil Engineering, University of Leeds, where he is Senior Lecturer. He is a chartered Civil Engineer, a Member of the Institution of Civil Engineers and of the Institution of Highways and Transportation.

Anthony D. May

Anthony D. May graduated with first class honours in Mechanical Sciences from Pembroke College, Cambridge, and subsequently studied at the Bureau of Highway Traffic at Yale University. He spent ten years with the Greater London Council, with responsibilities for policy on roads, traffic and land use, before joining the University of Leeds in 1977 as Professor of Transport Engineering. While at Leeds he has been awarded over fifty research grants and contracts, and has served as Director of the Institute for Transport Studies, Head of the Department of Civil Engineering and Dean of the Faculty of Engineering. He is currently Pro Vice Chancellor for Research.

Christopher A. Nash

Christopher A. Nash has a BA in Economics with First Class Honours from the University of Reading and a PhD in Transport Economics from the University of Leeds. He joined the staff of Leeds University in 1974 and is currently Professor of Transport Economics in the Institute for Transport Studies. He is author or co-author of four books and more than eighty published papers in the fields of project appraisal and transport economics.

Coleman A. O'Flaherty

Coleman A. O'Flaherty worked in Ireland, Canada and the USA before joining the Department of Civil Engineering, University of Leeds in 1962. He was Foundation Professor of Transport Engineering and Foundation Director of the Institute for Transport Studies at Leeds University before being invited to Canberra, Australia as Chief Engineer of the National Capital Development Commission. Since retiring as Deputy Vice-Chancellor, University of Tasmania in 1993 he has been made a Professor Emeritus of the University and granted the honorary degree of Doctor of Laws.

Preface

In 1967 I wrote a basic textbook entitled *Highways* that was aimed at undergraduate civil engineers who were interested in centring their careers on highway planning, design and construction. The book was well received and subsequently two further editions were prepared. These later editions were each divided into two volumes, one dealing with those aspects of particular interest to the young traffic engineer, and the other with the physical location, structural design, and materials used in the construction of highways.

When I was invited by the Publisher to prepare a fourth edition, I resolved instead to invite some of the top engineering educationalists in Britain to collaborate with me in the preparation of two new books. In this first volume *Transport Planning and Traffic Engineering* I am very fortunate that Mike Bell, Peter Bonsall, Gerry Leake, Tony May and Chris Nash agreed to participate in this endeavour. All are recognised experts in their fields and I am honoured to be associated with them in this book.

Transport Planning and Traffic Engineering is essentially divided into four parts.

The first part (Chapters 1–11) deals with planning for transport, and concentrates on the historical evolution of the transport task; transport administration and planning at the governmental level in Britain; principles underlying the economic and environmental assessment of transport improvement proposals, and of transport analysis and forecasting; contrasting traffic and travel demand-management strategies; a basic approach to the development of a town centre parking plan; planning for pedestrians, cyclists and disabled persons; roles and characteristics of the various transport systems in current use; and introductory approaches to the planning of public transport and freight transport systems.

Planning of any form is of limited value unless based on sound data. Thus the second part (Chapters 12–15) is concerned with issues in survey design; observational and participatory transport surveys; and studies relating to the prevention, investigation and reduction of road accidents.

The third part (Chapters 16–23) deals with practical road design for capacity and safety. It covers an introduction to traffic flow theory; the US highway capacity manual and British design-standard approaches to road design; road accident considerations; the geometric design of roads (including intersections) for both safety and capacity; an introduction to computer-aided design; road lighting; and the design of off-street parking facilities.

The final part (Chapters 24–28) is concerned with the management and control of traffic in, mainly, urban areas. As such it concentrates on regulatory methods of traffic management; *in situ* physical methods of traffic control; traffic signal control at intersections and in networks; and the role and types of driver information systems.

Whilst this book is primarily aimed at senior undergraduate and postgraduate university students studying transport and traffic engineering I believe that it will also be of value to practising engineers and urban planners.

Coleman O'Flaherty
July 1996

Acknowledgements

My colleagues and I are indebted to the many organisations and journals which allowed us to reproduce diagrams and tables from their publications. The number in the title of each table and figure indicates the reference at the end of each chapter where the source is to be found. It should be noted that material quoted from British government publications is Crown copyright and reproduced by permission of the Controller of Her Majesty's Stationery Office.

I am personally indebted to Ms Judy Jensen, Librarian-in-Charge at the State Offices Library, Hobart for her very considerable unassuming, professional help in obtaining reference material for me. I must also acknowledge the courteous, as well as professional, help provided to me at all times by Ms Eliane Wigzell, Arnold's Civil and Environmental Engineering Publisher; it was and is a pleasure to work with her.

Last, but far from being least, I thank my wife, Nuala, whose patience and forbearance have helped me immeasurably in my writing enjoyment.

Coleman O'Flaherty

PART I

Planning for transport

CHAPTER 1

Evolution of the transport task

C.A. O'Flaherty

Everybody travels whether it be to work, play, shop or do business. All raw materials must be conveyed from the land to a place of manufacture or usage, and all goods must be moved from the factory to the market place and from the staff to the consumer. Transport is the means by which these activities occur; it is the cement that binds together communities and their activities. Meeting these needs has been, and continues to be, the transport task.

Transport, because of its pervasive nature, occupies a central position in the fabric of a modern-day urbanised nation. To understand this it is useful to consider how today's land transport system, and particularly its road system, has developed over time. In Britain, as in most countries, this has been a story of evolutionary change with new transport developments replacing the old in response to perceived societal and economic needs. How people live and work has also changed as a consequence of improvements in lifestyle and in transport capabilities. What can be said with certainty about the future is that these interactive changes will continue, and that it will be the task of the transport planner and traffic engineer to cope with them.

Because of the pervasiveness of transport, 'solutions' to transport problems can have major influences upon people's lives. These influences are reflected in the constraints which society currently places on the development and evaluation of road proposals; that is, generally, they must be analytically based, economically sound, socially credible, environmentally sensitive, politically acceptable and inquiry proof. Meeting these needs has resulted in the development in relatively recent times of a new professional area, transport engineering.

Transport engineering applies technological and scientific principles to the planning, functional design, operation and management of facilities for any mode of transport in order to provide for the safe, rapid, comfortable, convenient, economical, and environmentally compatible movement of people and goods. Traffic engineering, a branch of transport engineering, deals with the planning, geometric design, and traffic operation of roads, streets, and highways, their networks, terminals, abutting land, and relationships with other modes of transport.

1.1 The road in history

The birth of the road is lost in the mists of antiquity. However, with the establishment of permanent settlements and the domestication of animals some 9000 years ago, the trails deliberately chosen by people and their animals were the forerunners of the first recognised travelways which, in turn, evolved into today's streets, roads and highways.

Although the wheel was invented in Mesopotamia *ca* 5000 BC, it did not come into wide usage as a carrier of humans or goods until well into the second millennium AD. For thousands of years, therefore, the transport task was carried out by humans and pack animals walking to their destinations. On long trips people rarely walked more than 40 km in a day and, consequently, settlements tended to develop about well-used resting places 15 to 40 km apart. Typically, these were at sites which had reliable water supplies and were easily defended. These settlements, in turn, reinforced the establishment of recognised travelways between these sites. Many settlements, especially those at crossings of streams and/or travelways, or on dominant sites adjacent to waterways, eventually grew into villages and towns. Until the Industrial Revolution these settlements rarely exceeded 45 minutes' travel by foot from the outskirts to their centres.

The first manufactured roads[1] were the stone-paved streets of Ur in the Middle East (*ca* 4000 BC), the corduroy log paths of Glastonbury, England (*ca* 3300 BC), and the brick pavings in India (*ca* 3000 BC).

The oldest extant wooden pathway in Europe, the 2 km Sweet Track, was built across (and parts subsequently preserved in) marshy ground near Glastonbury. Corduroy road sections have also been found in marshy ground in continental Europe. Many of these formed part of a comprehensive network of trade routes, the Amber Routes,[2] which developed over the period 4000 BC to 1500 BC.

The oldest extant stone road in Europe was built in Crete about 2000 BC. About 50 km long, its function was to connect the then capital Knossos in the north of the island with the southern port of Leben, thereby gaining access to the Mediterranean trade. However, notwithstanding the many examples of stone roads which have been found in various parts of the world, it is the early Romans who are now credited with being the first real roadmakers.

The Roman road system was based on 29 major roads, totalling 78 000 km in length, which radiated from Rome to the outer fringes of the Empire. The pavements were usually constructed at least 4.25 m wide to enable two chariots to pass with ease and legions to march six abreast. These roads were constructed by the military, using slave labour, to aid administration and enable the legions to march quickly to quell rebellion after an area had been occupied. They had long straight sections to minimise travel time, and often followed firmer and safer old travelways along the sides of hills. Many pavements were constructed on embankments up to 1 m, sometimes 2 m, high (for defence reasons) in locales where attacks were likely. Soil for the embankments was mostly obtained by excavating longitudinal drains on either side of the road; in soft soils foundations were strengthened by driving wooden piles. Stone pavements were laid with crossfalls to aid drainage, widened at bends to accommodate the unwieldy carts and wagons of the day, and reduced in width in difficult terrain.

Following their invasion of Britain in 55 BC, the Romans constructed some 5000 km of major road in 150 years. This road system radiated from their capital, London (located at the first upstream crossing of the Thames) and extended into Wales and north to beyond Hadrian's Wall. The withdrawal of the last legion from Britain in 407 AD marked the final decline of the Roman Empire — and the breakdown for centuries of the only organised road system in Europe.

After about 100 years the Roman roads fell into decay from the wear and tear of natural and human forces. During the Dark Ages Britain was split into small kingdoms whose rulers' needs were parochial rather than national, so they exerted little effort to preserve the through-roads. When sections became untraversable, trackways were simply created around them. New roads consisted of tracks worn according to need, with care usually being taken to avoid cultivated land and private property. These practices largely account for the winding nature of many of Britain's present-day roads and lanes.

Throughout the Middle Ages through-roads were nothing more than miry tracks, and the rivers and the seas tended to be relied upon as the main trade arteries. This was in contrast with the situation in continental Europe, especially in France whose centralist rulers built main roads radiating from Paris as a means of holding the country together. The only significant commitment to road works in Britain was by the medieval religious authorities, who saw road repairs as meritorious work similar to that of caring for the poor and the sick. The suppression of the monasteries by Henry VIII removed these road maintainers, however, and the new owners of the ecclesiastical estates were not inclined to continue their road obligations.

1.1.1 Emergence of passenger transport

A feature of the Middle Ages was the growth of many prosperous villages into towns. Consequently, lengths of stone-paved street were constructed within some of the larger towns. The building of these roads was often associated with the need to provision towns from their rural hinterlands, i.e. good access roads were needed to withstand the high wheel pressures created by wagons and carts (and eventually, coaches with passengers).

The first non-ceremonial coach to be seen in Britain appeared in London in 1555. However, Milan led Europe in the development of urban coach travel, with 60 coaches in use in 1635.[1] Long wagon-coaches were in use in Spain as early as 1546 to provide for long-distance passenger travel.

The first British stagecoach to stop at regularly-spaced posthouses to change horses operated between Edinburgh and Leith in 1610. The development (in Austria) in the 1660s of the Berliner coach with its iron-spring suspension system led to the rapid expansion of coach-type travel so that, by 1750, four-wheeled coaches and two-wheeled chaises (introduced from France) had superseded horseback-riding as the main mode of intertown travel for Britain's wealthy and the growing middle class. This expansion of coach travel was facilitated by major initiatives in road-making, initially in France and then in Britain.

At the turn of the eighteenth century Britain's roads were so abominable that Parliament passed in 1706 the first of many statutes which created special bodies known as Turnpike Trusts. These Trusts, which eventually exceeded 1100 in number and administered some 36 800 km of non-urban roadway, were each empowered to construct and maintain a designated length of road and to levy tolls upon specified kinds of traffic. The development of the toll road system, particularly in the century following 1750, was important for a number of reasons: first, it resulted in the emergence of skilled road-makers, e.g. John Metcalf, Thomas Telford, John Loudon McAdam; second, it established that road-users should pay road costs; third, it determined the framework of the present-day main road network; and fourth, it made coach travel quicker, easier, more comfortable, and more attractive.

By the turn of the nineteenth century the value of road drainage and of firm roads with solid surfaces was widely accepted and many thousands of kilometres of good quality main road had been built between towns. In urban areas heavily-trafficked main streets were surfaced with stone setts. Wooden blocks were often used instead of stone setts to alleviate the noise and unhygienic pollution engendered by animal-drawn iron-tyred vehicles. Whilst these block roads were relatively easy to sweep, they were slippery when wet, smelly, and had fairly short lives. Thus it was not until steel tyres were replaced by pneumatic tyres (patented by Robert Thompson in London in 1845 and made workable by John Dunlop in 1888) and streets were surfaced with 'artificial' bituminous asphalt (mainly from the turn of the twentieth century) that these early environmental pollution problems were alleviated.

1.2 Railways, bicycles and motor vehicles

With the advent of the industrial revolution, there was a great wave of migration from the countryside and villages became towns and towns became cities. This was accompanied by a population explosion resulting mainly from improved living and health conditions; for instance, in 1800 the population of England and Wales was less than 10 million (*ca* 17 per cent lived in towns of more than 20 000 population) and 100 years later the population was over 30 million (*ca* 55 per cent in such towns). The material successes of the Industrial Revolution were the catalysts for major changes in both intra-urban and inter-urban transport.

1.2.1 Initiation of rail transport

The opening of the Surrey Iron Railway on 26 July 1803 – this, the first public railway, was horse-drawn and operated between Wandsworth and Croydon in South London – marked the onset of the rail age. When the steam-powered Stockton–Darlington railway was opened in 1825, it was the beginning of the end for long-distance horse transport. The transfer of long-distance passengers and goods from road to rail was practically instantaneous whenever towns were accessed by a railway. What McAdam called the calamity of the railways fell upon the Turnpike Trusts between the years 1830 and 1850. The relative advantages of rail travel over coach travel were such that many Trusts were quickly brought to chronic insolvency and they began to disappear due to lack of traffic. The final Trust collected its last toll on 1 November 1885 on the Anglesey portion of the London–Holyhead road. In 1850 British railways carried some 67 million passengers; by 1910 this had risen to nearly 1300 million.

In continental Europe, steam traction was introduced into Germany (between Nüremberg and Fürth) in 1835, and quickly spread into Russia (1836), Austria (1838), the Netherlands (1839), Italy (1839), Switzerland (1844), France (1844), Hungary (1846), Denmark (1847) and Spain (1848).[3]

At the same time as the railway was being developed for mainly long-distance transport purposes, a number of other transport modes were also being developed for mainly intra-urban travel purposes. The horse-drawn omnibus, first used in Bordeaux in 1812, was introduced into London in July 1829. The horse-bus and the hansom cab were the main movers of the middle class in towns – the fares were too high for usage by the working classes – until the turn of the twentieth century. August 1860 saw the initiation of the first horse-tram service to be operated in England (in Birkenhead); it could carry more people more comfortably and more quickly than the horse-bus. January 1863 saw the opening of the world's first 'underground' (steam) railway in London and, five years later, the first 'elevated' urban railway was opened in New York.

Whilst the railway was a major connector for suburban towns in the late 1800s, thereby encouraging an existing commuting tendency, it was the overhead-powered electric tram, particularly the American electric streetcar of the late 1880s, that really changed urban travel and urban form internationally. Twenty years after the introduction of the first electric tramway to use an overhead-wire conductor (in Leeds in 1891) nearly every town in Britain had its own network of low-fare electric tramways. Typically, these radiated outward from the central areas of towns and flexibly serviced lower-density residential areas along their routes. Thus low-cost efficient transport began to be clearly associated with the development of residential suburbia. This relationship was firmly established following the introduction of the municipally-operated motor bus (in Eastbourne in 1903) and trackless tram, i.e. the rubber-tyred electrically-powered trolley bus (in Leeds in 1911).

1.2.2 Bicycle and motor vehicle beginnings

While the above developments in public transport for the masses were taking place throughout the nineteenth and early twentieth centuries, developments with even greater potential were happening in respect of private transport on the road. These involved the motor vehicle, which eventually caused the demise of the tramway systems (except in Blackpool) as well as significant reductions in the usage of the other public transport systems, and the bicycle.

The Macmillan *bicycle* of 1839 is generally credited with being the first true bicycle, i.e. its forward motion was obtained with pedals without the rider's feet touching the ground. However, it was not until the development, some 35 years later, of the low bicycle that the cycling boom really began. As it was developed technologically, and road surfacings were improved, the bicycle was accepted as a cheap alternative to public transport and, with the growing emancipation of women, cycling became socially acceptable for both sexes. In the economically depressed 1920s and 1930s, the bicycle became the private vehicle of the mass of the populace. Its use then grew to the extent that, at its peak in 1952, about one quarter of all of Britain's then vehicle-kilometres were attributed to pedal cyclists. By 1993, however, the cycling traffic had dropped to just over 1 per cent of the total road traffic.

Whilst there is some argument regarding the identity of the inventor of the first internal combustion engine, and when it occurred, there is no doubt but that the history of the *motor vehicle* really began in 1885 and 1886 when Karl Benz of Mannheim and Gottlieb Daimler of Constatt, respectively, working independently and unbeknown to each other, produced their first motor vehicles: a petrol-driven tricar by Benz and a four-wheel petrol-driven coach (minus shafts) by Daimler. These vehicles heralded a transport revolution and the start of a return of four-wheeled traffic to the roads.

The motor vehicle had little effect upon rural travel or town development for some considerable time. Until the introduction of the low-cost mass-produced Model-T car (by Henry Ford, in the USA) prior to World War I, the motor vehicle was only enjoyed by the wealthy for touring and pleasure-driving. Much more important to the then general populace were the low bicycle and public transport. Thus the first concerted pressure for the improvement of rural roads came from the cyclists, who formed a Road Improvement Association in 1886.

The considerable advantages which internal combustion engined vehicles had over horse-drawn vehicles were recognised during the 1914–18 War, as the military invested heavily in motor vehicles, especially lorries. After the war, the ready availability of surplus lorries, and of trained personnel to drive and service them, was the catalyst for the start of the transfer of freight from rail to road – a process which has continued since then. At their peak in 1923 the railways moved 349 million tonnes of freight; 10 years later this had dropped to 255 million tonnes.

1.2.3 Rise of the motor vehicle

After World War II, the 1950s saw personal incomes begin to climb and with increasing affluence came a worldwide growth in the numbers and use of the motor vehicle. The extraordinary usefulness and convenience of the motor vehicle and especially of the *private car*, are reflected in the data in Fig. 1.1. In general, when people could afford a car they bought it – and subsequently developed a car-dependent lifestyle.

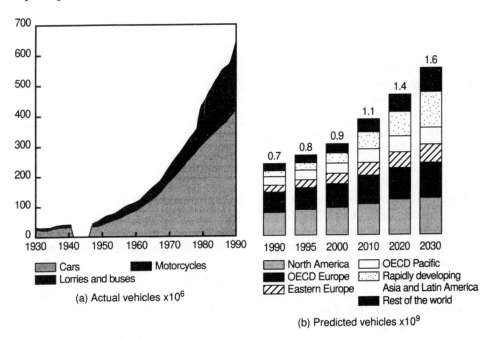

Fig. 1.1 Global trends in motor vehicle numbers: (a) actuals, 1930–90, and (b) predicted by region, 1990–2030[14]

In 1903 there were 17 000 motor vehicles registered in Britain: 8000 cars (0.2 per 1000 population), 4000 goods vehicles, and 5000 public transport vehicles. Ninety years later the total number had risen to 24.83 million (comprising 20.10 million private cars, 2.19 million light goods, 0.43 million (heavy) goods vehicles, 0.65 million two-wheelers,

0.11 million public transport, 0.32 million agricultural, 0.98 million Crown exempt, and 0.05 million others). 84 per cent of the total vehicle growth and 90 per cent of the private car growth took place after 1950.

In 1951 some 86 per cent of Britain's 14.5 million households did not have a car, while 13 per cent and 1 per cent had one and two cars, respectively. In 1992 the number of households had grown (by 54%) to 22.3 million, while the number without a car dropped to 32 per cent; 45 per cent had one car, 20 per cent had two cars and 5 per cent had three cars. Figure 1.2 and Table 1.1 show, however, that the proportions of households with and without cars vary considerably according to economic region and household structure. Approximately 96 per cent of the households in the employer, managerial and professional groups have direct access to a car; this compares with just over half the households in the unskilled manual group.

Table 1.1 Households with regular use of a car, by household structure[15]

Household structure	Percentage with			
	no car	1 car	2 cars	3+ cars
1 person under 60	42	55	3	–
1 person 60 or over	79	21	–	–
2 adults under 60	14	51	33	2
2 adults, 1/both over 60	32	55	11	1
3 or more adults	13	31	36	21
.1 adult with child(ren)*	61	39	–	–
2 adults with child(ren)*	12	53	33	2
3+ adults with child(ren)*	12	35	36	17
Total	32	45	20	5

Note: Due to rounding the totals do not equal 100 per cent
* A child is ≤ 17 years and living in the parental home

Since 1952 passenger-kilometres by private transport have increased by about 3 per cent per annum while the number of cars has grown by 2.2 per cent annually. The availability of a car increased the amount of travel from a household, particularly for non-work purposes; further, the greater the car availability per household, and the wealthier the household, the greater the increase.

The Government has published[4] upper and lower forecasts of traffic and vehicles on all roads; these, rebased to 1993, are shown in Table 17.13. They assume that Britain's gross domestic product will grow by between 3 per cent (high forecast) and 2 per cent (low forecast) each year to 2025, and that no technical change will upset the dominance of road transport in that time-frame. Note that, in 1993, the total motor-vehicle traffic was 410.2×10^9 vehicle-kilometres, of which cars and taxis accounted for 82.1 per cent, heavy goods vehicles 6.9 per cent, two-wheeled motorised vehicles 1.0 per cent, light goods vehicles 8.9 per cent, and large buses and coaches 1.1 per cent.

Commercial (road) vehicles dominate the transport of freight in Great Britain. Table 1.2 shows the breakdown between the different transport modes for tonnes of freight lifted and tonne-kilometres of freight traffic. Factors which favour the use of the road for freight transport in Britain include:[5]

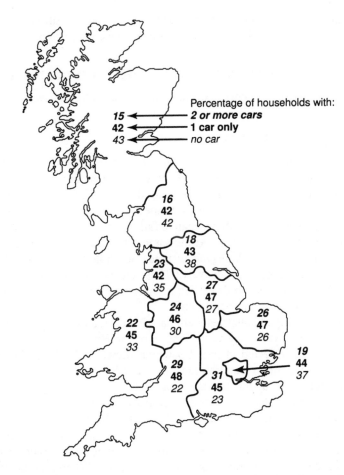

Fig. 1.2 Households with regular use of a car, by economic planning region, in 1992 (based on data in reference 15)

1. population and industry are concentrated in a triangle with London, Leeds and Liverpool at each apex, providing few lengths of haul long enough for rail to gain a competitive advantage
2. there are no long navigable (internal) waterways which can carry heavy barge traffic
3. the motorway system, although heavily congested at times, is extensive
4. deregulation of the haulage sector in 1968 freed it from capacity control and enabled it to expand readily to meet new demand.

Table 1.2 Freight transport by mode, 1993[15]

	Road		Rail		Water		Pipeline	
	(t)	(t-km)	(t)	(t-km)	(t)	(t-km)	(t)	(t-km)
Amount ×10⁹	1.62	134.5	0.10	13.8	0.14	52.0	0.13	11.6
Per cent	82	63	5	6	7	25	6	5

Also (until the Channel Tunnel was opened) Britain's island location retarded the development of a through rail link for the international movement of goods.

1.3 Some changes associated with the motor vehicle

Social change, growth in affluence, and changes in personal travel have always been linked.

1.3.1 Demographic changes

Until the advent of mechanised transport the size of a town was usually limited by how far people could walk to/from work and shops and, in many larger towns, this led to high residential densities and unhealthy living conditions. With the advent of increasing affluence and urban transport in the mid-nineteenth century people began to spread themselves and residential densities began to fall, although towns continued to grow due to migration from the countryside. At the same time household sizes began to reduce; for example, in England and Wales the average household sizes in 1901, 1951 and 1981 were 5, 3.6 and 2.7 persons, respectively.[6] These changes are mainly due to young people forming their own households earlier, a lowering of the birth rate, and the trend for the elderly not to live with their children.

In almost all developed countries there has also been a drift of population from large cities to small towns over many years. The trends in Britain (which has a long history of urbanisation) are reflected in Fig. 1.3 which shows that (a) London peaked in the early 1950s and the main conurbations in the 1960s, (b) medium-sized cities are approaching their peak at the present time, and (c) smaller towns are still growing strongly. In older cities also, many of the traditional manufacturing industries are dying so that the inner city areas surrounding the central business districts are left with a declining number of job opportunities. New industries, especially high technology ones, are preferring to locate in smaller towns or in the outer areas of larger towns where land is cheap and access to the national road network is good. When both the origins and the destinations of trips are in the outer areas trip lengths are increased, the provision of adequate public transport services becomes more difficult and expensive but the use of cars is made easier.[7] Service and retail jobs are still mainly central area-based and growing, but not at a sufficient rate to counter the loss of public transport trips caused by declines in the manufacturing industry in the area about the central business district.

As well as becoming more decentralised the organisation of work is also becoming more flexible. For example, there have been significant increases in the numbers of self-employed, part-time workers, women workers (rising from 25 per cent in the 1930s to more than 40 per cent in the 1980s), and workers with more than one job, and these are reflected in changing transport needs.

Although the population of Great Britain (56.40 million in 1992) is increasing only relatively slowly, the increase in size of the driving-age cohort is much more rapid. 81 per cent of men and 53 per cent of women now have driving licences, and the proportions are still growing. People are now living longer and it can be expected that car usage amongst the elderly will also increase as time progresses and more of today's working generation enter retirement. At the other end of the scale the proportion of children taken to school by car is increasing; it was *ca* 30 per cent in 1990 whereas it was only *ca* 1 per cent a generation previously.[8]

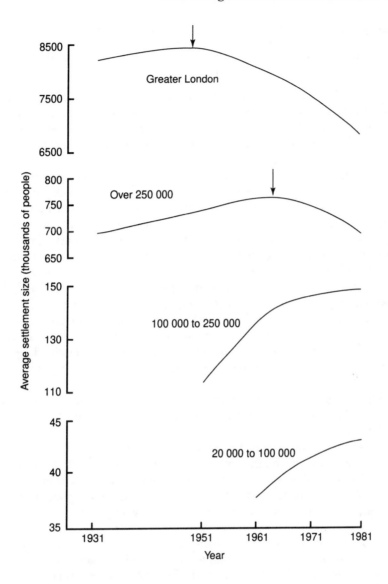

Fig. 1.3 Changes in population in towns of different size in England and Wales, 1931–81[7]

There have been major changes in relation to where people shop. Only 5 per cent of retail sales took place at out-of-town stores in 1980, but by 1992 this proportion had grown to 37 per cent. Most out-of-town shoppers use cars on their shopping trips; the great majority of people without cars do not shop at out-of-town centres and stores because of difficulties in getting to them. It is reported[7] that the catchment area of a major edge-of-town store is 40 minutes' driving time, while that for minor high street stores is *ca* 20 minutes.

1.3.2 Some trip patterns

An analysis[6] of travel data from a number of British towns has revealed a number of important trends in relation to the changing patterns of urban travel.

1. As the number of cars in a household increases, the total number of trips made per person per day by all transport modes increases as a result of an increase in car trips. Roughly, the effect of adding a car to an urban household is that about 4 additional trips are made by car per day for a first car and about 3 for a second car; at the same time trips by other modes are lost.
2. For all modes and all households taken together (Table 1.3) the average person makes 0.77 work trips per day (29 per cent of all trip-making), 0.41 trips/day to and from educational facilities (15 per cent), 0.33 shopping trips/day (12 per cent), 0.51 social or recreational trips/day (19 per cent), and 0.67 trips/day for all other purposes including non-home-based trips for whatever purpose (25 per cent). Note that the data on which this table is based indicate that (a) even households without cars make *ca* 10 per cent of their trips by car, mostly as passengers, while one-car and two-car households use a car for 50–60 per cent and 70 per cent of trips, respectively; (b) the most important single journey purpose is work, possibly because the journey to work tends to be long and less likely to be walked; (c) home-based shopping trips are proportionally only half as many in 2+ car households as in households without cars, and (d) non-home-based trips are more important in car-owning households, possibly because having a car makes it easier to chain together a series of trips for different purposes.
3. In general, use of public transport tends to be higher in larger cities while cycle and motorcycle travel are more important in smaller cities or where the use of public transport is low.
4. The rate at which car ownership has risen in a country is largely determined by national economic growth, and the ranking of car ownership in different countries is generally reflective of their relative economic standings. Car ownership is only very marginally affected by the level of public transport provision, i.e. good public transport in any area seems to reduce car ownership levels by only about 0.04–0.06 car per person.
5. Higher operating costs do depress car ownership, but only to a small degree, i.e. a fuel price increase of 10 per cent (in real terms) probably reduces car ownership by about 2 per cent. However, a short-term rise in fuel cost of 10 per cent causes car traffic to fall by 1–3 per cent, while the long-term fuel price elasticity is likely to be larger.
6. If a car is available for an urban journey, the roads are not too congested and parking is easy, it tends to be used irrespective of the level of public transport fares because of its greater door-to-door speed and comfort. Public transport tends to be more competitive with the car in dense urban areas where road congestion reduces speeds and makes driving less pleasant and where finding a parking space at the central destination is more difficult and expensive.
7. While the car is convenient for those who have one, the car-based society is making life more difficult for those households without a car. These mainly comprise the poor, the elderly, and those who cannot drive or are unable to drive because of some disability.

Table 1.3 Trip rates for different purposes in Britain (trips per person per day)[6]

Purpose	Car ownership	Walk (over 5 min)	Pedal cycle	Moped and motor cycle	Car driver	Car passenger	All car	Public transport	All modes
Work	0	0.166	0.069	0.031	0.006	0.068	0.073	0.253	0.592
	1	0.106	0.041	0.024	0.425	0.114	0.540	0.143	0.861
	2+	0.060	0.024	0.018	0.618	0.090	0.708	0.085	0.936
	All	0.123	0.049	0.026	0.287	0.092	0.379	0.181	0.768
Education	0	0.229	0.015	0.001	0.000	0.008	0.009	0.071	0.319
	1	0.279	0.024	0.003	0.028	0.052	0.080	0.079	0.458
	2+	0.198	0.028	0.006	0.083	0.117	0.200	0.074	0.503
	All	0.249	0.021	0.003	0.025	0.044	0.069	0.076	0.411
Shopping	0	0.194	0.011	0.003	0.001	0.011	0.012	0.153	0.365
	1	0.097	0.009	0.002	0.092	0.064	0.155	0.059	0.319
	2+	0.057	0.007	0.003	0.134	0.053	0.187	0.029	0.282
	All	0.124	0.009	0.003	0.064	0.042	0.106	0.092	0.330
Social/ recreation	0	0.194	0.017	0.009	0.003	0.067	0.070	0.131	0.418
	1	0.110	0.018	0.009	0.208	0.161	0.368	0.046	0.552
	2+	0.081	0.021	0.011	0.318	0.183	0.502	0.031	0.649
	All	0.137	0.018	0.009	0.142	0.126	0.268	0.079	0.512
Non-home- based and other	0	0.194	0.020	0.008	0.010	0.042	0.052	0.117	0.390
	1	0.150	0.016	0.008	0.406	0.126	0.532	0.059	0.776
	2+	0.107	0.011	0.010	0.727	0.133	0.860	0.044	1.069
	All	0.160	0.017	0.009	0.300	0.095	0.396	0.079	0.671

1.3.2 Impact on public transport

Car availability is the biggest single factor affecting public transport usage. For example, a first car in a household results in a drop in public transport trip-making of roughly 40 per cent while a second car removes a further 30 per cent.[7]

Buses are the most important form of public transport in most urban areas and it is upon the (mostly) relatively short trip-making by bus that the car has had its greatest detrimental effect. Figure 1.4 shows that between 1952 and 1993 the total bus and coach passenger-kilometres declined by 54 per cent. The main events which contributed to this decline are identified in Fig. 1.5. Identifying characteristics of frequent and infrequent bus users are summarised in Table 1.4.

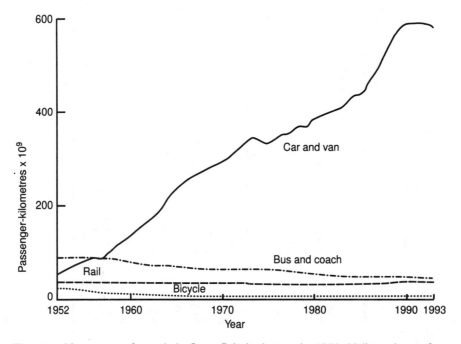

Fig. 1.4 Movement of people in Great Britain, by mode, 1952–93 (based on reference 15)

The research upon which Table 1.4 is based also indicates that the current market for local bus travel is likely to decline. The national projections given in Table 17.13 suggest that no growth is expected in bus and coach travel in the foreseeable future.

The increase in car and van travel (+895 per cent) between 1952 and 1993 also had a minor impact on rail passenger travel in that total rail passenger-kilometres declined by 5 per cent. Closer examination of these data also shows, however, that the total number of passenger journeys made via British Rail over this time period declined from 1017 million to 713 million (−29.9 per cent) while the mean journey length increased substantially; also the number of journeys via London Underground increased from 670 million to 735 million (+9.7 per cent).

The main usage of urban rail is normally for journeys to work. However, it is reported[9] that London Underground's market includes about 40 per cent commuters, 30 per cent off-peak radial trips (e.g. shoppers), and 15 per cent visitors to London; of the visitor journeys

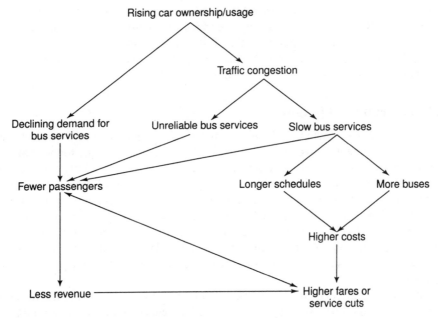

Fig. 1.5 Mechanism of the decline in bus services[16]

Table 1.4 Examples of market segmentation for public transport[17]

Characteristic	Most frequent users	Most infrequent users
Local bus		
Geographic	Residents of London and major cities	Residents of rural areas
Demographic	Teenagers; women; pensioners	Middle-aged men
Economic	Unemployed; pensioners; low-paid; students; schoolchildren; housewives	Professional and managerial
Car availability	No car and no licence	Multi-car and licence
Long-distance rail: business		
Geographic	Areas well served by rail	Areas with poor rail services
Demographic	Younger working age; men	Teenagers; pensioners; women
Economic	Professional and managerial	Semi- and unskilled; pensioners; housewives; unemployed
Car availability	Multi-car and licence	No car and no licence
Long-distance rail: non-business		
Geographic	Areas well served by rail	Areas with poor rail services
Demographic	Under 24; pensioners; women	Older working age; men
Economic	Professional; students; servicemen	Semi- and unskilled; house-wives; unemployed
Car availability	No car and licence *or* multi-car and no licence	One or more cars and licence

about 90 million are made by persons from overseas and 20 million by visitors from other parts of Britain. Identifying characteristics of long-distance rail users are also given in Table 1.4. The research on which Table 1.4 is based suggests that the market for long-distance rail travel for non-business purposes is likely to decline in the future while that for business purposes is reasonably robust.

1.3.3 Impact on rail freight

Notwithstanding the growth in Britain's economy, rail freight tonne-kilometres declined by about 63 per cent between 1951 and 1993, while road freight tonne-kilometres increased by over 365 per cent. The growth in road freight is undoubtedly due to the efficiency and flexibility associated with using larger lorries to travel greater door-to-door distances more quickly. New developments in manufacturing, like just-in-time production with its demand for punctual delivery, require fast and frequent transport facilities and are therefore likely to intensify the demand for road-based freight transport in the future (see the large projected growth in Table 17.13). This is happening at a time when the tonnes-lifted of bulk commodities (which are most suitable for transport by rail) are in decline.

Nearly three quarters of the rail tonne-kilometres are now devoted to the movement of only four commodities: solid mineral fuels (28 per cent), minerals and building materials (20 per cent), petroleum products (14 per cent) and metal products (12 per cent). However, the Channel Tunnel (operationally opened 1993) has the potential to generate at least a partial renaissance for intermodal rail transport by providing significant journey-time savings *vis-à-vis* heavy commercial road vehicles involved in direct international freight haulage.

1.3.4 Impact on the environment

The growth in the numbers and usage of the motor vehicle has taken its toll of the environment in many ways, not least of which is through road congestion and vehicle noise and emissions. Road accidents (see Chapter 18) are also a major cost in both monetary and human terms.

Congestion

As well as imposing high costs on industry and road users through wasted time and fuel, delayed deliveries and reduced reliability, congestion increases air pollution, global warming and the usage of (scarce) oil resources. In urban areas, it encourages traffic to use unsuitable residential roads, thereby endangering the quality of life of adjacent householders. Overall, congestion costs in OECD countries are equivalent to about 2 per cent of GDP.[10]

Congestion in urban areas in Britain is currently being tackled by government policies which, mainly, seek better integration of land use and transport planning, greater use of public transport in towns, and packages of traffic management measures aimed at easing traffic flows on main roads; proposals for major road building proposals in urban areas are not being encouraged. However, there is also considerable interest in Britain at this time (and elsewhere[11]) into the feasibility/desirability of using road pricing to reduce congestion on urban roads. Congestion on inter-urban roads is being tackled by the provision of an expanded road programme (which emphasises the widening of exist-ing major roads) and by the use of improved technology (e.g. variable message systems and signalised access-metering) to maximise ease of movement. The promotion of increased rail usage as a means of reducing congestion on inter-urban roads is consid-ered to have limited practical value; for instance, it is estimated[12] that a 50 per cent increase in rail passenger traffic would reduce road traffic by less than 5 per cent.

Traffic noise

Noise from vehicles disturbs sleep, impairs job performance, impedes the learning process (especially in schools close to busy roads), hinders social activity and verbal communication, and affects health through stress generated by frustration from lack of sleep and a general deterioration in the quality of life. Many studies[13] suggest that to comply with desirable limits for well-being indoors, the representative outdoor noise level experienced during a 24-hour day should not exceed 65 dB(A). About 11 per cent of Britain's population are known to have been regularly exposed to outdoor noise lev-els in excess of 65 dB(A) in the early 1980s while a further 39 per cent lived in 'grey' areas of 55–65 dB(A).

The traffic noise problem is being tackled to a limited extent by both legislative mea-sures and traffic engineering ones. For example, from 1995 any new design of car must not exceed 74 dB(A) when accelerating in low gear at full throttle; the new limit for the heaviest type of commercial vehicle is 80 dB(A) and new motorcycles must not exceed 75–80 dB(A) according to engine size.[8] However, the full effects of these standards will not be felt until the vehicle fleet is replaced over, say, 10–12 years. It is also government policy for existing buildings and housing in black-spot noise areas to be protected with anti-noise screens and sound-proofing, and for greater account to be taken of noise abatement objectives in the location, design and operation of roads.

Emissions

The main emissions from motor vehicles are carbon dioxide (CO_2) and what are termed the air pollutants: carbon monoxide (CO), nitrogen oxides (NO_x), oxides of sulphur (SO_x), hydrocarbons (HC), and lead (Pb) and other particulate matter. Two major con-cerns arise regarding vehicle emissions: their impact on human health (estimated at about 0.4 per cent of GDP in OECD countries[10]) and on global warming.

It is now accepted that vehicle emissions can be the cause of ill-health, e.g. irritation of cardio-respiratory, eye or other systems, acute toxic systemic effects, mutagenic or carcinogenic action, and adverse effects upon the defence mechanism against common infections. However, while the qualitative linkages are well established, quantification is still difficult and controversial. A measure of the potential problem wherever vehicles congregate can be gathered from the fact that in the centre of an average town, motor vehicles usually account for 100 per cent of all CO and Pb levels, at least 60 per cent of NO_x and HC levels, about 10 per cent of the SO_x level, and 50 per cent of particulate levels.[13]

Certain gases which are in balance in the atmosphere create a natural greenhouse effect that keeps the Earth's surface temperature at a level suitable for life. However, as a result of the world's industrialisation and population increases of the past 200 years, changes have taken place in the composition of the atmosphere which are causing global warming. The rate at which additional warming will occur in the future is very uncertain as the mechanisms involved are complex, and future emissions depend on economic and social factors. The principal gases contributing to global warming are also generated by the motor vehicle; for instance, carbon dioxide contributes about half of the estimated annual increase in warming and road transport currently contributes 14–16 per cent of this gas. Thus, if vehicle emissions continue at their present levels and if predicted global numbers of vehicles are achieved, road traffic's contribution to global warming could help threaten the planet in the long term.

The need to control emissions from vehicles is recognised and Britain is one of 155 nations which signed a treaty (the Rio Declaration) establishing guidelines for this purpose. Steps currently being taken to alleviate this problem include:

- improving vehicle and engine design to reduce emissions from and usage of hydrocarbon fuels
- using state-of-the-art technology to improve traffic flow and reduce congestion
- improving usage of public transport and minimising penalties associated with inter-modal freight transfers
- imposing traffic restraint measures to relieve congestion and using improved communication technology to replace person-movements with information flow.

1.4 Britain's road network

The first 40 years of the twentieth century were a time of evolutionary development for roads, with the main emphasis being on 'laying the dust', reconstructing existing roads, and providing work for the unemployed. In addition, there was opposition to the expansion of road transport by a strong rail lobby. Consequently, in 1938 Britain's road network was essentially unchanged in total length (289 086 km) from that at the turn of the century, and only 44 km of 4953 km of trunk road were dual carriageway. This was in contrast with the USA (first Parkway in 1906), Italy (first Autostrada in 1924), and Germany (first Autobahn in 1932) which had substantial lengths of motorway-type road at the start of World War II.

The 1920s and 1930s also saw much inner-city slum clearance, a strengthening of the flight of home-dwellers to low-density suburbs – helped by improved public transport which provided access to cheap land away from crowded city centres – and the movement by industry to sites adjacent to good roads. Traffic congestion in towns was mainly tackled by road widenings to increase capacity.

Britain's first motorway, the 13 km Preston bypass, was opened in 1958. In 1970 when completion of its then target of 1600 km of motorway was in sight, the Government announced plans to improve 6750 km of road to at least dual carriageway standard to form a strategic inter-urban network for England which would tie in with previously published strategic networks for Scotland and Wales. This was a fundamental development in that it marked a national commitment to the comprehensive planning and construction of a strategic network of high-calibre highways which would enhance road safety, be easily connected to all major centres of population, promote economic growth and regional development, and divert through traffic from unsuitable roads in towns and villages.

The national road system in 1993 totalled 364 477 km and carried a traffic load of 410×10^9 vehicle-kilometres. Trunk roads (3062 km motorway, 1408 km urban trunk, and 10 822 km non-urban trunk) carried nearly one third of this traffic even though they comprised just 4.2 per cent of the total length; the motorway network carried nearly one half of the trunk road traffic.

Over 90 per cent of the trunk road network is non-urban. The total length of non-motorway dual carriageway trunk road is 3522 km; the balance of the trunk network (8707 km) is single carriageway. In other words, the trunk road classification does not imply a guaranteed level of service; rather the service provided in any given locale is a function of the road's national and regional importance and can range from that provided by a motorway to that provided by an all-purpose road less than 5.5 m wide.

The majority of urban areas in Britain have cartwheel-type road patterns with central business districts located at the centre. The spokes of the wheel are the radial routes which developed historically to link town centres and suburbia to central areas; they have high densities of development alongside and are heavily used by buses as well as cars. Radial routes are normally single or dual carriageway all-purpose roads, depending upon traffic demand and parking availability in the central area. The hub of the wheel is the inner ring road; its function is to promote the convenient use and amenity of the central area by deflecting through traffic while affording convenient access to essential traffic. The location and design of the inner ring road is bound up with the size, layout and usage of the central area. In practice, this ring road may be round, square or elongated, and may be incomplete on one or more sides.

Towns with populations of 20 000 or more tend to have a single inner/outer ring road, whereas cities of more than 0.5 million may have inner and outer ring roads. The outer ring road is the rim of the cartwheel. While it is now also used by through traffic to bypass a town, its original purpose was to link outer communities and promote development infill by acting as a distributor between radials. Thus these ring roads are generally located within the lower-density outer fringes of urban development, and they tend to be more circumferential than inner ring roads. Their quality of design and completeness depend upon needs at specific locations. Outer ring roads are not heavily used by public transport.

In large urban areas intermediate ring roads may be located between the inner and outer ring roads. Very often these intermediate rings incorporate existing local streets with all their diversity of use.

1.5 A final comment

As well as providing a brief history of some major developments in land transport, a major objective of this chapter has been to indicate that, as people's lifestyles and social needs have changed over the centuries, the means by which their transport needs have been met have also changed. Social developments and transport provision are inextricably related, and current transport problems are in many ways simply reflections of today's social needs. As times and needs change into the future, it can be expected that new transport problems and 'solutions' will also emerge. These challenges present exciting opportunities for the transport planners and traffic engineers of today – and tomorrow.

1.6 References

1. Lay, M.G., *Ways of the world*. Sydney: Primavera Press, 1993.
2. Franck, I.M., and Brownstone, M.D., *To the ends of the Earth: The great travel and trade routes of human history*. Oxford: Facts on File Publications, 1984.
3. Marshall, J., *The Guinness railway fact book*. Enfield: Guiness Publications, 1994.
4. *National road transport statistics (Great Britain) 1989*. London: HMSO, 1989.
5. Cooper, J., *Freight needs and transport policy*. Discussion Paper No. 15 in the 'Transport and Society' research project. London: Rees Jeffreys Road Fund, 1990.
6. Webster, F.V., Bly, P.H., Johnston, R.H., Paulley, N., and Dasgupta, M., Changing patterns of urban travel. *Transport Reviews*, 1986, **6** (1) 49–86.
7. Webster, F.V., and Bly, P.H., Policy implications of the changing pattern of urban travel. *Municipal Engineer*, 1987, **4**, 15–23.
8. *Transport and the environment; Command 2674*. London: HMSO, 1994.
9. Mackett, R.L., Railways in London. *Transport Reviews*, 1995, **15** (1) 43–58.
10. *Urban travel and sustainable development*. Paris: OECD and EMCT, 1995.
11. *Curbing gridlock: Peak-period fees to relieve traffic congestion*. Transport Research Board Special Report 242, Vols 1 and 2. Washington, DC: National Academic Press, 1994.
12. *Roads for prosperity; Command 693*. London: HMSO, 1989.
13. *Transport and the environment*. Paris: OECD, 1988.
14. Gwilliam, K.M. (ed.), *Transport policy and global warming*. Paris: European Conference of Ministers of Transport, 1993.
15. *Transport statistics Great Britain 1994*. London: HMSO, 1994.
16. Buchanan, M., *Urban transport trends and possibilities*. Discussion Paper in the 'Transport and Society' research project. London: Rees Jeffreys Road Fund, 1990.
17. Hill, E., and Rickard, J.M., *Forecasting public transport demand: The demographic dimension*. Discussion Paper 16 in the 'Transport and Society' research project. London: Rees Jeffreys Road Fund, 1990.

CHAPTER 2

Transport administration and planning

C.A. O'Flaherty

Transport infrastructure and administration have been rapidly expanded in every developed country to meet people's expectations and commercial and industrial needs. The extent, rate and manner in which this is now occurring varies considerably from country to country, however. The following is a brief overview of the British approach to transport administration and planning.

2.1 Transport administration in Great Britain

Responsibility for transport administration in Britain is divided between the central Government and local authorities. At the national level the responsibilities are shared between the Secretary of State for Transport (for England), the Secretary of State for Scotland, and the Secretary of State for Wales, each of whom is supported by permanent civil servants via their departments. At the local level they are shared between district councils, London boroughs, and county councils.

2.1.1 The Department of Transport

Of the three national bodies, the Department of Transport (DOT) is the most authoritative in respect of its responsibilities for land, sea, and air transport. These include: sponsorship of the rail and bus industries; motorways and trunk roads; airports; domestic and international civil aviation; shipping and the ports industry; and navigational lights, pilotage, HM Coastguard and marine pollution. The Department also has oversight of road transport, including: vehicle standards; registration and licensing; driver testing and licensing; bus and road freight licensing; regulation of taxis and private hire cars; and road safety.

Because of the relationship between land use and transport the Department of Transport interacts closely with the Department of the Environment, which has responsibility in England for functions relating to the physical environment in which people live and work, e.g. planning, local government, new towns, housing, inner city matters and environmental protection.

In April 1994 a new executive agency of the Department of Transport, the Highways Agency, was created to manage, maintain and improve the strategic motorway and trunk road network in England on behalf of the Secretary of State and his supporting

Ministers, who are the Minister of State for Railways and Roads, and the Parliamentary Under-Secretary of State/Minister for Transport in London/Minister for Local Transport and Road Safety. (There is also a supporting Minister of State for Aviation and Shipping.) With this arrangement Ministers retain responsibility for overall Government transport policy, for policy on trunk roads in England, and for determining the strategic framework within which the Highway Agency is required to operate, including:

- the scale of the motorway and trunk road network
- the content and priorities of the new construction programme including decisions about schemes entering the programme, the choice of preferred route, and final decisions following public inquiries
- the methodology to be used for traffic and economic appraisal of motorway and trunk road schemes
- the policy for charging for the use of inter-urban roads and private finance for roads.

The Ministers also determine the Highway Agency's key objectives and targets and allocate resources to it. Its current key tasks are to:

- deliver the programme of trunk road schemes to time and cost
- reduce the average time taken to deliver trunk road schemes
- maintain the trunk road network cost effectively by delivery year on year of a structural maintenance programme
- improve the information supplied to road users through improved signing and better information re roadworks.

In addition to its direct responsibilities the Department of Transport also influences the transport decisions of local authorities through a variety of mechanisms. Key amongst these is the provision (in conjunction with the Department of the Environment) of planning policy guidance on transport matters which local authorities must take into account when preparing their development plans. The Department of Transport also oversees the provision of Transport Supplementary Grants to local authorities to assist with capital expenditure on, for example, non-trunk roads of more than local importance, public transport improvements (including light rail) and significant street lighting, road safety and traffic management schemes. It also publishes the results of research (mainly through the Transport Research Laboratory) and provides advice notes, circulars, and other planning and design information to local authorities. The Department may also provide rural, public transport and innovation grants.

Rail transport

The main railway companies in Great Britain were nationalised in 1948. The British Railways Board, which was established in 1962 and has its members appointed by the Secretary of State for Transport, until recently operated the national mainline railway network in accordance with Department of Transport policy and finance considerations. Its services were divided into six business sectors. InterCity, Network SouthEast, and Regional Railways were responsible for the passenger services: InterCity operated the mainline passenger trains between major urban centres, Network SouthEast operated the commuter rail services into London (excluding the services run by London Underground), and Regional Railways operated the rural routes and services in urban areas outside the south east of England. The other three sectors were Railfreight

Distribution, Trainload Freight, and Parcels. A subsidiary company, European Passenger Services Ltd, was also established to operate international passenger rail services through the Channel Tunnel. Another subsidiary business, Union Railways, was set up to take forward proposals for a new high-speed rail link between London and the Channel Tunnel.

British Rail's role is changing rapidly as a result of the Government's decision to privatise the railway network. The Government's policies (enshrined in the Railways Act 1993) involve:

- the franchising of all passenger services to the private sector
- the transfer of the freight and parcels operations to the private sector
- the provision of a right of access to the rail network for the new operators of the privatised passenger and freight services
- the separation of track from train operations, so that a new track authority, 'Railtrack' (a government-owned company that is shortly to be privatised), is responsible for timetabling, operating signal systems, and track investment and maintenance, while passenger services continue to be operated by British Rail until they are franchised
- the establishment of a 'Rail Regulator' to oversee the fair application of arrangements for track access and charging, and for promoting competition and the interests of consumers
- the setting up of an 'Office of Passenger Rail Franchising' responsible for determining minimum service standards and for negotiating, awarding and monitoring franchises
- the creation of opportunities for the private sector to lease stations
- the development of subsidy arrangements for individual rail services or groups of services that are socially necessary, albeit uneconomic.

British Rail restructured its passenger services into 25 train-operating units as a basis for the privatised rail network, and over half of the services were to be franchised by April 1996.[1] Freight users have had rights of access to the rail network since April 1994; this allows the introduction of services by rail freight operators from the private sector. The Channel Tunnel (which is estimated to have cost *ca* £10 000 m) was formally opened on 6 May 1994, and a British-French group has been granted a 65-year operating concession for the tunnel by the British and French Governments. Proposals have been developed for a private consortium (which will be aided with government grants) to build a new high-speed line from London to the Channel Tunnel; this consortium will also take over both Union Railways and European Passenger Services.

London Transport (LT), which is the body responsible for the provision of public transport (including rail transport not provided by British Rail) in Greater London, became responsible to the Department of Transport in 1984. In 1985 London Transport set up two wholly-owned major operating subsidiaries, London Underground Ltd (LUL) and London Buses Ltd (LBL). London Underground Ltd operates high-speed train services on (currently) 383 km of railway of which 41.5 per cent runs underground in the centre before rising to the surface in the suburbs.

The third railway system in London is the Docklands Light Railway (DLR). This is a separate 23 km long surface route that currently connects the City of London with Docklands, Beckton, and Stratford. In 1992 the ownership of the DLR was transferred from London Transport to the London Docklands Development Corporation.

2.1.2 Local authorities

The structure of local government in Britain is currently under review. At the time of writing the local authorities for non-trunk roads (i.e. roads that are not the responsibility of the Department of Transport) are: in England the metropolitan district councils, the London borough councils, and the county councils; in Scotland the regional and the islands councils; and in Wales the county councils.

Metropolitan counties

In addition to London there are six conurbations that are styled metropolitan counties in England: Greater Manchester, Merseyside, South Yorkshire, Tyne and Wear, West Midlands and West Yorkshire. Each of these metropolitan counties is divided into districts (36 in all). Each metropolitan district has a single tier of local government, the metropolitan district council, which is the planning authority for its area and is therefore responsible for development control and the granting of planning permissions. Each district council is also responsible for non-trunk roads and associated functions such as traffic management and parking within its area.

Public transport policy in a metropolitan county is the responsibility of a Passenger Transport Authority (PTA). Each PTA is composed of representatives from the county's constituent district councils, and has a Passenger Transport Executive (PTE) to carry out its executive and administrative tasks. Prior to the 1985 Transport Act, PTAs/PTEs had very considerable powers and operational responsibilities in relation to public transport services within the conurbations. However, by mid-1994 their bus operations had been privatised and they had lost their power to provide financial support to determine the overall level of fares. Their main emphasis is now on policy matters, including the co-ordination of public transport; for instance, they may enter into agreements for the provision of passenger rail services, administer concessionary fare schemes, ensure school transport services, and identify and (after competitive tender) subsidise socially desirable but non-profitable bus routes to secure specific fare and service levels. In Tyne on Wear, Greater Manchester and South Yorkshire, they still have responsibilities in respect of the Metro, Metrolink and Supertram light rail systems, respectively. The PTAs are also the bodies which deal with the licensing of taxis and private hire cars in the metropolitan counties.

London

Greater London, which has an administrative area of about 1580 km^2 and a resident population of *ca* 7 million, comprises 32 boroughs and the City of London. Each London borough council effectively acts as a unitary authority with land use planning and transport powers along the lines of a metropolitan district council. In the case of public transport and traffic management, however, there are some significant differences.

The members of the London Transport Board are appointed by the Minister and, as noted previously, the Underground rail system is operated by the Board through its subsidiary London Underground Ltd. Privatisation of London Transport's twelve area-based bus subsidiaries (established under London Buses Ltd) was completed at the end of 1994; however, comprehensive network planning and competitive tendering for socially necessary but uneconomic routes was retained by London Transport. (The

London-wide 'travelcard' concessionary travel scheme that is a major incentive encouraging the use of public transport was also continued.)

London also has an independent Traffic Director with London-wide traffic management powers. The Traffic Director is currently responsible for the coordination, introduction and operation of a *ca* 500 km 'Red Route' network of priority roads in Greater London (composed of 300 km of trunk roads, with the balance being local authority roads) by the end of 1997 which are marked by red lining and special signs, and are subject to special parking, loading and stopping controls and other traffic management measures. The Traffic Director is required to ensure specified reductions in accidents, improvements in bus journey times, and bus reliability; to improve facilities for cyclists, pedestrians, and people with disabilities; and not to encourage further car commuting into central London or more traffic to cross the central area. The Government has instructed the Traffic Director to prepare the Local Plans for the trunk roads (see Section 2.2 for discussion of the land use planning process).

Non-metropolitan counties and regions

Outside of London and the six metropolitan counties, England is currently administered through a two-tier system of local government based on non-metropolitan counties which are subdivided into districts. Each English county and each district within a county is run by a separate council. Wales also has a two-tier local government system, as does mainland Scotland (except that in Scotland the senior tier is composed of regional councils instead of county councils). There are also three unitary local authorities in Scotland, i.e. the Orkney, Shetland and Western Isles island groups, each of which has a single all-purpose council.

The Government has announced its intention to change this local government structure. Thus from April 1996 all local authority areas in Scotland and Wales will be governed by unitary councils. The two-tier system of local government will generally continue in England, except that some 20 of the larger towns and cities will have unitary councils.

Currently, however, the county/regional councils are the land use planning and transport planning (except for trunk roads) authorities for their areas, having roles generally similar to those of the metropolitan district councils. A significant difference, however, is that there are no passenger transport authorities; instead the PTA-type responsibilities are directly exercised by the councils. The functions of a PTA are undertaken by Strathclyde Regional Council in respect of the Greater Glasgow conurbation.

Following the 1985 Transport Act the local bus operations owned by the district and regional councils were formed into 'arms length' Passenger Transport Companies (PTCs) and encouragement was given to their privatisation. At the time of writing about 20 PTCs, i.e. less than half the number in 1985, still remain in local authority ownership.[2]

All long-distance bus and coach services are now operated by private enterprise. Privatisation of the nationalised National Bus Company (the largest single bus and coach operator in England and Wales) and of the Scottish Bus Group (the largest bus operator in Scotland) were completed in 1988 and 1992, respectively. There are now no restrictions on the routes served, or on the number of vehicles operated on each long-distance route, by private operators.

2.2 The statutory land use planning process

The town and country planning system in Great Britain is designed to regulate the development and use of land in the public interest. In the non-metropolitan areas it is driven by a development planning process which involves the preparation of structure plans which set out strategic policies in non-metropolitan counties/regions, and of complementary local plans, waste local plans and mineral local plans which set out detailed development policies for non-metropolitan districts. In the London boroughs and the metropolitan districts the planning authorities prepare unitary plans which combine the functions of structure and local plans.

Each non-metropolitan county council is currently required to prepare and continually update a Structure Plan for the area over which it has jurisdiction. This is a major statement of the key strategic policies that are deemed structurally important to the development of land over the subsequent 15 years, and which can be used as a framework for local planning by district councils and National Park authorities. Its strategic nature is emphasised by the limitation that the structure plan must contain a key diagram rather than a map – and this key diagram cannot be reproduced on an Ordnance Survey base.

Policies included in structure plans[3] relate to housing, conservation of the natural and built environment, the rural economy, the urban economy (including major employment-generating and wealth-creating developments), strategic transport and road facilities and other infrastructure requirements, mineral workings and resources, waste treatment and disposal, tourism, leisure and recreation, and energy generation.

Each district council/National Park authority is required to prepare and continually update a local plan for, usually, the following 10 years which develops the strategic policies and proposals of the structure plan and relates them to precise areas of land defined on an Ordnance Survey base map. As such it provides the detailed basis for the control, coordination and direction of land use development, whether it be publicly or privately owned. In practice, the local plan is also the main means by which detailed planning issues are brought before the public.

The Local Government Act 1985 which established the London borough councils and the metropolitan district councils as the planning authorities also ushered in a new system of development planning for the new authorities: Unitary Development Plans (UDPs). These plans replace the structure, local and other development plans for the areas for which they are prepared. Each UDP has two parts: Part 1 contains the strategic policies previously included in a structure plan, and Part 2 contains the more detailed information that would normally be included in a local plan. In this instance, however, both parts are prepared by the same planning authority.

Development plans must contain land use policies and proposals (including time frames and priority) relating to the road and rail network and to related services, e.g. rail depots, public transport interchange facilities, docks and airports. These must also reflect regional and national policies. Currently, for example, local authorities are required[1] to adopt planning and land use policies that:

- promote development within urban areas, at locations highly accessible by means other than the private car
- locate major generators of travel demand in existing centres which are highly accessible by means other than the private car
- strengthen existing local centres – in both urban and rural areas – which offer a range

of everyday community, shopping and employment opportunities, and aim to protect and enhance their viability and vitality

- maintain and improve choice for people to walk, cycle or catch public transport rather than drive between homes and facilities which they need to visit regularly
- limit parking provision for developments and other on- or off-street parking provision to discourage reliance on the car for work and other journeys where there are effective alternatives.

The aim underlying these national policies is to reduce the need to travel, especially by car, by influencing the location of different types of development relative to transport provision (and *vice versa*), and by fostering forms of development which encourage walking, cycling and public transport use.

2.3 Finance

Prior to the 1980s transport was generally regarded as a public good, most transport infrastructure costs came from the public purse, and it was accepted as the norm that public transport revenues would be supplemented by governmental authorities. Following deregulation, the financing of roads, railways and public transport changed as components of the transport industries were sold or franchised.

Factually, there is no relationship between the money raised by government through its taxes on passenger and freight vehicles using the road and rail systems and the expenditures which it lays out on these systems. For example, in 1992 the taxes raised from the road system totalled £21.46 billion whereas the recurrent and capital expenditure (including subsidies and capital grants) on roads in the same year amounted to £7.46 billion.[4]

The Department of Transport receives most of its funding directly from the Treasury as part of the national budgetary process. An important aspect of the European Community's policy in regard to regional development is the provision of transport infrastructure to link the member states; as a consequence the Government is also the recipient of funding ($18 million in 1993/94) from the EEC for trunk road, rail and harbour developments that meet their criteria.

Capital expenditure on rail infrastructure (which is now the responsibility of Railtrack) is now provided from the access charges paid by train-operating companies. It is now government policy to seek to expand the investment by private enterprise in rail infrastructure; thus, for example, the proposed £2700 million high-speed rail link between London and the Channel Tunnel is expected to involve financial commitments by the public and private sectors.

In 1992/93 the British Railways Board received £1354 million in capital grants and subsidies; most of this (85 per cent) was from the central government for the public service obligation of operating sections of the rail passenger network that would otherwise not cover their costs, while 7.6 per cent was from Passenger Transport Executives.[4] (Subsidies have greatly increased since then, however, as a result of train operators having to pay commercial charges for track access and to lease rolling stock under the new organisational arrangements.) In 1992/93 the Government also subsidised London Transport to the extent of £956 million; some two thirds of this was for capital works (most for London Underground and the Docklands Light Railway), 17.5 per cent was revenue support for the buses, and 11.7 per cent was for concessionary fare reimbursements (for passengers).

In 1992/93 some £711 million was also paid by the Government as grants and subsidies for bus and tram operations outside the London region. Of this amount 33 per cent was for revenue support, 44 per cent was for concessionary fare reimbursement (for passengers), and the balance related to fuel duty rebate.

Of the £4336 million spent on road construction in Great Britain in 1993/94 some 59 per cent was spent on the national road network (for new road construction/reconstruction, major maintenance, etc.) while 41 per cent was spent on local roads and car parks.

The Government is 100 per cent responsible for capital expenditures on trunk road schemes. Local authority expenditure on roads and public transport comes mainly from the central government in the form of block rate support grants (RSGs) and transport supplementary grants (TSGs). (Note, however, that the use of TSGs is now being confined to road schemes.)

The block rate grant is an amount calculated by the Department of the Environment which is supposed to enable each local authority to provide its services to a required standard. A local authority may use some of its block funding for transport purposes, e.g. for road maintenance or to support borrowing for capital works on roads.

A local authority with transport responsibilities is also required to submit annually to the Department of Transport a transport policy and programme (TPP) document which sets out the aims of its transport policies (which must be consistent with the development plans) and provides a costed 5-year programme of capital works for road and traffic regulation. Each TPP is considered on its merits, and the Department of Transport may agree that a transport supplementary grant (TSG) covering a share of the cost be allotted for road construction (e.g. for a bypass, or a road on the primary route network, or a major urban road) or for major traffic regulation purposes (e.g. for an urban traffic control system) on road proposals that are of more than local importance.

In recent years significant amounts of private capital have also been invested in local roads by private developers, in return for being granted planning permission by local authorities for related development projects. With Government approval a local authority may also borrow money for transport infrastructure projects; it may also obtain grants under Section 56 of the 1968 Transport Act toward the cost of approved public transport schemes. The remainder of a local authority's expenditure on transport normally comes from a Council Tax which is levied on the local community on the basis of business and private property values.

2.4 Some transport planning considerations

For many years the main focus of transport planning related to the provision of roads. Figure 2.1 shows in a simplified way the steps involved in the governmental process relating to the development of a trunk road scheme in a rural or urban area. A similar process is followed in relation to major roads sponsored by local councils except that structure/local/unitary plan approvals are involved. These processes can take 10–15 years (between when a scheme is first conceived and its actual opening to traffic) depending on the size, location and complexity of the schemes and their acceptability to the public.

While all the stages in Fig. 2.1 are important, the traffic/transport study, the economic and environmental assessments, and the public consultation processes deserve some overview comments here.

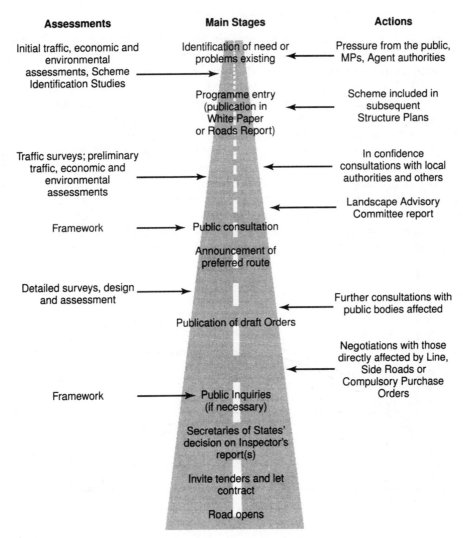

Assessments	Main Stages	Actions

Fig. 2.1 Stages in the development of a trunk road scheme in Great Britain[14]

2.4.1 Transport/traffic study process: an overview

At some stage(s) in the planning of a road or road system it will be necessary to carry out traffic studies to estimate the volume(s) of traffic that will have to be considered in a design year, as well as to satisfy statutory obligations relating to noise. Traffic data are also required for economic and environmental assessments in relation to the justification, scale and location of scheme alternatives. The collection and analysis of data can be a complicated process, particularly in urban areas. Why this should be so can be illustrated by examining the traffic components that constitute the design-year volume for a new/improved road.

Components of the design volume for a road

Traffic volumes for some future design year are derived from measurements of current traffic and estimates of future traffic. In Britain the design year is normally taken as 15 years after the opening of the road/road improvement. Given that 10–15 years may elapse before a road scheme is open to traffic, the actual design period may therefore be up to 30 years into the future. Therefore, when carrying out prediction exercises it needs to be appreciated that their credibility declines with increasing time into the future.

The basic constituents of the design volume for an individual road are shown in Fig. 2.2. By *current traffic* is meant the number of vehicles that would use the new road if it were open to traffic at the time the current measurements are taken. Current traffic is composed of reassigned traffic and redistributed traffic. *Reassigned traffic* is the amount of existing same-destination traffic that will immediately transfer from the existing road(s) that the new road is designed to relieve. *Redistributed traffic* is that which already exists on other roads in the region but which will transfer to the new road because of changes in trip destination brought about by the new road's attractiveness.

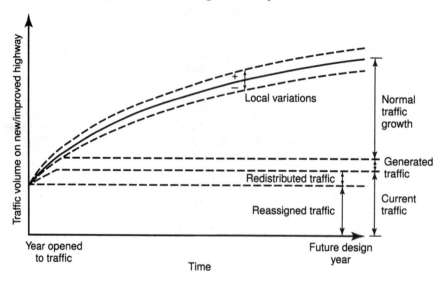

Fig. 2.2 Constituents of a road's design traffic volume

On low-volume roads in rural areas, classified traffic count data alone may be sufficient to evaluate the current traffic volumes. In this case (and provided that the implications of overestimation/underestimation are not important) the numbers of vehicles attracted to the new/improved road may be estimated adequately by an experienced traffic planner having a thorough knowledge of local traffic and travel conditions. However, with high-volume rural roads or bypasses around smaller urban areas the situation becomes more difficult and more rigorous techniques are required to validate the estimates of current traffic patterns. Information regarding journey times is normally also needed to estimate the traffic likely to be attracted to the new/improved road; i.e. the greatest number of vehicles will be attracted when the travel time and/or distance savings are significant.

Normal traffic growth is the increase in traffic volume due to the cumulative annual increases in the numbers and usage of motor vehicles. In this respect care needs to be taken when deciding the extent to which national projection figures (see Table 17.13) should be applied to particular local situations, to ensure that the figures finally selected reflect the local growth rates. For example, one detailed study[5] suggested that for a national saturation level of 0.45 car/person it is likely that saturation levels of 0.25–0.30 would apply to the central cities of conurbations (including Inner London), 0.30–0.45 to other large cities, and 0.45–0.60 to other areas. Local differences are also reflected in Table 2.1 which shows the variations in the annual distances travelled per person by all modes for different regions in the period 1991/93 *vis-à-vis* 1985/86. Also, between the periods 1985/86 and 1991/93 the travel distances by car in all regions grew to 8332 km (80 per cent of the total personal travel) from 6502 (76 per cent).

Table 2.1 Average distance, km, travelled per person per year by region of residence, 1991/93 (based on reference 15)

	London	Other English metropolitan areas	South East (excluding London)	Rest of England and Wales	Scotland
All modes 1991/93	8723	8652	13 142	10 689	9315
All modes 1985/86	8148	7033	10 618	8718	7485

By *generated traffic* is meant future vehicle trips that are generated anew as a direct result of the new road. Generated traffic is generally considered to have three constituent components: induced, converted, and development traffic. (See reference 6 for an excellent recent examination of generated traffic.)

Induced traffic consists of traffic that did not exist previously in any form and which results from the construction of the new facility, and of traffic composed of extra journeys by existing vehicles as a result of the increased convenience and reduced travel time via the new road. *Converted traffic* is that which results from changes in mode of travel; for instance, the building of a motorway may make a route so attractive that traffic which previously made the same trip by bus or rail may now do so by car (or by lorry, in the case of freight). *Development traffic* is the future traffic volume component that is due to developments on land adjacent to a new road over and above that which would have taken place had the new road not been built. Increased traffic due to 'normal' development of adjacent land is a part of normal traffic growth and is not a part of development traffic.

If the journey time by the new road divided by the time by the quickest alternative route, i.e. the travel time ratio, is high it can be expected that the amount of induced traffic will be low. The amount of converted traffic is mainly dependent upon relative travel costs, convenience and journey times. Experience with highly improved/new major roads suggests that adjacent lands with ready access to them tend to be subsequently developed more rapidly than normal; consequently the amount of development traffic generated depends upon the type of development and the extent to which the planning authority encourages/allows it to take place.

Comprehensive transport demand studies

The estimation of traffic flows on new/improved roads for some future year is at its most complex within large urban areas where the influences upon the road and rail transport systems are immense and, often, the data required to plan can be obtained only by carrying out a comprehensive study.

These studies can take many forms, but the classical *land use transport study process* is as follows:

1. carry out inventories and surveys of: goal and objectives; present travel activities of persons and freight; present traffic facilities, public transport services, and transport (including parking) policies; present and future land uses and populations; and appropriate present and future economic, environmental and employment data
2. determine existing interzonal travel patterns, and derive and calibrate mathematical models to represent them
3. develop and evaluate transport options to meet future needs
4. use the models developed at 2 to predict future trips for the scenarios outlined at 3
5. select the optimum acceptable option, and develop this in detail
6. continue replanning of the transport system to the extent that the available funds (and the techniques available to analyse further the ageing database) allow.

Most of the land use transport studies of the 1960s and 1970s put a major proportion of their resources into data gathering and the development of the representational models and gave only limited consideration to the generation of transport solutions to be tested by these models. Some of these models also gave an illusion of accuracy and objectivity which subsequent analyses showed not to be valid. Very few of the studies made any real attempt to gauge public attitudes to their proposed solutions, or to establish updateable analytical systems to support review and revision. Thus, while much good technical work and in many instances good outcomes were achieved by many of these studies, their credibility came into dispute in the 1970s and 1980s.

In recent years the term 'Integrated Transport Studies' has begun to be applied to various transport studies carried out in large urban areas in Britain. As shown in Table 2.2 one version of the new approach begins with a statement of vision for the city and ends with detailed evaluation of the preferred strategy. Typical objectives for such a transport study are:[7] efficiency in the use of resources; encouragement of economic regeneration; accessibility within the city and to regional, national and international facilities; environmental quality, including noise, air pollution, severance, townscape, safety and security, global impacts, and sustainability; and practicability, in particular financial feasibility. The integrated transport planning approach outlined in Table 2.2 emphasises the use of simplified models for sketch planning and experimentation purposes. Other versions of the integrated transport study approach use more complex models to test alternative policies/proposals.

The scale of operation involved in the collection of data for a comprehensive transport study is most easily illustrated by considering some of the information-gathering carried out in 1991 as part of the London Area Transport Survey (LATS).[8] This is the fourth comprehensive transport study carried out in London; the previous three were carried out in 1962, 1971 and 1981, respectively. The 1991 LATS, which updated and extended these earlier studies, is the largest programme of data collection ever undertaken

Table 2.2 An integrated transport study process[7]

- A statement of vision for the city

- A set of transport policy objectives consistent with the vision which serves as the basis for the appraisal framework

- The definition of a series of performance indicators, both quantitative and qualitative, of the objectives to include in the appraisal framework

- The development of a strategic forecasting model and associated processing tools to provide information for the appraisal framework

- Definition of a range of future economic and land use scenarios for the city

- Identification of future transport and environmental problems, typically for a 20 year horizon, for each of the scenarios, on the assumption that only currently committed transport policies are implemented (the do-minimum strategy)

- Compilation of a long list of transport policies capable of addressing the identified problems

- At this stage, in some studies, a process of local consultation to obtain views on the identified problems and to elicit suggestions for further policies to be added to the long list of measures

- Packaging all the transport policy measures into a set of illustrative strategies representing different concepts for addressing the problems (e.g. a road-based strategy, a rail-based strategy and a management and pricing-based strategy)

- Appraisal of the initial strategies (generally against the background of a single economic and land use scenario) and extensive sensitivity testing to identify the policy measures which contribute most effectively and which are supportive of one another

- Preliminary assembly of the measures on which to base a preferred strategy

- Potentially, a second round of consultation to present the results of strategy tests and to obtain views on the preliminary composition of the preferred strategy

- Further testing and refinement of combinations of the selected policy components to produce a preferred strategy in which the measures are combined to achieve as much synergy as possible

- Robustness testing of the preferred strategy against alternative planning scenarios and re-adjustment of the strategy for different scenarios in the light of the findings

---

for travel in London.[4] It involved carrying out major household and roadside interview surveys, and a number of smaller surveys designed to fill gaps in the major survey data, e.g. surveys of coach passengers and Inter-City rail passengers, a cordon count in Inner London of bus passengers, and a diary survey of drivers at the outer cordon. Data on weekday rail travel between 7 a.m. and 9 p.m. were obtained from British Rail (Network South East) and London Transport Underground.

The study area defined by LATS included 3.07 million households (70 000 in Central London, 1.02 million in Inner London, and 1.97 million in Outer London) and 7.47 million people. It was bounded by and included the M25 London Orbital motorway, and parts of boroughs that extend outside the M25. Outer London, which comprised nearly two thirds of the population and 86 per cent of the spatial area, was defined as the outer London boroughs and the surrounding area between the Greater London boundary and the M25.

The household survey involved some 60 000 households, each of which provided information on the round-the-clock travel by its residents on a weekday and a Saturday. The roadside interview survey provided information regarding the origins and destinations of weekday vehicular movements in and around London by residents and non-residents. Vehicles entering the LATS area were interviewed at sites on the outer cordon (just outside the M25) or on slip roads joining the radial motorways for traffic travelling toward London. Traffic movements in the wider M25 corridor were determined from interviews on radial screenlines that were extended outside the LATS boundary. Vehicle movements within the study area were determined from interviews on an internal network of screenlines and cordons.

The various surveys gave a comprehensive picture of travel in London by all modes and people (whether residents or not). It determined, for example, that over 20 million trips were made either wholly or partly within the LATS area on a typical weekday in 1991, and that some 4 million of these were walking trips by London residents. Table 2.3 shows that 58 per cent of the 15.8 million non-commercial vehicle trips had their origins in Outer London, and that 78 per cent of these also had their destinations in Outer London.

Table 2.3 Non-walk trips × 10^6 in London, by area of origin and destination[4]

Origin	Destination				
	Central London	Inner London	Outer London	External	All
Central London	0.5	0.5	0.5	0.3	1.8
Inner London	0.7	2.0	0.7	0.2	3.5
Outer London	0.5	0.7	7.1	0.7	9.1
External	0.3	0.1	0.7	0.3	1.4
All	2.0	3.4	9.0	1.4	15.8

2.4.2 Economic assessments

At various stages in the developmental process associated with a major road it is usual to carry out economic assessments of the route alternatives being considered so that they can be compared with the 'do-minimum' case that is the basis for comparison with all options. Economic evaluations can be carried out in many ways. That which is commonly used in Britain, particularly for trunk roads, involves the use of a computer program known as COBA 9[9] to determine the present values of the benefits and costs associated with each option under consideration.

Figure 2.3 sets out the means by which the COBA program calculates the present values of the benefits and costs, to obtain the net present value of a preferred scheme. In practice, COBA analyses are used to contribute to various decisions, for example:

- assessing the economic justification for a corridor improvement prior to its inclusion in the Department of Transport's preparation pool
- determining the priority that might be accorded to an individual scheme, by comparing its rate-of-return with those from other proposed schemes in the region/country
- estimating the timing of a scheme, including the merits of staged construction
- helping in the selection of a short list of alignment options to present at public consultations

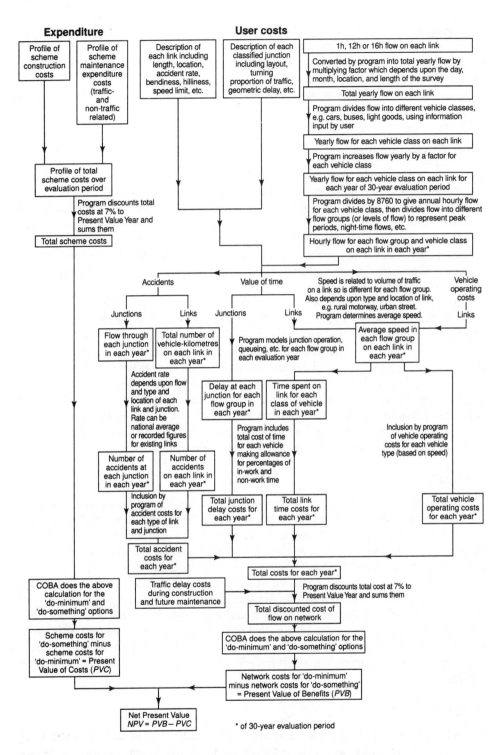

Fig. 2.3 Overview of the COBA method of economic evaluation[9]

- helping in the selection of a preferred option to recommend to the Minister after public consultation and public inquiry
- assisting in the selection of economically optimal link design standards to use with the various options under consideration
- assisting with the initial economic assessment of the optimal intersection design to use with the various options.

2.4.3 Environmental assessments

Economic assessments are major contributors to decision-making in relation to new/ improved trunk roads; however, they are not normally the sole contributor. In Britain, the economic assessments are combined with a bundle of twelve environmental assessments to form an assessment framework which is normally the basis for selecting a preferred route, for choosing between options within a route corridor, for public consultation and (if needs be) public inquiry and, ultimately, for decision-making through the political process (i.e. the Secretary of State for Transport) as to whether the proposed scheme should proceed.

The twelve environmental impacts which are currently assessed in relation to the framework are:[10] air quality; cultural heritage; disruption due to construction; ecology and nature conservation; landscape effects; land use; traffic noise and vibration; pedestrian, cyclist, equestrian and community effects; vehicle travellers; water quality and drainage; geology and soils; and policies and plans. The extent to which any one of these is more or less important depends upon the proposal in question. Some of these impacts can be measured quantitatively; other can only be assessed qualitatively. Reference 10 provides details of how these assessments are carried out, including the preparation of environmental impact tables for public consideration in conjunction with the economic consequences. The following is a brief overview of the environmental factors.

Air quality

This evaluation involves making a comparison between current air quality levels and those anticipated in the design year on the basis that the scheme is proceeded with, and that it is not. The vehicle pollutant levels assessed under this measure include carbon monoxide (CO), oxides of nitrogen (NO_x), hydrocarbons (HC), particulate matter, lead (Pb) and carbon dioxide (CO_2). These are normally measured/estimated in relation to their impacts at buildings or areas within 200 m on either side of a proposed route. Particular attention is paid to the likely impacts at sensitive locales, e.g. at buildings with susceptible populations (such as hospitals and old people's homes) and at locations where high pollutant levels can be expected (such as tunnel portals and major intersections). As it is not possible to take measurements in relation to the future, use is made of established relationships (e.g. see reference 10) to make predictions regarding future air quality.

Cultural heritage

Nowadays, communities are very sensitive to the preservation of their past and many otherwise worthy road proposals have foundered due to inadequate attention being paid to their likely impact on such features as areas of archaeological remains and ancient monuments, and listed buildings and conservation areas of special architectural or historic importance. Impacts which require particular attention (preferably by avoidance or, at least, by mitigation) include the demolition of buildings or the destruction of ancient remains, increased visual intrusion, increased noise and vibration (especially where settings are open to the public), severance from other linked features (e.g. gardens and outbuildings), changes to related landscape features, and general loss of amenity.

Disruption due to construction

This describes the impacts on people and the natural environment between the start of the proposed roadworks (including related advance works by public utility organisations) and the end of the contract maintenance period. Construction nuisances which are particularly felt by people living/working/shopping within about 50 m on either side of the site boundary include increased noise, dust, dirt, and vibration, as well as a general loss of amenity due to the operations of heavy construction equipment and vehicles. Travellers may have their journey distances and/or times increased, while traffic that is diverted temporarily may be the cause of undesirable environmental effects in adjacent localities.

Ecology and nature conservation

Ecology is the scientific study of the interrelationships between living organisms and their environment, e.g. climate, soils and topography. Nature conservation is concerned with maintaining the diversity and character of the countryside's wildlife communities, viable populations of wildlife species, and important geological and physical features. Road schemes can have many detrimental impacts upon ecology and nature conservation.

- They are the cause of a direct loss of wildlife habitats.
- Embankments and cuttings create barriers across wildlife habitats, and animals are killed crossing roads to traditional foraging areas.
- Changes to the local hydrology may affect wetland sites, both locally and some distance away.
- Local watercourses may be polluted by oil, deicing salts, particulates, and spillages.
- Road lighting can adversely affect invertebrates and disorientate birds.
- Certain flora specimens are at risk from particular emissions from vehicles while (in harsh winters especially) saline spray from the road surface can damage some plants up to 15 m from the carriageway.

Landscape effects

The impact of a road upon the landscape is dependent upon the extent to which the proposed alignment is integrated/in conformity with the character of the surrounding terrain. Visual assessments need to take seasonal differences into account, as well as likely changes as vegetation develops and matures. The existing pattern of settlement is also important, e.g. will the proposed road detract from existing attractive man-made features or will it follow existing roads, railways, power lines, etc.?

Land use effects

Factors normally taken into account in any impact assessment of agricultural land use include land-take, type of husbandry, severance, and the major accommodation works for access, water supply and drainage. As far as practicable assessment should also cover the impacts upon potential land use developments that might be prevented from happening if the proposed scheme is implemented; the local authority's development plans are important in this latter evaluation. Property blight in the form of reduced land and property values can affect an area as soon as alternative routes are announced, and will continue until the line is finally defined.

Traffic noise and vibration

When cruising at the low speeds experienced in urban areas the *noise* from both light and heavy vehicles is dominated by that from the vehicles' power-plants; at speeds above 40 km/h the noise from light vehicles is dominated by tyre noise, whereas with heavy lorries it is not until speeds of 80–100 km/h are achieved that tyre noise nearly equals power-train noise. At high speeds open-graded bituminous surfacings are less noisy than cement concrete ones. A doubling of traffic volume typically leads to an increase of *ca* 3 dB(A) in noise level, if other factors remain unchanged. Hill-climbing also results in more noise. Factors which therefore affect the noise level at any given locale include traffic volume and composition, tyre condition, carriageway type and texture, gradient and imposed speed limits.

A vehicle generates ground *vibrations* when it travels over irregularities in the carriageway surface and, dependent upon the characteristics of the underlying soil, these can cause differential settlements beneath adjacent structures that are not well founded. In practice, however, this vibration effect is not normally a problem with new carriageways. Of more importance are the vibrations from air-borne sound waves that are generated by the engines or exhausts of vehicles. These can couple into a building via windows and doors and cause elemental vibrations and rattlings which disturb and annoy people, even though they may cause no structural damage.

Pedestrian, cyclist, equestrian, and community effects

This evaluation is very often concerned with the severance impacts of a proposed scheme upon the local scene. A major aspect of severance is the dividing effect that it has upon people, through the breaking of established connections with friends and the services which they normally use, e.g. schools, shops, recreation facilities and workplaces. Severance may also be caused by the demolition of a community facility or the loss of land used by the public. Community severance effects are not evenly spread amongst people in the area about a proposed road: elderly people, children, and the disabled are particularly affected. In rural areas, farms that are of small size and/or existing close to the economic margin, and/or operate an intensive crop/livestock system can be very detrimentally affected by the severance effects of a new road.

Vehicle travellers

The two main impacts upon vehicle travellers that are considered as part of an environmental assessment are view from the road and driver stress. *View from the road* is the extent to which vehicle travellers can see and enjoy the different types of scenery through which a route passes; the quality of the landscape is obviously also important in this respect. *Driver stress* is defined[10] as the adverse mental and physiological effects experienced by a driver traversing a road network. Factors which influence a driver's stress level include road layout and geometry, carriageway characteristics, intersection frequency, and traffic speed and flow per lane. Research indicates that as frustration, annoyance and discomfort increase drivers become more aggressive and inclined to take risks, and that commuters and professional drivers are less stressed than other drivers for a given combination of road and traffic conditions.

Water quality and drainage

Water is essential for the sustenance of life by humans, animals and plants, as well as being critically important in relation to agriculture and industry, waste disposal, transport, recreation and some organised sports. In practice, design standards for run-off drainage and treatment (e.g. grit/silt traps, oil interceptors, French drains, sedimentation tanks/lagoons, grass swales, aquatic/vegetative systems, pollution traps), and flood relief measures in flood plains, normally ensure that pollution associated with run-offs from carriageways or with spillages of hazardous materials have relatively insignificant impacts on water quality. However, particular attention may need to be paid to proposed schemes that are close to ecologically valuable watercourses and/or where the anticipated traffic composition contains a high proportion of road tanker vehicles (e.g. on routes from chemical plants or ports).

Geology and soil considerations

Major road schemes can also have impacts on geology and soils. For example, geological or geomorphologic sites of scientific interest and importance may be exposed, and these will need to be protected from natural or artificial degradation. Surcharging resulting from the imposition of a road embankment may accelerate the collapse of underground mine workings or natural caverns. Important mineral deposits can be buried by a new road. Valuable agricultural soils and rare natural seedbanks can be lost by covering; if the topsoil is to be removed for reuse elsewhere, care has to be taken with its handling and storage.

Development policies and plans

Structure/local/unitary development plans are now in place for all areas in Great Britain. The extent to which a particular proposal is compatible with these plans, and is in accord with national and regional policies, is an important consideration in the component of the environmental assessment which deals with policies and plans.

2.4.4 Public participation in road planning

Prior to the late 1960s transport planners could generally assume that their professional expertise and 'unbiased' presentation of the facts would find reasonable acceptance at the decision-making stage in the development of a road proposal. The 1970s saw this comfortable feeling challenged by the public, often in very vociferous and determined ways. This desire by people to have more direct influence on road decisions that affected their lives resulted in the development of processes whereby the public could become involved in the plan development prior to its presentation for final decision-making. These processes came to be known as public participation.

Although the participation process can be time-consuming, it has the following advantages.

1. It provides a mechanism whereby planners can get a direct and timely understanding of community concerns and values, and perceptions of acceptable alternative alignment options.
2. Potential opponents of a proposal are involved in the decision-making process, thereby strengthening the planner's ability to implement the decision when it is eventually taken.
3. The participation process is educational for the public and the transport planner.
4. It reassures the final decision-maker, i.e. the political process, that the democratic ideal of involving people affected by a proposal in its development has been pursued.

In concept, there are three essential components of any successful public participation programme: first, the members of the public to be involved in the planning exercise must be identified; second, timely two-way communication flows must be established whereby the planner provides information on the planning process, requirements to be met, public participation events, the public's role in the decision-making, and alternative proposals and their impacts, while the public participants give information on community conditions and concerns, contribute ideas and opinions on transport problems and potential solutions, and express their values pertinent to the planning issue; and third, positive interaction must be encouraged, using techniques which get people to work together in a positive way on a shared concern.

In practice, most successful public participation programmes result from careful planning, a supportive governmental climate, simple techniques, and the basic skills and common sense of practitioners who are sensitive to people. There are no formulae guaranteed to work in every instance; however, openness and honesty, flexibility, and a willingness to deal with people in a constructive non-defensive way are key attributes.

Techniques used in the participation process[11] may include: group discussions, e.g. with small groups, public meetings, search conferences, or workshops; contributions from individuals, e.g. via individual discussions, oral or written submissions, questionnaire surveys, or participant observation through a planner residing in the area; and publicity designed to provide feedback, e.g. public displays, providing a staff source of information/counselling in the area, or media releases.

British practice[12] in respect of involving the public in the selection of the most appropriate route alignment from the alternatives being considered is called *public consultation*. It involves: (a) a comprehensive professionally-mounted exhibition of the alternative schemes that is displayed locally and staffed by representatives of the road-sponsoring agency, e.g. the Department of Transport in the case of a trunk road; (b) a widely-distributed consultative document in the form of an attractive brochure containing a statement of the purposes that the scheme is intended to satisfy, maps showing the alternative alignments, a description of each alignment and an assessment of its relative economic costs, efficiencies, and environmental impacts; and (c) a prepaid questionnaire included with the consultative document, for feedback regarding the alternative routes. The absolute need for the scheme or other ways of meeting the need without road construction are not considered in this exercise; this is taken to be a matter for governmental decision.

The advice received from the Department of Transport's public consultation process is provided to the transport planners who combine this with all the other planning information available, and arrive at a choice of a preferred alignment for the proposed scheme. If there are objections to the preferred line that are not subsequently withdrawn, a Public Inquiry[13] is usually held.

Although not normally considered as part of the consultative process, the public inquiry is in one sense the ultimate stage of the public's involvement in the preparation of a road proposal. The inquiry is carried out by an independent Inspector whose brief is to conduct the hearings in an open, fair and impartial way, record the relevant facts, present his or her views regarding the arguments mounted, and make recommendations to the Secretary of State as to whether or not the proposal(s) should be approved, with or without modification.

2.5 References

1. Department of Transport and Department of the Environment, *Transport: Planning policy guidance 13*. London: HMSO, 1994.
2. White, P., Deregulation of local bus services in Great Britain: An introductory review. *Transport Reviews*, 1995, **15** (2), 185–209.
3. Department of the Environment, *Development plans and regional planning guidance: Planning policy guidance 12*. London: HMSO, 1992.
4. Department of Transport, *Transport statistics Great Britain*. London: HMSO, 1994 edition.
5. Tanner, J.C., *Saturation levels in car ownership models: Some recent data*. Report SR669. Crowthorne, Berks: Transport Research Laboratory, 1981.
6. Coombe, D. (ed.), Special issue on induced traffic (7 papers). *Transportation*, 1996, **23** (1).
7. Roberts, M., Conurbation transport policy: Making the most of what we may yet spend. *Proceedings of the Institution of Civil Engineers: Transport*, 1995, **111** (3), 205–212.
8. Department of Transport and London Research Centre, *Travel in London: London area transport survey 1991*. London: HMSO, 1994.
9. *COBA: A method of economic appraisal of highway schemes*. London: The Department of Transport, 1982 (and subsequent amendments).
10. *Design manual for roads and bridges: Environmental assessment, Vol. 11*. London: HMSO, 1993.
11. Sinclair, A., *Public participation in transport planning (in Australia)*. Bureau of Transport Economics Occasional Paper No. 20. Canberra: Australian Government Publishing Service, 1978.
12. *Trunk road planning and the public: The procedures outlined*. London: The Department of Transport, 1991.
13. *Public inquiries into road proposals: What you need to know*. London: The Department of Transport, 1990.
14. Standing Advisory Committee on Trunk Road Appraisal, *Urban road appraisal*. London: HMSO, 1986.
15. *National travel survey 1991/93*. London: HMSO, 1994.

CHAPTER 3

Transport policy

A.D. May

3.1 Introduction

Transport problems, such as congestion, pollution and accidents, are a cause of widespread public concern. Recent surveys have indicated that 70 per cent of UK residents rate peak traffic conditions as bad, and 60 per cent of EU residents consider the consequences of car traffic in urban areas unsatisfactory.[1,2] Delays in transport are also a serious problem for industry. A study by the Confederation of British Industry suggested that congestion on the roads costs the UK some £15 billion each year.[3]

Not surprisingly, there are a wide range of suggested solutions to these problems: from building new roads to banning cars, and from improving bus services to the use of telecommunications as an alternative to travel. Many of these 'solutions' are expensive, and may not be very effective; moreover they may introduce new problems. New roads, for example, consume precious land; bans on cars may result in a loss of trade.

It is the task of politicians, and of the skilled professionals who advise them, to identify the most appropriate solutions to today's, and tomorrow's, transport problems. These solutions form the basis of a transport policy, which can be designed for a nation,[4] an individual city or town or a rural area.[5] This chapter describes the approach to the formulation of a transport policy. It starts by outlining a logical approach to transport policy formulation. It then discusses the objectives of transport policy, and the ways in which problems can be identified and assessed. Finally it describes briefly the range of solutions which are available (many of which are considered more fully later in this book) and outlines the case for an integrated approach to transport policy formulation. Some of the material in this chapter was originally prepared for guidelines published by the Institution of Highways and Transportation.[6]

3.2 A logical approach to transport policy formulation

There is a danger in the formulation of transport policy that politicians and, in some cases, professionals immediately assume that a particular solution is needed. New roads have often been promoted on this basis without considering their wider implications; so have measures such as traffic restraint and bus priorities. It is essential that professionals are clear on the reasons for such solutions: that is, that the objectives which are to be achieved can be specified.

If it is known that the aim is to reduce the costs of congestion, improve accessibility, and enhance conditions for those dependent on public transport, there is a case for considering, for example, bus priorities. Even if a further aim is to avoid the environmental impact of longer queues, one might still be prepared to accept bus priorities because the benefits against one set of objectives outweigh the losses against another. But if one simply pursues bus priorities because of a belief that they are a good thing, then it is all too easy for a subsequent group of policy-makers to take the counter view and remove them.

Figure 3.1 presents a structure for strategy formulation in which objectives are the starting point. They are used initially to identify problems, both now and in the future, as indications that the objectives are not being met. Possible solutions are then identified, not as desirable measures in their own right, but as ways of overcoming the problems which have been identified. The potential solutions are then compared, often by means of a predictive model of the transport system (see Chapter 5), by appraising them against the objectives which they are designed to meet. As measures are implemented, their impact is assessed, through before and after studies, again in terms of achievement against objectives. On a regular basis, too, conditions are monitored and the current conditions and problems reassessed, in terms of the overall objectives. This process may seem somewhat idealised and remote from standard practice, but it has several virtues. First, it offers a logical basis for proposing solutions, and also for assessing any proposals offered by others. If the answer to the question 'what problems would this solution solve?' is unconvincing, the solution is probably not worth considering. Second, it ensures that the appraisal of alternatives is conducted in a logical, consistent

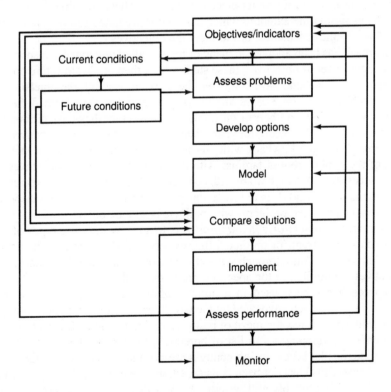

Fig. 3.1 An objectives-led structure for strategy formulation

and comprehensive way against the full set of objectives. This has been the basis for the development of the *Common Appraisal Framework*, now recommended by the UK Department of Transport.[7] Third, assessing the performance of the implemented measures improves the ability to judge the potential of similar measures elsewhere, and to predict their impact. Fourth, regular monitoring provides a means of checking not just on the scale of current problems, but also, through attitude surveys, on the perception of those problems. In this way the specification of objectives, and of their relative importance, can be modified to reflect changing attitudes and concerns. The last two of these are represented by the feedback loops in Fig. 3.1.

A clear statement of objectives, and of the related problems, thus provides a logical structure for identifying possible solutions, appraising their potential contribution, assessing measures which are implemented and hence learning by experience, and regularly monitoring the state of the transport system.

3.3 Problem-oriented planning and the objectives-led approach

There are in practice two different types of approach which can be adopted to identifying objectives and related problems.

The first is the true *objectives-led approach*, as described in Fig. 3.1, in which broad (or more detailed) objectives of the kind described in Section 3.5 are first specified, typically by the local authority or its elected members. These are then used to identify problems by assessing the extent to which current, or predicted future conditions, in the absence of new policy measures, fail to meet the objectives.

This approach has been adopted in many of the so-called integrated transport studies.[8,9] Having specified a set of objectives, these studies have then predicted future conditions if nothing new were done, and have compared these conditions with the objectives to identify future problems. In some cases this list of problems has then formed the basis for discussions with elected members or the public to see whether they have different perceptions of the problems. If they do, these are then used to re-define the objectives to match their concerns. The main drawback with this approach is that many elected members and members of the public are less familiar with the abstract concept of objectives (such as improving accessibility) than they are with concrete problems (such as the nearest job centre being 50 minutes away). It is to bridge this gulf that some integrated transport studies have checked the predicted problems with politicians and the public.

The alternative *problem-oriented approach* is to start by defining types of problem, and to use data on current (or predicted future) conditions to identify when and where these problems occur. This approach starts at the second box in the flow chart in Fig. 3.1. The objectives are implicit in the specified problems, and may never actually be stated.

This approach has been used in a number of recent studies of smaller conurbations.[10,11] It is the approach advocated by the UK Department of Transport in its *package approach* guidance.[12] It has the merit of being easily understood. However, it is critically dependent on developing a full list of potential problems at the outset. If particular types of problem (like access to job centres) are not identified because the underlying objective (accessibility) has not been considered, the resulting strategy will be partial in its impact. It is thus probably still wise to check with elected members and the public that the full set of problems has been identified.

3.4 Types of objective

An objective is a statement of a desired end-state. However, that statement can range from the very general, such as a successful urban economy or a high standard of quality of life, to the very specific, such as avoiding pollution levels above a specified threshold. Both are helpful, the first in providing the context for the strategy, and a direction to it; the second in providing a basis for assessing whether the objective is being met.

3.4.1 Statements of vision

The most general specifications often appear in Statements of Vision: broad indications of the type of area which politicians or the public wish to see. These serve to identify long-term goals to which more detailed transport policy objectives can contribute.

The London Planning Advisory Committee, for example, has specified a 'Fourfold Vision for London' in which the capital would provide a strong economy; a good quality of life; a sustainable future; and opportunities for all.[13] Similarly, the Birmingham Integrated Transport Study started with a vision of Birmingham as:

- having a national and international standing equivalent to that of other European provincial capitals
- maintaining its special and high level role as a regional centre
- providing a social and cultural environment in which its diverse groups of residents could each play a satisfying and distinctive part.[9]

These broad statements often say nothing about transport itself; instead they raise the question: 'how best can transport help to realise this vision?'. The answers to this question help to specify the higher level transport policy objectives.

3.4.2 Higher level objectives

These higher level objectives, sometimes referred to as aims or goals, identify attributes of the transport system, or its side effects, which can be improved as a means of realising the vision. Typical among them are the desire to reduce congestion, protect the environment, avoid accidents and improve accessibility.

Birmingham's five transport objectives within its overall vision were:

- efficiency in the use of resources
- accessibility within and outside the city
- an enhanced environment, including townscape and safety
- economic regeneration
- practicability, including financial feasibility.[9]

These broad objectives indicate the directions in which strategies should be developed. They are sufficient to indicate that the appraisal procedures should predict and assess the level of congestion, noise, pollution, accidents and access. They also provide a means of assessing the relative performance of different strategies in reducing pollution or accidents. They do not, however, indicate whether a particular solution is adequate in its impact. To do this more specific, quantified objectives are needed.

3.4.3 Quantified objectives

Quantified objectives may indicate a requirement, for example, to avoid frontage noise levels in excess of 68 dBA, or residents without cars being more than 30 minutes from the nearest superstore. They provide a clearer basis for assessing performance of the strategy, but they do require careful definition if the specified thresholds are to be realistic. The study of Bury St Edmunds, for example, specified upper thresholds for delay of 60 seconds at signalised junctions and roundabouts, and 25 seconds at other junctions, and developed a series of quantifiable indicators for other types of objective.[10] Once this is done, quantified objectives provide a direct basis for identifying problems, for current or future conditions, on the basis that a problem occurs wherever the quantified objective is not met.

3.4.4 Solution-specific 'objectives'

It is important to avoid specifying solutions within the objectives, since this constrains the search for solutions, and may well lead to an overall strategy which is less appropriate to the area's needs. Where politicians, or interest groups, wish to introduce general objectives such as to impose physical restrictions on car use, it is preferable to ask why this solution is being proposed and what it is designed to achieve. Answers to such questions should lead to a clearer specification of the true underlying objectives. The solution-specific 'objective' can then be replaced by the set of underlying objectives, and the proposed solution can be tested alongside others in the strategy formulation process.

3.5 A possible set of objectives

While the specification of objectives, and of priorities among them, is the responsibility of the politicians concerned, a number of objectives regularly appear in such statements. This chapter lists and defines a number of possible objectives. They are expressed as higher level objectives rather than as detailed quantified objectives, since the thresholds required for the latter will to a large extent be location-specific. They are not necessarily listed in priority order.

3.5.1 Economic efficiency

Much economic analysis is concerned with defining 'efficient' allocations of scarce resources. Economic efficiency is achieved when it is impossible to make one person or group in society better off without making another group worse off. In such a situation, it is impossible to find any measures for which – if they were undertaken – the gainers would be able to compensate the losers and still be better off themselves. In other words, seeking economic efficiency means taking all measures for which the 'willingness to pay' of the beneficiaries exceeds the 'required compensation' of the losers.

Such a definition, applied to transport, would involve comparing benefits to travellers such as faster travel time with disbenefits such as increased noise and pollution. This would subsume virtually all of the objectives listed below.

In practice, in transport, the efficiency objective is defined more narrowly. It is concerned primarily with maximising the net benefits, in resource terms, of the provision of transport. Efficiency defined in this way is central to the principles of social cost-benefit analysis, and a higher net present value from a cost-benefit assessment represents a more efficient outcome. However, it is based directly on the values which individuals assign to their journeys, and there has been some concern recently that the resulting emphasis on increases in the amount of travel, and in speed of travel, may not be wholly consistent with the needs of society. These concerns are probably better treated under other objectives such as that of sustainability.

While some cost-benefit analyses focus solely on the costs and benefits for motorised travel, and treat the impacts on pedestrians and cyclists, such as pedestrian delay, as an environmental impact, it is more logical to consider economic efficiency for all travellers together, whatever their mode of travel.

3.5.2 Environmental protection

The environmental protection objective involves reducing the impact of transport facilities, and their use, on the environment of both users and non-users. Traditionally, the environmental impacts of concern are those listed in the UK Design Manual for Roads and Bridges.[14] They include noise, atmospheric pollution of differing kinds, vibration, visual intrusion, severance, fear and intimidation, and the loss of intrinsically valuable objects, such as flora and fauna, ancient monuments and historic buildings through the consumption of land.

While some of these can be readily quantified, others such as danger and severance are much more difficult to define and analyse. Attempts have been made, with impacts such as noise and pollution, to place money values on them, and hence to include them in a wider cost-benefit analysis, but it is generally accepted that it will be some time before this can be done reliably even for those impacts which can be readily quantified.

More recently, particularly following the Rio Summit, the environmental protection objective has been defined more widely to include reduction of the impact of transport on the global environment, particularly through emission of carbon dioxide, but also by consumption of scarce and unrenewable resources.[15] These issues may be better covered under a broader sustainability objective, as discussed below.

3.5.3 Safety

The safety objective is concerned straightforwardly with reducing the loss of life, injuries and damage to property resulting from transport accidents. The objective is thus closely associated with the concerns over fear and intimidation listed under environmental protection above, and these concerns could as readily be covered under either heading.

It has been common practice for some time in the UK to place money values on casualties and accidents of differing severity, and to include these within a social cost-benefit analysis. These values include the direct costs of accidents, such as loss of output, hospital, police and insurance costs, and replacement of property and, more controversially, an allowance for the pain, grief and suffering incurred. The latter are valued using 'willingness-to-pay' techniques.[16] To this extent, the safety objective has been subsumed within the efficiency objective. However, there are some misgivings about some elements of the valuation of accidents, and it is probably therefore helpful to estimate accident numbers directly as well.

3.5.4 Accessibility

Accessibility can be defined as *ease of reaching*, and the accessibility objective is concerned with increasing the ability with which people in different locations, and with differing availability of transport, can reach different types of facility.[17] In most cases accessibility is considered from the point of view of the resident, and assessed for access to activities such as employment, shopping and leisure. By considering accessibility separately for those with and without cars available, or for journeys by car and by public transport, the shortcomings of the existing transport system can be readily identified. It is possible also to consider accessibility from the standpoint of the employer or retail outlet, wanting to obtain as large a catchment as possible in terms of potential employees or customers. In either case, access can be measured simply in terms of the time spent travelling or, using the concept of generalised cost, in terms of a combination of time and money costs.

A reduction in the generalised cost of a journey will contribute to both the accessibility and efficiency objectives, and consideration of accessibility separately may appear to involve double counting. However, it is in practice a useful concept in its own right, since it helps to identify opportunities to travel, whether the journeys are made or not.

3.5.5 Sustainability

The sustainability objective has been defined as being the pursuit of *development that meets the needs of the present without compromising the ability of future generations to meet their own needs.*[18] It can therefore be thought of in transport terms as a higher level objective which considers the trade-off between efficiency and accessibility on the one hand and environment and safety on the other. A strategy which achieves improvements in efficiency and accessibility without degrading the environment or increasing the accident toll is clearly more sustainable.

However, the definition of sustainability also includes considerations of the impact on the wider global environment and on the environment of future generations. Issues to be considered under this heading include the reduction of carbon dioxide emissions, which are a major contributor to the process of global warming,[15] controlling the rate of consumption of fossil fuels, which are non-renewable, and limiting also the use of other non-renewable resources used in the construction of transport infrastructure and vehicles.

It is easy to argue, at the level of the strategy for an individual urban area, that there would be no significant impact on the global environment, and hence that this wider objective can be discounted. The flaw in this argument is that global consumption of fuel and emission of carbon dioxide are the result of a myriad of such local decisions, and need to be treated at this level. It was for this reason that the Rio Summit agreed to impose targets on all industrialised nations; the UK government has since reflected this in its own policy documents.[15,19]

3.5.6 Economic regeneration

The economic regeneration objective can be defined in a number of ways, depending on the needs of the local area. At its most general it involves reinforcing the land use plans of the area. If these foresee a growth in industry in the inner city, new residential areas or a revitalised shopping centre, then these are the developments which the transport strategy should be supporting. At its simplest it can do so by providing the new infrastructure and services required for areas of new development. But transport can also contribute to the encouragement of new activity by improving accessibility to an area, by enhancing its environment and, potentially, by improving the image of the area. The economic regeneration objective therefore relates directly to those of accessibility and environmental protection.

3.5.7 Equity

While all of the above objectives can be considered for an urban area as a whole, they also affect different groups of people in society in differing ways. The equity objective is concerned with ensuring that the benefits of transport strategies are reasonably equally distributed, or are focused particularly on those with special needs. Among the latter may be included lower income residents, those without cars available, elderly and disabled people, and those living in deprived areas. The equity objective will also be concerned with avoiding worsening accessibility, the environment or safety for any of these groups. One way of considering these equity, or distributional, issues is by reference to an impact matrix, which identifies the impact groups of concern to decision-makers (among both residents and businesses) and the objectives and indicators which are of particular concern to them. Fig. 3.2 is an impact matrix taken from the Edinburgh study.[8]

3.5.8 Finance

Financial considerations act primarily as constraints on the design of a strategy. In particular, they are a major barrier to investment in new infrastructure, or to measures which impose a continuing demand on the revenue account, such as low fares. In a few cases, the ability to raise revenue may be seen as an objective in its own right, and it is clearly the dominant objective for private sector participants in a transport strategy. The finance objective can therefore be variously defined as minimising the financial outlay (both capital and revenue) for a strategy or as maximising revenue.

Some elements considered under the financial objective will also be included under efficiency. In particular, the capital costs of new infrastructure and the resource costs of operating vehicles will appear in both. However, the resource value of time will not appear as a financial consideration, since no money is involved. Conversely, the payment of fares by passengers will not appear in a full social cost-benefit analysis, since the fares are simply a transfer payment from passenger to operator, but they will appear in the financial assessment of the operation of the strategy.

3.5.9 Practicability

The other major constraints on strategy design and implementation are practical ones. Issues under this heading include the availability of legislation; the feasibility of new technology; the ability to acquire land; and the simplicity of administration and enforcement of regulatory and fiscal measures. Public acceptability can also be considered under this heading. Flexibility of design and operation, to deal with uncertainties in future demands or operating circumstances, may also be important. The practicability

	Impact group				
Objective/indicator	People in different areas	People travelling for different purposes	People travelling by different modes	Different economic sectors	Different transport operators
Efficiency					
Capital					*
User time		*	*		
Operating cost					*
User money		*	*		*
Accessibility					
Generalised time	*	*	*	*	
Environmental					
Noise	*			*	
Pollution	*			*	
Fuel Consumption			*		
Environmental Quality	*		*	*	
Safety & Security					
Danger	*		*	*	
Insecurity	*		*	*	
Accidents			*		
Economic					
Development	*			*	
Practicability					
Finance					*
Planning					*
Land Availability					*
Operation					*
Enforcement					*

Fig. 3.2 A possible impact matrix[8]

objective can therefore be defined as ensuring that policies are technically, legally and politically feasible, and adaptable to changing circumstances. Table 3.1 provides a useful checklist.[6]

3.5.10 Conflicts, constraints and double counting

While the objectives listed in this Section may all be aspirations of a particular urban area, they will almost certainly not all be able to be achieved. Some will be in conflict with others. For example, the requirements of the efficiency objective may well, by encouraging faster or more frequent travel, run counter to those of sustainability. Equally, means of improving accessibility, by car or by bus, may contribute to increased intrusion into the local environment. The equity objective represents an area in which many of these conflicts are focused. It will almost certainly not be possible to achieve similar improvements in the environment in all areas of a city, or similar increases in safety or accessibility for all modes of travel. For these reasons it is particularly important to be able to specify priorities among objectives (for example that protection of the environment is more (or less) important than economic development) as well as among impact groups.

Table 3.1 Checklist of practicability issues[6]

- *Degree of control* – is the policy or proposal directly under the control of the local authority, or do other decision-makers have to be influenced?

- *Feasibility* – what is the likelihood of the decision being implemented?

- *Scale of resources* – what is the scale of resources, such as land or finance required?

- *Enforcement* – does the proposal require other, supporting enforcement measures to ensure that it is effective?

- *Complexity* – does the policy or proposal involve numerous factors? Most transport policy decisions are, of course, complex but the extent varies.

- *Time-scale* – what is the time-scale for the implementation and the effects of that policy or proposal?

- *Flexibility* – is the decision final or merely a preliminary one that can be revised later?

- *Dependence* – how many and what types of other decisions will be affected, and is the policy or proposal dependent on or supportive of others?

- *Complementarity* – are the proposals complementary, such as light rail and park and ride facilities, or are they independent?

- *Conflicts* – does the policy or proposal conflict with others that have been or are likely to be made?

- *Partitioning* – can the policy or proposal be broken down into a series of simpler, discrete components?

- *Political nature of policies and proposals* – how should the policy or proposal relate to the way that political choices are made, and in what form should information about it be communicated to decision-makers?

Once this is done, the other objectives serve as constraints. The environment can be enhanced subject to there not being too great an adverse impact on economic activity. The safety of cyclists can be improved subject to there not being too great a restriction on bus users. In this way all the objectives, and not just those concerning finance and practicability, can be specified also as constraints: as outcomes which the strategy should avoid.

A final consideration under this heading is that of double counting. If all the objectives are to be considered together in one aggregated assessment, as might occur in a full cost-benefit analysis, it is obviously important to avoid letting any one element count twice. For example, impacts on fear and intimidation, if they could be costed, might be included under environmental or safety issues, but not both. However, for the purposes of strategy development, the objectives and impacts on them are best considered separately as illustrations of the impact of a strategy, and to inform decision-makers. If this is done, then double-counting may well help in assimilation of the strategy. It is thus unnecessary to take steps to avoid it.

3.6 Quantified objectives and targets

While the higher level objectives defined above indicate the directions in which a strategy should aim, they say nothing about the amount which it would be appropriate to achieve. As a result, it may be difficult to judge whether a proposed strategy is successful, or whether more could be achieved. More quantified objectives can be specified in terms of a series of targets, which can be either general or specific.

For some objectives general targets can be readily defined. The two best examples in the UK at present are the government's target of a one third reduction in casualties between 1985 and 2000[20] and the Rio target of carbon dioxide emissions in 2000 of no greater than those in 1990.[15] It would be perfectly feasible to extend these to other attributes of the environmental protection and sustainability objectives, but it is much harder conceptually to suggest overall targets for objectives such as efficiency, accessibility or finance.

The UK Royal Commission on Environmental Pollution[21] suggested in addition targets for noise, local pollutants and carbon dioxide involving:

- reducing day- and night-time exposure to road and rail noise to not more than 65dBA and 59 dBA respectively at external facades
- achieving full compliance with WHO health-based air quality guidelines for transport-related pollutants by 2005
- reducing the emissions of carbon dioxide from surface transport in 2020 to no more than 80 per cent of the 1990 level.

It also proposed targets for reductions in car use and increases in public transport use and cycling, including:

- reducing the proportion of urban journeys undertaken by car in London from 50 per cent to 45 per cent by 2000 and 35 per cent by 2020, and outside London from 65 per cent in 1990 to 60 per cent by 2000 and 50 per cent by 2020
- increasing the proportion of passenger-kilometres by public transport from 12 per cent in 1993 to 20 per cent by 2005 and 30 per cent by 2020
- increasing cycle use to 10 per cent of all urban journeys by 2005, compared to 2.5 per cent in 1993.

There are three problems with setting targets of these kinds. First, there is no guarantee that they are achievable. It may be in practice that performance against other objectives, such as efficiency, finance or practicability, would be much worse if the target were met in full. Second, and conversely, they may be all too readily achieved; if targets are set which are easy to meet they may result in under-achievement against the underlying objective. Third, targets which have to do with the strategy rather than the objectives, such as seeking a given increase in the level of cycling, beg the question as to whether this is the most appropriate strategy for achieving the objectives.

It may well be preferable to develop an overall strategy which goes as far as possible to achieving the priority objectives, without producing unacceptable disbenefits against the lower priority objectives, and then to convert that strategy into a series of targets which can be used to monitor performance and achievement. In other words, the strategy ought to determine the targets, rather than the targets being allowed to define the strategy.

At a more detailed level, it may be appropriate to define specific targets for the achievement of specified objectives in particular locations. These can be produced for each of a series of indicators for each objective. These might include delay under the efficiency objective; noise, pollution and severance under environment; accident rates under safety; and access times under accessibility. Table 3.2 provides an example of this style of treatment.[10]

Table 3.2 Thresholds for problem identification[10]

Issue	Threshold
Access for private vehicles, commercial vehicles, buses, taxis, emergency vehicles and cyclists	The excess of the actual distances over the minimum distances were graded in bands of 0 to 10%, 11 to 25%, 26 to 50%, and more than 50%
Delays to cars and goods vehicles	Lower threshold: 25 s for signals and roundabouts, and 10 s for other junctions Upper threshold: 60 s for signals and roundabouts, and 25 s for other junctions
Delays to buses	As for cars and goods vehicles
Delays to pedestrians crossing roads	Lower threshold: 10 s – minimum average delay equivalent to that at a zebra crossing assuming a safe stopping distance for vehicles Upper threshold: 40 s – maximum average delay equivalent to that for pedestrians at a signalised junction
Parking	Threshold taken as 85% of theoretical supply
Safety	Personal injury accidents per year at junctions involving: all road users – 0.93; pedestrians – 0.16; cyclists – 0.14; motorcyclists – 0.33. Personal injury accidents per kilometre per year on links involving: all road users – 1.12; pedestrians – 0.19; cyclists – 0.17; motorcyclists – 0.40
Noise	Noise levels were graded as follows up to 65 dBA – not a problem; 66 to 70 dBA – slight problem; 71 to 75 dBA – moderate problem; more than 75 dBA – severe problem

Fear and intimidation	Degree of hazard	Average traffic flow over 18-hour day, veh/h	18-hour heavy vehicle flow	Average speed over 18-hour day, mile/h
	extreme	1800+	3000+	20+
	great	1200 to 1800	2000 to 3000	15 to 20
	moderate	600 to 1200	1000 to 2000	10 to 15

This level of treatment makes it easy to identify problems (as discussed in the next section) and to demonstrate achievement either with a predicted strategy or with implementation. However, once again it is essential to avoid targets which are too demanding, or insufficiently so. In some cases the targets can be derived from scientific studies which demonstrate that an impact, such as carbon monoxide levels, causes a serious problem above a given level. An alternative approach is to use attitude surveys, which indicate that people are much more dissatisfied when problems such as delays or noise exceed a given level. However, this presupposes that the target, or threshold, uniquely defines the point at which problems occur. In practice, this will often not be the case, as illustrated in Fig. 3.3, which shows the relationship between facade noise level and percentage of occupants likely to be annoyed. Superimposed on this is the standard 68 dB threshold above which action is required in the case of road construction. It is clear that failure to deal with noise levels below 68 dB will leave a large proportion annoyed. Equally there will be benefits to be gained from a reduction from, say, 75 dBA to 70 dBA, which will not be identified by such a target. A more defensible approach is to define a series of thresholds, including one which it is desirable to attain, and one above which action is definitely necessary. An example of this, taken from Table 3.2 (which adopts this approach) is also shown in Fig. 3.3.

There is no correct set of targets, or thresholds, which can be applied uniformly in all areas. While it is therefore inappropriate to be prescriptive as to the thresholds, it is feasible to provide guidance on the types of indicator which might be used. These will be similar to those for problem identification and are considered in the following section.

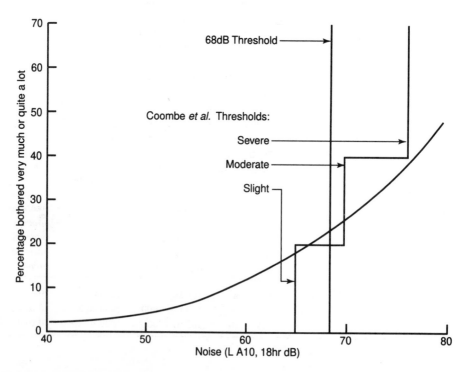

Fig. 3.3 Noise thresholds and attitudes[14,10]

3.7 Problem identification

As noted in Section 3.3, the problem-oriented approach to transport planning starts by identifying problems and developing solutions to them. The objective-led approach defines problems in terms of specified objectives. Both methods converge at the stage of problem identification and then use these as a basis for identifying solutions and strategies (Fig. 3.1). In either case it is essential to be comprehensive in the list of types of problem. This may be difficult to achieve with the problem-oriented planning approach in which there is no pre-defined set of objectives to prompt the question 'how do we know that we have a problem?'

With the objective-led approach the situation is simpler. Once quantified objectives have been defined and defensible targets and thresholds specified, it is a straightforward process to use these for problem identification. This approach was advocated in a study for the UK Department of Transport.[22] Table 3.3 is taken from that report. It lists the objectives in the left-hand column, and uses them to define thresholds, beyond which problems occur, in the right-hand column. Table 3.2 adopts a similar approach, but uses multiple thresholds for some issues. Studies which have adopted this approach have tended to focus on traffic-related issues and it is difficult to find an example of a study which has addressed the full set of objectives listed in Section 3.5 in this way. Table 3.4 suggests a set of possible indicators which could be used for each of these objectives with the exception of practicability, which is covered in the checklist in Table 3.1. As noted earlier, specific thresholds have not been suggested since these will depend on the conditions in a given location, and the expectations of its politicians and public.

When thresholds are defined, they can be used, with current data, to identify the locations, times of day, and groups of traveller or resident for which problems occur. Given an appropriate predictive model, a similar exercise can be conducted for a future year. The model can also be used to assess whether a strategy will overcome these current or future problems, and whether it will induce new ones. As an extension of this approach, it is also possible to compare the current or predicted conditions with the threshold, and assess severity of the problem. A problem which exceeded the upper threshold would be considered as more severe than one which only exceeded the lower one. Equally, a problem which exceeded the threshold by a greater margin could be identified as more severe.

Table 3.3 Objectives and problem indicators for urban road appraisal[22]

Issue group	Issue headings	
Efficiency	Delay	
	● private vehicles	● cyclists
	● commercial vehicles	● pedestrians
	● public transport	
Safety	Road accidents	
Human environment	Occupiers/users of facilities	
	● noise	● visual impacts
	● vibration	
	Pedestrians	
	● noise	● visual impacts
	● pedestrian delay	● severance
	● air pollution	● fear and intimidation

Table 3.4 Suggested indicators for different transport policy objectives

Objective (See Section 3.5)	Indicators
Economic efficiency	Delays for vehicles (by type) at junctions Delays for pedestrians at road crossings Time and money costs of journeys actually undertaken Variability in journey time (by type of journey) Costs of operating different transport services
Environmental protection	Noise levels Vibration Levels of different local pollutants (CO, HCs, NO_x, particles) Visual intrusion Townscape quality (subjective) Fear and intimidation (see Table 3.2) Severance (subjective)
Safety	Personal injury accidents by user type per unit exposure (for links, junctions, networks) Insecurity (subjective)
Accessibility	Activities (by type) within a given time and money cost for a specified origin and mode Weighted average time and money cost to all activities of a given type from a specified origin by a specified mode
Sustainability	Environmental, safety and accessibility indicators as above CO_2 emissions for the area as a whole Fuel consumption for the area as a whole
Economic regeneration	Environmental and accessibility indicators as above, by area and economic sector
Finance	Operating costs and revenues for different modes Costs and revenues for parking and other facilities Tax revenue from vehicle use
Equity	Indicators as above, considered separately for different impact groups (see Fig. 3.1)
Practicability	See Table 3.1 for a useful checklist.

3.8 The instruments of transport policy

Transport planners have available to them, at least in principle, a wide range of instruments of transport policy. These are the means by which the objectives described above can be achieved, and problems overcome. These instruments can be categorised in several ways. This chapter considers them under the headings of infrastructure provision; management of the infrastructure; information provision; pricing; and land use measures. Where relevant it considers in order, under these headings, measures which provide for the private car; public transport; cyclists and pedestrians; and freight.

The key question with each of the measures is its ability to achieve one or more of the objectives. Unfortunately, this is an area of transport policy in which information is often sparse. Experience with some measures, such as bus priorities and cycle lanes, has been well documented through a series of demonstration projects. In other cases, of which road construction is the most glaring example, there have been very few before and after studies to provide the evidence on the impact of the measure. Even where the evidence exists, it may be difficult to generalise from it, since results in one urban context are not necessarily transferable to another. In the absence of real life trials, the most obvious source of evidence is desk studies, typically using computer models. This is, for example, still the only source of guidance on the impact of road pricing in UK cities. Such results are, of course, only as reliable as the models which generate them. Sections 3.9 to 3.13 provide a brief summary of the evidence on each of the main types of measure available, and provide references to particularly useful sources of such evidence.

3.9 Infrastructure measures

3.9.1 Provision for the car

New road construction

This has traditionally been justified in terms of the savings which it generates in travel time, primarily for cars and commercial vehicles, and which typically account for around 80 per cent of the predicted economic benefits. However, these economic benefits have been called into question by the 1994 SACTRA Report[23] which indicates the extent to which they can be eroded by the generation of additional traffic. New roads can, by bypassing particularly sensitive areas, achieve substantial environmental improvements there, as evidenced by a series of studies.[24] In this way, orbital roads may have a different impact from radial ones. However, these are only likely to be sustained if steps are taken to avoid traffic growth on the roads which are relieved of traffic. The key environmental issue, however, is the impact of the new road on the urban fabric and adjacent communities. New roads should almost certainly contribute to a reduction in accidents, by transferring traffic to purpose-built roads whose accident rates should be much lower than those of typical urban streets. To some extent this effect, too, will be eroded by the generation of new traffic. New roads are likely to have an adverse impact on sustainability. They focus particularly on the car, and are likely to encourage its use for faster and longer journeys. This in turn will make public transport, cycling and walking relatively less attractive, and increase fuel consumption and carbon dioxide emissions.

New roads will contribute positively to accessibility for their users: that is for cars and commercial vehicles travelling in that corridor. Relief of other roads may also help local accessibility. However, these benefits disappear rapidly if the new road, and its enhancement of accessibility, attract new traffic. Moreover, new roads may well, if not carefully designed, worsen accessibility across the alignment, particularly for pedestrians and cyclists. The impact of new roads on economic regeneration is far from clear. If they improve accessibility to an area, they may well encourage new development there but, as noted above, the impact on accessibility may well be short-lived. Moreover, accessibility is a two-way process, and a new road to an inner urban area may well encourage development on the periphery of the urban area, from which the inner areas can now be served more readily. An early report concluded that it was only where roads crossed major barriers to movement that the economic regeneration benefits would be significantly greater than those measured through time savings alone.[25] New roads clearly fail to contribute to an equity objective, since they provide in the main for those who already have the greatest advantages in terms of transport.

New roads in urban areas are extremely expensive; costs of £20 million per kilometre are not uncommon and provision for environmental protection may result in figures substantially above this. Even significant time and accident savings may be difficult to justify when set against such costs. The main practical constraints on road building are the time taken between design and implementation, which is often made more protracted by the strength of public opposition; and the restrictions imposed by existing built form and services.

New car parks

These are the other main way in which infrastructure is provided for cars. There is even less evidence of their impact, but much will depend on the measures which complement such provision. Additional parking provision can contribute to economic efficiency by reducing the need to search for parking space. Although there is little hard evidence, it does appear that a significant part of town centre traffic is made up of cars searching for available parking space. However, lack of parking also acts as a control on car use, and expansion may simply encourage additional car use. New off-street parking is probably therefore best combined with a reduction in on-street parking. This should reduce searching traffic (since parking locations are clearer), improve the environment and increase safety. It may, however, aggravate accessibility problems, particularly for those who need to park close to their destination. As with new roads, the cost of parking provision, time-scale and land availability are likely to be significant constraints.

3.9.2 Provision for public transport

Conventional rail provision

This involves both the opening of new rail lines and provision of new stations. There are several well documented studies of the impact of such measures, and procedures for predicting their effects. Such schemes contribute to efficiency both by reducing travel time for existing users and by attracting users from other modes. Several studies have shown that, while around 60 per cent of new usage comes from bus, around 20 per cent is transferred from car use, and 20 per cent newly generated.[26] Transfer from bus may contribute positively if bus mileage, and its contribution to congestion, can be reduced. The transfer from car will also contribute positively to the environment, while the reopening of closed lines and stations, if carefully designed, should have little negative environmental impact. It will also contribute positively to safety. So, potentially, can the reduction in bus use. The Tyne and Wear Metro generated a 17 per cent reduction in accidents in the city centre, largely through reductions in bus movements,[27] but unfortunately this was lost once deregulated buses were able to compete with the Metro. The impact of rail infrastructure projects on sustainability is uncertain. By reducing levels of car use they reduce energy consumption and hence CO_2 emissions; however, they may encourage longer distance travel and more decentralised patterns of land use.

Rail infrastructure measures contribute positively to accessibility, by reducing access distances to public transport, by reducing waiting times and, particularly, by increasing in-vehicle speeds, since the trains are protected from road congestion. Those schemes which increase penetration of a city centre are particularly effective in reducing walking distances. The impact can be substantial; for example the Tyne and Wear Metro increased by 35 per cent the population within 30 minutes of the city centre. However, these benefits are specific to the areas immediately around the new facilities. The effect on economic regeneration is disappointing.[27] While accessibility has been much improved, and rail infrastructure presents a positive image and enhanced environment for an area, there is little evidence from detailed studies of Glasgow and Newcastle that it has attracted significant investment by industry. The one positive impact has been on maintaining the vitality of the city centre. Rail infrastructure projects are likely to have positive equity implications, since they offer a service which can be used by all. However, these benefits are limited to the corridors directly served, and any resulting reduction in bus services may disadvantage certain groups of travellers.

Rail infrastructure projects vary substantially in cost. A single new station may be able to be constructed for under £0.2 million, and a line reopened for as little as £3 million per kilometre (1995 prices). At the other end of the spectrum, the cost of the proposed Crossrail in London is estimated at over £100 million per kilometre. Cost may therefore be a substantial barrier to implementation. The main practical constraints, as for other infrastructure projects, are time-scale for implementation and availability of land. Schemes of this kind will only be of relevance where journeys in excess of 5 km can be made by rail. Below this, bus services, with their more frequent stops and better town centre penetration, will provide shorter access times. This in turn limits application to urban areas with a population of over 250 000.

Light rail

Light rail has become a very popular alternative to conventional rail provision in recent years. In many ways it can be expected to have a similar impact, although there is as yet little documented evidence. Its main differences are that it can operate on street, have more frequent stops, and achieve better penetration of town centres.[28] Its impacts on efficiency, the environment, safety, accessibility, sustainability, economic regeneration and equity are thus likely to be similar to those of conventional rail, with a few exceptions. While its efficiency impact will generally be positive, light rail may potentially have adverse impacts on efficiency if capacity for other traffic has to be reduced. It can also have an adverse impact on the environment, and this has been a significant barrier to implementation in some cases. Conversely, it is likely to provide greater accessibility, by having more frequent stops, and thus be appropriate in slightly smaller urban areas. Even so, the minimum population to justify such provision will be in excess of 150 000. Finance is again a major barrier. Light rail schemes are expensive, not least because of the requirements of street running, and typically exceed £5 million per km.

Guided bus

This vehicle (see Chapter 9) provides a lower cost alternative to light rail. While totally separate rights of way can be provided, as in Adelaide, most UK proposals envisage providing guideways solely where buses need to bypass congestion. This can be achieved with minimal space requirements; the guideway need only be 3 m wide, and is only needed in the direction in which congestion is experienced. Specially equipped buses can then operate normally on the rest of their routes, hence providing much more extensive suburban coverage than light rail.[29] The impact of guided bus is as yet uncertain. It should have less adverse impact on efficiency than light rail, by requiring less space, but its positive impacts depend critically on its ability to attract patronage. If it is perceived by car users as a slightly improved bus it will be unlikely to contribute significantly to the reduction of congestion, environmental impact and accidents, and will perform much as bus priority measures do. If it is seen as a higher quality service approaching that of rail, its impact will be much greater. While its impacts are uncertain, the costs of provision are clearly much lower than those of light rail.

Park-and-ride

This extends the catchment of fixed track public transport into lower density areas, by enabling car drivers to drive to stations on the main line. It has also been used successfully in smaller cities such as Oxford and York in conjunction with dedicated bus services.[30] The parking facility itself provides a low cost way of extending the benefits of public transport, by increasing the numbers able to use public transport, and hence reducing congestion, environmental intrusion and accidents in inner urban areas. It does not, however, offer significant improvements in accessibility and equity since, by definition, only car users can use the facility. Some doubt has been cast on the true benefits of park-and-ride; it has been suggested that it may in practice generate longer journeys by rural residents, and hence increase car use. However, the beneficial impact on inner urban areas is not in dispute. The net effect will depend on where the facility is located. Land availability and cost are likely to be the main practical barriers although several recent schemes have been financed as part of new retail developments.

3.9.3 Provision for cyclists and pedestrians

Cycle routes

These provide dedicated infrastructure for cyclists, and hence extend the range of cycle priorities. As well as making cycling safer, they have been designed to attract more people to cycle in preference to driving, hence achieving the benefits of reduced car use. In this, in the UK at least, they have so far proved unsuccessful.[31] It appears that cycle routes can achieve travel time benefits for cyclists, but will not attract more people to cycle in the absence of other measures. Moreover, any reduction in cyclist casualties on main roads appears to have been balanced by an increase in casualties on minor roads. This being the case, the costs of provision need to be offset solely against the travel time benefits. The costs themselves will depend very much on the availability of suitable corridors and land availability.

Pedestrian areas

These provide a dramatic improvement in the environment for pedestrians, and have proved very successful in enhancing trade in many town and city centres. There is little evidence to support traders' claims that pedestrian streets cause a loss in trade, provided of course that they are well designed.[32] As well as achieving environmental and safety benefits, they may well therefore have positive impacts on the urban economy and on land use policy. However, they present some accessibility problems for car and bus users and, particularly, for goods deliveries and for disabled people. Exemptions for some of these, whether permanently or, as with deliveries, at certain times of day, inevitably reduce the benefits somewhat. Pedestrian areas almost inevitably reduce efficiency, by reducing road capacity. They also potentially cause disbenefits in the surrounding area, both by diversion of traffic and by attracting trade to the protected area. These potential adverse impacts can be reduced by careful design. Aesthetic design is of crucial importance in maintaining trade and will in turn inevitably add to the costs of the measure.

3.9.4 Provision for freight

Lorry parks

These facilities provide a means of reducing the environmental impact of on-street overnight parking of lorries. Dedicated provision in a well designed and secure parking area, together with on-street restrictions, may well be beneficial.

Transshipment facilities

These aim to provide a means of transferring goods from the larger vehicles needed for efficient line haul to smaller, less environmentally intrusive vehicles for distribution in town centres. Some proposals have also envisaged trolleying of goods over short distances and, at the other extreme, underground freight distribution. Experience to date in the UK suggests that such facilities are unlikely to be attractive to freight operators, and hence to be cost effective, at least until much greater restrictions on existing practices can be justified.[33]

Encouragement of other modes

Attempts to encourage usage of other freight modes are likely to focus primarily on railborne freight, but in appropriate cases could extend to water and pipeline. Such modes are only competitive over longer distances and for bulk freight, given the additional costs for handling and road-based distribution. Moreover, the road-based distribution is likely still to take place in urban areas; the main beneficiaries are likely to be communities on inter-urban routes.

3.10 Management measures

3.10.1 Provision for the car

Conventional traffic management

Included under this heading are such a wide range of measures that it is impossible to cover them all. In general, measures such as one-way streets, redesign of junctions, banned turns and controls on on street parking have been shown to have beneficial impacts on travel time and on accidents, and typically to repay the costs of implementation within a matter of months. It is, however, necessary to bear in mind their possible adverse impacts. If such measures cause some traffic to reroute, journey lengths may increase and, in the extreme, this could more than offset the benefits of any increase in speed. The efficiency benefits are particularly sensitive to this process. Such rerouting may also introduce environmental intrusion into previously quiet streets. Accessibility may be reduced for certain users: one-way streets pose problems for bus services and deliveries; parking restrictions affect local frontages; and, in the extreme, a gyratory system can make access to the 'island' very unattractive.[34] Finally, any measure which reduces the cost of car use may encourage it to increase. There is as yet little evidence of this effect, which will be smaller in scale than that now attributed to new roads,[23] but it could potentially offset many of the resulting benefits. A major practical consideration with all traffic management is that of enforcement. Unless measures are self-enforcing, the costs of enforcement action need to be included in any appraisal.

Urban traffic control (UTC)

These systems are a specialist form of traffic management (see Chapter 27) which extends the principles of traffic signal control by integrating the control of all signals over a wide area, using the control parameters of split (of green times in a cycle), cycle time, and offset (of the start of green at a given junction) to optimise a given objective function such as minimising travel time or stops. Widespread trials have demonstrated the benefits of such systems. An up to date fixed time system can achieve savings in travel time of up to 15 per cent, although this may be degraded by as much as 3 per cent per annum as traffic patterns change. A vehicle-responsive system may achieve as much as a 20 per cent saving, which should be able to be maintained.[35] Such efficiency gains also improve the environment (since there are fewer stops and queues) and safety (with typical reductions in accidents of the order of 10 per cent). However, the potential for these benefits to be eroded by generated traffic, as mentioned above, needs to be borne in mind.

Advanced transport telematics (ATT)

This is the current title for the range of applications of information technology to transport which have been developed under the EU DRIVE programme.[35] This includes more recent developments in UTC to encompass, for instance, motorway access control and queue management techniques, for which results are beginning to be reported. It also includes the extension of UTC to provide priorities for buses, and their integration with information systems such as dynamic route guidance, which are covered in more detail in Sections 3.10.2 and 3.11.1. While results are beginning to be made available, it is too early to judge the effectiveness of these techniques in widespread application. Some technologies are still at the prototype stage, and will require further development before their practical limitations are clear.

Accident remedial measures

Accident reduction measures[36] cover a wide range of possibilities; for example, see Chapters 15 and 18. Most blackspot treatment and mass action measures (such as skid-resistant surfacing) will have few impacts other than a reduction in accidents; their effects on other objectives can therefore be considered minimal. Area-wide measures are likely to have other impacts, and are considered below under the general heading of traffic calming.

Traffic calming measures

These measures are designed to reduce the adverse environmental and safety impacts of car and commercial vehicle use. They have traditionally focused on residential streets, for which Buchanan, in *Traffic in Towns*, proposed an environmental capacity of 300 vehicles per hour,[37] and have involved two types of approach: segregation, in which extraneous traffic is removed; and integration, in which traffic is permitted, but encouraged to respect the environment. More recently they have also been extended to main roads, where integration is the only possible solution.[38]

Segregation can be achieved by the use of one-way streets, closures and banned turns, which create a 'maze' or 'labyrinth', which makes through movement difficult, and hence diverts it to surrounding streets. The extra traffic on surrounding streets can add to congestion and environmental intrusion there, and this trade-off needs to be carefully considered at the design stage. However, the maze treatment also reduces accessibility for those living in the area, and this loss of accessibility has often led to the rejection of such measures by the residents whom they are designed to benefit.[39] An alternative approach, more often used in city centres, is the traffic cell, in which an area is divided into cells, between which traffic movement, except perhaps for buses and emergency vehicles, is physically prohibited. This can also cause some access problems, particularly where parking supply and demand in individual cells is not in balance, but experience suggests that these are outweighed by the environmental benefits.[40]

Integration measures include low speed limits, speed humps, chicanes, pinch points, resurfacing and planting, all designed to encourage the driver to drive more slowly and cautiously. Experience in the UK with measures other than speed humps is still limited, but it is clear that these can achieve significant reductions in speed and accidents. Moreover, by making routes through residential areas slower, they can also induce rerouting, and hence a reduction in environmental impact. Such benefits may, of course, be offset by increases in congestion and environmental impact on the diversion route. This apart, such measures are likely to generate significant environmental, safety and equity benefits, without adversely affecting accessibility. The key issue currently is that of balancing cost of provision with effectiveness and visual quality.[38]

Physical restrictions on car use

These have been proposed as ways of reducing car use in urban areas. Possibilities include extensive pedestrian areas and traffic calming, and also the use of bus lanes to reduce capacity at junctions and give clear priority to buses. There has in practice been little experience of such measures, with the one notable exception of the Nottingham 'zones and collar' experiment.[41] Moreover, the reduction in road space is likely to increase the total time spent travelling in the network, thus reducing efficiency, and to increase queues, thus adding to the environmental impact. There was also some evidence in Nottingham of the queues encouraging unsafe driving practices. A serious practical problem is the need for road space in which to store the queues generated.

Regulatory restrictions on car use

These have been used as an alternative way of reducing car use. In several Italian cities, permits are allocated to those who can justify needing their cars in the centre, and others are banned. In Athens and Lagos an 'odds and evens' system operates, in which cars with odd number plates can enter on alternate days, and those with even numbers on the other days. Jakarta has its own regulatory system ('three in one') in which only cars with three or more occupants are permitted on certain roads. Permit systems are likely to prove expensive in terms of the resources required to check the validity of applications and to issue permits,[42] and there will inevitably be an element of rough justice in the way that they are allocated. Those without permits may in some cases experience a serious reduction in accessibility. That apart, such a system should in practice be able to achieve any required reduction in car use. Experience with the Lagos system suggests that, while it is easier to operate, it is less effective, since drivers can respond by owning two cars, and some who would not otherwise have chosen to drive may elect to do so.[43]

Parking controls

Parking controls potentially provide a more effective way of controlling car use. Controls can be by reducing the supply of spaces, restricting duration or opening hours, and regulating use through permits or charging. Local authorities are able to impose any of these controls on on-street space and in publicly operated car parks. The main problems, however, are that controls cannot readily be imposed on private non-residential parking, which typically accounts for 40 per cent to 60 per cent of all town centre space, or on through traffic. As a result, even the harshest controls on public parking may simply result in an increase in traffic parking privately or driving through the area.[44] Where private non-residential space is small, or already fully used, and through traffic can be controlled, parking controls can be effective in reducing car use. This in turn should reduce congestion, environmental impact and accidents. However, performance will depend very much on the way in which controls are applied. Simply reducing space is likely to increase the amount of time spent searching for parking space, which may have adverse impacts on congestion. However, one study in The Hague reports a 20 per cent reduction in car use as a result of car park closure.[45] Targeted restrictions on duration or on categories of parker will avoid this problem, but will introduce inequities similar to those from other regulatory restrictions. Inevitably accessibility will be reduced for some categories of parker with all types of parking control. Controls are generally inexpensive to implement, but may require continuing expenditure on enforcement if they are to be effective.

Car sharing

The sharing of cars offers a means of reducing car traffic while retaining many of the advantages of private car travel. Several experiments have aimed to encourage drivers to share their cars with others or to 'car pool' by taking it in turns to drive. Unfortunately, experience suggests that the numbers sharing voluntarily, even with incentives, are unlikely to exceed 5 per cent of car users, and that their passengers are as likely to transfer from bus use as from other cars.[46] Such schemes are thus likely to have a minimal impact in urban areas although, at the margin, they may offer some improvement in accessibility.

3.10.2 Provision for public transport

Bus priorities

The provision of priority for buses enables them to bypass congested traffic and hence to experience reduced and more reliable journey times. The most common measures are with-flow bus lanes; others include exemption from banned turns, selective detection at signals, and UTC timings weighted to favour buses. Contra-flow bus lanes and bus access to pedestrian areas are designed specifically to reduce the adverse impact on buses of certain traffic management measures. Conventionally, bus priorities in the UK are designed to keep loss of capacity to other traffic to a minimum. With with-flow bus lanes this is done by providing a setback at the stop line. Provided that this is done, efficiency is usually improved; travel time savings to buses can exceed 25 per cent while there are few losses to other traffic. The segregation of traffic also appears to enhance safety. The main disadvantages are to frontage access, if parking is restricted, and to the environment, since queues will be longer, and traffic diversions may be induced.[47] Unfortunately there is little evidence from the UK that such priorities induce a switch from car to bus travel; thus the potential wider efficiency and environmental benefits are not achieved. It appears that more continuous application of bus lanes, as practised in Paris, may be more beneficial in this regard. The main practical limitations with bus lanes are the lack of sites with suitable space for the extra lane and storage of the longer queue of other vehicles, and the need for effective enforcement. A more recent development in bus priorities has been the use of 'Red Routes' in London, in which bus lanes are combined with intensive, well enforced, parking restrictions. Travel time savings on the pilot Red Route have been dramatic, while the evidence on effects on frontage access and trade is mixed.[48]

High occupancy vehicle lanes

HOV lanes extend the use of with-flow (and potentially contra-flow) bus lanes to other vehicles which make more effective use of scarce road space. These can include car sharers, taxis and commercial vehicles. There is as yet little experience of such measures in the UK. Experience elsewhere suggests that these can provide greater benefits than conventional bus lanes, provided that the delays to buses are not great. However, enforcement is more difficult.

Bus (and rail) service levels

The level of public transport service can be modified to increase patronage, and hence to attract diversion from car use. For bus services the main options are to increase route density or to increase frequency on existing routes. The first of these reduces walking time, while the second affects waiting time. Since both of these have a greater impact on passengers than does a similar change in time on the bus, they can be expected to be more effective in increasing patronage.[47] The most appropriate allocation of a given fleet of buses between denser and more frequent routes will depend on local circumstances. Other bus service measures include the use of minibuses, which are more expensive per seat-km to operate, but can achieve greater penetration and may be more attractive; and demand-responsive bus services, such as dial-a-bus, whose operating costs have tended to exceed benefits. There is also a wide spectrum of paratransit measures involving unconventional bus and taxi services; their impacts are too varied to summarise here. With rail services, the only option available is usually to increase service frequency.

The effects on patronage can be measured by elasticities of demand. Elasticity of demand with respect to bus-km run is typically around +0.6, suggesting a 6 per cent increase in patronage for a 10 per cent increase in service level. However, under deregulation, response to bus service increases has been much lower. Unfortunately cross-elasticities for car use are typically an order of magnitude lower, at around −0.08, suggesting less than a 1 per cent reduction in car use for a similar service improvement.[47] Cross-elasticities are typically somewhat higher for rail service improvements. These elasticities will apply, of course, only to the corridor in which the improvements are introduced. Thus service improvements have, as their primary impacts, improvements in accessibility and equity; they are unlikely on their own to achieve significant efficiency or environmental benefits except, to a limited extent, within the specific corridor.

The main practical barriers to such measures are costs of operation and the fact that, in the UK, responsibility lies with private operators. In theory, at least, the cost of service increases can be met from increased fares revenue. Indeed, it has been suggested that an optimal operating strategy within given financial constraints may be to increase both service levels and fares.[47]

Bus service management measures

Operational measures can be designed to improve the reliability of bus services, to enhance their quality of service, or to reduce operating costs. Measures which improve reliability, including real time garage and fleet management procedures, are likely to be particularly beneficial, since uncertainty in travel time, and the extra waiting time resulting from irregular services, are major disincentives to travel. Such measures should generate significant efficiency benefits, and can potentially contribute to reduced car use.

3.10.3 Provision for cyclists and pedestrians

Cycle lanes and priorities

These serve the same function as cycle routes (Section 3.9.3) and experience with them is similar. They reduce accidents for cyclists, and may encourage some increase in cycle use, but have yet, in the UK, to achieve a transfer to cycling from other modes.[31] They are easier to implement than bus lanes, because they require less road space, but still pose problems of frontage access.[31]

Cycle parking

The provision of parking facilities for bicycles, to increase availability and security, may also be beneficial, but its impacts have rarely been studied.

Pedestrian crossing facilities

These are primarily a safety measure but may also reduce travel time for pedestrians. Indeed, it is not uncommon to find that total delay to pedestrians at city centre junctions exceeds that for vehicle users. In such circumstances, reallocation of signal time and linking of pedestrian phases may achieve efficiency benefits. There are relatively few other ways in which pedestrians' travel time can be improved, but other measures such as parking controls and footway widening may improve their environment and safety. Unfortunately there has been little or no analysis of the effects of such measures on pedestrian activity.

3.10.4 Provision for freight

Lorry routes and bans

This form of traffic management is primarily designed to reduce the environmental intrusion of heavy lorries, rather than to improve their operating conditions. Lorry routes can be mandatory, but are more frequently advisory, and thus avoid serious reductions in freight accessibility. Bans can be area-wide (for example in the cells between lorry routes) or limited to particular roads, or applied solely to short lengths of road forming a screenline or cordon. They can be complete, or limited to certain times and certain sizes of vehicle, or with exemptions for access. Such exemptions avoid problems of lost accessibility, but are difficult to enforce. Generally, restrictions on lorries are likely to result in reduced efficiency, and will require increased enforcement costs. Conversely they should, if well designed, improve the environment and safety. There have been relatively few studies of such measures, although that for the Windsor cordon demonstrated that any environmental benefits may be more than offset by increased operating costs, and by environmental losses on the diversion routes.[49]

3.11 Information provision

3.11.1 Provision for the car

Conventional direction signing

Good direction signing can provide benefits to car users, and other traffic, by reducing journey lengths and travel times; evidence suggests that around 6 per cent of travel time may be accounted for by poor routing, and that inadequate destination signing may as much as double the time spent searching for unfamiliar destinations.[50] Conversely, direction signing can be used to divert traffic away from environmentally sensitive routes; however, familiar drivers are unlikely to respond to such measures.

Variable message signs

These enable drivers to be diverted away from known, but unpredictable, congestion. They are very location-specific in their application, and hence in their benefits. Potential benefits are primarily in terms of efficiency; drivers are unlikely to be willing to divert in significant numbers to avoid environmental and safety problems.

Real-time driver information systems and route guidance

These are the most rapidly developing form of Advanced Transport Telematics (ATT) application.[35] Information from equipped vehicles or traffic sensors is used to provide radio or in-vehicle display messages of delays, or to indicate preferred routes to avoid congestion. Evidence suggests that familiar drivers are more likely to prefer information, and to choose their own routes, while unfamiliar drivers prefer guidance.[51] Several studies have predicted reductions in travel time of around 10 per cent from such systems, when applied in urban areas, together with reductions in accidents.[52] Unfortunately, the only documented field trial of dynamic route guidance, in Berlin, has suggested that the benefits may be much lower than this.[53] Most benefits will, of course, accrue to equipped vehicles in the form of improved accessibility; benefits for other private traffic and for buses may well be very small, thus raising important equity considerations. It has also been suggested that improved information may generate additional travel. There is, however, little evidence to support such claims.

Parking information systems

These are a further application of ATT principles, designed to reduce the high level of traffic searching for parking space in urban centres. Detectors identify car parks which are full or almost full, and trigger signs indicating the route to the nearest available space. Studies have demonstrated a significant reduction in time spent finding a parking space, but it has proved more difficult to estimate the resulting reduction in vehicle-km.[54] The efficiency and accessibility benefits from reduced searching may be associated with some reductions in environmental intrusion and accidents, but these will depend on the local circumstances.

Telecommunications

Modern telecommunication technology provides an alternative to travel for all, but studies have focused particularly on their use as an alternative to car travel. Teleworking, through which employees can work at home, has been more extensively studied, but there is interest also in teleshopping and teleconferencing. Studies in the US and Holland suggest that teleworking can reduce car use; typical teleworkers work from home two days a week, and their cars are used much less on the days when they are at home.[55] Such reductions may have efficiency benefits, and will also contribute to environmental and sustainability objectives; they will not reduce accessibility, since teleworking provides an alternative to travel. However, their impact is limited by employees' and employers' willingness to permit working from home.

Public awareness campaigns

Campaigns have been developed recently by several local authorities as ways of making residents, and particularly car users, more aware of the effects of their travel behaviour on the environment, and to alert them to the alternatives available, including use of other modes and changes in destination and frequency of travel. Early surveys have highlighted the conflict between individuals' preferences for car use and their concern over the impacts of car use.[56] It may therefore be possible to channel the resulting sense of guilt into more environmentally appropriate travel behaviour; however, it is too early to assess the impact of such schemes.

3.11.2 Provision for public transport

Service information

Timetabling, which is the basic form of service information to public transport users, has become degraded in many areas in the UK since bus deregulation. Indeed, studies of deregulation have identified lack of service information, aggravated by more frequent timetable changes, as one of the main causes of lost patronage[57] and surveys have indicated the potential for improved information to generate additional patronage. This, in turn, could have accessibility and equity benefits and, potentially, help to reduce car use. One problem in a deregulated environment is the reluctance of any operator to contribute to information which includes competitors' routes.

Real-time passenger information

This is now being provided, not just at major terminals, but at individual stations and bus stops, and on trains and in buses. Such information, on delays and alternatives, may on occasion enable travellers to save time by taking alternative routes. Its main impact, however, is in reducing the uncertainty and stress associated with late running services. Studies on the London Underground have attempted to estimate the benefits of such information, and have indicated the potential for increasing patronage.[58] There is some evidence that larger bus operators are prepared to invest in such information systems, in conjunction with local authorities, in order to increase market share.

Operation information systems

These use ATT-based fleet management facilities to identify locations of buses and to reschedule services to reduce the impact of unreliability. Such systems were studied intensively in the 1970s and have been tested more recently under the EU DRIVE programme[35] but there has been little interest as yet in applications in the UK's deregulated environment. Once again, equipment costs are still high.

3.11.3 Provision for cyclists and pedestrians

Static direction signs

Conventional signs are virtually the only measure available under this heading, but can be used to enhance the use of cycle priority routes and to improve access within pedestrian areas for disabled pedestrians. Tactile footways are a further facility providing specifically for visually handicapped pedestrians.

3.11.4 Provision for freight

Fleet management systems

These systems have been introduced widely for freight vehicles, enabling them to respond more rapidly to the changing demands of 'just in time' delivery schedules, and reducing the number of empty return journeys. They can also extend to dynamic route guidance to avoid congestion. Most such systems are, however, introduced by freight operators, and local authorities have little role in their implementation or operation.

3.12 Pricing measures

3.12.1 Provision for the car

Vehicle ownership taxes

These taxes are the most obvious direct charge on the private car. However, there is little evidence that they, or taxes on car purchase, have a significant impact on car ownership. Even if they were to do so, they would have no effect on car use. Indeed, by increasing the proportion of car use costs which are fixed, they could potentially have the opposite effect. They are, however, a major source of revenue which can potentially be used to finance transport investment.

Fuel taxes

These taxes have a more direct effect on vehicle usage. The UK government is committed to a 5 per cent p.a. real increase in the tax rate as a contribution to its sustainability objective, and the Royal Commission on Environmental Pollution has advocated more rapid increases than this.[21] Studies at the time of the 1970s fuel crises suggested a short-run elasticity of around −0.2 to fuel price, suggesting a 2 per cent reduction in car use in response to a 10 per cent increase in price. However, more recent research suggests that around half the response is in terms of increased fuel efficiency; short-run mileage elasticities are nearer to −0.1.[59] Moreover, it appears that most reductions occur in evening and weekend leisure travel. The impact on congestion would thus be very small. Moreover, in the longer term, drivers are more likely to switch to more fuel-efficient vehicles. This would still contribute to fuel savings and hence to the efficiency, environmental and sustainability objectives, but would have little effect on congestion or safety. Fuel taxes bear most heavily on low income drivers and rural residents, whose accessibility they may adversely affect. Fuel taxes are again a major source of revenue.

Parking charges

Charging for parking is one of the most widely used forms of parking control. Uniquely among parking control measures, they enable demand to be kept below the supply of parking space, thus reducing time spent searching. Elasticities with respect to parking charges vary depending upon the availability of alternatives, but figures in the range −0.2 to −0.4 have been quoted.[60] The wider impacts depend on the alternative used by the car driver; parking on the fringes of the controlled area, or in private parking spaces, will inevitably have less impact on efficiency and the environment than switching to public transport. A recurring concern with the introduction, or increasing, of parking charges is that it will encourage drivers, and particularly those shopping, to go elsewhere, thus adversely affecting the urban economy. There is some evidence that this may happen, although it has proved difficult to collect reliable data. Parking charges will affect low income drivers more, and thus have equity implications. They may as a result have some minor effects on accessibility. They are a source of finance, although the potential for profits is usually small. As with parking controls, parking charges can readily be applied to publicly controlled parking space. Parking charges cannot be imposed at private car parks and, by definition, do not apply to through traffic. As noted earlier, these represent major loopholes in the effectiveness of any form of parking control.

Congestion charging

Charging for road use has been proposed in a number of forms. Most envisage charging to cross screenlines or cordons, using paper licences (as in Singapore), toll gates (as in Norway) or fully automated electronic charging. Systems which charge continuously in a defined area, based on time taken, distance travelled or (as in Cambridge) time spent in congestion, have also been proposed. The only direct evidence of the impact of such schemes comes from Singapore[61] and Norway.[62] However, the latter schemes were designed to raise revenue rather than to control traffic levels. Other results are available from the large number of desk studies conducted in the UK and elsewhere.

It is clear that congestion charging could significantly reduce car use in the charged area, and hence reduce environmental impact and accidents. Traffic would divert to boundary routes, other times of day and other modes; much of the transfer would be to bus, which would benefit from the reduced congestion. Careful design is needed to ensure that these alternatives do not themselves become congested. Subject to this, congestion charging can achieve significant efficiency, environmental and safety benefits; it should also increase accessibility, although the time and money costs of private travellers will increase. It will also generate substantial revenue, which can potentially be used to finance other elements of a transport strategy.[8]

The equity impacts are more uncertain. Bus users, pedestrians and cyclists will benefit; rail users will be little affected except, perhaps, by increased crowding; car users, and particularly those on low incomes, will suffer. This is one of the major concerns with congestion charging. A second is the potentially adverse impact on the economy of the charged area if charging encourages drivers to travel elsewhere, on which there is little or no evidence. A third concern relates to the practicability of the technology, which is largely untested, and the enforcement procedures. The underlying barrier to progress, however, is that new legislation is needed to permit congestion charging.

3.12.2 Provision for public transport

Fare levels

Fares can be adjusted on all public transport services, and will have a direct effect on patronage and on car use. Evidence suggests a fares elasticity of around −0.3 for buses and slightly higher for rail. Cross-elasticities for car use are around +0.05.[62] Thus a 10 per cent reduction in bus fares could increase patronage by around 3 per cent, but would only reduce car use by 0.5 per cent. However, unlike service level changes, fares changes apply throughout an urban area, and may thus have a greater absolute impact on car use. The 'Fares Fair' campaign in London in the early 1980s, which reduced fares by 32 per cent, was estimated to have reduced cars entering central London by 6 per cent. Fares reductions can, therefore, contribute to efficiency and environmental objectives, as well as improving accessibility for public transport users and hence equity. There is also some evidence that they can reduce accidents.[63] Their major drawback is their cost. There is also some evidence that low fares may encourage longer distance travel, and hence land use patterns which are in the longer term less conducive to sustainability.

Fares structures

Variations in fare structure include the introduction of flat and zonal fares as alternatives to conventional graduated fares; lower off-peak fares; and travelcards and season tickets which allow unlimited travel within a defined area. There is some evidence that simplification of fares structures may do more than fares reductions to increase patronage.[64] Changes in structure may thus also contribute positively to efficiency, environmental and safety objectives, as well as improving accessibility by reducing the cost of marginal journeys. If appropriately designed, they may not impose a significant additional financial burden. However, many such structures rely on the ability to offer a common set of charges, and free interchange, to all services in an area. Deregulation in the UK has made this more difficult, and many innovative fare structures have been abandoned or drastically curtailed as a result.

Concessionary fares

These provide lower fares or free travel to identifiable categories of passenger with special needs. These may include schoolchildren, elderly people and people with disabilities. Their main objective is equity-related, in enabling people who would otherwise find public transport too expensive, or who cannot use cars, to travel. They probably have no significant efficiency or environmental benefits, but they do improve accessibility for the target population.[65] They do, however, impose a substantial financial burden on the local authorities which support them.

3.12.3 Provision for cyclists and pedestrians

Pricing is rarely an issue for cyclists (except, possibly, for secure cycle parking) or pedestrians.

3.12.4 Provision for freight

The fiscal measures described for cars are relevant for freight as well. Current UK government policy involves charging higher vehicle ownership taxes for larger vehicles to reflect the additional costs which they impose, particularly on road maintenance. There is, however, some evidence that the higher licence charges fail to do this. Fuel taxes cannot readily distinguish between types of vehicle, but parking charges typically vary with vehicle type, and some congestion charging proposals have also envisaged doing this.

3.13 Land use measures

3.13.1 Application to different modes

Land use measures cannot in the main be focused on a particular mode, and are therefore considered together in this section. However, most of them are designed to encourage the use of public transport, cycling and walking, shorter distance journeys (which in turn further favour cycling and walking), and less frequent travel.[66] Land use measures are less likely to significantly influence freight movements, with the obvious exception of encouragement of development near to rail and waterborne freight facilities.

3.13.2 Types of measure

Flexible working hours

These can be considered as a form of land use measure designed to reduce demand for peak travel and the resulting congestion. True flexible hours working provides the employee with flexibility in hours of arrival and departure, while specifying a required core time and number of hours per week or month. In many cases they were introduced by employers to retain employees rather than for transport policy purposes, and the scale of their operation, and impact, is thus not well understood.[67] Staggered hours, in which employers are encouraged to change the fixed working hours of all or a proportion of their employees, were popular in the US in the 1970s, and were designed specifically to reduce peak loadings on the transport system.[68] Another variant which has been discussed, but rarely tested, is the four-day week in which employees work the same hours per week, but travel on one fewer day. Studies of flexible working hours and staggered hours suggest that the efficiency benefits have been small. In some cases they have enabled peak public transport services to be withdrawn, thus saving operating costs,[67] but in the main they have simply transferred travel to slightly less congested times. However, they have achieved significant time savings for those who participate, and thus have accessibility benefits. It had been feared that flexible working hours would discourage car sharing and public transport use. In practice US experience suggests the reverse; some car users switch either to car sharing or to public transport because they can adjust their working hours to match the schedules imposed.

However, all such measures are the direct responsibility of the employer, and can be changed without consultation with transport providers. Local authorities therefore have no real power to influence them, other than through encouragement. There are now a few examples of local authorities working with selected employers to develop more sustainable travel patterns.

Development densities

A local authority can specify densities for new development. Higher densities enable more opportunities to be reached within a given distance, and hence may encourage shorter journeys and use of cycling and walking. By increasing population and employment densities, they may also make public transport more viable. However, there is very little evidence of the scale of these effects, except for cross-sectional comparisons which demonstrate that residents in lower density areas are more likely to use the car and to travel longer distances.[69] As with other land use measures, increased densities are only likely to be effective if developers respond positively and if users elect in practice to travel to opportunities nearer to towns and workplaces. There is evidence that developers are prepared to accept denser development, not least because it increases the return on land investment, but there is little evidence on the response of users.

Developments within transport corridors

These provide a way of concentrating denser development, and activities which can more readily be served by public transport, in those areas where public transport is readily available. This can lead to a corridor-style development, and has been used to considerable effect in cities such as Toronto.[70] The Dutch ABC policy is an extension of this concept. Developments are categorised in terms of their ability to use public transport and their need for road-based freight transport, and allocated to different zones in an urban area.[4] While such strategies are intuitively sensible and should reduce journey lengths, improve accessibility and have some efficiency and environmental benefits, it is difficult to assess their impact on overall travel levels.

Development mix

This can be specified in development plans in such a way that houses are closer to places of work and to other attractions such as schools, shops and leisure facilities. Several US states have introduced a 'jobs-housing balance' into their planning controls for this purpose.[71] However, while such policies should improve accessibility, there is little evidence that users do in practice travel to the jobs and leisure facilities which are nearer to their homes. Generally there is insufficient understanding at present of the impact on travel of different policies for development density and mix. It also needs to be borne in mind that any such policies will only apply to new development and redevelopment, and will thus take some considerable time to have a significant impact.

Travel reduction ordinances

These are a further variant, introduced in the US, in which developers are given permission to develop on the condition that they, or their tenants, produce a plan specifying ways in which they will reduce car use to a level below that which would normally be expected from such a development. As with so many of these measures, there is little evidence of their impact.[71]

Parking standards

Standards for parking probably offer the single most direct impact on levels of car use among land use measures. Conventionally these have required developers to provide at least a minimum number of parking spaces per unit floor area to ensure that all parking generated takes place off street. The resulting parking adds to the stock of private non-residential space, and further reduces the ability of local authorities to use parking controls as a restraint tool. Several local authorities have now proposed much more restrictive standards.[72] Such measures can limit the growth in parking space and hence in car use. As properties with larger car parks are redeveloped, this can lead to efficiency and environmental benefits. There may be some localised reductions in accessibility. However, there is always the danger that, under pressure from developers, local authorities will relax these new standards to attract valuable development. This is particularly the case in areas with competing centres with different parking standards.

3.14 Integration of policy measures

3.14.1 Need for integration

It will be clear from the previous sections that no one measure on its own is likely to provide a solution to urban transport problems. Most have at least one positive contribution to make, in reducing travel time, environmental impact or accidents, but also have adverse impacts on, say, accessibility or equity. Some, such as traffic calming, can achieve benefits in one location at the expense of deterioration elsewhere. Some, such as bus and cycle priorities, would be more effective if they could influence mode choice; without such an impact they only benefit the users of the affected mode.

For all of these reasons, a package of measures is likely to be more effective than selecting any one measure on its own. A set of measures is likely to tackle more problems;

Table 3.5 A matrix of interactions between strategy measures

	Highways	PT infrastructure	Park and ride	Parking supply	Traffic management	Bus priorities	Traffic calming	Parking control	PT service levels	Information systems	Parking charges	Road pricing	Fuel prices	PT fares	Development control
Highways						C	C					C			C
PT infrastructure			C					C/P			C/P	C/P	C/P		C
Park and ride								C			C	C			
Parking supply					C			C							C
Traffic management						C	C		C	C		C			C
Bus priorities					C		C		C	C					
Traffic calming	C							C/P			C/P	C/P	C/P		C
Parking control						C			C	C					C
PT service levels							C	C/P			C/P	C/P	C/P		C
Information systems					C		C	C	C		C	C			
Parking charges	C/F	C/F	C/F	C/F			C/F	C	C/F					C/F	
Road pricing	F	C/F	C/F	F		C	C/F		C/F					C/F	
Fuel prices	F	C/F	F	F		C	C/F		C/F	C					
PT fares	C/F	C/F	C/F	C		C	C/P	C	C		C/P	C/P			
Development control	C	C						C	C						

Key: Measures in the left-hand column can reinforce the measure in the appropriate column by:
C: complementing it
F: providing finance for it
P: making it more publicly acceptable

one measure can offset the disadvantages of another or avoid the transfer of problems to another area; a second measure can reinforce the impact of the first, hence, for example, inducing a change of mode and generating greater benefits.

In these ways, synergy can be achieved between measures; that is, the overall benefits are greater than the sum of the parts. The identification of measures which might achieve such synergy is at the core of successful urban transport planning. This is the principle behind the UK Department of Transport's 'Package Approach' for local authority funding, in which authorities are encouraged to identify packages of complementary measures which can be financed and influenced together, and which should be more successful in meeting the authority's objectives.[12]

3.14.2 Potential benefits from integration

Integration can potentially achieve benefits in several ways. The first involves measures which complement one another in their impact on users. Obvious examples are the provision of park and ride to increase rail or bus patronage; the use of traffic calming to reinforce the benefits of building a bypass; the provision of public transport, or a fares reduction, to intensify the impact of traffic restraint; and the encouragement of new developments in conjunction with rail investment.

The second involves measures which make other elements of the strategy financially feasible. Parking charges, a fares increase or road pricing revenue may all be seen as ways of providing finance for new infrastructure.

The third concerns public acceptability, and the need to package measures which are less palatable on their own with ones which demonstrate a clear benefit to those affected. Once again an example is to be found in road pricing, which attitudinal research demonstrates is likely to be much more acceptable if the revenue is used to invest in public transport.[1]

Table 3.5 shows in matrix form, for a selection of those measures described in previous sections, those which are particularly likely to complement one another in one of these ways.[73]

3.15 References

1. Jones, P., Gaining public support for road pricing through a package approach. *Traffic Engineering and Control*, 1991, **32** (4).
2. Brog, W., Behaviour begins in the mind: possibilities and limits of marketing activities in urban public transport. *ECMT Round Table 91*. Paris: ECMT, 1991.
3. Confederation of British Industry, *Trade routes to the future: Meeting the infrastructure needs of the 1990s*. London: CBI, 1989.
4. De Jong, M.A., National transport policy in the Netherlands. *Proceedings of the Institution of Civil Engineers: Transport*, 1995, **111** (3).
5. May, A.D., Transport policy: a call for clarity, consistency and commitment, *Proceedings of the Institution of Civil Engineers: Transport*, 1995, **111** (3).
6. Institution of Highways and Transportation, *Guidelines for the development of urban transport strategies*. London: IHT, 1995.
7. The MVA Consultancy, The Institute for Transport Studies and Oscar Faber TPA, *A common appraisal framework for urban transport projects*. London: HMSO, 1994.

8. May, A.D., Roberts, M. and Mason, P., The development of transport strategies for Edinburgh. *Proceedings of the Institution of Civil Engineers: Transport*, 1992, **95** (1).

9. Wenban-Smith, A., May, A.D., and Jones, A.D., Integrated transport studies: lessons from the Birmingham study. *Traffic Engineering and Control*, 1990, **31** (11).

10. Coombe, R.D., Goodwin, R.P., and Turner, D.R., Planning the transport system for the historic town of Bury St Edmunds. *Traffic Engineering and Control*, 1990, **31** (1).

11. Ramsden, J., Coombe, R.D., and Bamford, T.J.G., Transport strategy in Norwich. *Highways and Transportation*, 1994, **39** (2).

12. Department of Transport, *Transport policies and programme submissions for 1996–9: Supplementary guidance on the package approach*. London: Department of Transport, 1995.

13. London Planning Advisory Committee, *Advice on strategic planning guidance for London*. London: LPAC, 1994.

14. Department of Transport, Environmental assessment. In *Design manual for roads and bridges*, Chapter 11. London: HMSO, 1993.

15. Department of the Environment, *Climate change: the UK programme*. London: HMSO, 1994.

16. Jones-Lee, M., *The value of preventing non-fatal road injuries*, CR330. Crowthorne, Berks: Transport Research Laboratory, 1993.

17. Jones, S.R., *Accessibility measures: a literature review*. LR 987. Crowthorne, Berks: Transport Research Laboratory, 1981.

18. Brundtland, G., *Our common future: Report of the 1987 World Commission on Environment and Development*. Oxford: Oxford University Press, 1987.

19. Department of the Environment, *Sustainable development: the UK strategy*. London: HMSO, 1994.

20. Department of Transport, *Road safety: The next steps: Interdepartmental review of road safety policy*. London: Department of Transport, 1987.

21. Royal Commission on Environmental Pollution, *Transport and the environment*, 18th Report. London: HMSO, 1994.

22. Coombe, R.D., Urban road appraisal: transport planning and assessment. *Proceedings of Seminar on Highway Appraisal and Design*. London: Planning and Transport Research and Computation (International) Co. Ltd, 1985.

23. SACTRA, *Trunk roads and the generation of traffic*. London: HMSO, 1994.

24. Mackie, A.M., and Davies, C.H., *Environmental effects of traffic changes*. LR 1015. Crowthorne, Berks: Transport Research Laboratory, 1981.

25. Advisory Committee on Trunk Road Assessment (ACTRA), *Report of the Advisory Committee on Trunk Road Assessment*. London: Department of Transport, 1977.

26. Nash, C.A., and Preston, J.M., Appraisal of rail investment projects: recent British experience. *Transport Reviews*, 1991, **11** (4).

27. Tyne and Wear PTE, *The Metro report*. Newcastle: TWPTE, 1985.

28. Howard, D.F., The characteristics of light rail. *Highways and Transportation*, 1989, **36** (11).

29. Read, M.J., Allport, R.J., and Buchanan, P., The potential for guided busways. *Traffic Engineering and Control*, 1990, **31** (11).

30. McPherson, R.D., Park and ride: progress and problems. *Proceedings of the Institution of Civil Engineers: Municipal Engineer*, 1992, **93** (1).

31. Harland, G., and Gercans, R., *Cycle routes*. PR 42. Crowthorne, Berks: Transport Research Laboratory, 1993.
32. Hass-Klau, C., Impact of pedestrianisation and traffic calming on retailing: A review of the evidence from Germany and the UK. *Transport Policy*, 1993, **1** (1).
33. Collis, H., The lorry management study. *Traffic Engineering and Control*, 1988, **29** (11).
34. Pearce, K., and Stannard, C., *Catford traffic management study*, Vol. 1, Planning and Transportation Research Report 17. London: GLC, 1973.
35. Keen, K., European Community research and technology development on advanced road transport informatics. *Traffic Engineering and Control*, 1992, **33** (4).
36. Institution of Highways and Transportation, *Urban Safety Management*, IHT Guidelines. London: IHT, 1990.
37. Buchanan, C., *Traffic in towns*. London: HMSO, 1963.
38. Hass-Klau, C., Nold, I., Böcker, G., and Crampton, G., *Civilised streets: A guide to traffic calming*. Brighton: Environmental and Transport Planning, 1992.
39. McKee, W.A., and Mattingley, M.J., Environmental traffic management: The end of the road. *Transportation*, 1978, **6** (2).
40. Elmberg, C.M., The Gothenburg traffic restraint scheme. *Transportation*, 1972, **1** (1).
41. Vincent, R.A., and Layfield, R.E., *Nottingham zones and collar experiement: The overall assessment*. LR 805. Crowthorne, Berks: The Transport Research Laboratory, 1978.
42. Greater London Council, *Area control*. London: GLC, 1979.
43. Ogunisanya, A., Improving urban traffic flow by restraint of traffic: The case of Lagos. *Transportation*, 1984, **12** (2).
44. May, A.D., Parking control: Experience and problems in London. *Traffic Engineering and Control*, 1975, **16** (5).
45. Gantvoort, J.T., Effects upon modal choice of a parking restraint measure. *Traffic Engineering and Control*, 1984, **25** (4).
46. Bonsall, P.W., Spencer, A.H., and Tang, W.S., Its Yorkshire: Car sharing schemes in West Yorkshire. *Traffic Engineering and Control*, 1981, **22** (1).
47. Webster, F.V. *The demand for public transport*. Crowthorne, Berks: Transport Research Laboratory, 1980.
48. Wood, K., and Smith, R., Assessment of the pilot priority Red Route in London. *Traffic Engineering and Control*, 1992, **33,** (7 and 8).
49. Christie, A.W., Hornzee, R.S., and Zammit, T., *Effects of lorry controls in the Windsor area*. SR 458. Crowthorne, Berks: Transport Research Laboratory, 1978.
50. Jeffery, D.J., *The potential benefits of route guidance*. LR997. Crowthorne, Berks: Transport Research Laboratory, 1981.
51. Bonsall, P.W., The influence of route guidance advice on route choice in urban networks. *Transportation*. 1992, **19** (1).
52. Jeffery, D.J., and Russam, K., Information systems for drivers. *Transport Planning and Technology*, 1984, **9** (3).
53. May, A.D., Bonsall, P.W., and Slapa, R., Measuring the benefits of dynamic route guidance. *Proceedings of the Seminar on Advanced Transport Telematics*. London: Planning and Transport Research and Computations (International) Co. Ltd, 1991.

54. Polak, J.W., Hilton, I.C., Axhausen, K.W., and Young, W., Parking guidance and information systems: Performance and capability. *Traffic Engineering and Control*, 1990, **31** (10).

55. Hamer, R., Kroes, E.P., and van Ooststroom, H., Teleworking in the Netherlands: An evaluation of changes in travel behaviour. *Transportation*, 1991, **19** (4).

56. Ciaburro, T., Jones, P., and Haigh, D., Raising public awareness as a means of influencing travel choices. *Transportation Planning Systems*, 1994, **2** (2).

57. Association of Metropolitan Authorities, *Bus deregulation: The metropolitan experience*. London: AMA, 1990.

58. Silcock, D.T., and Forsyth, E., Real time information for passengers on the London Underground. *Proceedings of the Seminar on Public Transport Planning and Operation*. London: Planning and Transport Research and Computation (International) Co. Ltd, 1985.

59. Virley, S., The effects of fuel price increases on road transport CO_2 emissions. *Transport Policy*, 1993, **1** (1).

60. Feeney, B.P., A review of the impact of parking policy measures on travel demand. *Transport Planning and Technology*, 1989, **14** (2).

61. Holland, E.P., and Watson, P.L., Traffic restraints in Singapore: Measuring the effects of the Area Licence Scheme. *Traffic Engineering and Control*, 1978, **19** (2).

62. Larsen, O.I., The toll ring in Bergen: The first year of operation. *Traffic Engineering and Control*, 1988, **29** (4).

63. Allsop, R.E., London fares and road casualties. *Traffic Engineering and Control*, 1993, **34** (12).

64. Gilbert, C.L., and Jalilian, H., The demand for travel and travelcards on London Regional Transport. *Journal of Transport Economics and Policy*, 1991, **25** (1).

65. Goodwin, P.B., Hopkin, J.M., and McKenzie, R.P., *Bus trip generation from concessionary fares schemes: A study of six towns*. RR 127. Crowthorne, Berks: Transport Research Laboratory, 1988.

66. Department of the Environment, *Planning policy guidance 13: Transport*. London: Department of the Environment, 1994.

67. Daniels, P.W., Flexible hours and journey to work in office establishments. *Transportation Planning and Technology*, 1981, **6** (1).

68. O'Malley, B., and Selinger, C.A., Staggered work hours in Manhattan. *Traffic Engineering and Control*, 1973, **14** (1).

69. Department of the Environment and Department of Transport, *Reducing transport emissions through planning*. London: HMSO, 1993.

70. Knight, R.L., and Trygg, L.L., Evidence of land use impacts of rapid transit. *Transportation*, 1977, **6** (3).

71. Wachs, M., Learning from Los Angeles: transport, urban form and air quality. *Transportation*, 1993, **20** (4).

72. Sanderson, J., A matrix approach to setting parking standards. *Transportation Planning Systems*, 1994, **2** (1).

73. May, A.D., and Roberts, M., The design of integrated transport strategies. *Transport Policy*, 1995, **2** (2).

CHAPTER 4

Economic and environmental appraisal of transport improvement projects*

C.A. Nash

In Chapter 3 reference was made to a range of objectives that transport policy may adopt, and some of the indicators that may be used as evidence of the success of any particular policy decision in meeting those objectives. Assessing whether a proposal is worthwhile clearly involves forecasting the effect it will have on these indicators and weighing them up to decide whether overall the proposal is beneficial. This process is known as appraisal, and methods for doing this are the subject of this chapter. The emphasis throughout is on the concepts involved; those wishing to see in more detail how it is carried through into practice are recommended to look at the British COBA 9 manual.[1]

When reference is made in this chapter to a project, the term is being used in the broad sense, and defined as:[2]

> A set of interrelated expenditures, actions and policies designed to achieve a country's specific objectives for economic and social development within a specific time-scale.

Consequently the term 'project' could refer to the construction of new transport infrastructure, but it could equally well mean a management measure, such as change in fares policy or the introduction of a traffic management scheme, or indeed an integrated strategy as a whole.

Techniques of project appraisal generally rest, wholly or partly, on the concept of economic efficiency and at the outset that concept will be explained. Tests of economic efficiency rest on the valuation of all costs and benefits of a project in money terms, and methods of doing this in respect of all the objectives of transport policy will then be considered. A project is economically efficient if the benefits measured in money terms exceed the costs; the most efficient project is that for which the difference is greatest. A method is also required for dealing with the fact that costs and benefits of transport projects are spread over many years. Conventionally this is handled by the technique of

---

* I am indebted to Tony May and Abigail Bristow of the Institute for Transport Studies, University of Leeds, for helpful comments on an earlier draft of this chapter.

discounting for time, which will be explained first in the context of financial appraisal and then of social cost-benefit analysis.

Finally discussed is the role of indicators that are not expressed in money terms and readily aggregated into a single measure of the net benefit of the project. These may arise for two reasons: first, the difficulty of finding satisfactory methodologies for valuing some benefits and costs in money terms, and second, that decision-takers may wish to look at a broader range of criteria than economic efficiency. In particular, equity, and the distribution of costs and benefits, is an objective that cannot be viewed simply as a part of the search for economic efficiency.

4.1 Economic efficiency

A key aspect of transport appraisal is usually the search for an 'efficient' allocation of scarce resources, and it is important at the outset, therefore, to have a clear understanding of what is meant by the concept of economic efficiency. An economically efficient allocation of resources is achieved when it is impossible to make one person or group in society better off without making another group worse off. In other words, if projects could be found and undertaken which would make everyone better off, those projects would serve to promote economic efficiency.

Now it may reasonably be argued that few, if any, projects in the transport sector or any other sector of the economy have such an effect. Almost anything one does is bound to hurt somebody, if only the taxpayer who pays for it. However, this is ignoring the possibility of compensation. Suppose for instance that a project is undertaken which will give a certain group of people a faster journey to work. Suppose also that collectively they are willing to pay £100 000 for this faster journey, but that implementation of the project will cost only £50 000. It is clear that, in principle, it would be possible to make everyone better off by undertaking the project. Those who experience the faster journey would be willing to compensate fully those who paid for the project and still be better off themselves. The principle can be extended to encompass other effects of transport projects, for instance on safety or the environment. Suppose that the project in question increases the risk of accidents and leads to environmental damage in the form of noise or air pollution. Provided that those experiencing these costs of the project would be willing to accept that a payment of less than £50 000 would fully compensate them for bearing these costs, then the project could still in principle make everyone better off.

Thus, seeking economic efficiency means taking all measures for which the 'willingness to pay' of the beneficiaries exceeds the 'required compensation' of the losers.

Two important points should be noted. The first is that it is rare for compensation actually to be paid, so that projects have gainers and losers. While repeated application of the compensation test should achieve an economically efficient situation, there is a whole range of possible economically efficient allocations of resources, each with its own distribution of costs and benefits. In some of them, a few people may be very well off and everyone else very poor; in others everyone might be equally well off. Many people will have preferences between alternative economically efficient positions in terms of their distribution of costs and benefits, but such positions cannot be compared in terms of economic efficiency; they are all equally efficient. To deal with this issue, it is necessary to follow the example of Chapter 3 and introduce equity as a further goal.

Second, it is necessary to define more clearly what is meant by the welfare of an individual. It has been implied that if a change occurs for which an individual is willing to pay more than they have to in order to secure it, or for which they receive more than the minimum compensation they require, then that individual can be regarded as being better off with the change than before it occurred. In other words, welfare can be judged in terms of individuals' own preferences as revealed in their choices about how much to pay or how much compensation they require. There are clearly cases where it might be thought that individuals do not act in accordance with their own best interests, and a more paternalistic view may be taken. For example, it should be clear that under the definitions so far, both drug trafficking and prostitution could be considered welfare promoting activities. Perhaps the strongest argument for a more paternalistic approach is where individuals are not fully informed of the long-term consequences of their actions; this is an issue that arises frequently in the case of environmental effects.

4.2 Economic efficiency and markets

In a free market, under certain conditions, an economically efficient allocation of resources will be achieved simply by the activities of producers and consumers freely buying and selling goods and services. Basically this is because firms competing to supply products will supply them as long as the amount that people are willing to pay for them exceeds the cost of production. In a perfectly operating market economy all goods will be produced for which consumers are willing to pay the costs, and prices will be equal to the marginal costs of production (or in other words the cost per unit of a small increase in production). This is an argument for leaving resource allocation up to the free market.

This may be illustrated with a simple example. Figure 4.1 shows the market mechanism at work. Suppose that the demand for a particular good (Q) is determined by the equation:

$$Q = 100 - 2P \tag{4.1}$$

where P is the price. This is one of the fundamental concepts of economics – the demand curve – and shows the relationship between the demand for a product and its price, all other factors (such as the price of competing goods and the income of the consumers) being held constant. Suppose also that the cost of producing the good is equal to 10 per unit (i.e. for simplicity it will be assumed that this cost is constant no matter how much of the good is produced). In other words, the marginal cost (MC) of an extra unit of the good is equal to the average cost (AC) of all units produced. Thus $MC = AC$ = 10 regardless of the value of output in the diagram. In a free market, producers will be willing to supply the good as long as they receive a price of at least 10 per unit. In other words, the line $MC = AC = 10$ may be regarded as a supply curve which shows that producers will be willing to supply any amount of the good at a price of 10 per unit. Thus consumers will buy as much of the good as they are willing to pay at least 10 units to obtain. This quantity is found by looking at the quantity given by the demand curve when price equals 10; in the diagram this turns out to be 80. Thus price will be established at 10 and the quantity sold will be 80. Consumers will receive as much of the good as they are willing to pay the costs of producing, and the outcome will be economically efficient.

In other words, an economically efficient allocation of resources implies that the price of all goods should be equal to the marginal cost of producing them. Therefore, if a demand curve is defined which relates the quantity demanded (*Q*) to price (*P*) and a marginal cost curve which relates the marginal cost (*MC*) to volume (*Q*), then:

$$Q = Q(P) \qquad\qquad (4.2)$$

and
$$MC = C(Q) \qquad\qquad (4.3)$$

The optimal level of output for any good can be found by setting $P = MC$ and solving this pair of simultaneous equations.

However, there are a number of objections to the presumption that a free market will lead to an efficient allocation of resources in the transport sector:

1. Natural monopoly exists where it is impossible or inefficient to have competing firms, and consequently the single operator has the power to raise prices above marginal costs. It has generally been considered that transport infrastructure forms a natural monopoly, because it is very expensive to have competing networks of roads or of railway lines serving the same points. The implication of this is usually taken to be that the state needs either to own it outright or, if it is privately owned, to regulate the price charged for its use.
2. Transport imposes costs on third parties (e.g. accidents, delay, environmental pollution). These costs are known to economists as *external costs* because they are not borne by the individual consumer or the firm supplying the transport facilities, and will be ignored unless the government intervenes (e.g. by charging a tax equal to these costs, or by introducing direct controls to limit its effect).
3. As was implied in Section 4.1, the free market may produce an efficient outcome, but it is not necessarily equitable. For instance, some people (particularly poorer people or people with disabilities) may be left without transport. Central or local government may intervene to provide services for such people.

Thus in practice consumers (freight and passengers), private producers, state owned producers, central government and local government all have a part in the transport system.

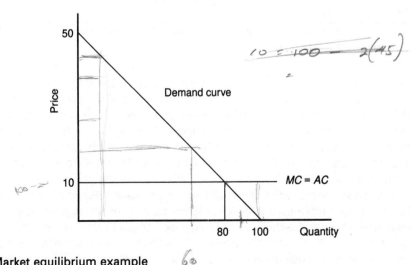

Fig. 4.1 Market equilibrium example

4.3 Valuing costs and benefits

4.3.1 Transport efficiency

This chapter is concerned primarily with appraisal criteria for those circumstances where the state does intervene in the transport sector. This section considers the alternative methodologies that are available to find the willingness to pay of beneficiaries and the compensation required by losers from the effects of transport projects.

Consider the appraisal of a typical road scheme (Table 4.1). Like any other project, it will involve capital, maintenance and operating costs. Unlike many other projects, the bulk of the operating costs will be incurred by people other than the agency undertaking the project – namely motorists, bus companies and road hauliers. Moreover, to the extent that in the absence of the scheme they would have used poorer quality, more congested roads, then operating costs will typically be reduced by provision of a new road; it is only in respect of any traffic generated by the road itself that costs will increase. There may also be some maintenance cost savings on existing roads as a result of the reduced level of traffic using them.

Table 4.1 Costs and benefits of a road scheme[24]

	Traffic growth (alternative assumptions)	
	High (£000s)	Low (£000s)
Costs		
Construction cost	2491	2491
Maintenance cost	72	72
Delays during construction	32	32
Total	2595	2595
Benefits		
Time and operating cost savings	4218	2658
Accident savings	417	304
Total	4635	2962
Net Present Value	2040	367

Note: These values are in fact present values, in 1979 prices, discounted at 7%. The calculation of present values is explained below

So far, the cost-benefit analysis appears very straightforward. All the above items are readily measured in money terms. By the arguments of the previous section, it follows that, wherever competitive markets are operating freely, market prices are appropriate as measures of costs or benefits of transport projects. Consequently, for construction, maintenance and operating costs of projects, market prices are usually taken as representing the true cost of the resources. Where markets are distorted, market prices may need adjustment and this leads to the process of shadow pricing – substituting for the market price a value which is thought more appropriate than the market price. Shadow pricing is most often used in developing countries with severe problems of a surplus of unskilled labour and a shortage of foreign exchange. To consider it here would be outside the scope of this text; the interested reader is referred to reference 2 for a simple explanation in a transport context. The one adjustment that is commonly undertaken even in developed countries is to remove any element of taxation from these costs, as this is a transfer to the government rather than a cost to the economy as a whole (i.e. the cost to the taxpayer of paying the tax is offset by the availability of the tax revenue provided that it is wisely spent).

But many costs and benefits of transport projects – time savings, pain and grief resulting from accidents, environmental effects – do not have a market price. In this case, a variety of methods have been used to try to establish what those affected would be willing to pay for the benefits or would require in compensation for the costs.

In the case of time savings, there is a distinction to be made between time spent travelling during working hours (which includes bus and lorry drivers as well as business travellers), and time spent travelling during one's own time. In the former case, it is usual to value the time at the wage rate of the employee concerned plus a mark-up to allow for overhead costs of employing labour (such as social insurance charges). This assumes that the time saved can be gainfully employed, and that the gross wage represents the value of the marginal product of labour in its alternative use. Doubts, however, may be raised on a number of grounds. Is the time saving large enough to be of use, or will it simply be wasted as idle time? (Individual transport projects often yield savings of less than a minute, although these may be aggregated with savings from other schemes to form more useful amounts of time.) Will the labour released find alternative work, or add to unemployment? If it does find alternative employment, does the gross wage really reflect the value of its marginal product in the new use?[3]

For non-working time, the problem of valuation is greater. The approach here has been to try to discover what people are willing to pay to save time, either by 'revealed preference' or by 'stated preference' methods. Revealed preference methods rely on studying people's behaviour in situations in which they reveal an implicit value of time. The most popular case is that of the choice of travel mode, where people may have a choice between two modes one of which is faster and more expensive than the other. If a model is estimated which forecasts the probability that someone chooses one mode rather than the other as a function of journey time, money cost and any other relevant quality differences, then the relative weight attached to time and money can be used to estimate their *value of time*.

This approach was used for many years, but it suffered from some problems. It is necessary to find cases where such trade-offs really exist and are perceived by a representative cross-section of the population. To estimate the value of time to a reasonable degree of accuracy, samples running into thousands are needed, and the data usually have to be collected specifically for this purpose by means of a questionnaire survey. An alternative is to use stated preference methods, in which the respondents to the survey are asked what they would choose given hypothetical alternatives (an example is given in Table 4.2). This enables the individual trade-offs to be designed to reveal the maximum information about the value of time; moreover, each respondent can be asked about a number of different choices. This allows great economies in sample size. After piloting and testing to ensure that the results were similar to those produced by revealed preference methods, this approach was used extensively in the studies that determined the values of leisure time currently used by the British Department of Transport (Table 4.3).[4]

Table 4.2 Example of a stated preference question*

Please compare the following alternative combinations of train fare and service level

	A				
LONDON, dep	2.50	3.20	3.50	4.20	4.50
Stockport ...	5.10	5.40	6.10	6.40	7.10
Manchester, arr	5.20	5.50	6.20	6.50	7.20

Fares: one way £12, return £24
Scheduled journey time: 2 h 30 min
Reliability: up to 10 min late

	B				
LONDON, dep	2.50	.	3.50	.	4.50
Stockport ...	5.40	.	6.40	.	7.40
Manchester, arr	5.50	.	6.50	.	7.50

Fares: one way £10, return £20
Scheduled journey time: 3 h
Reliability: up to 30 min late

Do you:

Definitely prefer A	Probably prefer A	Like A and B equally	Probably prefer B	Definitely prefer B

*Source: Institute for Transport Studies, University of Leeds questionnaire

Table 4.3 Resource values of time per person (pence per hour)[1]

	Average 1988 prices and values (pence/h)
a) *Working time*	
Car driver	849.7
Car passenger	705.3
Bus passenger	701.2
Rail passenger	1066.1
Underground passenger	1050.0
Bus driver	647.6
Light goods vehicle occupant	660.8
Heavy goods vehicle occupant	622.5
All workers	841.6
b) *Non-working time in-vehicle*	
Standard appraisal value	207.5
c) *Walking, waiting and cycling*	Double the in-vehicle values in (b)

4.3.2 Generated traffic

The above discussion suggests that the cost of making a particular journey may be taken as

$$G = M + vT \qquad (4.4)$$

where G = generalised cost, M = money cost, v = value of time, and T = journey time.

(Obviously this is a simplification, and other variables may need to be added to generalised cost. For instance, time is often split into walking, waiting and in-vehicle time.) Thus the price in a standard demand curve may be replaced by this notion of generalised cost, which allows for the fact that transport users pay for their journey in a combination of time and money devoted to it. It would be expected that, as the generalised cost of journeys fell, so the number undertaken would rise, so that a demand curve can be drawn relating this generalised cost to the number of journeys made, or $Q = Q(G)$. Assuming for simplicity a straight line demand relationship, this may be drawn as in Fig. 4.2.

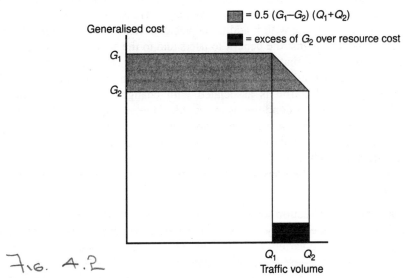

Fig. 4.2

Fig. 4.2 Benefits from reduced transport costs

BENEFITS from reduced transport costs.

Now suppose that a transport project reduces the generalised cost from G_1 to G_2, and causes an increase in demand from Q_1 to Q_2. The generalised cost saving to the existing users is straightforwardly measured as $(G_1 - G_2) Q_1$ (but note that there may be some offsetting loss of tax revenue to the government to be taken into account). New users are actually paying G_2 for their journeys. They must value these journeys on average at least this highly. On the other hand they were not willing to travel when the cost was G_1. In fact the demand curve tells the maximum that each user is willing to pay for the journey, and if it is linear then on average this is halfway between G_1 and G_2. They are actually paying G_2. Thus on average the difference between what they are willing to pay and what they actually pay is $0.5 (G_1 - G_2)$. This is termed the *consumers' surplus* on the additional journeys. Overall the benefit to users from the generalised cost change, or the additional consumer surplus to users, is therefore $0.5 (G_1 - G_2) (Q_1 + Q_2)$.

(Strictly speaking there is also another factor to consider. If the generalised cost for the generated traffic included a payment of tax, then – as explained above – this should be seen as a transfer rather than a true resource cost. The excess of G_2 over resource cost should be seen also as a net benefit of the generated traffic; it is a part of the willingness to pay of the user unmatched by any resource cost to the economy concerned.)

The implication of this analysis so far is that if generated traffic is ignored, then the benefits of a transport project will be understated, and indeed that the greater the generated traffic the more beneficial the project. This may clash with current widely accepted views that generated traffic is a problem rather than a benefit. There are several reasons why generated traffic may be seen as a problem.

1. If the road will be congested even after the project, then the presence of generated traffic will raise the generalised cost over what it would otherwise be. In other words the final generalised cost will not be G_2, but somewhere between G_1 and G_2. If future benefits have been estimated assuming the speeds that would exist without this increase in traffic, then the benefits may have been seriously overstated.[5]
2. Generated traffic will bring with it additional accident risks and environmental costs which have to be set against the benefit to users and to the state.
3. It should be remembered that the simple two-dimensional representation of demand used so far assumes that all other variables affecting demand are held constant. If the presence of generated traffic is associated with changes in other variables, for instance land use, then this condition is infringed. If the extra (or more likely longer) journeys arise because the project in question has led to a relocation of facilities at more remote sites then the above logic cannot be used to impute a benefit to users from these trips. In such cases, more complex measures beyond the scope of this text are needed to measure the consumer surplus from the simultaneous transport and land use changes.[6]

The case dealt with in this section so far is very simplified. In principle the transport system is modelled as a single journey with a unique generalised cost. In fact, it comprises many journeys over a wide variety of routes and by a number of modes. These interact in that together they determine the level of congestion on the network, and in that a change in the generalised cost on one route or mode may change demand on others. Examining the complications this leads to for the appraisal process is beyond the scope of this text. But it turns out that in many circumstances, a good approximation to the change in consumer surplus on the system as a whole is given by simply summing the above measure over all links and modes, i.e.

$$\sum_m \sum_l 0.5(G1 - G2)(Q1 + Q2) \tag{4.5}$$

where m = mode, l = link, and G_1, G_2, Q_1 and Q_2 are as defined previously.

The reader should be warned, however, that for very large changes, and in particular where totally new modes of transport are being introduced, this measure is not adequate and more complicated methods are needed.[6]

4.3.3 Safety

Turning to accidents, the costs may be divided into those that are readily valued in money terms, and those that are not. The former include damage to property and vehicles, health service, ambulance and police costs, and loss of production due to victims being unable to work (this again is typically valued at the gross wage). What is more difficult is to place a money value on the pain, grief and suffering caused by death or injury in an accident. For many years, in Britain, this value was determined by the political process rather than the preferences of those directly involved. However, it is possible to apply both revealed preference and stated preference techniques to this issue as well. The way to do this is to recognise that transport improvements do not save the lives of specific known individuals; rather they lead to a reduced probability of involvement in an accident for all users. Thus real or hypothetical trade-offs between safety and cost may be used to derive the *value of a life*. Such a stated preference study[7] is indeed the basis of the value currently used by the British Department of Transport (Table 4.4), although there may be doubts as to how well people are able to respond to questions involving changes in very small probabilities.

The British Department of Transport utilises a computer program (COBA) to calculate the value of its trunk road projects. This program includes the money values of all the items so far discussed, as indeed do the methods used in all the major countries of Western Europe (although until recently it did not generally allow for the existence of generated traffic). But it does not value any of the other effects of road schemes, of which by far the most important and controversial are the environmental effects of such schemes. The issue of how to take these into account in the appraisal process is described next.

Table 4.4 Average cost per casualty by severity[1] (1988 prices)

Type of casualty	Cost (£)
Fatal	565 900
Serious	58 920
Slight	4 810

4.4 Valuing environmental effects

So far the valuation methods discussed have taken into account two of the objectives referred to in Chapter 2 – transport efficiency and safety. This section deals with the issue of environmental evaluation. First this is discussed in terms of the ability to value environmental effects in money terms. (For a more detailed discussion of methods see reference 8. A recent attempt at comprehensive monetary valuation is contained in reference 9). Later in the chapter, alternative approaches are considered.

Transport projects have many important environmental effects, both at the local and global level. At the local level, they lead to property demolition, noise nuisance, visual intrusion and air pollution. At the same time, by taking traffic off other, perhaps more environmentally sensitive roads, projects may offer environmental benefits. More globally, schemes require inputs, such as limestone for construction purposes; to the extent that they generate additional traffic they also require oil production and produce pollutants with more than purely local effects, such as nitrogen oxides (implicated in acid rain) and carbon dioxide (a greenhouse gas).

4.4.1 Revealed preference methods

These methods rely on finding a market in which members of the population reveal the value they attach to the attribute in question in terms of willingness to pay for it or to accept compensation for its loss. This concept has already been discussed in relation to valuing time savings and accidents.

Turning to environmental issues, there are two long-standing revealed preference approaches in use – hedonic pricing, and the Clawson approach. *Hedonic pricing* is most often used in the context of house price models, although it is also used for instance in the estimation of the value placed on accidents via wage rate studies. The approach is to estimate the relationship between house prices and the environmental characteristics of houses, controlling for other factors such as the physical characteristics of the house and its accessibility. The estimated effect on price of environmental factors, such as the level of noise and air pollution, is often taken as some sort of mean valuation of the character-istics in question, although it is possible to apply more complex methods.

The house price approach has been subject to many criticisms. For instance, as usu-ally applied, it assumes a perfect market in which buyers with perfect knowledge can obtain any combination of characteristics they wish. At best it can obviously only be used to value attributes experienced in the home, and where people correctly perceive the effect on themselves. Thus it is likely to be more appropriate as a way of valuing noise nuisance than of valuing the health impacts of air pollution.

The *Clawson*, or travel cost, approach by contrast is only applicable for valuing the benefits of visiting facilities (e.g. country parks, nature reserves, forests, beaches). It relies on estimating a demand curve relating the frequency of visit to the travel cost involved. Again there are many practical problems involved. The most common is prob-ably the fact that most leisure trips are multi-purpose, and it is not clear to what extent the cost of them is incurred to visit a single facility, such as a beach or a stretch of wood-land, as opposed to the other components of the trip. Again it is a very partial technique; at best it can only measure the benefit people get from visiting a site rather than the ben-efit the site may hold in terms of scientific research, in terms of forming the subject of books and films or simply because people are willing to pay to preserve it.

In general then it appears that revealed preference methods are likely to be of value only in measuring amenity values of benefits readily perceived and understood and experienced in a limited number of locations.

4.4.2 Stated preference methods

In recent years, stated preference methods have taken over from revealed preference as the dominant method used in overcoming valuation problems in cost-benefit analysis. Again, as discussed in relation to valuing time savings and accidents, the approach is usually to ask respondents to choose between a number of discrete choices. However, stated preference methods also exist in a number of different forms. The one which has become most popular in the environmental field is the *contingent valuation method*. This involves actually asking a straight 'willingness to pay' to achieve or avoid some particular result, such as to protect a forest from destruction or to prevent building of a power station in a particular location. The attraction of this approach is that in principle it can be used to value anything, whether it can readily be quantified or not, whether it has actually been experienced or not. With the aid of this technique it would appear pos-sible to quantify all the externalities involved in the transport sector.

But again there are problems. There has been concern ever since the approach was first formulated with the likely biases that may creep in[10] when asking hypothetical questions about issues that may be unfamiliar. There is conflicting evidence on the seriousness of all these problems, and in part this reflects the fact that they are likely to be much more serious in a badly designed survey than in a well designed one. But it is also the case that despite the apparent universality of the technique some issues are easier to deal with than others. For instance it is likely to be easier to tackle the preservation of a feature which is well known (so people understand the issue) but about which feelings do not run high (so they do not have a strong incentive to bias their answers) than either a very controversial issue or one which is complicated and poorly understood. In practice, the fact that environmental effects of transport projects are typically both complicated and very controversial may make the technique difficult to apply.

4.4.3 Opportunity cost

The opportunity cost approach is quite different from both stated and revealed preference approaches in that it does not attempt to estimate willingness to pay for the benefit or to avoid the cost. Rather it asks what expenditure would be needed to offset it. In general the problem with this approach is that if the value placed by the population on the effect in question is not known, then it is not known whether in fact it is worth offsetting it. In some circumstances, however, it may be clear that it is worthwhile offsetting it (for instance, where air pollution damages a building and it is cheaper to repair it than to replace it, or where it destroys crops and it is known that their market value is at least as great as the cost of replacing them).

A version of this approach has become much more common in recent years as a result of two developments. The first is a tendency in the face of uncertainty about the true damage costs caused by different pollutants to adopt a 'precautionary principle' of limiting the level of the pollutant to what is considered a safe level. In this situation, any project which pushes pollution above the limit must be balanced by another (shadow) project to offset this effect. For instance if the limit for greenhouse gas emissions that can be tolerated is known, and emissions from transport are to be allowed to rise, then other emissions must be reduced elsewhere. In this context, the cost of reducing greenhouse gas emissions elsewhere by one unit becomes the opportunity cost of allowing them to rise in transport. Quantification of this opportunity cost is also not without its problems; strictly it requires examination of all possible ways of reducing greenhouse gas emissions elsewhere in the economy in order to identify the one with the least cost.

4.4.4 Comment on the various economic methods and the Leitch framework

In principle, then, methods exist which may be used for valuing all the costs and benefits of transport projects, but all have their problems and the reliability of all is open to doubt. For instance, the effects of property demolition could be studied by means of a contingent valuation survey, asking people the minimum compensation they would need to willingly sell their existing house (this was undertaken as part of the studies of the proposed third London airport in the early 1970s; see reference 10, Chapter 9). Noise, visual amenity and local air pollution have all been valued by means of studies of house price differentials, which are one way in which people indirectly reveal their willingness to pay for a superior environmental quality.[11] For global pollutants, the opportunity cost approach may be taken.

In the current British methodology, no attempt is made to provide a monetary value to the environmental effects of road schemes. Rather, the environmental effects are set out in a matrix known as the *Leitch framework* (after the chairman of the committee which devised it[12]), subsequently updated as Chapter 11 of the Department of Transport *Design Manual for Roads and Bridges*.[13] A summary of the elements included in the original Leitch framework is shown in Table 4.5. From this it will be seen that there is a wide variety of measures, in different units, viz. physical measures, numbers of houses, rankings, and verbal descriptions, as well as the financial ones; the revised version contains a similar mixture of financial, quantitative and verbal information. At the same time, no measures are included of non-local environmental effects of schemes. This is because, with traffic assumed constant regardless of what road schemes are built, the level of these pollutants hardly varies.

Table 4.5 Summary of the measures used in the Leitch framework[12] for assessing the costs and benefits of road schemes

Incidence group	Nature of effect	Number of measures	
		Financial	Other
Road users	Accidents	1	3
	Comfort/convenience	6	
	Operating costs	5	
	Amenity		2
Non-road users directly affected	Demolition disamenity (houses, shops, offices, factories, schools, churches, public open space)		37
	Land take, severance, disamenity to farmers		7
Those concerned	Landscape, scientific, historical value, land use, other transport operators		9 (+ verbal description)
Financing authority	Costs and benefits in money terms	7	
Total		19	58

By contrast, a number of other countries, including Germany and Sweden, do explicitly put economic values on certain local environmental effects in their appraisal of transport projects. For instance both of these countries value noise and local air pollution. It must be said, however, that the values used are based on somewhat shaky evidence, and are derived from estimates of the opportunity costs of achieving environmental standards by alternative means (such as double glazing, or fitting catalytic converters to vehicles).

4.5 Equity considerations

As noted above, seeking economic efficiency involves determining those projects for which the 'willingness to pay' of the beneficiaries exceeds the 'required compensation' of the losers. This rule, if applied throughout the economy, ensures that a position of economic efficiency is reached. That position could, however, involve a distribution of income with a small very rich group and a large very poor group within an economy.

If one is concerned with the distribution of income in the economy, then it is necessary to know not just whether a project contributes to economic efficiency but also who gains and who loses from it. This can be undertaken by identifying groups in terms of function (motorists, consumers, bus users, government, transport operators, residents) and income level, and disaggregating costs and benefits to these groups. Then it is possible to attach, explicitly or implicitly, higher weight to poorer sectors of society.

However, this makes the analysis much more complicated, as the repercussions of projects have to be traced through all those affected. For instance, suppose that a project reduces the costs of freight transport to a city centre. The results obtained may be a mixture of higher profits for freight operators, higher profits for retailers, higher rents for property owners, higher taxes for central and local government, and lower prices for consumers. As can be seen, it is much easier to measure the transport cost saving than to attempt to trace through who ultimately gains from it.

The Leitch framework goes some way towards allowing for equity, by dividing costs and benefits into those falling on road users, occupiers of property, those interested in the historical or scientific value of the site, etc. But there are some strange inconsistencies surrounding the treatment of public authorities. There is no clear separation of the cash flow to and from public authorities from the benefits to users – for some reason, both are treated as benefits to the financing authority. This certainly reinforces the view that the financing authority appears only to be interested in time savings for users but not in environmental costs and benefits. Amongst other groups for whom benefits are not clearly isolated are companies (regarding freight and business travel), public transport operators, cyclists and pedestrians.

A more consistent approach to the specification of and measurement of effects on incidence groups has recently been published.[14] Initially developed as a Planning Balance Sheet[15] and later extended and renamed to form what is now known as a Community Impact Evaluation, this approach clearly specifies all the relevant groups and consistently identifies costs and benefits to them, either in whatever the natural units of measurement of the effect in question are, or in money units where valuations are deemed to be sufficiently reliable to be of value. A more recent extension of this approach is the common appraisal framework developed for the Department of Transport.[16]

One final point may be made on the distributional issue. When evaluating methods of valuing time and accident savings, there is – not surprisingly – clear evidence that these are related to ability to pay. Thus if one were applying cost-benefit analysis purely as an efficiency test, one would need to disaggregate benefits by income group and apply higher values of these benefits to the better off. This would systematically bias decisions towards improving roads used more by the affluent, for instance those in wealthier parts of the country.

In practice, this has never been seen as politically acceptable in Britain. Thus it is usual simply to apply average values to all road users. This in itself could introduce some curious biases to decisions, however. For instance, it may lead authorities to spend money on securing time savings for travellers in poor areas on the basis of the average value of time, when in fact those travellers value the time savings at less than the cost of the scheme, and would rather have received the cash as a tax reduction. This illustrates the problem that effectively uprating one item of benefit for the poor while not applying similar weights to all others distorts the relative values of different types of cost and benefit. It is more consistent to value all costs and benefits at people's own willingness to pay or to accept compensation, and then to apply extra weight to all costs or benefits experienced by poorer groups if this is desired.

4.6 Economic regeneration considerations

In many countries, a main motivation behind transport projects is the encouragement of economic development and the promotion of particular patterns of land use. Thus for instance better roads to remote areas may be built to reduce their disadvantage in terms of transport cost; improved public transport to a city centre may be used to try to reduce decentralisation of jobs.

It is clear from the above that the approach in Britain has generally been to concentrate on the direct transport benefits of projects, on the assumption that these are overwhelmingly the most important factors. Part of the reason for this is that, in a small country with an already well developed transport system, even a major transport project will only have a small effect on the total costs of production and distribution of most industries in a particular location. Typically, in Britain, transport costs amount to some 8 per cent of total production and distribution costs, and even major projects will change total cost by less than 1 per cent.[17] Moreover, even if one could reduce the disadvantage of remote areas, they will still not be favoured unless they have some advantages which outweigh the fact that their transport costs will still be higher than those in more accessible locations.

Nevertheless, there clearly are cases where transport improvements do affect land use and economic development. A major estuary crossing, for example, may enable firms to concentrate their distribution facilities (or even production) on one side of the estuary, with consequent exploitation of economies of scale.[18] A major motorway development close to a major conurbation will tend to attract distribution and retailing activities, particularly at junctions with other motorways.[19] Improved rail services to the city centre may well trigger house building for commuter purchase,[20] and perhaps also make the city centre a more attractive location for firms. It should also be noted that these developments are not always beneficial. In the case of the M25 motorway around London, much new development has been attracted to a green belt area, at considerable environmental cost. Improved roads to remote areas may promote tourist travel, but they also enable firms to serve those areas from major centres, leading to the closure of local facilities such as bakers and distribution depots. Improved rail services may lead to the growth of long-distance commuting and urban sprawl. These considerations are likely to become of greater importance in the future if transport policy, while restraining the car in urban areas, continues to sanction its unrestrained use elsewhere.

Although models do exist to try to predict these sorts of repercussions of transport projects, they are complex to use and they are not yet at the stage that they can be relied on to give more then general guidance on likely effects.[21] Thus weighing up of these factors still has to be mainly a matter of judgement. In practice, an 'opportunity cost' approach based on the cost of creating jobs by alternative means is commonly used for valuation. Nevertheless, it appears that many European countries pay more attention to regional and land use factors when reaching a view on the overall desirability of a scheme than does Britain.

4.7 Budget constraints

It is very rare for a transport agency to have the financial resources to undertake all projects that would otherwise be justified. Thus it is usually important to look at the financial implications of the projects in question. This means not just estimating capital costs, but also operating and maintenance costs and sources of revenue over the life of the project. The presence of a predicted revenue stream may give rise to the possibility of additional borrowing, to be repaid from the revenue, or of the direct involvement of private capital, in either case easing the current budget constraint. Given a limited budget, the aim is to identify that set of projects which will yield the greatest net benefits per pound of government finance.

4.8 Appraisal criteria

What has been considered so far is a list of relevant costs and benefits and how they may be measured and valued. But how then may all the items be brought together to reach a decision?

When a commercial enterprise analyses whether it will be profitable for it to undertake a project, it undertakes what is commonly termed a *financial appraisal*. A financial appraisal of a project involves measuring all the effects of the project on the cash flow of the agent undertaking it. These are then 'discounted' back to the present to find its net present value in financial terms. In a social appraisal, one is not just concerned with cash and not just concerned with the agent undertaking the project: the objective is to measure the benefits and costs whoever receives them and whatever form they take.

In order to undertake an appraisal it is necessary to identify: (a) the *base case* (i.e. what will happen without the project), and (b) the *option* (what will happen with it).

For a financial appraisal, one simply seeks to identify the change in cash flow between the above two cases. However, in considering cash flows it is necessary to allow for the fact that firms would rather have cash now than in the future, because of the interest they could have earned if they had the money immediately.

To a firm which can lend money at a rate of interest of r per cent per annum, £1 now is worth $(1 + r)$ after 1 year, $(1 + r)^2$ after 2 years and $(1 + r)^t$ after t years. Simply reversing the procedure, it may be said then that the present value of £1 in 1 year's time is

$$\frac{1}{(1+r)} \tag{4.6}$$

in 2 years' time $\qquad = \qquad \dfrac{1}{(1+r)^2} \tag{4.7}$

and in t years' time $\qquad = \qquad \dfrac{1}{(1+r)^t} \tag{4.8}$

This, therefore, is the basis of the method known as *discounting for time* to calculate the *Net Present Value* (NPV) of the project (see Table 4.6). Note that the costs and benefits arising in each year of the life of the project are simply multiplied by the discount factor which converts them into present values. The net present value is simply the difference between the sum of the discounted costs and the discounted benefits.

Table 4.6 Example of an investment appraisal, showing effect of the project

Year	Change in costs C_i (£)	Change in revenue R_i (£)	Net cash flow (£)	Discount factor at 10%	Discounted cash flow (£)
0	+100	0	−100	1	−100
1	−10	+20	+30	0.909	+27.27
2	−10	+20	+30	0.826	+24.78
⋮	⋮	⋮	⋮	⋮	⋮
20	−10	+20	+30	0.149	+4.47

$$NPV = 155.3$$

$$\text{Or } NPV = \sum_{i=1}^{t}\left[\frac{R_i - C_i}{(1+r)^i}\right] - K$$

(PV)

It is usual even in a social cost-benefit analysis to adopt the same approach of discounting for time, although the justification is less obvious. Three different justifications, with differing implications for the appropriate choice of discount rate, are to be found in the literature.

1. That because people are generally becoming better off over time, they place less weight on a given benefit the further it is into the future (i.e. the social time preference argument).
2. That any financial resources devoted to the transport sector could alternatively have been invested elsewhere in the economy, where they would have yielded an economic return. The aim of discounting then is to ensure that they earn at least as high a return in the transport sector as they would have done elsewhere (i.e. the opportunity cost argument).
3. To ration scarce financial resources within the transport sector (i.e. the capital rationing argument).

Of these, it is argument (2) that usually wins with governments and lending agencies. Thus an opportunity cost rate of discount is usually adopted. In Britain at the present time this is taken to be 8 per cent; in many countries, particularly developing countries, it is 10–12 per cent. It is important to realise the degree to which this process reduces the weight given to future costs and benefits; for instance after 10 years, the weight attached to costs and benefits when discounting at 5 per cent is 0.61 and at 10 per cent is 0.39. After 30 years, the relevant values are 0.23 and 0.06 respectively.

A number of decision rules have been proposed, but the simplest to use is to undertake all projects for which the net present value is positive. This is only valid, however, when there is no shortage of funds to undertake all the projects in question. If a number of projects are competing for scarce resources, a simple value for money index can then be derived by dividing the net present value of the benefits minus costs of the project by the net present value of the financial requirement, and then ranking the projects in order of this indicator.

If all costs and benefits could be valued in money terms, this would lead to a simple criterion on which the value of a project could be judged. In practice, it is usual for an appraisal to contain some factors which have not been valued in money terms.

In the British Department of Transport's approach to appraisal, no formal method is used for trading off these various measures against each other and against the 'economic' costs and benefits. At the local level, the Leitch framework is used to reach a judgement as to which of a number of local variants of the scheme is the best overall. But in setting national priorities between schemes for funding this leads to problems. How could one possibly use the Leitch framework approach to rank schemes on a national level, for instance to set priorities between bypasses, motorway upgrading, new urban roads and development roads in remote areas?

So far, the examples of transport appraisals used in this chapter have been based largely on methods devised by the British Department of Transport for central government use in the evaluation of motorways and trunk roads. Local authorities have tended to take a somewhat different approach. Perhaps because they are more concerned with planning and environmental issues rather than narrow transport benefits, they have been much more ready to adopt objectives achievement or other multi-criteria appraisal methods. It is also the case that appraisals in many other European countries are more explicitly multi-criteria in orientation, with economic efficiency being seen as just one of a number of objectives.

Multi-criterion approaches require three stages: firstly definition of a set of objectives, which may for instance relate to accessibility, the environment, safety, economy and equity; next, measurement of the extent to which each project contributes towards the desired objective; finally, weighting of the measures in order to aggregate them and produce a ranking of projects. An extensive literature exists on these methods: for a computerised transport application see reference 22.

As it stands, this method would be quite consistent with the principles of cost-benefit analysis if the following conditions held:

1. all the objectives related to factors that affect the welfare of the population concerned
2. the measures of achievement and weighting of them are based on the preferences of the people affected by the projects, subject possibly to some form of equity weighting.

In practice, the first condition probably generally holds but the second does not. Measurement of the degree of contribution to objectives is often based not on detailed measurement but on the judgement of the professional staff planning the projects. This might be defended either on grounds of convenience (it is easier to ask professional staff than the public at large) or on the grounds that professional staff know better what matters than does the public at large.

Whether these weightings are expressed in money terms or not, they are essentially performing the same function as money values in expressing relative valuations. Moreover, to the extent that at least one of the performance measures – cost or economy – is expressed in money terms, they can readily be transformed into money values.

There is therefore less difference between this approach and traditional cost-benefit analysis than might at first sight be supposed.

It appears then that, as currently practised, multi-criteria decision-making techniques are essentially concerned with aiding and ensuring consistency in this latter stage of weighting by the decision-taker. This is a separate role from that played by the cost-benefit analysis, and should be seen as complementary rather than competing. To the extent that the information provided by a cost-benefit analysis is seen as relevant to the decision-taker, it still needs to be provided. But what is clear is that it must be provided in a sufficiently disaggregate form for the decision-taker to apply, explicitly or implicitly, his or her own weights. This again argues for the Planning Balance Sheet or Community Impact Evaluation form of presentation as outlined briefly above.

4.9 Appraisal of pricing policies

The outcome of a project appraisal will be influenced by the pricing policy followed, even where pricing policy is not explicitly one of the factors being considered in the appraisal. It has been argued above that the economically efficient pricing policy for a transport system, as for any other good or service, is to price at marginal cost. However, this should be the marginal cost imposed on all members of society by consumption of the good in question. This must include safety, environmental and congestion costs. It may be that environmental and safety costs are imposed on third parties by the decision to make a trip, but what about congestion? Surely congestion is already suffered by the trip maker, so why should they also pay for it through the price they pay to use the facility?

The argument is a very old one.[23] The entry of an additional vehicle on to the road system will, if traffic is already sufficiently dense to prevent free flow conditions, lead to a further reduction in speed for all traffic. Thus, as well as the delays suffered by the additional vehicle, all other vehicles on the road system will suffer delays. At the margin, the external congestion cost caused by one more vehicle is the additional delay to all other vehicles it causes. This must be distinguished from the average delay compared with free flow conditions, which may be a much greater number.

In an ideal world, vehicles would be charged in accordance with the externalities they created. This would require a pricing structure in which a price per kilometre was charged which varied with:

1. the characteristics of the vehicle, which determine the noise, emissions, delay to other vehicles and accident risk involved. Strictly these obviously depend not just on the characteristics of the vehicle when new, but also on its condition and the way it is driven
2. the characteristics of the road it is being driven on, including physical features (width, gradient, curvature) of the road itself and the surrounding land use (housing, countryside etc.)
3. the time at which it is being driven (which is important, for instance, in terms of the degree to which noise and local air pollution are a nuisance) and the traffic conditions on the road at that time.

Electronic road pricing, which if universally applied would offer the capability to implement such a pricing structure, in which the price per kilometre is adjusted in accordance with continuous monitoring of the location and condition of the vehicle and the road conditions in which it is being driven, is discussed in Section 3.12. Suffice it to say here that, although a small number of cities are examining its implementation, its extensive application is still clearly some way off. Even if it is now technically feasible to do so, one would still need to consider whether it was worth the cost of implementation, and whether people would actually adjust more effectively to a simpler, more understandable tariff.

What most countries have at the moment is a very different structure consisting of a fuel tax, which may vary with the type of fuel (diesel, leaded/unleaded petrol) and an annual license fee which varies with the type of vehicle. This offers some possibility for influencing both the type of vehicle people buy and the extent to which it is used, but can only charge for external costs on the average in each case. The case remains then for using a variety of other means to influence the way in which vehicles are used in specific circumstances. These means might include pricing measures (e.g. electronic road pricing in particular areas) and physical measures (bans on particular types of vehicles, parking controls, traffic management). There is no prospect in the foreseeable future of being able to handle transport externalities solely through pricing measures even if that were clearly seen as the most efficient approach. Nevertheless, having information on the value attached to the externality in question is an essential element in the appraisal of any measure to overcome the problem of transport externalities.

4.10 Public transport appraisal

So far discussion has focused on road projects as the typical transport application of cost-benefit analysis. However, the technique can be readily applied to rail infrastructure projects, air and seaports etc. It can also be used to appraise pricing policies and levels of service on public transport.

When applied to public transport, broadly the same list of issues arises as considered above in the case of roads. However, it should be noted that, given the absence of appropriate pricing to cover the external costs of congestion, accidents and environmental effects on roads, benefits often arise from public transport projects in terms of relieving these problems by diverting traffic from car or lorry.

However, there is one major difference. In the case of public transport, usually a fare is charged for the journey, and often the fares and service decisions are left up to the operator, acting on a commercial basis. This is only possible in the case of road schemes if a toll is charged, or some other form of road pricing is implemented, and this is the exception rather than the rule.

At one level, all the presence of commercial public transport operators does is to add a further complication to the analysis. If, for instance, one was examining provision of a facility such as a bus priority system, and the agency undertaking the project had no control over the service decisions of the bus operator, it would simply be necessary to predict the reactions of the operator, just as one needs to predict the behaviour of road users in any road scheme.

However, the question arises as to whether it is desirable to change that behaviour, either by provision of grants or subsidies or by direct ownership and control. Consider, for instance, the case of a rail project, where the rail operator is usually an integrated provider of track and services. If the operator is simply set commercial objectives, it will obviously appraise investment in rail infrastructure solely on financial terms. Many of the items included in a full cost-benefit appraisal – user benefits (except inasmuch as

Table 4.7 Comparison of net present values of two rail investment programmes[25]

	West Yorkshire (6 new stations on existing services), (£)	Leicester–Burton (new service serving 14 new stations*), (£)
Gain in public transport revenue from new users	997	8897
Loss in public transport revenue due to increased journey time	−166	–
Recurrent costs	−147	−9154
Capital costs	−656	−5806
Financial NPV	28	−6063
Time savings to new rail users	515	4582
Time savings to existing rail users	−472	–
Time savings to road users	113	3304
Accident savings	277	2612
Tax adjustment	−282	−2326
Social NPV	179	2109

* Revised in subsequent work for Leicestershire County Council

they may be recouped as fare revenue), benefits of relief of road congestion, accidents and environmental degradation, and other environmental effects – will be left out of a purely financial appraisal.

Thus the use of cost-benefit analysis for public transport projects inevitably involves some sort of replacement of purely financial objectives with social ones, and usually some sort of grant or subsidy. For governments which believe in leaving decisions to the market wherever possible – perhaps because of a belief that grants or subsidies automatically lead to inefficiency – this is an unwelcome message.

Table 4.7 illustrates the difference between a purely financial appraisal and a social cost-benefit analysis as actually determined for two separate local rail projects in Great Britain. (Note that each comparison is based on a 30-year project life, 7 per cent interest rate, and 1986 prices.) In the first case, which involved the establishment of new stations in West Yorkshire, both criteria show positive net present values, although the financial net present value is very small. In the Leicester–Burton example, a large negative financial net present value becomes positive when other benefits are taken into account.

4.11 Final comment

In this chapter, various methods of appraisal of transport projects have been examined. It was demonstrated that the concept of economic efficiency requires that benefits be measured in terms of what recipients are willing to pay for them, and costs in terms of the compensation those bearing them require. If all benefits and costs were valued in money terms, then one could simply calculate the net present value of the project using the technique of discounting for time. The value for money offered by any particular project would then be judged by the ratio of this net present value of benefits to the present value of finance the project required.

In practice, many items are not valued in money terms, whether because of practical difficulties in ascertaining appropriate valuations or because of the inclusion of objectives other than economic efficiency. In this situation, some sort of 'framework' layout of costs and benefits by incidence group is the most popular approach to appraisal, whether or not it is accompanied by a formal multi-criteria weighting system.

4.12 References

1. *Coba 9 manual: A method of economic appraisal of highway schemes*. London: The Department of Transport, 1989 revised edition.
2. Adler, H.A., *Economic appraisal of transport projects*. Washington: Johns Hopkins Press, 1987.
3. Marks, P., Fowkes, A.S., and Nash, C.A., Valuing long distance business travel time savings for evaluation: A methodological review and application. *Proceedings of the Seminar on Transportation Planning Methods*. London: Planning and Transport Research and Computation, 1986.
4. The MVA Consultancy, Institute for Transport Studies, University of Leeds and Transport Studies Unit, University of Oxford, *The value of travel time savings*. Newbury: Policy Journals, 1987.
5. Williams, H.C.W.L., and Yamashita, Y., Travel demand forecasts and the evaluation of highway schemes under congested conditions. *Journal of Transport Economics and Policy*, 1992, **26** (3), 261–282.
6. Williams, H.C.W.L., Travel demand models, duality relations and user benefit analysis. *Journal of Regional Science*, 1976, **16** (2), 147–166.
7. Jones-Lee, M., *The value of transport safety*. Newbury: Policy Journals, 1987.
8. Mitchell, R.C., and Carson, R.T., *Using surveys to value public goods*. Washington, DC: Resources for the Future, 1989.
9. Mauch, S.P., and Rothengatter, W., *External effects of transport*. Paris: Union International des Chemins der Fer, 1995.
10. Dasgupta, A.K., and Pearce, D.W., *Cost benefit analysis: Theory and practice*. London: Macmillan, 1972.
11. Pearce, D.W., and Markyanda, A., *Environmental policy benefits: Monetary valuation*. Paris: OECD, 1989.
12. Leitch, Sir George, Chairman, *Report of the Advisory Committee on Trunk Road Assessment*. London: HMSO, 1977.
13. *Design manual for roads and bridges*. London: The Department of Transport, 1993.
14. Lichfield, N., *Community impact evaluation*. London: UCL Press, 1996.
15. Lichfield, N., Kettle, P., and Whitbread, M., *Evaluation in the planning process*. Oxford: Pergamon Press, 1975.
16. The MVA Consultancy, Oscar Faber TPA and The Institute for Transport Studies, University of Leeds, *Common appraisal framework for urban transport projects*. London: HMSO, 1994.
17. Dodgson, J.S., Motorway investment, industrial transport costs and sub-regional growth: a case study of the M62. *Regional Studies*, 1974, **8**, 75–91.
18. Mackie, P.J., and Simon, D., Do road projects benefit industry? A case study of the Humber Bridge. *Journal of Transport Economics and Policy*, 1986, **20** (3), 377–384.

19. McKinnon, A.C., Recent trends in warehouse location. In Cooper, J. (ed.), *Logistics and distribution planning: Strategies for management*. London: Kogan Page, 1988.
20. Harman, R., *Great Northern Electrics in Hertfordshire*. Hertford: Hertfordshire County Council, 1980.
21. Webster, F.V., Bly, P.H., and Paulley, N.J., *Urban land-use and transport interaction: Policies and models*. Aldershot: Avebury, 1988.
22. Mackie, P.J., May, A.D., Pearman, A.D., and Simon, D., Computer-aided assessment for transportation policies. In S.S. Nagel (ed.), *Public administration and decision-making software*. New York: Greenwood Press, 1990, 167–177.
23. Pigou, A.C., *The economics of welfare*. London: AMS Press, 1924.
24. *Economic evaluation short course notes*. Leeds: Institute for Transport Studies, University of Leeds.
25. Nash, C.A., and Preston, J., Appraisal of rail investment projects: Recent British experience. *Transport Reviews*, **11** (4), 295–309.

CHAPTER 5

Principles of transport analysis and forecasting

P.W. Bonsall

5.1 The role of models in the planning process

Models are simplified representations of reality which can be used to explore the consequences of particular policies or strategies. They are deliberately simplified in order to keep them manageable and to avoid extraneous detail while hopefully encapsulating the important (determining) features of the system of interest.

The reason for using models is that estimates can be made of likely outcomes more quickly and at lower cost and risk than would be possible through implementation and monitoring.

Models can be used in a variety of ways.

1. To predict future conditions in the absence of policy intervention – thus providing an assessment of the extent to which conditions will deteriorate, or ameliorate, and giving an indication of the conditions which are likely to prevail at some future date.
2. To predict future conditions on the assumption that each of a series of specified policies or designs is implemented – thus helping to establish the extent of any benefits which can be attributed to each one and thus in turn providing the basis for an appraisal of their relative costs and benefits.
3. To test the performance of a given policy intervention in each of a series of imagined futures – thus indicating its 'robustness' in the face of future uncertainty.
4. To produce very short-term forecasts as part of an on-line management or control system such as might be found in a sophisticated area traffic control system.

The model may be specified to represent a small area, such as an individual junction, in considerable detail or a large area, such as a city or region, in rather less detail. The former scale might be appropriate when considering alternative designs (e.g. different sized roundabouts). The latter would be more appropriate to a strategic question such as whether or not to build a new LRT system or implement urban road pricing.

The purpose to which a model is to be put will determine not only its geographical coverage and detail, but also its output indicators. Thus if the model is to be used as part of the safety audit of two junction designs it will need to produce estimates of various types of vehicle conflicts and accidents, whereas if the model is being used as part of a strategic assessment of alternative car park pricing policies it would need to produce estimates of revenue flows, impacts on the demand for space in each of several car parks and so on.

Most modelling work is concerned with the appraisal, in the broadest sense of the word, of alternative options or designs. This appraisal will require estimates of the costs and benefits of each option and will use models to estimate their performance in terms of factors such as travel times, vehicle operating costs, accidents, environmental impacts, and revenues.

5.2 Desirable features of a model

A model will ideally produce an accurate forecast, at minimum cost in terms of data and computing resources. The art of modelling consists fundamentally of trading off accuracy requirements on the one hand against resource constraints on the other.

To effect this trade-off properly the analyst would need to understand what level of accuracy is really required in the current context (recognising that, for some purposes, a ball-park estimate may be all that is really required), what data already exist, at what cost additional data could be provided, what computing resources are available and at what cost. Only then can the analyst make an informed judgement as to the most appropriate form of model. This is the ideal; in practice, however, the decision will be heavily influenced by the analyst's familiarity with different forms of model and access to different modelling software packages.

There are several properties to a good model.

Accuracy and precision

The degree to which this is important is determined by the context in which the model is to be used, bearing in mind that extra accuracy is usually obtained at a cost and that apparent precision is often spurious. A model's ability to produce accurate results, is, of course, dependent on its specification, calibration and validation (see below).

Economy in data and computing resources

Although ongoing development in automatic data capture and computing make it possible to contemplate relaxing these constraints somewhat, they continue to be important.

Ability to produce relevant indicators at appropriate level of disaggregation

The actual indicators required will of course depend on the context.

Ability to represent relevant processes and interactions

For example, it may or may not be necessary for the model to allow for choice of mode, choice of time of travel or choice of destination. It is important that the model should include a representation of any processes which may be influenced by the policy measures being tested. Only in this way can it be said to be policy sensitive.

Appropriate geographical spread

Unless the model includes the whole area in which the effects of a policy might be felt, it cannot possibly indicate the totality of impacts and may produce a distorted estimate of likely benefits. For example, the effects of a new bypass may be felt at some considerable distance from the road itself due to the diversion of traffic from previously congested roads and the possibility of new traffic being generated by the new availability of fast journeys.

It is also desirable, but not essential, that a model should be *transparent* and *user friendly*. Transparency implies that the workings of the model should be apparent to the user – as for example in the case of a simulation model which predicts junction performance by tracing the behaviour of individual vehicles passing through a junction. Transparency can provide a useful check on the plausibility of the results, but unfortunately most models are not transparent and so the user has to take it on trust that they are correctly specified. A user-friendly model would be one that enabled a novice user to interact with it, quickly run the desired set of tests and interpret the output. Some models are user friendly but all too many have rigid requirements as to data formats and use specialist definitions which can be off-putting to the novice user.

Good modelling practice dictates that it is better for a model to be slightly over-specified rather than slightly under-specified because, although the former risks redundancy, the latter risks bias or error. Resource constraints, however, will discourage anyone from *deliberately* over-specifying their model.

5.3 Specification, calibration and validation

Most models are based on the premise that, by observing past or current behaviour of systems or individuals, one can infer rules which determine that behaviour and can then use those 'rules' to predict as yet unobserved behaviour. The process of deciding what 'rules' to include in the model is known as *specification*. For example, it might be specified that the number of trips produced in an area is some function of the population in that area. The process of quantifying the rule such that it is able to reproduce observed behaviour is known as *calibration*. For example, data from a town in Yorkshire might be used to calibrate the rule about trips and population and might yield the result that the number of trips per 24-hour period is given by multiplying the population by 2.6. This calibrated rule might now be used to estimate the number of trips per 24-hour period in that same town in future years (when the population may have changed) or to predict the number of trips per 24-hour period elsewhere in Yorkshire or further afield. Before doing so, however, it would be wise to check that the calibrated rule holds true for other towns or other years. This is known as the *validation* of a model and should be conducted on an independent database.

Only when a model has been validated in a wide range of situations can it be said to be truly *transferable* and *causal* rather than simply *correlative*. (The ability of a model to produce the right answer in a limited range of situations may simply be due to coincidence or correlation between the input and output variables, but if a model performs well in a wide range of circumstances, the chances are that it is representing underlying cause and effect and is thus generally applicable.)

Unfortunately many models in fairly widespread use have not been subject to rigorous validation tests.

5.4 Fundamental concepts

The following concepts lie at the heart of most, but not all, models in widespread use.

5.4.1 Utility maximisation and generalised cost

Many models make use of the idea that behaviour is a result of individuals attempting to maximise their net gain, or minimise their net loss, from each decision. Thus, for example, they will choose to make a trip if it gives them 'benefits' (e.g. access to employment, shops or leisure facilities) which outweigh the 'costs' (the time and money involved in making the trip) and they will choose the mode of transport and route which yields the lowest net costs (e.g. the one with the lowest travel time, greatest comfort and lowest money cost). This process of choosing the best option is known as utility maximisation (i.e. the individual is choosing the option with the greatest net 'utility' to himself or herself). But of course the process involves trade-offs between money costs, time, comfort and all the other relevant attributes. This trade-off is usually represented through the concept of *generalised cost* which allows combination of the various elements of cost by putting appropriate weights on the different attributes.

Equation 5.1 is a typical definition of the generalised cost of travel and incorporates a 'value of time' to translate time costs into generalised 'money' units.

$$C_{ij}^{\ k} = IVT_{ij}^{\ k}X_1 + OVT_{ij}^{\ k}X_2 + OPC_{ij}^{\ k} \qquad (5.1)$$

Here $C_{ij}^{\ k}$ is the cost of travelling from origin i to destination j by mode k, $IVT_{ij}^{\ k}$ is the in-vehicle time required to travel from i to j by mode k, $OVT_{ij}^{\ k}$ is the total out-of-vehicle time (e.g. walking and waiting) involved in travelling from i to j by mode k, $OPC_{ij}^{\ k}$ is the out-of-pocket costs (e.g. fares, petrol, parking charges) associated with travelling from i to j by mode k, X_1 is the value of in-vehicle time, and X_2 is the value of out-of-vehicle time.

The 'value of time' is clearly an important coefficient in the generalised cost equation. It is one of the coefficients whose value is estimated as part of the calibration process and can be inferred by careful study of the decisions which people make when faced with options with different times and costs. Currently, a typical value of time might be 6 p/minute for in-vehicle time or 12 p/minute for out-of-vehicle time.

5.4.2 Equilibrium

Most modelling work is based on the idea that real-world systems are made up of a series of equilibrating forces which, if allowed to operate, will bring the system into equilibrium. For example, there might be an equilibrium split of traffic between two parallel routes: if one route had an increase in flow, the extra traffic would create additional delay on that route which would persuade some drivers to switch to the other route, producing a new equilibrium. Similarly there might be an equilibrium queue length at bus stops: if, over time, the queues got longer the passengers would get fed up and go elsewhere or the bus company would increase their service frequency, but if, over time, the queues got shorter this would attract additional passengers or cause the bus company to cut its service frequency – all of which would tend to leave the queue length much as it was.

Clearly there will be cases where no such equilibrium exists either because the system is never stable for long enough for these forces to act or because other, disequilibrating, forces may be at work. A well-known example of a disequilibrating force is the so called 'vicious spiral' of public transport decline whereby an increase in fares leads to a loss in patronage, which in turn causes loss in revenue, which leads to increases in fares or reduction in service, which leads to further loss in patronage ... and so on.

The equilibrium concept is sometimes adopted in models for purely practical reasons – it being argued that the equilibrium condition is at least as likely as any other result and that it therefore represents a possible and unique outcome in a given set of circumstances which can be compared with the similarly plausible and unique outcome of a different set of circumstances. The existence of such reference points allows comparisons to be made which would be more difficult if there were no unique result of any given set of circumstances. This issue is further addressed later in the chapter.

5.4.3 Aggregation of individuals' decisions to produce population estimates

The performance of a transport system is the result of decisions by many individual travellers but most analyses are in practice concerned with abstract totals rather than with individuals. Thus the analyst might wish to know the total flow or the average speed on a link, the total emission of a particular pollutant or the average noise level rather than the conditions experienced or pollutants emitted by each and every individual. This is not to say that the detailed results might not be useful – merely that they are not normally required. The only disaggregation usually sought would be in terms of space (conditions on individual links or in specific zones within the study area), time (e.g. morning peak, interpeak, evening peak, evening) and possibly person type (e.g. the conditions experienced by people with different incomes, different levels of car ownership, different journey purposes and so on). Most models achieve this by representing groups of travellers (from a particular origin to a particular destination for a particular purpose) as more or less homogeneous units but, as will be seen, there is an increasing use of models which simulate each individual traveller separately and then aggregate the results as necessary.

5.4.4 Ignoring irrelevant dimensions of response – limiting the model domain

There are numerous ways in which travellers can respond to changes in the transport system. For example, imagine that an increase in the attractiveness of a peak period bus service might affect the following:

- choice of departure time (e.g. to travel in the peak rather than the off-peak period)
- choice of service (e.g. to use the improved bus rather than an unchanged one)
- choice of mode (e.g. to use bus instead of rail)
- choice of destination (e.g. to travel to the city centre rather than in the suburbs)
- choice of trip frequency (e.g. to travel daily rather than once a week, or not at all).

Over a longer time period it might even influence:

- car purchase decisions (perhaps the second car is not necessary after all)
- residential location (perhaps it is worth relocating to be on the improved bus route)

and so on.

Clearly it is not possible to represent all these decisions in every model. Nor is it necessary, because the effect of some of these responses might be so small as to be negligible and certainly of less significance than the general noise and uncertainty inherent in any model.

It must be recognised, however, that any limitation on the range of responses represented in the model will necessarily limit its areas of applicability or 'domains'. For example, the domain of a model which excludes choice of departure time would not include the investigation of a peak period pricing policy. Similarly, the use of models which only represent drivers' choice of routes to appraise new road schemes has recently been criticised on the grounds that such models ignore the potential of such schemes to induce (or generate) additional traffic. One alternative to the explicit representation of all choices is to subsume all the 'minor' effects into a catch-all 'other' response. This approach can extend the domain of a model somewhat, but does not, of course, allow any investigation of potentially important differences between the consequences of the separate responses involved.

5.5 Selecting a model

5.5.1 The range of models available

Models range in sophistication from simple equations encapsulating empirical relationships which can be worked out on the back of an envelope, on a calculator or in a spreadsheet, through to suites of computer programs each involving hundreds of lines of code to perform sophisticated mathematical functions or detailed simulations. Generally speaking, the more complicated the process being represented and the more detailed the predictions sought, the more complex will be the model, the more data it will require and the more demanding it will be of computing resources.

Most modelling work is conducted with programs drawn from one or other of the commercially available model suites. These programs will normally have been designed only to require data which is likely to be readily available and to produce results of a type and complexity which is suited to most practical planning purposes. Another advantage of the standard software is that the very fact of its being standard makes it unnecessary for the user to justify a departure from normal practice; a tailor-made model may give a better prediction but its use will invite a scrutiny and scepticism which a standard model, for better or worse, will usually escape.

5.5.2 Specifying the requirements

When selecting a model for a particular purpose the analyst should begin by clarifying the requirements of the exercise and the resources available. The requirements should be expressed in terms of:

- what input variables are required, and at what level of detail?
- what output variables are required, at what level of detail and in what form?
- what responses should be allowed for?

Specification of the input and output variables will usually be self-evident; for example, if the task is to test the effect of different pricing regimes on bus use then bus fares will need to be among the input variables and number of passengers will need to be among the outputs. The level of detail required may not be so obvious; is it, for example, necessary to consider the range of fares experienced by different groups of travellers with different concessions available, travelling different distances, at different times of day and with different frequencies? Is it necessary to distinguish changes in bus use on different routes or at different times of year? The answers to these questions will help determine the level of detail at which the model must operate.

The question of which responses are to be allowed for in the model relates to the issue of model domains. Analysts should be aware that, if their model excludes any potentially important responses, the results may be of limited value and, perhaps almost as important, the only way to demonstrate that such and such response is *not* significant is to allow for it anyway. Following this argument, an analyst may be wise to err on the side of including too many dimensions of response.

Some years ago the only form of output available was hard-copy printout but, although there is still a role for this medium, it is increasingly common to require model output in the form of on-screen results and graphics which can facilitate interactive use of the model using the results of one test to help specify the next. There is also a growing demand for models which can interface with spreadsheets and desk-top publishing so that model output can rapidly be incorporated into reports and documents. In some circumstances it can also be useful to have model predictions output as an animated sequence – for example the predicted performance of a proposed road junction can be made to 'come alive' for decision-makers (and analysts!) by having a simulation model's output displayed as an animated sequence.

5.5.3 The constraints of time and money

Against the above requirements must be set the constraints of time and money. These will determine whether new software can be bought or developed, whether any new data can be collected, what computing facilities could be made available, the amount of time for data preparation and model runs. Most modelling work is now done using software designed to run on standard desk-top computers which are now so cheap and ubiquitous that the cost of the software and data is likely to be more significant than that of the hardware. Indeed, current trends in computing power and costs are soon likely to make it feasible to contemplate using models employing advanced computing techniques (such as neural networks, expert systems, parallel computing and memory-intensive simulation) which have hitherto been restricted by the relatively high cost of the required hardware.

5.5.4 Marginal change modelling versus ab initio modelling

It is always wise, when making a forecast, to make maximum use of relevant data. Thus if one is engaged in short-term forecasts, or even long-term forecasts in a slowly changing scenario, it will be wise to consider whether the forecast can be seen as a marginal change from the existing picture. If it can, and if that picture is known, then one should use one of the marginal change models described below. On the other hand, if the present-day data are inadequate or the factors which are expected to influence the forecast year are fundamentally different from those currently at work, one might do better to use an *ab initio* model – one which makes a forecast from scratch without reference to the current situation. An exception to this latter rule would be when the future pattern is so very clearly the result of an evolutionary process of development that it becomes necessary to consider the dynamic evolution of the new pattern from the old.

5.6 Classes of model available to the transport analyst

This section will review some of the types of model available – starting with the simplest equations and going up to complex simulations. The review is not intended to be comprehensive but will cover all the commonly used models.

5.6.1 Simple formulae

An equation of the general form:

$$y = f(x) \tag{5.2}$$

can be used to represent an empirically observed relationship between one variable (x) which can be taken as given, and another (y) which is to be predicted. Commonly used examples of this type of model include: (a) work trips per household per day = f (employed residents); (b) shopping centre car parking requirement = f (retail floorspace); and (c) speed on a link = f (volume of traffic on that link).

In each case the value of f can simply be deduced by processing data on the x and y variables from a range of sites such that the average value of y is divided by the average value of x. It will normally be noticed, when doing this, that the value of f varies from site to site depending on local circumstances. This variation can either be treated as 'noise' in the relationship or an attempt can be made to discover what characteristic of a site causes the variation (e.g. for the car parking requirement, the value of f might be lower at city centre sites than at suburban sites). The result might then be a set of f-values each appropriate to a given set of circumstances. The resulting formulae will be applicable in the defined circumstances but cannot be expected to give accurate predictions if other factors change.

Particular care must of course be exercised in the application of any of these functions when the x value lies outside the range on which f was calibrated. For example, even though one might expect the number of accidents involving cyclists to be a function of the number of bicycle-kilometres travelled, it might be that, if ever the use of bicycles were to increase substantially, car drivers would become more familiar with them, more skilled at anticipating them and hence less likely to collide with them, thus decreasing the risk per bicycle-kilometre. An f-value calibrated against low or medium levels of bicycle use might therefore be inappropriate at higher levels.

Simple formulae are widely used in transport planning as rules of thumb or ball-park estimates for use during sketch planning and as the basis for setting capacity standards. The ease with which they can be estimated and applied makes them popular in a wide range of applications and anyone with even the most basic understanding of the way that transport systems work will have little difficulty in imagining potentially useful formulae. For example: how about road accidents as a function of vehicle-kilometres, pollutants emitted as a function of petrol sold, or bus trips as a function of the number of non-car owners? The list is endless but it must be recognised that the formulae are only representing correlations in the data and do not indicate causality. They can be used to predict future values of certain variables only on the assumption that other influences remain unchanged.

Simple formulae are of little use in predicting the effects of policy intervention unless separate values of f have been derived from sites with and without the policy in question. For example, one might derive separate values for the relationship between road accident casualties and vehicle-kilometres from years in which there was a national safety campaign and from years in which there was not.

An important group of formulae models are the *trip rate models* which are used to predict the number of trips 'generated' by a given type of household or land use. A trip rate model may be concerned with trip origins (in which case it is called a 'trip production model') or with trip destinations (in which case it is called a 'trip attraction model') and may seek to predict the total number of trips during a given time period or a subset such as trips by a particular mode (e.g. by car or bus) or trips associated with a particular purpose (e.g. work trips or leisure trips). Thus one might have a trip production model to predict the number of work trips by car produced by a typical household per day and one might have a trip attraction model to predict the number of shopping trips by public transport attracted per day by a given amount of city centre retail floorspace.

Clearly the number of trips generated by a household will depend on the characteristics of the household; for example, a household with more members, or higher car ownership, is likely to generate more car trips. This fact is explicitly recognised in a well-known trip generation model known as the *category analysis model* or the *cross classification model*. The model was developed in the 1960s[1] and is simply based on the idea of determining the trip rate of various categories of household in a base year and then applying these trip rates to the expected future distribution of households among the same categories and summing the resulting trips to produce a forecast total number of trips.

The original category analysis model allowed for 108 categories of household; defined as the product of three levels of car ownership (0, 1, 2+), six levels of income and six household structures (defined in terms of the number of employed and non-employed members). Subsequent versions of the model have used other methods of categorisation, sometimes using techniques such as cluster analysis to define them, but the basic concept is the same. The fundamental assumptions behind the category analysis model are firstly that the trip rate associated with each group will remain the same, which is obviously dependent on there being no significant change in the factors influencing trip rates; and secondly that the future distribution of households among the various categories can be predicted. This prediction is in practice often based on what is known about the planned housing stock in the area, assumptions or forecasts of changes in household structure, and forecast growth in car ownership and income. Each of these, although most obviously the car ownership and income forecasts, are likely to take into account any known trends (see below).

5.6.2 Time series models

Many aspects of transport demand vary over time. For example, car ownership and total vehicle-kilometres have been rising almost continuously, except for blips coinciding with fuel crises or economic recessions, for several decades. By plotting the relevant data over time (see, for example, Fig. 1.4 or 18.2) it becomes possible to deduce an underlying trend. A prediction of future levels can then be made by extrapolating this trend into the future. Trend-based forecasting is very popular because of its simplicity and because it is so obviously based on past evidence.

A time series model is clearly an example of a marginal charge model (as defined above) and is therefore limited to those situations where the current picture is known and where the factors affecting the future are not expected to be significantly different from those that have affected the past and present. More strictly, any prediction from a time series model is based on the assumption that the trends, and any factors influencing them, will continue unabated.

A trend may be determined by fitting a line, 'by eye', to data on a graph or, more formally, by regressing the observed data against time. This is simply done using the regression function in one's calculator or spreadsheet. If the trend is apparently linear (as in Fig. 5.1(a)) it is common practice to calculate the gradient of the line and then express the trend as an annual growth rate of a given absolute amount. For example an increase of x units per annum could be expressed by a formula such as:

$$\text{Flow in year } (t + n) = \text{Flow in year } t + (nx) \tag{5.3}$$

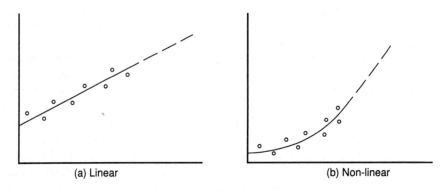

(a) Linear   (b) Non-linear

Fig. 5.1 Time trends

If the trend has a marked non-linearity (as in Fig. 5.1(b)) then a more complex mathematical function may be found to represent it. For example, if growth appears to be exponential one might have a formula such as:

$$\text{Flow in year } (t + n) = \text{Flow in year } t \times \exp^{n} \tag{5.4}$$

Alternatively, particularly if the trend does not fit one of the standard functions, the forecast can be read off directly from the graph each time one is needed or, more conveniently, values for each year can be read off and stored in a table.

Examination of time series data may suggest that there is some seasonal or cyclical fluctuation with or without a clear trend (Fig. 5.2). If this fluctuation is significant it would clearly be useful to be able to incorporate it into the forecast. If there is an obvious periodicity in the data (e.g. an annual cycle) then this is achieved quite straightforwardly by: (1) calculating an average monthly flow for each year; (2) measuring, for each year, each month's deviation from that year's average monthly flow; and then (3) calculating the average deviation for all the Januaries, the average deviation for all the Februaries, and so on. The annual trend, if any, would be determined by examining each year's average monthly flow. Assuming that a linear trend of x per cent growth per year is found, the January flow for a given year would then be given by the following formula:

January flow in year $(t + n)$ = (average flow in year t)
+ $((nx/100) \times$ average flow in year t)
+ (average January deviation) (5.5)

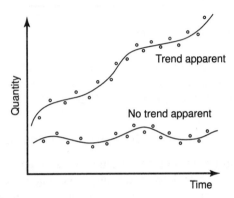

Fig. 5.2 Cyclical fluctuation

Clearly the situation may not be as simple as the above. For example; the monthly deviations may themselves follow a trend (e.g. the usual January decrease in traffic levels may be becoming less marked as cars get more reliable and comfortable for winter driving). It is more likely, however, that the data will contain variations which cannot be explained by any trend or cyclical process. One option in such circumstances would be to treat these variations as 'noise' in the forecast; another would be to seek to explain them and, if this fails, to use a quite different form of model.

There are, of course, limits beyond which it would be foolish to extrapolate a trend line however strong it appears in the historic data. It is obviously necessary to avoid theoretical absurdities such as extrapolating a decreasing trip rate until it goes negative or extrapolating an increasing proportion of income devoted to transport until it exceeds 100 per cent; but other limits may be less obvious. For example, it would be reasonable to assume that growth in car ownership, which has historically exhibited an accelerating trend in the early stages before settling into a linear growth rate, must one day show some signs of slowing down due to saturation of the market for new cars. A long-term extrapolation of car ownership trends should therefore include an indication of the point at which saturation is expected to take over from the period of more or less linear growth (see Fig. 5.3).

More generally, one should be aware that the extrapolation of a trend presupposes no change in the underlying factors which have previously supported that trend. This may not be justified if there is an important shift in policy which might affect the trend or if the extrapolation takes the forecast into an area where new constraints and forces may begin to apply; for example, an extrapolation of traffic growth would be invalid if the

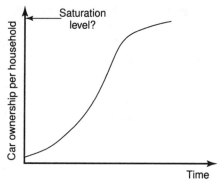

Fig. 5.3 Expected trend in car ownership

implied growth were to exceed the capacity of the road network to cater for it. Provided that such issues are taken into account, trend-based forecasts can have an important role in predicting the 'do-nothing' or 'do-minimum' future.

5.6.3 Averaging and smoothing – forecasting without a trend

One way of making short-term forecasts when there is no apparent trend in the data is to use simple arithmetic to calculate an average, a moving average or an exponentially smoothed prediction. The *average forecast* simply uses the mean value of all previous observations; this is simple but will give a distorted picture if data from the beginning of the series were influenced by different factors from those operating now. A *moving average forecast* attempts to overcome this by taking an average of only the most recent time periods (e.g. if one were using a 15-year moving average the 1997 forecast would be based on the average of the years 1982–1996); the problem with this method is that if the time period is too long it may suffer the same problems as the unconstrained average method and if it is too short it takes no benefit from less recent data and may be too subject to recent random events. An *exponentially smoothed forecast* seeks to avoid these pitfalls by making use of all available data while giving increased weight to the most recent. The formula is:

$$E_t = O_{t-1} + \alpha\,(E_{t-1} - O_{t-1}) \tag{5.6}$$

where E_t is the expected value in year t, O_{t-1} is the observed value in year $(t-1)$, and α is the smoothing coefficient (typically taking a value between 0.3 and 0.5).

Table 5.1 shows results for all the above three methods.

5.6.4 Regression analysis

As has already been mentioned, simple regression analysis can be used to derive the gradient of a trend line. It does, however, have many other roles in transport modelling. Regression is the process of identifying the mathematical function (perhaps a straight line or perhaps a more complex curve) which best fits the observed data. The general form of the regression equation is:

$$y = a + b_1x_1 + b_2x_2 + b_3x_3 \ldots + b_nx_n \tag{5.7}$$

Table 5.1 Forecasting by moving averages and exponential smoothing

Year	Observed data	Forecast				
		Uncon-strained average	5-year moving average	10-year moving average	Exponen-tially smoothed with $\alpha=0.3$	Exponen-tially smoothed with $\alpha=0.5$
1982	123	–	–	–	–	–
1983	125	123.00	–	–	(123)	(123)
1984	124	124.00	–	–	124.40	124.00
1985	125	124.00	–	–	124.12	124.00
1986	125	124.25	–	–	124.74	124.50
1987	124	124.40	124.4	–	124.92	124.75
1988	122	124.33	124.6	–	124.28	124.38
1989	121	124.00	124.0	–	122.68	123.19
1990	125	123.63	123.2	–	121.50	122.09
1991	128	123.78	123.2	–	123.95	123.55
1992	124	124.20	123.8	124.2	126.79	125.77
1993	121	124.18	123.8	124.3	124.84	124.89
1994	125	123.90	123.6	123.9	122.15	122.94
1995	123	124.00	124.4	124.0	124.15	123.97
1996	129	123.90	124.0	123.8	123.34	123.49
1997	126	124.27	124.2	124.2	127.30	126.24

where y is the dependent variable whose value is to be predicted, $x_1 - x_n$ are the independent variables whose values help to determine the value of y, and a and $b_1 - b_n$ are calibrated coefficients which define the mathematical functions which best explain y in terms of the x variables.

Various packages exist to calibrate a regression model but the user will need to specify the dependent variable, the potential independent variables and the potential functional form of the b coefficients (e.g. linear, square, square root, exponent). The user will also need to be sure that the x variables are truly independent because, if there is any correlation among them, the resulting regression model will be flawed. The quality of the regression model is usually measured by the value of the correlation coefficient (r^2) whose value should be as close to 1.0 as possible; a value of less than 0.6 is usually regarded as disappointing. It is, however, a mistake to assume that a regression model with a high r^2 is necessarily a good model – an apparently good fit can be created artificially by specifying a large number of independent variables. A really good model will have a high r^2 and a high ratio of observations to variables (degrees of freedom). Further information on the correct use and interpretation of regression analysis may be found in statistical texts (see, for example, references 2 and 3).

The x variables in a regression model may be continuous variables (such as income, population or time elapsed) or can be conditional 'dummy' variables which take a default value of zero, but assume a value of 1 when a particular condition is true for a particular observation. For example, a 'maleness' dummy would have value 1 if the subject/driver/tripmaker was male and a 'bad weather' dummy would have a value of 1 if the observation was made during bad weather. The flexibility of the regression model's structure means that it can be used for a wide range of purposes in transport modelling.

The most common applications are in the prediction of car ownership, trip ends (numbers of origins or destinations at a given location) and trip volumes (in an area, between a given pair of zones or along a particular link, or using a particular facility), but there is no reason in principle why regression models should not also be used to predict such things as waiting times at a junction, numbers of accidents or levels of pollution.

Table 5.2 lists some regression models which have been, or might be, developed. Note the use of different types of dummy variables in equations (i), (ii) and (v) in this table. Note also that checks would need to be made to ensure that there was no correlation between x_1 and x_2 in model (ii), between x_2 and x_3 in model (iv) or between x_1 and x_2 in model (v). Finally, note the distinction between equations (iv) and (v); (iv) is seeking to predict trip origins in an area (a 'zonal' regression model) while (v) is predicting trip origins by particular types of household (a 'household' regression model). Model (v)'s prediction could be aggregated to produce an area total by summing the predicted number of trips for each type of household for all such households in the area. This may sound more laborious than the zonal regression but is likely to give a more accurate result because the household model, which was based on data for individual households, will have been able to identify relationships in the data which are lost at the zonal level. The general rule is that, for the most accurate forecasts, regression models should be built with the most detailed data available.

A major difference between the regression models and the trend models or simple formulae described earlier is that it is possible, within one regression equation, to represent the effect of policy variables as well as other factors. It therefore becomes possible to use the equation to predict the effect of policy instruments as well as background factors outside the control of the policy decision-maker (compare x_6 and x_4 in equation (i)).

An equation, such as (ii) in Table 5.2, which is used to predict the volume of trips between a given pair of zones, is known as a *direct demand model* because it produces a forecast of demand without explicit consideration of the various subsidiary choices (e.g. about destination or mode) which the traveller might have gone through. Direct demand models are useful for forecasting travel demand along a specified route or service and are often used for preliminary assessment of a proposed new route or service. They are not, however, recommended for forecasting numerous individual flows in a complex network because they take no account of the interactions between different flows or of other factors likely to affect the overall pattern of demand. If a forecast of the overall pattern is required, it becomes necessary to consider the individual cells in the demand matrix; this is the subject of the next section.

5.6.5 Matrix estimation models

The cells of a demand matrix (otherwise known as the *trip matrix* or *origin–destination matrix*) indicate the number of trips between each origin–destination pair, and the row and column totals indicate the total number of origins and destinations respectively in each zone. The matrix provides a fundamental picture of travel demand in a study area and estimation of its cell values is a key component of many transport analyses.

If a matrix already exists, and if there is reason to believe that the basic pattern of demand within it is likely to remain unchanged, it is normal practice, in line with the general principle of using a marginal change model wherever possible, to seek to update the old matrix rather than to estimate an entirely new one. The simplest way of updating

Table 5.2 Some regression models

The general model is: $y = a + b_1x_1 + b_2x_2 + b_3x_3 \ldots + b_nx_n$

(i) y = volume of traffic (veh-km p.a.) on holiday routes in Scotland
  x_1 = years elapsed since 1980
  x_2 = dummy variable (=1 if average May daytime temperature exceeded 15°C)
  x_3 = dummy variable (=1 if August average daytime temperature exceeded 20°C)
  x_4 = dummy variable (=1 if previous years' August average daytime temperature exceeded 20°C)
  x_5 = average price of 2-week holiday in Spain (£ adjusted for inflation)
  x_6 = expenditure on advertising by Scottish Tourist Board in 12 months from preceding October (£ thousand)
  x_7 = average price of petrol (£/km adjusted for inflation)

(ii) y = volume of car trips (per 24 h) between a given pair of settlements in an agricultural area
  x_1 = distance between the settlements (km)
  x_2 = time required to travel between the settlements in typical traffic conditions (h)
  x_3 = population of one settlement
  x_4 = population of the other settlement
  x_5 = dummy variable (=1 if either settlement has a weekly agricultural market)
  x_6 = car ownership (cars/head) in the two settlements
  x_7 = total value of agricultural production in the area (£ p.a. adjusted for inflation)

(iii) y = car ownership (cars/head) in the UK in a given year
  x_1 = years elapsed since 1950
  x_2 = GDP/head (adjusted for inflation)
  x_3 = population over 17 years of age
  x_4 = average price of petrol (£/km adjusted for inflation)
  x_5 = average price of local bus travel in previous 5 years (£/km adjusted for inflation)

(iv) y = work trip origins in a given area (24 h)
  x_1 = employed population in the area
  x_2 = car ownership in the area (cars/head)
  x_3 = average income per employee in the area (£ thousand p.a. adjusted for inflation)
  x_4 = average price of bus travel in the area in the previous 5 years (£/km adjusted for inflation)

(v) y = work trip origins from a household (24 h)
  x_1 = household income (£ thousand p.a. adjusted for inflation)
  x_2 = number of cars owned in the household
  x_3 = number of full-time employed residents in the household
  x_4 = number of part-time employed residents in the household
  x_5 = number of unemployed residents in the household
  x_6 = dummy variable (=1 if house is within 500 m of bus stop with regular service to city centre)

an old matrix is to apply a *uniform growth factor* multiplying all cells by the same amount, as shown in Equation 5.8.

$$T_{ij\hat{t}} = T_{ijt} \times G_{t\hat{t}} \tag{5.8}$$

where $T_{ij\hat{t}}$ = trips from zone i to zone j in forecast year $\hat{t}$, T_{ijt} = trips from zone i to zone j in observed year t, and $G_{t\hat{t}}$ = expected growth in trip numbers between years t and $\hat{t}$.

Usually, however there will be reason to believe that some cells in the old matrix are likely to grow faster than others; for example, if a particular zone is the site of a new shopping centre one would expect shopping trips to that zone to grow more than proportionately. The *singly-constrained growth factor* approach uses such evidence to apply different growth factors to different rows (or columns) in the matrix. Equation 5.9 shows the origin-constrained growth factor formula and Equation 5.10 shows the destination-constrained growth factor formula.

$$T_{ij\hat{t}} = T_{ijt} \times G_{it\hat{t}} \tag{5.9}$$

where $G_{it\hat{t}}$ is the expected increase in trips originating in zone i

$$T_{ij\hat{t}} = T_{ijt} \times G_{jt\hat{t}} \tag{5.10}$$

where $G_{jt\hat{t}}$ is the expected increase in trips destined to zone j.

The growth factors, $G_{it\hat{t}}$ and $G_{jt\hat{t}}$, might be taken directly from policy statements or land use data (e.g. a 10 per cent increase in households in zone x) or, preferably, would make use of growth in the number of trip ends as predicted by a model such as a category analysis model, regression model or zonal trend forecast described earlier in the chapter.

The singly-constrained growth factor approach can be applied either to origins or to destinations but not to both. If data are available on growth in origins and in destinations it is necessary either to ignore one of them or to use a *doubly-constrained growth factor* approach. The best of several available methods of implementing this is the so-called *Furness* procedure which involves iteratively factoring to the new origin total (O_i) and the new destination total (D_j). The steps in the procedure are shown in Table 5.3.

If a trip matrix does not already exist for a particular study area or if the changes in land use or transport have been, or are likely to be, so great as to render any previous matrix irrelevant it becomes necessary to estimate a matrix from scratch. There are various ways of doing this but the most widely used is the *gravity model* which derives its name from its use of the gravity analogy – that the attraction (trips) between two bodies (zones) will be directly proportional to their mass (trip ends) and indirectly proportional to their separation (travel cost between them). At the heart of the gravity model is a 'deterrence function' which represents the decrease in trip making associated with increased travel cost. Various forms of function have been used but the most common is the negative exponential – as in Fig. 5.4.

As with the growth factor model described above, the gravity model can be origin-constrained, destination-constrained or doubly-constrained. The equations for the three cases, as proposed by Wilson,[4] are given in Equations 5.11, 5.12 and 5.13 respectively.

$$T_{ij} = O_i A_i D_j \exp^{-\beta C_{ij}} \tag{5.11}$$

$$T_{ij} = O_i D_j B_j \exp^{-\beta C_{ij}} \tag{5.12}$$

$$T_{ij} = O_i A_i D_j B_j \exp^{-\beta C_{ij}} \tag{5.13}$$

where T_{ij} = predicted number of trips from i to j, O_i = origins at zone i, D_j = destinations at zone j, C_{ij} = cost of travel from i to j, β = calibrated coefficient (the deterrence function), and A_i and B_j are balancing factors.

In Equation 5.11

$$A_i = 1 / \left(\sum_j D_j \exp^{\beta C_{ij}} \right)$$

In Equation 5.12

$$B_j = 1 / \left(\sum_i O_i \exp^{-\beta C_{ij}} \right)$$

In Equation 5.13

$$A_i = 1 / \left(\sum_j D_j B_j \exp^{-\beta C_{ij}} \right) \quad \text{and} \quad B_j = 1 / \left(\sum_i O_i A_i \exp^{-\beta C_{ij}} \right)$$

Table 5.3 Steps in the Furness procedure

---

1. Calculate origin growth factors G_i as

$$G_i = O_i / \left(\sum_j T_{ij} \right)$$

    where O_i is the new total of origins in zone i and T_{ij} is the old matrix

2. Test for convergence (are all G_i close to 1.0?* if yes, stop)

3. Multiply the cells in each column† of the current matrix by its G_i to produce a revised matrix, $\hat{T}_{ij}$

4. Calculate destination growth factors G_j as

$$G_j = D_j / \left(\sum_i \hat{T}_{ij} \right)$$

    where D_j is the new total of destinations in zone j and $\hat{T}_{ij}$ is the matrix produced at the previous step

5. Test for convergence (are all G_j close to 1.0?* if yes, stop)

6. Multiply the cells in each row† of the current matrix by its G_j to produce a revised matrix, $\hat{T}_{ij}$

7. Calculate origin growth factors G_i as

$$G_i = O_i / \left(\sum_j \hat{T}_{ij} \right)$$

8. Go to step 2

---

* The process is normally assumed to have converged satisfactorily when the G_i or G_j factors are in the range 0.9 to 1.1.
† The assumption here is that each column relates to a separate origin and each row relates to a separate destination.

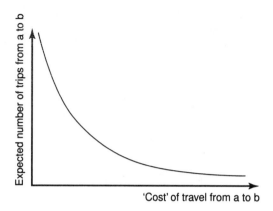

Fig. 5.4 The use of the negative exponential as a deterrence function in a gravity model

The singly-constrained model can be estimated in one pass whereas the doubly-constrained version requires an iteration, equivalent to that in the Furness procedure starting out with estimates of A_i's assuming the B_j's are all set equal to 1.

It will be appreciated that the deterrence function (β) controls the shape of the matrix by altering the distribution of trip costs. A high value of β indicates that trip numbers are very sensitive to cost and hence that there will be few long trips whereas a low value of β indicates an insensitivity to cost and hence a relatively large number of long trips.

The value of β is estimated from observed data by finding that value which is best able to reproduce the observed distribution of trip costs (or more simply, in some cases, is able to reproduce the same mean value). It has been shown that satisfactory results can be obtained using a technique known as 'entropy maximisation' which gives the most likely value consistent with specified constraints. A detailed description of this technique is beyond the scope of this chapter and interested readers should consult the original work[4] or more recent references.[5, 2]

If one is seeking to estimate a current matrix rather than to predict a future one then an alternative method can be employed making use of evidence from contemporary traffic counts. This approach, known rather prosaically as *matrix estimation from traffic counts* has a number of variants (see for example references 6 and 7). A detailed description of these methods is beyond the scope of the present chapter; however, the rationale behind them is that a given observed pattern of flows in a network is consistent only with a finite number of possible underlying matrices, that some of these possible matrices are inherently more likely than others, and that mathematical theory can be used to identify the most likely matrix. The confidence with which this most likely matrix can be identified is increased if some previous matrix is available to provide a starting point for the procedure. These matrix estimation techniques are becoming increasingly popular because the required software is readily available and the required data are relatively cheap to collect.

5.6.6. Elasticity models

In the previous discussion of regression models it was indicated how policy variables could be included in the model specification; for example, in equations (iii) and (iv) in Table 5.2 a variable relating to the level of bus fares was included. By examining the calibrated value of the coefficient relating to this variable it is possible to get an immediate indication of the sensitivity of the forecast to the changes in the policy variable. If one is engaged in relatively short-term planning, such that other influences can be assumed constant, it is possible to specify a model which is solely concerned with the sensitivity of the forecast to key policy variables. Such a model is known as an *elasticity model* and it is one of the most widely used models of marginal change. Its general form is:

$$Y_t = Y_{t-1} \{1 + E(X_t - X_{t-1})/X_{t-1}\} \tag{5.14}$$

where Y_t = quantity demanded in year t, X_t = value of a supply variable (e.g. price or journey time) in year t, and E is the elasticity coefficient for Y with respect to X. The elasticity coefficient should be calibrated on past data containing evidence of the marginal effect on Y of a marginal change in X. Thus:

$$E = (\Delta Y / Y) / (\Delta X / X) \tag{5.15}$$

The most common application of elasticity models in transport has been to represent the effect on public transport usage of marginal changes in public transport price, but they can also be used to predict the effect of marginal changes in a wide range of supply variables. For example, one could use an elasticity model to represent the effect on demand of changes in petrol prices, car parking charges, road-user tolls, expenditure on advertising, frequency of bus service, and so on.

The basic elasticity equation shown in Equation 5.14 can be extended to have several terms; e.g.

$$Y_t = Y_{t-1} \{1 + E_1(X_{1t} - X_{1t-1})/X_{1t-1} + E_2(X_{2t} - X_{2t-1})/X_{2t-1} + E_3(X_{3t} - X_{3t-1})/X_{3t-1} \cdots$$
$$+ E_n(X_{nt} - X_{nt-1})/X_{nt-1}\} \tag{5.16}$$

Some of the Xs will relate to characteristics of the option for which the forecast of usage is being made (e.g. a particular bus service), while others might relate to characteristics of a competing option (e.g. a parallel bus service, a rail service or car journey between the same locations), this latter group are termed *cross-elasticity variables*. Thus, an elasticity equation to predict usage of a coach service between Leeds and London might be:

passengers in year 2 = passengers in year 1 × {1
  + E_1 × (year 2 fare − year 1 fare)/year 1 fare
  + E_2 × (year 2 frequency − year 1 frequency)/year 1 frequency
  + E_3 × (year 2 train fare − year 1 train fare)/year 1 train fare
  + E_4 × (year 2 cost of car trip − year 1 cost of car trip)/year
  1 cost of car trip)} $\tag{5.17}$

Note that the sign of each elasticity coefficient will have an intuitively correct value – own-price coefficients should be negative, own-level-of-service coefficients should be positive and cross-elasticity coefficients should have the opposite sign to ordinary elasticity coefficients.

Elasticity models are deceptively simple but they have the unfortunate effect of concentrating the analyst's attention on price and level of service variables to the exclusion of other relevant changes such as of population or income. Also, it is tempting to use an elasticity coefficient to predict the effect of changes on price which are bigger than those for which the coefficients were calibrated; the results in such cases can be misleading because it may well be that elasticities are higher (or lower) over different ranges. A 10 per cent increase in fares may have brought about a 3 per cent reduction in demand but a further 10 per cent increase might not cause a further 3 per cent reduction and a 50 per cent increase would probably not bring a 15 per cent reduction, nor might a 10 per cent decrease in fares bring about a 3 per cent increase in demand.

5.6.7 Demand allocation – modelling the choice between alternatives

Trips are made as a result of choices made by travellers; choices between alternative modes of travel, alternative times of travel and alternative routes. Some of these choices are heavily constrained (e.g. if the traveller has no car available or if a public transport timetable provides only one service per day) while others may be made from a very extensive choice set. Different groups in the population have different constraints and those who have a very restricted choice set may be described as *captives* of a particular option. Even if they are not captive to one option some groups may have a strong preference for one option over the others; for example, high income travellers are likely to have a high value of time which causes them to prefer a fast and expensive option over a slower and cheaper one. There is a wide and growing range of models which seek to predict travellers' choices among available options. They vary in sophistication from models which simply aim to reproduce an observed pattern to others which seek to replicate the underlying choice processes.

An important concept underlying many of these models is *market segmentation* whereby separate forecasts are made for subgroups within the total population who, by virtue of their particular characteristics of car ownership, income, journey purpose etc., may be expected to respond to the choices available differently from other subgroups. Having made a forecast, for each subgroup, of the proportion of group members selecting each option, the overall forecast is simply achieved by weighting the proportions by the expected size of each subgroup in the forecast year's population, as shown in Equation 5.18.

$$D_o = \sum_s N_s P_{so} \tag{5.18}$$

where D_o = predicted total allocation of demand to option o, N_s = expected size of subgroup s, and P_{so} = proportion of subgroup s expected to select option o.

If no change is expected in the relative attractiveness of the available options then the demand allocation model might simply use the observed proportions of each subgroup choosing each option as the basis of P_{so}; this would be a marginal change model simply reweighting the existing pattern of demand to reflect any increase in the size of particular subgroups within the population.

More usually, however, some change in the relative attractiveness of options will be expected – indeed the whole point of the modelling exercise may be to make an assessment of the impact of such changes. This obviously necessitates some consideration of ways in which the population, or subgroups within it, are likely to perceive the relative attractiveness of the options.

The most naive approach would be to assume that all members of a given group will take the same view as to these relativities and will therefore all choose the same option – the 'best' one from their point of view. This approach, known as *all-or-nothing allocation*, is set out in Equation 5.19.

$$A_s = \min \{P_{s1}, P_{s2} \dots P_{so}\} \qquad (5.19)$$

where A_s = option to which every member of subgroup s will be allocated, and P_{so} = cost of option o as perceived by subgroup s.

It has the virtue of simplicity but has little else to recommend it. It produces an unrealistically 'lumpy' pattern of demand with massive reallocations possible as a result of a minor change in one option which alters it from best (chosen by everyone) to second best (chosen by no one) or vice versa. The problem is disguised if a large number of subgroups, each with different perceptions of the relative costs, are defined. One way of achieving this in the absence of any data on which to base this disaggregation is to define groups more or less arbitrarily and then add a random element (positive or negative) to the costs of each option for that group prior to selection of the 'best' option for that group. This technique is known as a *stochastic all-or-nothing model* and is defined in Equation 5.20.

$$A_s = \min \{P_1 + e_{s1}, P_2 + e_{s2}, \dots P_o + e_{so}\} \qquad (5.20)$$

where A_s = option to which every member of subgroup s will be allocated, P_o = cost of option o, and e_{so} = random number added to the cost of option o for subgroup s.

The best known application of this model is the Burrel network assignment model[8] which allocates drivers to routes through a network by dividing the total population into subgroups each of whom have random elements added to the cost of each link before selecting the cheapest route. A stochastic all-or-nothing model is able to reproduce an apparently realistic distribution of demand among the alternatives but it has no real empirical justification.

The most straightforwardly empirical approach to the issue is to use the observed splits between pairs of alternatives of differing relative attractiveness as the basis of a model which predicts proportionate allocation of demand as a function of relative attractiveness. Such models are called *diversion curves* and a typical one is illustrated in Fig. 5.5 where a diversion curve for the choice between options *a* and *b* has been derived by plotting the use of *a* against some measure of *a*'s relative attractiveness and then drawing a best fitting curve. The gradient of the curve at any given point indicates the sensitivity of the choice to the cost ratio at the point. Note that the *x*-axis is here expressed as the cost ratio (*b/a*) but that, in some circumstances, a better fit might be found by plotting proportionate demand against the cost difference (*b–a*). Separate diversion curves can of course be defined for different groups within the population.

Diversion curves are rarely produced for situations where there are more than two options but there is no reason in principle why they should not be extended into extra dimensions – the limiting factor is perhaps the analyst's ability to imagine the shape of the multi-dimensional plane which is thereby produced, but if the calculations are being done by a computer this need not matter. Diversion curves, as their name indicates,

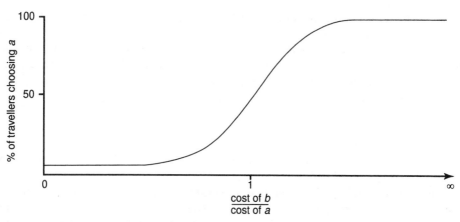

Fig. 5.5 Diversion curve

originated as tools with which to predict the amount of diversion of traffic which might occur from an existing road to a new or improved road running more or less parallel to it. They can also, of course, be constructed and used to predict the outcome of any choice – for example that between competing modes.

Diversion curves predate the widespread availability of computers and have largely been replaced by models which, while fulfilling essentially the same function, have a sounder mathematical, statistical and theoretical basis. The best known of these models is the *logit model of individual choice*. This model uses the relative attractiveness of each of a set of options to predict, for each option, the probability of an individual choosing it. These individual probabilities can simply be translated into proportionate allocations of a group of people faced with the same set of options. (If the probabilities of choosing options *a*, *b* and *c* were 0.5, 0.2 and 0.3 respectively then, if 10 people were making the choice, 5, 2 and 3 would be predicted to pick *a*, *b* and *c* respectively.)

The equation of the logit model of choice between *k* options is:

$$P_1 = \left(\exp^{-\lambda c_1}\right) \Big/ \left(\sum_k \exp^{-\lambda c_k}\right) \tag{5.21}$$

where P_1 = the probability of choosing option 1, c_k = a measure of the cost of using option *k*, and λ = a calibrated sensitivity coefficient.

This equation can deal with the choice between any number of independent alternatives (there must be no correlation between subsets of options) and produces an S-shaped probability curve similar to that shown in Fig. 5.5. This shape is interpreted as demonstrating that as soon as either option has an obvious advantage it will attract almost all the choosers and so cannot attract many more even if its advantage increases further; the real battleground between the options (where the probability curve is steepest) occurs where no one option has a particular advantage over the others.

The value of the λ coefficient determines the slope of the probability curve in this key area and hence determines the extent of the 'battleground', as can be seen in Fig. 5.6. The λ coefficient represents the sensitivity of the choice to the 'cost' variable – the higher the value of λ the greater the sensitivity. The theory which underlies this model can be summarised as the realisation that there is always some uncertainty about an

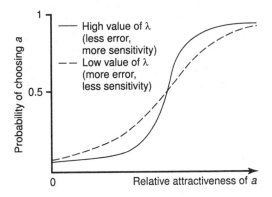

Fig. 5.6 Sensitivity of choice to value of λ in logit model

individual's perception of the attractiveness of an option and that this uncertainty can be seen as a random 'error' term (which may be positive or negative) added to the cost of each option such that, when two or more options are compared, some individuals will perceive their cost difference as being greater than it really is, some will see it as less than it really is and some will actually perceive the more expensive option as being the cheapest. The λ coefficient limits the contribution of the random error terms and so it is that, with high values of λ (and thus relatively small error terms) there is a more clear-cut preference for the option which is really cheaper.

Other models have been proposed that assume a different distribution of error terms but the negative exponential distribution which is used in the logit model has a number of practical advantages which have made it the most widely used. (Further details of the logit model and its derivation are beyond the scope of this book but may be found in references 5 and 9).

As has been mentioned, the logit model assumes that the random errors are independently distributed for each option and thus the model can only be used to predict the choice between uncorrelated options. This can sometimes be inconvenient; for example, when modelling the choice for a particular journey between three modes of transport – train, drive-alone, drive-with-passenger – it would be unreasonable to suppose that the perception of the costs of the two car modes would be independent. The solution in such circumstances is to use a *hierarchical logit model* which treats the choice as if it were in two stages – an 'upper' choice between train and car and a 'lower' choice, if car is selected, between drive-alone and drive-with-passenger. The upper and lower choices will each have their own λ coefficient and will be based on different cost comparisons. The choice between the two car modes in the example given above will simply be based on a comparison of the two car costs whereas the train/car choice will be based on a comparison of the train cost and a 'composite' car cost which should be defined as:[10]

$$CC = \frac{-1}{\lambda} \log\left(\sum_k \exp^{-\lambda c_k} \right) \tag{5.22}$$

where CC = composite cost, c_k = cost of options which are being combined, and λ = sensitivity coefficient used to choose between options which are being combined (i.e. from the lower choice)

Figure 5.7 demonstrates the essential difference between a standard logit model and a hierarchical logit model. In the above example, the 'nesting' of the logit model was self-evident but this is not always the case and a good part of the skill in using the hierarchical logit model is in determining the most appropriate nesting arrangement. Generally speaking the nests should represent expected correlations in perception and the hierarchy should be ordered such that the lowest order choices are most sensitive to cost differences.

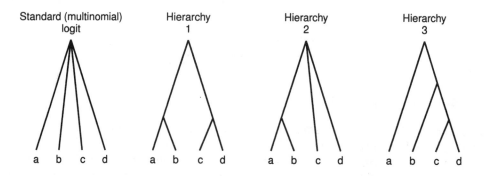

Fig. 5.7 Some possible hierarchical structures for a logit model of the choice between four options

The logit model, in its standard form or its hierarchical form, has become a very popular model of choice between competing alternatives. It has been particularly widely used to represent the choice between alternative modes but has also been applied to the choice between alternative routes, departure times, destinations, car ownership levels and even trip frequencies. Indeed, by making use of a hierarchical structure it is possible to put together a set of equations which represent all of these choices in one model suite. Although the logit model can be used to model the choice between destinations it is more common to do this via a matrix estimation model such as those described earlier in the chapter.

Before leaving the topic of demand allocation models, it is necessary to consider the special complications which are caused by the existence of capacity effects.

It is a characteristic of most transport systems that, above certain limits, the more heavily used a mode, route or service becomes the less attractive it will be to the users because of overcrowding, congestion and consequential delays. This effect, known as 'capacity restraint', may seriously limit the demand – not necessarily because there is simply no room for extra demand but because when a system becomes overcrowded it becomes less attractive. Some models take this idea one stage further to suggest that there will be an equilibrium level of demand which will come about over a period of time as users adjust their behaviour in the light of experience. If extra users begin to use an overcrowded facility the additional costs thereby imposed on all users will be such that, over a period of time, a number of them will voluntarily switch to some other option.

This kind of mechanism could apply to any choice where capacity effects may be important. For example, it might be used to explain commuters' choices between competing modes of public transport when one or more of them is overcrowded, to explain commuters' time of departure for work, or to explain shoppers' choice of supermarket or car park. The best known application, however, is to drivers' choice of routes in a congested network and the classic exposition of the result of the capacity equilibrium process is Wardrop's principle:[11] 'under equilibrium conditions traffic arranges itself in congested networks such that all used routes between an O–D pair have equal and minimum costs while all unused routes have greater or equal costs'. From which it follows that, under these conditions, no driver can unilaterally reduce his or her costs by switching to another route.

Few people would seriously suggest that the transport system is stable enough, or that individual travellers are knowledgeable enough, for this perfect equilibrium ever to come about in the real world, but the representation of this state has nevertheless been a goal of many modellers who recognised its value as a fixed point among a myriad of possible future states; a comparison of two such fixed points, relating to two different policy inputs, has considerable advantages over a comparison of two less tightly defined outcomes from the two policy inputs.

There are two main groups of approaches to achieving an equilibrium solution: the first group uses mathematics, statistics and algorithms to get as close as possible to perfect equilibrium while the other seeks in some way to represent the processes that might be involved. The mathematical/statistical/algorithmic approaches are mostly based on iteration with the alternate allocation of demand according to cost and recalculation of the cost in the light of demand. Clearly some form of damping is required to avoid infinite oscillation. Some models achieve this by averaging the costs from successive iterations while others seek to form a composite demand from a weighted sum of the intermediate estimates of demand.

The second group of approaches seek to represent more faithfully, if less efficiently in computing terms, the processes that might tend to lead towards equilibrium. One variant divides the demand into fractions and allocates each one in turn before recalculating the costs and allocating the next fraction – and so on until all the demand has been allocated. This model can be said to represent in simplified ways the informed decisions by successive groups of travellers during a single day with each group able to make its decisions in the light of those of all previous groups. (In practice it is found that, until capacity restraint sets in, all the fractions are allocated to the same choice and so the procedure has been made more efficient by starting with large fractions and reducing them progressively as 'time' goes on.)

Another approach is to represent the behaviour of individual travellers over several days during which time they experiment with new options and experience the consequences. Such an approach would be a form of simulation (see Section 5.6.8).

To date, the only area of modelling which has given significant attention to capacity restraint is that concerned with the assignment of trips to routes in a network, and it is from that field that most of the above examples have been taken. It can however, be expected that these issues will begin to be considered in other areas of modelling as other capacity ceilings in the transport system become more apparent.

5.6.8 Simulation models

Simulation models differ from abstract mathematical models in that they attempt to represent the dynamic evolution of some aspect of the transport system through an explicit representation of the behaviour of actors within it. Thus, at a detailed scale, the performance of a junction might be simulated by representing individual vehicles passing through it and the performance of a bus route might be simulated by representing the boarding and alighting of individual passengers at each stop. At a less detailed level, the development of a bus network might be predicted as the result of competition between different operating companies.

If a simulation is very detailed, particularly if it deals with individual decision-makers rather than groups, it may be termed a 'microsimulation' model, but the terms are often used interchangeably.

The decisions of actors within a simulation model may be represented by means of discrete choice probability models such as those described earlier (but with probabilities turned into discrete choices by means of techniques such as Monte Carlo sampling whereby a random number is used to select from a probability distribution), or by means of heuristic or algorithmic approaches drawn from detailed research into individual decision-making.

A major difference between simulation and abstract modelling is that simulation allows the modeller to develop a logical model of a very complex system by putting together a number of components which are themselves fairly well understood. Thus it becomes possible to develop a model of a complex system without having recourse to overarching theories about system dynamics or concepts such as equilibrium. It is therefore well suited to examining the way in which system performance depends on detailed aspects of the design or control strategy or is sensitive to external factors such as weather conditions or a variation in the pattern of demand or to the detailed behavioural assumptions underlying the model. Such testing can be expensive in computer time but is important if the results are to be generalisable.

The advent of very cheap computer memory and of parallel processors is now enabling simulation of large systems to be conducted at a very detailed level and this is undoubtedly the fastest growing area of modelling.

5.7 Transport modelling in practice

5.7.1 Model packages

Most transport planners and engineers prefer to make use of an existing model package rather than seek to develop a new model to meet their particular needs. Numerous packages are now available on a commercial basis and they range in price from a few hundred pounds to several thousand. Public sector planning organisations tend to be conservative and risk-averse and therefore tend to prefer a familiar and well documented model package whose predictions have generally been accepted to a new model whose capability is unproven. Thus it is that the planning profession has been slow to make use of new models and techniques.

Many models in widespread use can be traced back to the early days of transport modelling. A notable development in the USA in the 1960s was the so-called *four-stage (sequential) travel demand model* outlined in Fig. 5.8. This model was designed for use in transportation studies and used four submodels to predict, in turn, the trip ends (i.e. the numbers of origins and destinations), the trip distribution (the OD matrix), modal split, and assignment (the use of specific links or services). Each submodel made use of predictions for the previous one in the sequence but there was provision for feedback

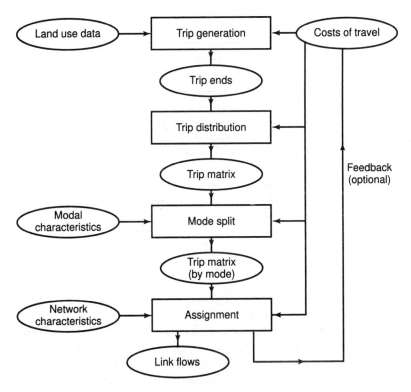

Fig. 5.8 Four-stage (sequential) travel demand model

effects such as the possibility that changes in costs, predicted by the assignment model due to congestion on individual links, might affect choice of mode or destination.

This sequential structure can still be found in many model packages even though, over the years, the individual submodels have become more sophisticated, the feedback mechanisms have been identified and a fifth submodel – choice of departure time – may have been added. Despite these improvements, the model is widely criticised for its sequential structure and the amount of data required to run the complete suite and it has become comparatively rare for all the submodels to be run in sequence. Modellers have instead come to select only those submodels which appear directly relevant to the task in hand. For example, junction design work is likely to be done using only an assignment model; if the network changes are thought significant enough to affect the travellers' choice of mode or destination, this is likely to be represented by an elasticity equation acting on the OD matrix (as in Fig. 5.9) rather than by making use of the distribution and mode split submodels from a sequential suite. The original rationale for the four-stage model was for strategic planning but this role is now often met by less cumbersome models utilising unified structures such as hierarchical logit.

5.7.2 Current trends in modelling

Current trends in modelling reflect firstly the current concerns of decision-makers, secondly the new opportunities being provided by the availability of cheap and powerful

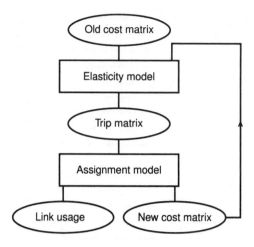

Fig. 5.9 Elastic assignment model

computers and new modelling techniques, and thirdly the ongoing concern of modellers to produce more accurate predictions.

In order to meet the current concerns of decision-makers, it has been necessary to develop models which seek to predict the impact of new policy options such as automatic toll collection, improved traveller information systems or more sophisticated traffic control systems (as described in Chapter 26 and Chapter 27). Also, in order to provide tools which might assist in the formulation and rapid assessment of integrated transport policies (see Chapter 2 and Chapter 3), it has been necessary to develop models which can quickly indicate the expected impact of a wide range of transport and land use policies. This has led to the development of sketch planning models using unified choice structures, highly aggregated zoning systems and skeletal networks. More controversially, some such models even go as far as replacing the link-based network by a zone-based representation of network capacity.[12]

The desire to capitalise on cheaply available computer power has spawned the development of detailed simulation modelling and the desire to exploit new techniques has led to a new breed of models which make use of the concepts of artificial intelligence, neural networks and fuzzy logic.

The ongoing concern of modellers to produce more accurate models has resulted in continued attempts to develop models which accord more closely with the insights revealed by behavioural researchers over the last several decades – insights such as the fundamental importance of an individual's activity schedule in determining travel demand, the existence of non-compensatory decision-making and the role of inertia and habit in determining daily behaviour.

5.7.3 A caution

It is easy for model enthusiasts to lose sight of the fact that models are useful only if they can deliver useful results on time and without undue demands on data or computing resources; this generally means that simpler models should be preferred to more complex ones. On the other hand the users of model predictions are apt to forget that models are simplifications of reality whose predictions should be regarded as estimates rather than precise forecasts. Modellers should have the courage to emphasise this point and ought to use the increased computer power now available to conduct sensitivity analyses and hence produce confidence limits for any predictions made.

5.8 References

1. Wooton, H.J., and Pick, G.W., Trips generated by households. *Journal of Transport Economics and Policy*, 1967, **1** (2), 137–153.
2. Taylor, M.A.P., Young, W., and Bonsall, P.W., *Understanding traffic systems*. Aldershot: Avebury, 1996.
3. Wonnacott, T.H., and Wonnacott, R.J., *Introductory statistics for business and economics*. New York: John Wiley, 1977.
4. Wilson, A.G., *Entropy in urban and regional modelling*. London: Pion, 1970.
5. Ortuzar, J. de D., and Williamson, L.G., *Modelling transport*. Chichester: John Wiley, 1994, second edition.
6. Williamson, L.G., Origin-destination matrix: static estimation. In M. Papageorgiou (ed.), *Concise encyclopaedia of traffic and transportation systems*, 315–322. Oxford: Pergamon Press, 1991.
7. Cascetta, E., and Nguyen, S., A unified framework for estimating or updating origin/destination matrices from traffic counts. *Transportation Research*, 1988, **22B** (6), 437–455.
8. Burrel, J.E., Multiple route assignment and its application to capacity restraint, In W. Leutzback and P. Baron (eds), *Beiträge zur Theorie des Verkehrsflusses*. Karlsruhe: Strassenbau und Strassenverkehrstechnik Heft, 1968.
9. Ben Akiva, M.E., and Lerman, S.R., *Discrete choice analysis: Theory and application to travel demand*. Cambridge, Mass: The M.I.T. Press, 1985.
10. Williams, H.C.W.L., On the formation of travel demand models and economic evaluation measures of user benefit. *Environment and Plannning*, 1977, **9A** (3), 285–344.
11. Wardrop, J.G., Some theoretical aspects of road traffic research. *Proceedings of the Institution of Civil Engineers*, 1952, **1** (36), 325–362.
12. Bates, J.M., Brewer, P., Hanson, McDonald, D. and Simmonds, D., Building a strategic model for Edinburgh. *Proceedings of the PTRC Summer Annual Meeting (seminar G)*. London: PTRC, 1991.

CHAPTER 6

Traffic planning strategies

C.A. O'Flaherty

Road and public transport problems are currently a significant issue in most countries, especially those with well-developed or burgeoning economies. However, resolving these problems requires changes in public attitudes and modifications in personal practices as much as technical solutions. Changes in desired lifestyle are hard to achieve in any society; they are particularly difficult in countries with democratically elected governments because of their impact on the political process. Thus timing is very important in terms of getting acceptance of controversial solutions and often it has to be recognised that 'things' may get worse before particular remedies can be applied.

Traffic congestion is now recognised as a major problem in Britain in respect of both work and non-work trips. Traffic delays, particularly in large urban areas, inter-city corridors, and on heavily-used tourist routes, are substantial and growing, and this is affecting people's convenience and business costs. There is also an increasing concern about road accidents, global warming, environmental pollution and fossil fuel depletion, and a growing acceptance that (for economic, social and environmental reasons) the building of new roads has only a limited role to play in respect to 'solving' today's (and tomorrow's) transport problems, particularly in urban areas.

In the early/mid-1980s the terms *congestion management* and *travel demand management* began to be identified with mechanisms which were being used to address problems in cities, inter-city corridors and substantial traffic-generating activity centres. Implemented individually or in concert with one another, these measures can help to achieve one or more of the following:[1]

- reduce the need to make a trip
- reduce the length of a trip
- promote non-motorised transport
- promote public transport
- promote car pooling
- shift peak-hour travel
- shift travel from congested locations
- reduce traffic delays.

Tables 6.1 and 6.2 summarise the applications to which various long- and short-term management measures (grouped according to strategy class) can be put in order to alleviate the above concerns. In the tables the strategy classes are also macro-grouped according to whether their major influences are supply-side or demand-side. *Supply-*

side measures generally aim at increasing the capacity of a road system so as to improve the traffic flow for all modes of transport using it. *Demand-side measures* are intended to reduce car demand by increasing the mode share carried by public transport, increasing vehicle occupancy, reducing the need to travel to a particular destination, and/or reducing the need to travel during peak traffic periods.

For discussion purposes it is useful to present the various demand- and supply-side measures in the context of five contrasting transport planning approaches in an urban area:

1. do-minimum
2. use land use planning to reduce trips and trip lengths
3. develop a transport network that is heavily car-oriented
4. develop a transport network that is heavily public transport-oriented
5. manage the demand for travel.

In practice, of course, most transport plans involve elements of all five ('transport packages'), with tilts and emphases in certain directions depending upon the political, social, economic and environmental conditions pertaining at the locale in question.

6.1 Do-minimum approach

At its extreme this approach assumes that traffic congestion, road accidents, and environmental degradation are inescapable features of modern-day life and, if left to itself, human ingenuity and self-interest will ensure that congestion will become self-regulating before it becomes intolerable. In practice, the do-minimum approach is used as a basis for comparison with 'do-something' proposals.

6.1.1 Congestion

In most large cities today traffic congestion is, in a sense, self-regulating. For example, despite enormous economic changes, and with only limited new road construction, the average speeds of traffic in London have not changed dramatically over many decades.[2]

Traffic congestion can be described as either recurring or non-recurring. *Recurring congestion* is associated with expected delays; it results from large numbers of people and vehicles travelling at the same time (e.g. during peak commuting or holiday periods) at the same places (e.g. at busy intersections). *Non-recurring congestion* is associated with unpredictable delays that are caused by spontaneous traffic incidents, such as accidents.

Both recurring and non-recurring congestion are associated with stop-start driving conditions which reduce fuel efficiency and increase air pollution, raise the cost of freight movement and distribution, hinder bus movements (thereby making transport by car appear more attractive), increase the numbers of accidents and delay emergency vehicles. Drivers who regularly encounter recurring congestion on main roads in urban areas often seek to reduce their delays by developing 'rat-runs' through adjacent neighbourhoods, thereby imposing considerable environmental hardships and social and economic costs on persons living in those areas. Recurring congestion also encourages the flight of home owners and business from inner cities to suburbia, leaving tracts of run-down areas in their wake. Minor road incidents during periods of recurring congestion can be the cause of severe non-recurring congestion, while several incidents within a short period of time at key points in a road network can result in the 'seizing' of the road system so that gridlock (i.e. the bringing of traffic to a standstill over a wide area) occurs.

Table 6.1 Applications of congestion management measures, divided according to strategy class[1]

Type of measure	Strategy class	Measures	Applications of measures							
			Urban	Inter-urban	Peak periods	Off-peak	Holiday periods	Construction/ maintenance	Special events	Incident management
Supply side	Road traffic operations	Entrance ramp controls on motorways	xx	x	xx	x	x	x	x	x
		Traveller information systems	xx	xx	xx	x	x	xx	xx	xx
		Traffic signalisation improvements	xx		xx	xx		x	x	x
		Motorway traffic management	xx	x	xx	x	x	x	x	x
		Incident management	xx	xx	xx	xx	x	xx	x	xx
		Traffic control at construction sites	xx	x	xx	xx	x	xx	x	
	Preferential treatment	Bus lanes	xx		xx	xx		x	x	
		Car-pool lanes	xx		xx	x		x		
		Bicycle and pedestrian facilities	xx		x	x			x	
		Traffic signal pre-emption for HOVs	xx		xx	x			x	
	Public transport operations	Express bus services	xx	x	xx	x		xx	x	
		Park-and-ride facilities	xx	x	xx	x		xx	x	
		Service improvements	xx	x	xx	xx	x	xx	xx	x
		Public transport image	xx	x	xx	x		xx	x	
		High capacity public transport vehicles	xx	x	xx	x		xx	xx	
	Freight movements	Urban goods movement	xx		xx	xx				x
		Inter-city goods movement	x	xx	xx	xx		x	x	x

Demand side

		Col 1	Col 2	Col 3	Col 4	Col 5	Col 6	Col 7	Col 8
Land use and zoning	Land use and zoning policy	xx	x	xx	x	x		xx	
	Site amenities and design	xx		xx		x		xx	
Communication substitutes	Tele-commuting	xx	x	xx	x			x	
	Tele-conferencing	x	xx	x	x			x	
	Tele-shopping	x	xx	x	x				
Traveller information services	Pre-trip travel information	xx	xx	xx	x	xx		xx	xx
	Regional rideshare matching	xx		xx		x		x	
Economic measures	Congestion pricing	xx	xx	xx	x	xx	x	xx	
	Parking pricing	xx	xx	xx	x			xx	
	Transport allowances	xx		xx	x				
	Public transport and rideshare financial incentives	xx		xx	x			x	
	Public transport pass programme	xx		xx	x			x	
	Innovative financing	xx		xx	x				
Administrative measures	Transport partnerships	xx		xx					
	Trip reduction ordinances and regulations	xx		xx	x				
	Alternative work schedules	xx		xx					
	Car restricted zone	xx		xx	xx			xx	
	Parking management	xx		xx	xx			xx	

xx = significant application, x = some application

Table 6.2 Impacts of congestion management measures, divided according to strategy class[1]

Type of measure	Strategy class	Measures	Impacts of measures							
			Reduce need to make trip	Reduce length of trip	Promote non-motorised transport	Promote public transport	Promote car pooling	Shift peak-hour travel	Shift trips from congested locales	Reduce traffic/traveller delays
Supply side	Road traffic operations	Entrance ramp controls on motorways				xx	xx	x	xx	xx
		Traveller information systems		x		x	x	xx	xx	xx
		Traffic signalisation improvements				x			x	xx
		Motorway traffic management		x		x	x	x	x	xx
		Incident management				x	x	x	xx	xx
		Traffic control at construction sites				x	x	x	xx	xx
	Preferential treatment	Bus lanes				xx		x	x	x
		Car-pool lanes				x	xx	x	x	x
		Bicycle and pedestrian facilities			xx					
		Traffic signal pre-emption for HOVs				xx				x
	Public transport operations	Express bus services				xx				x
		Park-and-ride facilities				xx	xx		x	x
		Service improvements				xx				x
		Public transport image				xx				x
		High capacity public transport vehicles				xx				x
	Freight movements	Urban goods movement						x	x	xx
		Inter-city goods movement	x					x	x	xx

Demand side	Measure	1	2	3	4	5	6	7	8
Land use and zoning	Land use and zoning policy	x	x	xx	x	x	x	x	x
	Site amenities and design	xx	xx	xx	xx	xx	x	x	x
Communication substitutes	Tele-commuting	xx	xx	x	x	x	x		x
	Tele-conferencing	xx	xx	xx	x	x		x	x
	Tele-shopping	xx	x						
Traveller information services	Pre-trip travel information	x	x	x	xx	xx	xx	xx	x
	Regional rideshare matching	xx			xx	xx	x	x	x
Economic measures	Congestion pricing	x	x	x	xx	xx	xx	xx	xx
	Parking pricing	x	x	x	xx	xx	xx	xx	xx
	Transport allowances			x	xx	xx	x		x
	Public transport and rideshare financial incentives				xx	xx	x		x
	Public transport pass programme				xx	x	x		x
	Innovative financing				xx	xx			xx
Administrative measures	Transport partnerships	xx		x	xx	xx	x	x	x
	Trip reduction ordinances and regulations		x	x	xx	xx	x	x	xx
	Alternative work schedules	x		xx	xx	xx	xx		x
	Car restricted zone	xx	xx	xx	xx	xx	x	xx	x
	Parking management	x	x	x	xx	xx	x	x	x

xx = significant impact, x = some impact

For the above reasons, therefore, it is now generally accepted that to let congestion find its own level unhindered would have the effect of causing irretrievable long-term damage to both urban and rural areas, and would ultimately have the effect of reinforcing car-dependent lifestyles.[3]

6.2 The land use planning approach

Transport plans which emphasise land use planning and the intricacies of individual site design recognise that the control of land use is to a large extent the key to the control of both the demand for transport and its impact upon the environment.

Table 6.3 Total distance travelled by mode and settlement population size in the United Kingdom, 1985/86 (reported in reference 4)

Area	All modes km*	Car km*(%)	Local bus km*(%)	Rail km*(%)	Walking km*(%)	Other[†] km*(%)
Inner London	141	76(54)	12(8.5)	34(24)	2.5(1.8)	17(12)
Outer London	167	113(68)	8.9(5.3)	23(14)	2.6(1.6)	19(11)
Conurbations with populations:						
(ave) > 0.25 m	117	74(63)	16(14)	5.2(4.5)	3.5(3.1)	18(15)
Urban areas with populations:						
0.1 m – 0.25 m	161	115(72)	8.6(5.4)	11(7.0)	3.2(2.0)	23(14)
0.05 m – 0.1 m	155	110(72)	7.2(4.7)	13(8.4)	3.7(2.4)	20(13)
0.025 m – 0.05 m	151	111(73)	5.7(3.8)	13(8.3)	3.7(2.5)	18(12)
0.003 m – 0.025 m	176	133(76)	7.2(4.1)	8(4.6)	3.0(1.7)	24(14)
Rural areas	211	164(78)	5.7(2.7)	11(5.2)	1.7(0.8)	29(14)
All areas	160	114(71)	9.3(5.8)	11(7.1)	3.2(2.0)	22(14)

* Kilometres per person per week (excluding trips < 1.6 km). [†]Other refers to 2-wheeled motor vehicles, taxis, domestic air travel, other public transport, and other types of bus travel.

Although the relationship between land use and transport is complex, it is well established that the decentralisation of people and jobs to suburbia, and to smaller free-standing towns in non-metropolitan areas, has been possible because of the increase in personal mobility brought about by the private car. This outward spread of people and jobs has been accompanied by an increasing physical separation of homes, jobs and other facilities, and has been reinforced by the concentration of shops, schools, hospitals and other services into fewer and larger units.[3]

Table 6.3 shows that in Britain the small towns and rural areas are associated with the greatest levels of travel and car dependency. However, there appears to be no systematic relationship between travel behaviour and the size of the urban area as far as the largest cities are concerned. Table 6.4 shows that trip lengths increased by some 31 per cent between 1965 and 1991.

6.2.1 Land use planning mechanisms

Transport plans that emphasise the land use planning approach generally seek to influence settlement patterns so as to increase the accessibility of jobs, shops, educational institutions, places of entertainment, etc., without the need to travel by car – or, when travel by car is essential, by minimising its usage and shortening the distance travelled. However, initiating area-wide changes in land use is not something that is easily done under a democratic system of government and in many instances considerable time must elapse before their full impact is realised. Consequently, the land use approach is normally accompanied by transport measures that aim to increase the competitiveness and attractiveness of urban centres *vis-à-vis* peripheral developments, and promote choice by increasing the relative advantage of travel other than by car.

Practical *land use control measures* that are now commonly related to transport plans are as follows.[4,5]

1. Limit the spread of cities so as to keep up residential densities and protect 'green belt' open land. Avoid the development of new settlements that are unlikely to achieve 10 000 homes within, say, 20 years.
2. Increase the supply of housing in existing larger urban areas (in market towns and above) where they are/can be easily accessible to existing facilities. Concentrate higher-density residential developments near public transport centres or alongside corridors well serviced by public transport.
3. Locate high-density offices and retail establishments at sites already well served by public transport, and in places easily reached from local housing by bus, cycle or walking. Facilities that attract large numbers of people from a wide catchment, e.g. universities and further education establishments, conference centres, hospitals, major libraries and governmental offices, should be located at sites in urban areas that are well served by public transport. University accommodation should be provided at locales with easy access to the campus.
4. Promote the juxtaposition of employment and residential uses ('urban villages') so that people have increasing opportunities to work at or near their homes.
5. Locate developments which attract significant movements of freight (e.g. large-scale warehousing and distribution depots, and some forms of manufacturing) away from residential areas and close to transport networks. Maximise the use of sites capable of being served from wharves, harbours or railway sidings.
6. Allocate sites unlikely to be served by public transport solely for uses which are not employment intensive.

Table 6.4 Growth in person travel, 1965–91 (Note: 1965 = 100)

	1965	1975/76	1985/86	1989/91
Population	100	102	104	106
Journeys/person	100	113	118	132
Km/person	100	114	120	131
Person-km	100	129	148	183

Source: National Travel Survey

Practical *traffic management measures* that are often associated with land use-based transport plans include:

● limiting the amount of car commuting to new employment developments that are well serviced by public transport, by imposing low maxima on the required number of parking spaces to be provided by the developers

- ensuring that parking provision at peripheral office, retail and similar developments is not set at such high levels that they significantly disadvantage areas that are more centrally located
- shifting a significant proportion of the supply of parking from central and inner-city areas to outer park-and-ride interchange locations
- limiting new road construction to that essential for the servicing of new developments
- implementing priority measures that promote the use of public transport, and safe cycling and walking
- implementing area-wide traffic calming schemes to protect existing housing developments from through movements by the private car
- establishing car-free zones at concentrations of shopping
- requiring large employers to introduce plans which reduce the demand for commuter parking.

6.3 The car-oriented approach

The post-World War II era saw very rapid developments in respect of the growth and usage of the private car (see Chapter 1). Associated with this growth was the worldwide production of transport plans in the 1950s and 60s which emphasised the urgent need to provide more and bigger roads (particularly roads linking city centre with city centre, and suburbia with central areas) if cities were not eventually to grind to a standstill. Associated with this car-oriented planning approach was usually a demand for huge numbers of parking spaces in and about town centres, and a tacit acceptance that the public transport system should be allowed to decline to some base level of service. Many of these 'predict and provide' plans were never implemented (for cost and environmental reasons); some were.

The City/County of Los Angeles which, over a 60-year period, has constructed many thousands of kilometres of freeway and high-quality dual carriageways in an effort to accommodate growth in car ownership and use (as well as providing for low-cost car travel), might be described as the epitome of the private car-oriented approach. In 1993 it was reported[6] that in Los Angeles:

1. there is now more than one registered vehicle per licensed driver
2. some 77 per cent of workers drive to work alone, 15 per cent go by car pool/van pool, 5 per cent travel by public transport, and about 3 per cent walk/cycle/jog
3. overall weekday peak hour vehicle occupancies are less than 1.2
4. there is still the heaviest traffic congestion of any city in the United States, and it is continuing to worsen.

It might be noted that cities that have adopted approaches which favour the private car (as well as those which have not, mainly since the mid-1960s) place considerable emphasis on the importance of road hierarchy and on the implementation of traffic operations which provide motorists with real-time information regarding congested locations and times, and improve traffic flows with the aid of traffic control technology (see reference 7).

6.3.1 The road hierarchy concept

Over the course of time many roads have become multi-functional, i.e. they now act as (a) carriageways for through vehicular traffic, (b) accessways to frontage properties, (c) routes for public transport, (d) parking spaces for vehicles, (e) passageways for pedestrians and cyclists, and (f) corridors for the location of water/sewerage/gas/electrical services. A major feature of the car-oriented transport plans of the 1950s in the United States was their clarification and prioritisation of the transport functions served by various types of road. In essence they divided roads into three main functional groups:

1. *arterial roads* which are primarily for longer-distance high-speed through-vehicle movements and, hence, provide minimal access to adjacent frontages
2. *local roads and streets* whose main function is to provide for frontage access and, thus, whose design and traffic management is intended to discourage through traffic
3. *collector roads*, i.e. the 'middle' group, which are intended to provide for both shorter through-vehicle movements and frontage access.

Arterial roads are subdivided according to whether they are limited access (e.g. freeways) with no direct frontage access whatsoever, major arterials (e.g. expressways) which have small amounts of frontage access, and minor arterials which have more frontage access. Local streets are generally subclassified according to the land use that they serve, e.g. residential, commercial, and industrial streets. Collector roads are described as major or minor, depending upon their relative amounts of through and access service.

In Britain non-urban roads are described according to function as being either primary routes or secondary roads. *Primary routes* − these are distinguished by direction signs with a green background − are mainly trunk roads (which include most motorways) and some local authority roads. Urban roads were designated in 1966 as *primary distributors* which are intended to serve a town as a whole by linking its business, industrial, and residential districts, *district distributors* which feed traffic from the primary network to environmental districts (e.g. town centres or industrial estates, or large residential districts) but do not traverse them, and *local distributors* which allow traffic from the major distributors to penetrate environmental districts and gain access to homes, businesses, factories, etc. by way of *access streets*. Table 6.5 summarises the various functions and features of these urban roads; also included in this table are details regarding a class of street which has come to prominence in more recent years − the *pedestrian street*.

6.4 The public transport-oriented approach

With this approach the transport plan emphasis is on proposals which result in significant improvements to the quality and quantity of road and rail public transport services. These are normally associated with land use, economic, administrative and improved traveller information services measures which encourage the usage of public transport.

Advocates of public transport-oriented plans emphasise that, irrespective of whether they are road- or rail-based, public transport systems are more energy efficient, emit less airborne pollutants, minimise the amount of land used for transport (including parking) purposes, and generally result in better physical environments in urban areas. Good public transport, especially rail transport, helps to retain employment and other activities in central areas, as well as facilitating pedestrianisation; in rapidly expanding towns it allows a greater proportion of new jobs and facilities to be located in the centre.

However, people are unlikely to turn to public transport unless it is provided at reasonable cost by clean, comfortable vehicles and unless services are regular, predictable and reliable.[3]

Table 6.5 Hierarchical classification system for urban roads in Britain, based on function[20]

	Pedestrian streets	Access roads	Local distributors	District distributors	Primary distributors
Predominant activities	Walking; meeting; trading	Walking; vehicle access; delivery of goods and servicing of premises; slow-moving vehicles	Vehicle movements near beginning or end of all journeys; bus stops	Medium distance traffic to Primary Network; public transport services; all through traffic with respect to environmental areas	Fast-moving long-distance traffic; no pedestrians or frontage access
Pedestrian movement	Complete freedom; predominant activity	Considerable freedom with crossing at random	Controlled with channelised (e.g. zebra) crossings	Minimum pedestrian activity with positive measures for their safety	Nil; vertical segregation between vehicles and pedestrians
Stationary vehicles	Nil, except for servicing and emergency	Some, depending on safety considerations	Considerable, if off-road facilities not provided	Some, depending on traffic flow conditions	Nil
Heavy goods vehicle activity	Essential servicing and frontage deliveries only	Residential-related activities only; other areas – delivery of goods and services	Minimum through trips	Minimum through trips	Suitable for all HGV movements, especially through trips
Vehicle access to individual properties	Nil, except for emergency vehicles and limited access for servicing	Predominant activity	Some, to more significant activity centres	Nil, apart from major centres, i.e. equivalent to local distributor level of vehicle flow	Nil, apart from sites of national traffic importance
Local traffic movements	Nil, but may include public transport	Nil	Predominant activity	Some – only a few localities may be severed, and junction spacing is important	Very little – junction spacing will preclude local movements
Through traffic movements	Nil	Nil	Nil	Predominant role for medium-distance traffic	Predominant role for long-distance traffic
Vehicle operating speeds/speed limits	Less than 5 km/h; vehicles enter on sufferance	Less than 32km/h (20 mile/h) with speed control devices	Subject to 48 km/h (30 mile/h) limit but layout should discourage speed	Subject to 48 or 64 km/h (30 or 40 mile/h) limit within the built-up area	More than 64 km/h (40 mile/h) depending on geometric constraints

6.4.1 Rail public transport

Rail systems are most effectively used to service densely populated cities with relatively long journey-to-work distances along radial corridors with congested roads which are central area-oriented. Because of the high cost of rail systems in urban areas, making the full use of existing rail infrastructure is the key to ensuring the success of most new rail schemes.[8]

In London, where some 70 per cent of those commuting to central London travel by rail, the full costs of operating the Underground and Network SouthEast heavy rail systems were only 58 per cent and 86 per cent, respectively, met by fares revenue in 1992. The average journey lengths on Network SouthEast and London Underground trains are 27.4 km and 7.9 km, respectively, and the average journey speeds are 56 km/h and 32 km/h, respectively.

The 29 km-long first phase of the South Yorkshire Supertram, which services Sheffield city centre, cost £240 million to construct, i.e. £8.28 million/km. This light rail system was opened in 1994.

New rail systems generally offer a more attractive alternative to car users than buses – as well as being more attractive to existing bus users. Table 6.6 shows that the majority of the users of new light rail systems came from bus patrons (some of whom still use the bus for part of their journeys). Since there is a suppressed demand for car travel in most large cities (which means that empty road space is filled as soon as it is created) this explains why it is that the introduction of a new light rail system often appears to have little impact on road congestion – and why there is a need for concurrent car-deterrence measures if the main objective is to reduce congestion. Information about routes, waiting times, and timetables is important in removing uncertainty and much effort is currently being placed on providing rail customers with real-time electronic information displays at stations.

Reliability is also critical, and many rail operators are now publishing charters which set out their target standards of service to passengers. For example, London Underground Ltd's charter has a compensation scheme that applies when delays of 15 minutes or more occur within its control. Compensation is also payable by British Rail when service falls by more than a small margin below its standards and under the privatisation proposals, franchisees will be required to produce charters that will at least match British Rail standards.

Table 6.6 Previous modes used by users of new metro and tram systems in French cities (reported in reference 4)

City	Passengers (%)		New journeys*
	Who formerly travelled by:		
	Bus	Car	
Marseilles	74	15	11
Lille	56	28	16
Lyon	71	12	17
Nantes	67	17	16
Grenoble	53	20	27

* May include those who changed home, job or school and now make different trips from previously.

'Travelcards' which allow patrons to change from one rail system to another, and from bus to rail and vice versa, without the need for re-ticketing have been shown to be a critical factor encouraging the use of public transport.

6.4.2 Bus public transport

Strategies used to favour bus public transport over the private car include: (a) land use planning which locates large traffic generators at sites which are capable of being well served by buses (see Section 6.2), (b) improving bus services, and (c) using traffic restraint to make car travel more difficult.

Improving services

London's 'Red Route' schemes, for example, have shown[9] that quicker, more reliable journeys can stimulate bus patronage as well as reducing overall journey times and costs. Measures used to assist in the improvement of services include limited stop services, the use of bus control systems, and schemes which give priority to buses over cars on congested roads.

Express bus services typically pick up passengers at a limited number of stops in an outer residential area and travel non-stop to a town centre or a major industrial area. A flat fare is usually charged for the trip. These services are most heavily used when applied to peak period travel over longer distances.

Bus control is concerned with maintaining regularity of service in traffic congestion by providing immediate information on traffic conditions to the bus operators which enables the bunching of buses to be reduced or gaps in the services to be filled. Methods used for this purpose range from fitting buses with telephones to allow direct communication between a control centre and drivers, to the use of real-time automatic vehicle location systems which promote more efficient scheduling and better service reliability.

Bus priority measures in widespread usage include:

- with-flow priority lanes reserved for buses that allow them to bypass congested traffic travelling in the same direction
- contra-flow priority lanes that allow buses to travel the 'wrong way' on one-way streets to avoid a detour or serve the street
- reserved bus lanes on, or that provide access to, urban motorways and which are sometimes reversible for tidal-flow bus operations
- bus-only streets, i.e. existing streets that are turned into bus roads, and busways, i.e. segregated roads that are designed and constructed for buses only
- allowing buses to have access to pedestrian precincts
- providing automatic or driver-initiated bus detection at traffic signals when bus flows are low and/or road widths are insufficient to accommodate a bus lane, or programming area-wide urban traffic control schemes to give favourable attention to buses
- using traffic regulations to give priority to buses when leaving bus stops, to impose parking restrictions on bus routes, and to provide exemptions to buses from prohibitions affecting turning movement at intersections.

In Britain the most common bus priority measure is the with-flow lane which allows buses to bypass queues of vehicles on the approaches to traffic signal-controlled intersections.

The majority of these lanes are less than 250 m long and require constant policing to ensure that they are allowed to operate effectively, without being interfered with by private cars. A significant proportion of these bus lanes operate only during morning and evening peak periods when traffic congestion is at its height.

Other conventional bus service improvements include route extensions and expanded hours of operation, all-day services operating through residential areas (generally using minibuses), and peak-period circulator services that operate between residential/employment areas and rail or bus stations.

Demand-responsive services are an innovative type of operation that has received much consideration in industrialised countries (e.g. see reference 10). Also known as *dial-a-bus services,* these are intended to provide a quality of service between that of a taxi and a stage-bus at a lower cost, and/or to service locales where the travel demand for buses is low. With *corridor* demand-responsive services (a) time flexibility is provided with the aid of communication devices provided at points along a fixed route which allow potential bus users to call the next bus and reduce their waiting time, while (b) space flexibility is provided by allowing bus drivers to undertake minor pick-up/drop-off deviations from an otherwise fixed route. With *zonal* dial-a-bus services an area is divided into sections, each of which is serviced by one or more minibuses which collect passengers at their homes following phone requests, and deliver them to a single destination (many-to-one operation) or to two or three possible destinations (many-to-few operation). The zonal minibuses may operate for all or only certain times of the day, have fixed or semi-fixed schedules, and/or operate on the basis of advance bookings only.

Dial-a-bus services are relatively expensive to operate because of the extra labour requirements at the central office and the relatively low utilisation rates of the minibuses. 'Successful' services tend to be those that have been kept in operation as paratransit services for the elderly and handicapped, and/or the service is subsidised and fares are deliberately kept low for political or marketing reasons.

In recent years particular efforts have also been made to raise the image of conventional bus transport by emphasising cleanliness and amenities in buses, by improving personal security and accessibility for the elderly and disabled at bus stops and bus stations, and by targeting particular market groups during advertising campaigns.

In large urban areas *park-and-ride schemes* (see Chapter 7), whereby motorists leave their cars at outlying car parks and travel by bus (or rail) to the central area or to other concentrated employee destinations, are usually a significant component of a public transport-oriented transport plan. However, if any park-and-ride scheme is to attract motorists from their cars then the perceived advantage of using the combination of car and public transport modes will need to significantly outweigh that of using the car alone, otherwise the motorists will stay in their cars for the entire trip.

Traffic restraint

There is much evidence to suggest that car traffic is not significantly reduced by measures which rely only upon improvements to public transport. Rather, the general consensus is that if bus (and rail) patronage is to be increased and road congestion lessened, then the public transport improvements must also be accompanied by measures that discourage the use of private vehicles in urban areas.

The restraint tool mostly used in Britain to promote public transport usage and reduce congestion, especially for the journey-to-work trip, is *car parking control*. Control measures currently utilised in areas well serviced by public transport include:

- minimising the number of spaces allowed in new developments, and attempting to reduce the total number of controlled spaces in existing developments, with the objective of ensuring that the access road system to the area under consideration (e.g. the central area) is kept within its capacity
- limiting the continuous length of time during which a car may stay in a controlled parking space so that the commuter is prevented from using choice parking places
- imposing a parking charge regime on controlled parking spaces which favours short-stay parkers (who are usually shoppers and travel off-peak) and discriminates against commuters.

Unfortunately, the effectiveness of parking control is often limited by its lack of influence over a large proportion of the existing parking spaces. For example, it is reported[3] that some 80 per cent of the people who drive to work in most large towns in Britain are provided with free parking at their workplaces. 60 per cent of the cars entering central London between 7 a.m. and 1 p.m. terminate there, while 40 per cent is through traffic. Three quarters of all car trips terminating in central London, and 87 per cent of car commuting trips, use an off-street parking space; most of the drivers of these cars are given the parking space free or are reimbursed for its cost.[11]

From the above it can be deduced that commuting by car to central London is now limited in the extent to which it is influenced by parking cost, as this can be easily offset by employers. Only a further reduction in car parking spaces is likely to affect the number of commuters who choose to drive to work, if the current reliance upon parking control as the principal traffic restraint technique is to be continued.

A high level of enforcement of on-street parking regulations, including the prohibition of illegal parking on footpaths and in laneways, is essential if the maximum benefits to public transport are to be achieved from parking control. In this context it might be noted that in 1990 some 80 per cent of the cars parked on-street in central London were reported[12] to be illegally parked and that 300 000 parking offences were committed daily with only 1 in 50 illegally parked vehicles receiving fixed penalty notices.

Traffic signal control has also been used[13] in a major demonstration project to restrain car travel and encourage bus usage into a town centre. If modal change is to be achieved using traffic signals as the restraint mechanism, then substantial delays have to be imposed on car drivers before the objective is achieved. This, in turn, means that very extensive queuing accommodation is required at the traffic signal control points – and this storage space is rarely available on radial roads in British towns. Also required is a high degree of (unpopular) police enforcement to ensure that the delayed cars do not use the bus priority lanes or try to go through the signals while they are showing red.

Traffic signals that operate according to pre-selected control strategies are also used on upstream entry slip roads of motorways to regulate the amounts of traffic joining the mainline carriageways. The most usual reason for using *entrance ramp control* (also known as *ramp metering*) is to maintain vehicle throughput at a (high) flow level that is within the downstream capacity of a mainline road while vehicle speeds are kept relatively high and uniform; this reduces stop-go traffic operation and the numbers of related accidents that would otherwise occur in congested conditions. This form of access control is easily adjusted to incorporate priority for buses at little additional implementation cost.[14] However, care must be taken in urban areas to ensure that the overall system is designed so that delayed vehicles are not able to divert into 'rat-runs' through adjacent local streets.

6.5 The demand management approach

In practice transport plans and strategies which emphasise managing the demand for travel tend also to promote anti-congestion measures which reduce the pressure on the road system, e.g. encouraging greater public transport usage (see Section 6.4) and car pooling (e.g. by giving priority at congested locales and times to high-occupancy vehicles), imposing traffic restraint via stringent parking control measures at destinations (Sections 6.4 and 7.1), and promoting the use of variable working hours and telecommunication working processes by large employers so as to shift travel from congested peak periods. More far-reaching proposals include the use of road pricing mechanisms that are based on the concept that road users who contribute to congestion are a cause of additional costs to society and if they were to be charged for these costs, some would travel at different times, by different routes or by different means, and congestion would therefore be reduced.

6.5.1 Car-pooling

The concept underlying car-pooling is simple: put drivers of single-occupancy vehicles into fewer vehicles and traffic congestion, vehicle-kilometres travelled, and air pollution will be reduced.

Car-pooling can be informal, i.e. formed by a group of people acting on their own who share the driving and thereby reduce the cost of driving alone, or organised, e.g. by an employer or by a governmental or private agency that co-ordinates the activities of a number of employers. Car-pooling is most effectively used when (a) trips are relatively long, so that the time spent gathering the occupants is small relative to the rest of the journey, and (b) the participants have the same travel schedule each day. Car-pools that continue for long periods of time have participants who are compatible, work in the same employment area, and have full-time jobs with regular hours.

Car-pooling appears to be most heavily promoted in the United States. The most successful employer-generated car-pooling involves large employers who have a continuing, highly visible, well-staffed co-ordination programme. However, for employers to make that commitment they must believe that their costs will provide them with direct benefits.[15]

Governmental bodies in the U.S. have sought to develop car-pools/van-pools by advertising for the names of potential users, and their telephone numbers, home and work locations, work schedules, and personal characteristics (e.g. smoker/non-smoker), and then matching them by computer. In the case of van-pools, the vehicle is larger (e.g. a minibus), is provided by the brokerage agency (which also provides insurance), the participants pay a monthly fee (which is often subsidised), and one person is designated as the driver and generally, as compensation, has the free use of the vehicle over the weekend.

Even though the travel cost is less for a van-pooler than a car-pooler, van-pools are not as popular because of the larger number of people involved. Keeping files up-to-date is more difficult because greater numbers of people change their jobs or homes, and because the most efficient way to arrange the morning pick-up is to have it happen at a single location – and this means that the many participants of a van-pool must normally drive to (and from) a convenient meeting place where they can also leave their own cars in safety. Experience in the U.S. also shows that van-pools are not attractive to participants unless the travel distance is long.

In very large cities, the encouragement of car-pools and van-pools in heavily travelled corridors often has the effect of seducing passengers from existing public transport services.

Car- and van-pooling are more likely to be effective if (a) poolers are given preferential treatment over other vehicles when travelling to their destinations on congested roads and at busy intersections, and (b) there is strict parking control at the destinations so that poolers can be granted the privileges of paying less (or nothing) and parking closer. Car- and van-pooling are most unlikely to be effective if there is free parking for all car travellers at the destinations, or when a fully/partially funded company car is regularly available to the driver. In the latter instance, the driver has no incentive to minimise the travel costs to work by joining a car- or van-pool. In some cities in the U.S.A. pooling is encouraged by local government ordinances which require employers with more than a specified number of employees to develop programmes that are designed to increase vehicle occupancies by their employees.

In respect of the above it might be noted that of the 20.8 million cars registered in Britain at the end of 1993, some 2.2 million (10.6 per cent) were registered in the names of companies. In 1991 the London Area Transport Survey (LATS) determined that the average car occupancies per trip in London were 1.14 for commuting trips, 1.18 for trips on employer business, and 1.67 for other trips. Also, the provision of high-occupancy vehicle (HOV) lanes on major roads to facilitate car pooling is not favoured in Britain.[3]

6.5.2 Varying working hours

The varying of working hours tends to take three forms in practice. With *staggered hours* a number of large employers agree to stagger their start/finish times by, say, 15 to 30 minutes, with no change to the number of hours worked each day. With *flexible working hours* the employees may start and finish work each day at times of their own choice, as long as a core period (necessary for meetings) is worked each day. With the *compressed working week* the employees work many more hours per day for fewer days per week or per fortnight. All three forms of operation reduce the commuting traveller's journey time, but depend for their success on the goodwill and support of the employer organisations.

Staggered hours generally have the effect of spreading the morning and evening peak periods and reducing the demand for public and private transport movements during the most heavily loaded 15 to 30 minutes of each peak period. Staggered hours are most appropriately applied to offices and piece-manufacturing establishments. Flexible work scheduling also spreads the morning and evening peak periods and is commonly applied to offices and to administrative and information workers. Compressed work-week programmes tend both to reduce the total number of commuter trips and to shift them to off-peak periods. Compressed work-week employees also make many short errand-type trips (usually during off-peak periods) during their 'free' day; overall, however, the programmes can reduce the average number of vehicle-kilometres travelled and also the levels of vehicle-source pollutants entering the atmosphere.[16] Experience suggests that compressed work-weeks are most appropriate for office and administrative functions, and for line- and piece-manufacturing processes.

6.5.3 Technological improvements

Since the 1960s most transport plans for urban areas have emphasised the importance of improving traffic operations at intersections. To a large extent it can be argued that the reason that already congested road networks have been able to cope with much of the tremendous growth in vehicle-kilometres in large towns over this time is because of *improvements in traffic signal operation* and the application of integrated *urban traffic control* (UTC) systems at intersections, which have enabled more and more of their potential capacity to be extracted from existing roads.

One of the expectations underlying the current technological thrust is that developments in telecommunication technology will reduce the amounts of travel, especially by car during peak periods, in future years. The three measures which are seen as having the greatest potential to make this happen are telecommuting, teleconferencing, and teleshopping. However, all of these are still in their infancy in terms of widespread application.[1]

Telecommuting, also known as *telework*, is the partial or total substitution of telecommunications for the daily journey-to/from-work by allowing employees to work at locations that are remote from the traditional office. Employees are usually linked to the main office by telephone and facsimile and, often, by computer and modem. Remote worksites include the employee's home, satellite work centres (run by single employers), or neighbourhood work centres (run by multiple employers).

Telecommuting is seen as being particularly important, because it is cheaply established and has the potential to dovetail with employer objectives such as improved employee morale and productivity and, from a transport aspect, it impacts mostly on peak-period travel. It is reported[3] that the 1990 national transport plan for the Netherlands saw the potential benefit of telecommuting as less congestion with the possibility of a five per cent reduction in peak-hour car traffic. At this time telecommuting is most well established in the computer services industry.

Teleconferencing is the substitution of telephone and television communication (i.e. audio, or audio and video) for trips normally taken to meet with several individuals or groups, usually for business reasons. It requires a substantial investment in equipment (e.g. studio, transmission lines) or the renting of facilities with the equipment available. At this time teleconferencing is seen as being most applicable to private businesses and education. No data are available regarding its likely impact on travel.

At this time *teleshopping* involves the use of the telephone and facsimile, and sometimes videotex, to shop and purchase items without physically travelling to the place of sale. It is a private, strictly commercial venture that is likely to spread quite slowly; its main travel impact will likely be on congestion at large shopping centres. More specialised applications such as telebanking and telephone bookings for sporting and art events and travel will achieve a much higher penetration but lead only to a small reduction in the number of trips made.

Another technological development foreseen for the long-term future is the use of *advanced electronic systems* to control vehicle movements on major roads. This could eventually involve fitting vehicles with intelligent cruise-control guidance and collision-avoidance systems that would increase road capacity (by enabling vehicles to operate at lower headways) and improve road safety in heavy traffic (by preventing them from colliding).

6.5.4 Charging for road use

Many countries are now considering the use of more radical mechanisms (as compared with parking control and traffic signal control) to constrain the use of the private car on congested roads. One of the more thoroughly examined measures (via desk studies) is that of *road pricing*. With a road pricing system the efficiency of operation on a congested road or road network is improved by employing the user-pay principle to make a car driver think very carefully about whether or not to use the controlled road(s). This charging process, which allows those who are willing to pay to travel under less congested conditions, is also known as *congestion pricing*.

The simplest way of levying the charge is at a toll collection point, provided that space is available for queuing vehicles. Alternatively, differential licences may be prepurchased which allow access onto the controlled road(s). Technology is also available which, analogous to telephone charging, uses a piece of equipment (e.g. an electronic number plate) on the vehicle to activate a central charging computer when the vehicle enters or leaves a controlled road system. A 'contactless smartcard' with prepaid stored value has also been developed which can be mounted on a vehicle's windscreen so that a charge can be automatically deducted each time that it passes a toll point. (An advantage of this smartcard is that personal privacy is protected since it requires no record of a vehicle's whereabouts.)

An early study[17] of the road pricing concept as it might be used in Britain suggested that the mechanism used to implement widespread usage should satisfy the following criteria:

1. The charges should be closely related to the use made of the controlled roads (e.g. by making the charges proportional either to the time or distance travelled on them).
2. It should be possible to vary the charges for different roads or areas, at different times of the day, week or year, and for different classes of vehicle.
3. Charges should be stable and readily ascertainable by drivers *before* they enter the controlled road system.
4. Payment in advance should be possible, although credit facilities might also be permissible under certain conditions.
5. Any equipment used should possess a high degree of reliability (and be robust and reasonably secure against fraud).
6. The method used should be:
   (a) simple for road users to understand and the police to enforce
   (b) accepted by the public as fair to all
   (c) amenable to gradual introduction commencing with an experimental phase
   (d) capable of being applied to the whole country (as well as to temporary road users from abroad).
7. The pricing control mechanism should, if possible, indicate the strength of the demand for road space in different places so as to give guidance to the planning of new road improvements.

Comment

Much has been written about the possible use of road pricing as a tool for congestion management, and a significant number of transport planners now argue that it should be a core element of a demand management strategy, particularly for a large city. However, relatively few schemes have been implemented (see Chapter 3) to date; politically, it has been too difficult. What has generally been accepted world-wide, however, is the concept of 'user-pay' in relation to motorways, i.e. toll roads. In Britain in late 1993, for example, the Government announced its intention to introduce electronic motorway tolling 'when the technology is ready'. Looking to the future, therefore, it is possible that the increasing use of toll roads will establish a tradition of paying for roads, which in due course may facilitate public acceptance of area-wide urban road charging – and lead eventually to a fully integrated road user charging system affecting all car journeys.[4]

6.5.5 Rationing road space

A major criticism of road pricing is that it favours the wealthy and discriminates against the less well off, because of its reliance upon non-willingness/non-ability to pay to reduce the number of vehicles on congested roads. Various 'fairer' schemes have been proposed which have ranged from those based on demonstrated need to selection by lottery. At this time the most favoured proposals appear to be the *'odds-and-evens' system* whereby vehicles with number plates ending in an odd number are allowed to use the controlled roads on certain days of the week while even-numbered vehicles can use them on the others, and the *'pooler' system* which allows vehicles with more than a certain number of occupants to be exempt from any road pricing/toll charges.

6.6 Transport packaging

As noted previously, current transport plans for urban areas do not rely solely on any of the contrasting approaches described above. Rather they rely on 'packages' of the above, tilted to meet the needs of the environs within which the urban area is located (see, for example, reference 18). For example, logically it can be expected that proposals for larger towns will differ from those for smaller ones, and that urban areas of the same-size population that are located in developed and newly-developing countries would place different emphases on the mix of land use policy, road and public transport infrastructure, and congestion and travel demand management measures proposed.

A package approach which may be considered as generally reflective of transport planning in many larger British towns at this time is that for the City of Leeds. Leeds sees itself as having the potential to become a leading European city, with increasing prosperity, an enviable environment and a city centre to be proud of. Transport is seen as a key to achieving this vision and in 1990 therefore the city commissioned a review of its transport strategy. The study was directed by a very broadly based group which included the leader of Leeds City Council, cross-party membership of city councillors, professional officers from the Council and the Passenger Transport Executive, and local business interests. The make-up of this steering group was important in that it resulted in two important early decisions,[19] (a) to seek a consensus solution, and (b) to subject the study to an extensive public consultation exercise.

New road schemes
- Schemes to take through traffic out of Leeds and the city centre
- Complete inner ring road
- Improve outer ring road
- Support the Government's M1–A1 link

Better bus services
- Bus lanes and priority signals
- More bus shelters with better information
- New bus station
- Experimental guided bus route

Better rail services
- City centre station improvements, allowing more trains into Leeds
- New stations – with park-and-ride
- Station improvements
- Electrification and new rolling stock

A new city centre
- High quality environment for pedestrians
- Easy access to public transport
- Electronic parking information

Leeds supertram
- Routes serving south, east, north-east and north-west Leeds
- Links with buses and trains in city centre
- Linked to park-and-ride at outer end

Traffic and the environment
- Traffic calming in residential areas and district and town centres
- Boost road safety schemes – more crossings
- Cycle routes
- More heavy lorry bans

Access for all
- Raised platforms at stations
- Budget for dropped pavements and crossings for visually impaired people
- Supertram fully accessible for all
- Wheelchair-accessible taxis

▭ Metro train	•••••• Seacroft–Crossgates bypass
▬ Supertram	– – – – Inner ring road
▪▪▪▪ M1–A1 link road and A1 upgrading	▭▭▭ Drighlington bypass
	•••••• Ring road improvements

Fig. 6.1 The Leeds transport strategy[19]

The strategy which emerged (see Fig. 6.1) identified roles and functions for each part of the transport system in meeting the vision of Leeds, public aspirations, and the city's economic and 'green' strategic aims. It stresses the need for:

1. new roads to help remove through traffic from the pedestrianised city centre, improve the environment and support economic development
2. better public transport, to cater for growing travel demands and provide an attractive alternative to the car, which includes (i) improvements to the existing mainstay bus and rail facilities and operations, (ii) a large-scale park-and-ride programme, and (iii) a high-quality environment-friendly electric 'Leeds Supertram' which will run generally on ground-level tracks between the city centre and the suburbs
3. city centre initiatives, including an extended pedestrian core, to make the heart of Leeds more attractive and accessible
4. demand management of traffic in support of these initiatives.

6.7 References

1. *Congestion control and demand management.* Paris: OECD, 1994.
2. *Transport statistics Great Britain.* London: HMSO, 1994.
3. Royal Commission on Environmental Pollution, *Transport and the environment.* Command 2674. London: HMSO, 1994.
4. *Urban travel and sustainable development.* Paris: ECMT/OECD, 1995.
5. *Planning policy guidance: Transport.* PPG13. London: Department of the Environment/Department of Transport, 1994.
6. Wachs, M., Learning from Los Angeles: Transport, urban form, and air quality. *Transportation,* 1993, **20** (4), 329–354.
7. Wood, K., *Urban traffic control, systems review.* Project Report 41. Crowthorne, Berks: Transport Research Laboratory, 1993.
8. Bayless, D., London's railway network. *Highways and Transportation,* 1994, **41,** (10), 25–27.
9. Turner, D., Special London traffic initiatives – Red Routes. *Highways and Transportation,* 1994, **4** (41), 21–24.
10. Sutton, J.C., Transport innovation and passenger needs – Changing perspectives on the role of Dial-a-Ride systems. *Transport Reviews,* 1987, 7 (2), 167–182.
11. Kompfner, P., Hudson, R.H.C., Shoarian-Sattari, K., and Todd, D.M., *Company travel assistance in the London area.* Research Report 326. Crowthorne, Berks: Transport Research Laboratory, 1991.
12. Martin, K.W., New initiatives in the management of urban congestion. *Municipal Engineer,* 1990, **7** (2), 59–68.
13. Vincent, R.A., and Layfield, R.E., *Nottingham zones and collar study – Overall assessment.* Report LR805. Crowthorne, Berks: Transport Research Laboratory, 1977.
14. *Travel demand management guidelines.* Sydney: Austroads, 1995.
15. Zupan, J.M., Transportation demand management: A cautious look. *Transportation Research Record 1346,* 1992, 1–8.
16. Ho, A., and Stewart, J., Case study on impact of 4/40 compressed workweek program on trip reduction. *Transportation Research Record 1346,* 1992, 25–32.
17. Smeed, R.J., *et al., Road pricing: The economic and technical possibilities.* London: HMSO, 1964.
18. May, A.D., and Roberts, M., The design of integrated transport strategies. *Transport Policy,* 1995, **2** (2), 97–105.
19. Steer, J., The formulation of transport policies in British conurbations, 1990–1995. *Proceedings of the Institution of Civil Engineers: Transport,* 1995, **111** (3), 198–204.
20. Institution of Highways and Transportation/Department of Transport, *Roads and traffic in urban areas.* London: HMSO, 1987.

CHAPTER 7

Developing the parking plan

C.A. O'Flaherty

The provision of parking is an essential consequence of the movement of people and goods into and within urban areas. However, parking is also associated with traffic congestion due to too many vehicles seeking to gain access to particular locales at one time and the inability of the road and parking systems to provide for them.

7.1 Parking policy – a brief overview

Parking demand is mainly influenced by the type and function of land use and the quality of the public transport system – and hence the parking policy developed for any particular area depends very much on the local situation.[1] In villages and the smaller towns and in the outer areas of larger towns parking policy is often based on providing enough parking spaces (supply) to meet the demand (vehicles). In large towns and cities, however, the overall transport objectives of reducing accidents and safeguarding (a) the accessibility of the central area (by reducing traffic congestion on the radial roads and in the town centre) and (b) the quality of life in and about the central area (by reducing air, noise and environmental pollution) usually become paramount, and parking policy may be aimed at controlling the parking supply so as to induce appropriate shifts in the modal split in favour of public transport (and, thereby, to reduce the parking demand). Priority for on- and off-street parking at the places provided in all town centres is normally given to selected groups, e.g. residents, people with disabilities, shoppers, and people on personal business.

The control of the parking supply and the enforcement of the parking regulations are normally regarded as the keys to the achievement of the above transport objectives. In the case of new developments local authorities are able to control parking supply through the application of parking standards which define the maximum or minimum numbers of parking spaces to be applied to any new land use activity. In the case of existing public parking spaces under their control the local authorities may decide to:

- reduce the total number of spaces (particularly on-street places)
- impose time restrictions on existing spaces so as to encourage short-term parking and discourage commuting vehicles
- use various mechanisms to give priority of parking to selected groups, e.g. by applying graduated pricing in off-street public car parks which favour short-term parkers, or by using parking meters, discs or vouchers to control on-street parking in shopping and business areas, and/or by providing pre-paid licences (or free permits) to residents

to enable them to park on-street adjacent to their homes
- provide appropriate amounts of peripheral parking for long-stay commuter parkers outside the town centre and perhaps on the outskirts of the town or city
- use variable message information systems to assist drivers in finding available parking spaces and, thereby, reduce their parking search time and consequent traffic congestion on central area streets.

The major weakness of any parking policy which is based on controlling parking supply is that a local authority has no control over the very significant numbers of existing off-street parking spaces that are privately owned. Further, any easing of traffic congestion on central area streets can result in an increasing usage of these streets by through vehicles which do not have destinations in the town centre.

7.2 Planning for town centre parking – the map approach

When all the policy decisions have been taken, and related to estimates of need/provision, the parking plan for a town centre must be devised. There are many ways in which this can be done. One of the more basic approaches, which simplifies the process and provides a perspective on the problems that have to be tackled, is the map approach devised some years ago as a guide for local authorities.

This approach concentrates on fundamentals, and describes how a series of maps can be compiled as a convenient and effective way of illustrating the interactive factors which, in most towns, have a bearing on parking problems and need to be taken into account when preparing a parking plan. It enables a measure of the parking problem as a whole to be taken quickly and with the minimum use of resources. However, its direct usage is now most applicable to smaller towns which have not carried out a comprehensive transport study and, hence, do not have detailed data at their disposal.

The map approach can be summarised as follows.

1. Appraise the effects of parking on the town centre, on the convenience, pleasantness and efficiency with which the central area and its peripheral areas are used and function, together with the effect on (and of) accessibility and environment.
2. Analyse the causes and pressures behind present parking usage and future parking demand.
3. Decide how to make optimum use of the facilities available now and within the planning period into the future.

In the following elaboration of these steps use is made of illustrative data (see pp. 158–65) from *Parking in town centres*.[2]

Map 1. This is prepared to show the main land use generators of parking demand, and the areas where competition for parking space is most intense. In Britain the areas used mainly by shopping parkers are most easily determined from a parking usage survey carried out on a Saturday, when offices are closed. The preferred office parking areas can be determined in a similar way on an early closing day for shopping. Residential parking is most easily isolated by a usage survey carried out at night. These data will suggest where time-limit and residential parking controls are obviously needed, and where multiple usage of public car parks might be appropriate.

Maps 2 and 3. These maps show the 'pinch-points' on the streets, where parking is a cause of traffic congestion and where parking and congestion impact on the operation of essential services. Data for Map 2 are generally available within a town's traffic department or, in the case of a small town, can easily be determined by inspection. Data for Map 3 are usually obtained via a parking usage survey.

Map 4. This map defines those areas where there is conflict between car parking and the physical environment.

Map 5. This map suggests how the hinterland of the central area can be divided into catchment areas, and the traffic from each related to the major approach roads carrying vehicles into the town centre. In this case, for example, the southern catchment has the highest car-availability potential. Since the two approach roads within the southern catchment converge at the outskirts of the central area it also suggests that congestion may be a problem on the final leg.

Map 6. This map shows the numbers of commuters and shoppers using cars which originate from each catchment area outside the town boundaries. This is useful in pointing out where peripheral park-and-ride car parks might be located.

Maps 7 and 8. These show the existing public transport services and suggest where car travel by the commuter is being encouraged through lack of good alternative means of journeying to work. These maps also emphasise that the town centre parking plan should always be developed in association with strategies for sustaining and/or encouraging good town centre access by public transport.

Map 9. This illustrates how the central area can be divided into convenient destination zones (in this case on the basis of figures for employment and floorspace) in order to give guidance as to where the competing demands for parking will be high.

Map 10. This map shows where parking controls (time-limit/pricing) need to be implemented to ensure that the short-time parker is given preference in the parking plan. In this example, the acceptable walking distance for shoppers is given as 180 m. A guide to present practice in any given town can be gained by observation and/or survey in and about the central area on a busy shopping day.

Map 11. This map is similar to Map 10 except that it identifies parking locations likely to be used by commuters. In this example the acceptable average walking distance from parking place to destination is also taken as 180 m while the outer limit is taken as about 275 m. It should be noted, however, that in practice the average walking distances that are acceptable to parkers vary according to both trip purpose and intensity of parking demand. For example, Table 7.1 shows that the larger the city the greater the average

Table 7.1 Average distances (metres) walked from parking place to destination[8]

Population group of urban area	Trip purpose			
	Shopping	Personal business	Work	Other
10 000–25 000	61	61	82	58
25 000–50 000	85	73	122	64
50 000–100 000	107	88	125	79
100 000–250 000	143	119	152	104
250 000–500 000	174	137	204	116
500 000–1 000 000	171	180	198	152

distance that parkers will walk to their destinations, and that journey-to-work parkers will accept greater walking distances than shopping parkers. Also (although not shown here), off-street parkers will normally walk further than kerb parkers.

Map 12. This map, which is an extension of the analysis initiated in Map 5, shows the existing traffic flows on the major approach roads, and indicates where there is spare capacity. (In this case, to simplify the illustration, the through traffic has been excluded; however, it should be appreciated that non-parking through traffic can often account for up to about 35 per cent of the vehicles entering a central area.) As the south and south-east approach roads are filled to capacity during the peak hours then, obviously, to provide additional long-stay parking in the central area for cars approaching from the south would simply add to the congestion on the already crowded road.

Map 13. All of the above analysis is intended to culminate in the development of the central area parking plan shown in Map 13.

The developmental sequence involves defining the following:

1. existing off-street car parks and land to be made available for off-street parking
2. where on-street parking is to be permitted, and where it is to be prohibited, e.g. (a) at access points to buildings, loading bays and service areas, (b) at intersections, pedestrian crossings, bus stops and taxi stands, and (c) on one side of narrow streets
3. those streets where uncongested flow is essential at specific times, e.g. during peak periods (*see Map 2*)
4. locales where only front service access to buildings is possible, and where rear service-access is available (*Maps 2 and 3*)
5. where priority should be given to service vehicle and other operational parking (*Map 3*)
6. where overnight parking for residents is essential (and thus, where resident-only parking permits should be issued and enforced) (*Map 4*)
7. those areas of historic and architectural significance where special attention needs to be given to the amount, layout and detailed design of any parking provided (*Map 4*)
8. the areas where very short-term parking is to be given preference, e.g. near post offices and banks
9. the areas where shoppers and short-term visitors to offices are to be given preference (*Maps 1, 9, 10 and 11*)
10. the extent of the likely displacement of long-stay commuter-type parkers into areas adjoining the central area, related to acceptable walking distances (*Maps 10 and 11*).

Note that in this instance the parking restrictions imposed in the residential areas adjacent to the town centre are denoted as experimental and subject to changes to ensure the best solution; if successful, similar provisions may be made at a later date in other parts of town.

Maps 14 and 15. Essential developments in the hinterland of the town cannot be isolated from central area parking proposals. Thus Map 14 is developed as an addition to Map 13 in order to show where special parking facilities are needed to serve hinterland (especially commuter-type) travellers using out-of-town railway stations, and Map 15 (which is derived from Maps 7 and 8) shows where express and improved bus services need to be provided. Note also that, both outlying and intermediate park-and-ride car parks are proposed on the southern approach route. An express bus service (without a park-and-ride car park) is proposed for the eastern approach road.

Map 1

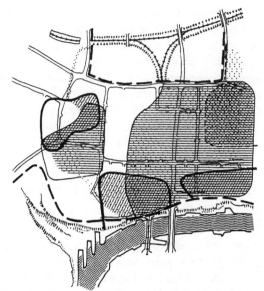

- – – Physical limit to parking areas
- ⋯⋯⋯ Extent of shoppers' parking
- Competition between shops and others
- ☐ Extent of office parking
- ≡ Competition between offices and others
- ☐ Extent of commercial and industrial parking
- ⫽⫽ Competition between commerce/industry and others
- Competition between residents and others

Map 2

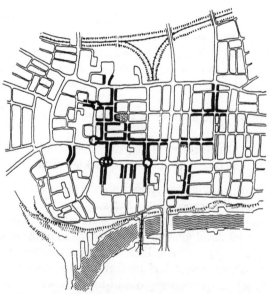

═ Minor streets requiring unobstructed flow during peak periods

Off-street car parks frequently causing congestion in adjoining streets and junctions

O Traffic flow along major roads frequently interrupted by congestion originating in minor streets

Map 3

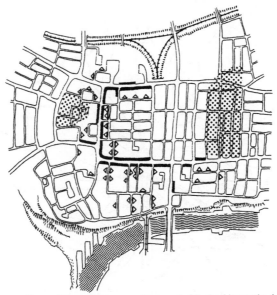

△ Essential service access frequently blocked by parked cars

▒ Areas where servicing is impeded by parked cars

▬ Service vehicles obstructing flow during peak periods

Map 4

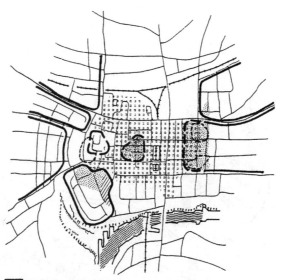

☐ Residential areas

⊏ ⊐ Residential use within mixed use areas

⋮⋮⋮⋮ Extent of primary central area uses

⌐¬ Areas of historic and architectural significance where
parked cars are an undesirable intrusion

▨ Town centre parking intruding into residential environment

▨ Residential areas where parked commercial vehicles are
detrimental to environment

Map 5

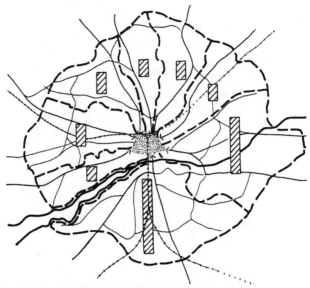

←— Major town centre approach roads

❑❑ Catchment areas within the town related to approach roads

▨ Potential parking demand from each catchment area

Map 6

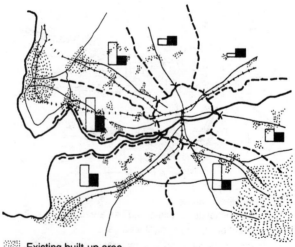

⬚ Existing built-up area

←— Principal approach roads

--- Limits of catchment areas outside the town

❑◼ Number of commuters using cars

◼ Number of shopping visitors using cars

Map 7

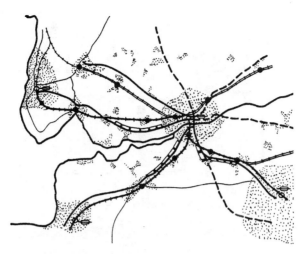

  ░ Existing built-up area
  Railways
  ┅╸ maximum wait less than 30 minutes at peak periods
  ╪╪╪ maximum wait more than 30 minutes at peak periods

  Railway stations
  ●╸ stopping trains
  ⊖╸ express and stopping trains

  Bus routes
  ─── maximum wait less than 20 minutes at peak periods
  ╴╴╴ maximum wait more than 20 minutes at peak periods
  ▭▭▭ where buses are often full during peak periods

Map 8

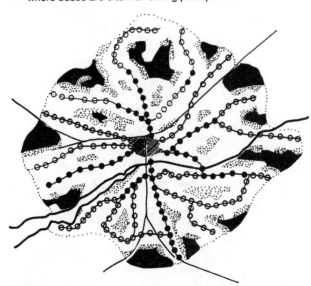

  ■ Residential areas more than 400 metres from bus stop
  ░ Residential areas between 180 and 400 metres from bus stop

  Bus routes
  ⊖⊖⊖ maximum wait more than 10 minutes at peak periods

  ●●● maximum wait less than 10 minutes at peak periods

Map 9

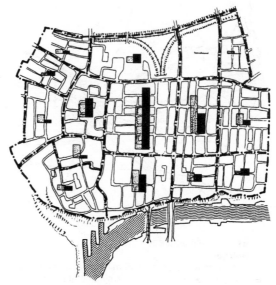

◨ Amount of employment
◼ Shopping floorspace

Map 10

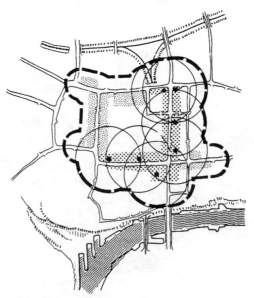

▦ Primary shopping areas
▒ Secondary shopping areas
• Principal shops
– – Acceptable maximum distance between car parks and shops
Areas within 180 metres of principal shops shown by circles

Map 11

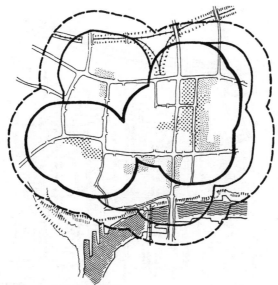

:::: Primary office areas
::: Secondary office areas
☐ Area within 180 metres of primary office areas
☐ Area within 180 metres of office areas
☐ Area within 275 metres of office areas

Map 12

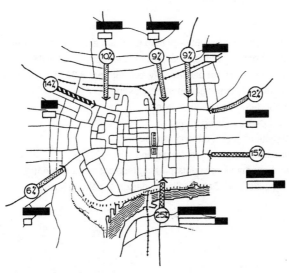

〈▨▨▨ Major approach roads
■■■ Capacity of roads
☐ Existing traffic flow to car parks from catchment areas
➤■ Frustrated element
〈▩▩▩ Major roads at capcity
〈ⵜⵜⵜ Major roads near capacity
Ⓩ Percentage of traffic to and from parking areas using each road

Map 13

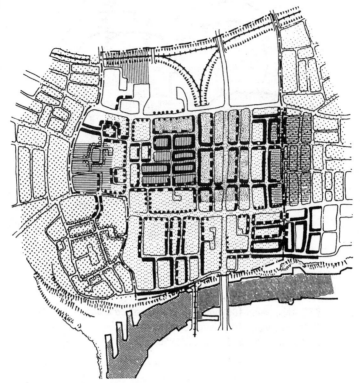

▬	On-street parking
▬◦▬	On-street parking except from 8.30 to 10 am and from 4 to 6 pm
■ ■	On-street parking with special provision for operational vehicles
• • •	Operational on-street parking only, but not from 8.30 to 10 am or from 4 to 6 pm
◦ ◦ ◦	Residents only overnight on-street parking
▦	Very-short-duration parking areas
⦂⦂	Short-duration parking areas
▥	Immediate off-street parking provision
—·—	Desirable limit of town centre parking
⦂⦂	Area of priority for residents' parking (subject to experiment)
▤	Area requiring special attention to design of parking layout

Map 14

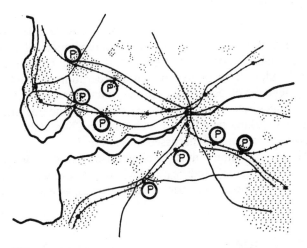

     Existing built-up areas
— Major roads
+++ Railways
—• Railway stations
Ⓟ Commuter car parks

Map 15

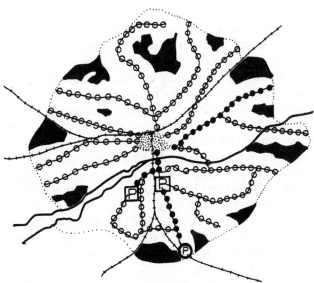

■ Residential areas more than 400 metres from bus stop
‑o‑o‑o Existing bus routes
‑•‑•‑• Proposed express bus routes
🄿 Proposed intermediate car parks
Ⓟ Proposed perimeter car parks

7.2.1 A final comment

The map approach just discussed has been used to provide a perspective of the inter-acting developmental features involved in the preparation of a basic town centre parking plan. A range of traffic management and parking control measures now need to be considered and, as appropriate, applied in order to implement the proposed park-ing plan and achieve the objectives underlying it; these measures are mainly discussed in Chapter 24. Issues relating to the use of parking standards and the location and design of off-street parking facilities are addressed in Chapter 22 (except for park-and-ride facilities, which are mainly discussed in Section 7.3).

To conclude this discussion it must be emphasised that the town centre parking plan should never be seen as static. Modifications will be necessary from time to time to meet changes in land use, traffic and environmental demands, as well as the needs of varying transport strategies.

7.3 Park-and-ride

Park-and-ride is commonly defined[3] as the act of parking at a custom-built car park and transferring to public transport to travel onward to one's destination. After hesitant starts in the 1960s and 70s it is now accepted that the provision of park-and-ride (including kiss-and-ride) facilities at peripheral locales that are away from a central area can be an integral part of a town centre's parking plan, and that the larger the urban area the more important this becomes.

In practice park-and-ride services are mostly bus-based and operated either as year-round services designed to relieve the central areas of large towns (and their radial access roads) of traffic congestion, or as seasonal services, e.g. during the summer hol-iday period at historic towns or the pre-Christmas period at towns noted for their shopping. In conurbations that are well served by rail systems park-and-ride is also rail-based. However, the potential for developing formal park-and-ride from rail sta-tions can often be limited by an inability to obtain suitable off-street parking space in built-up areas adjacent to stations – that is, unless accessible land can be made avail-able which is surplus to the railway's operating requirements, e.g. redundant goods depots, railway sidings.

In practice, also, many motorists park their cars on-street in locales adjacent to rail stations or at bus stops, and travel onward by public transport with no regard to the eco-nomic or environmental well-being of the areas where their cars are left. For example, at most London Underground and British Rail stations within Greater London informal on-street parking (known as 'railheading') is undertaken by at least twice as many motorists as use the formal station car parks. This railheading-type of operation, while assisting in the reduction of onward traffic congestion, can be a severe nuisance and very often justifies the introduction of local parking controls to protect the affected areas from unwelcome long-stay parkers.

7.3.1 Criteria for success

The successful operation of a significant park-and-ride scheme is dependent upon the following features (which are not in order of priority).

1. *The park-and-ride interchange must be serviced by a public transport system that offers reliable and frequent services in both the inward (to the central area) and outward directions.*
Generally, interchanges with reliable frequent services are more appealing to potential users since they maximise the travel-time choice, particularly for the return journey. Service intervals of half an hour have been proven to be preferred for long-distance rail park-and-ride to London.[4] Bus headways of 10 minutes have been recommended as being necessary for the successful operation of a park-and-ride scheme for a shopping centre.[5] Bus headways of not more than five minutes may be necessary during peak demand periods for journey-to-work travellers. (One widely used rule of thumb is that one minute spent waiting for a bus is roughly equivalent – in terms of influence on public transport usage – to three minutes on a bus.)

2. *The onward public transport mode must provide a reliable fast service from the interchange into the central area.*[6]
A park-and-ride interchange represents an additional leg in a motorist's journey. It affects the motorist's travel time and convenience by requiring him or her to leave the comfort of a car at an intermediate location and walk to a public transport stop and wait for a bus or train to complete the journey. To compensate for the uncertainty associated with this extra link it is essential that the public transport vehicle be provided, to a large extent, with reserved track running for the remainder of the journey, i.e. either by rail or by segregated busway/bus lane and priority at intersections when the express bus is travelling at-grade. When this segregation and priority are not provided the buses will be held up by the prevailing traffic congestion and their service reliability may be so reduced, and the travelling time so increased, that the motorist will perceive little advantage in using the park-and-ride service.

3. *The parking fee at the interchange plus the two-way public transport fare should be less than the perceived cost of travelling to the central area by car and parking there.*
For obvious reasons the round trip cost discernible to the user of the park-and-ride system must be less than the cost of travelling by car; if it is not, motorists will stay in their cars and hunt for parking in the central area or, alternatively, try to park free on the roads adjacent to the interchange so as to reduce their costs. There is ample experience to suggest that parking at the interchange should be either very cheap or, preferably, free so as to encourage park-and-ride usage. However, the cost of developing and operating the interchange has to be provided, and recommendations regarding this range from using central area parking fees for this purpose to levying developers for the additional value created as a result of having extra floorspace released at new developments through not having to provide land or resources for on-site parking.[7]

4. *Ample parking space must be provided at the interchange to ensure that parking is easily obtained at all times.*
If parking is not easily obtained at the interchange, potential long-stay users will either park on the adjacent roads or continue in their cars into the central area. There is no

simple way of determining the number of spaces required at a new interchange as this depends upon the success of the project as well as upon such variables as the size of the commuting area being serviced, the stringency of the parking control in the central area, and the degree of congestion experienced on the radial road(s). However, a guide to what might be required can be gained by looking at the needs created by a particular service interval.

Let it be assumed that a commuter park-and-ride service is to be provided by double-decker express buses (each capable of carrying 76 passengers) operating at a service interval of 5 minutes during the peak period (assumed in this instance, for ease of calculation, as 1 hour), and that the demand for this service will be such that each bus will be 90 per cent full on average when it leaves the interchange. The number of spaces required to meet the needs of this peak hour service is therefore

$$0.9 \times 76 \times \frac{60}{5} \times \frac{1}{1.15} = 714$$

assuming that the average car occupancy is 1.15. However, if significant numbers of patrons are 'kiss-and-ride' passengers the number of long-term spaces will be much lower.

The application of kissing to a park-and-ride operation bears no relationship to the degree of passion exhibited by commuters at an interchange; rather it is the term used to describe the practice whereby a person drives his or her partner to and from the interchange so that the car is not left there all day. Specially designated short-stay parking places must be provided for these users if congestion on the access road and/or adjacent roads is to be avoided. The number of pick-up spaces required will normally be much greater than the number of drop-off spaces, because passengers can be disembarked and a car driven away in approximately one minute whereas a waiting vehicle may stay 5–15 minutes before the arrival of the public transport vehicle. (It might be noted that British Rail looked to provide a total of some 50 short-stay spaces at a typical intercity station that is used for park-and-ride purposes.[4])

5. The park-and-ride interchange must be properly located.
The interchange should be sited so that the catchment area's road access allows for the free flow of traffic along 'natural' travel corridors. For urban areas with outer ring roads park-and-ride interchanges serviced by buses are often located where the ring road cuts the major radial roads leading into the town centre. Ideally, the car park should be located adjacent to the radial road, preferably on the left-hand side (in Britain) of the approach into the central area. The access road into the car park should be as short as possible, and the design of the intersection with the radial road (including any traffic control provided) should favour vehicles entering and leaving the interchange, particularly during peak periods.

A not uncommon locational mistake associated with both rail and bus park-and-ride involves siting the interchange too close to the central area. Obviously, the closer the interchange is to the town centre the greater the catchment from which it can draw. On the other hand if it is too close to the town centre the disadvantages to the motorist of diverting from the main route and changing vehicle will usually outweigh the advantage of travelling the rest of the trip by bus or rail – unless there is a major congestion point further in which can be bypassed easily by the public transport vehicle.

6. *The car park must be well designed and supervised.*

The interchange car park should be designed to at least the same standard (see Chapter 22) as a town centre car park, if it is to maintain its clientele under inclement weather conditions. Walking distances should be as short as possible between the cars and the waiting public transport vehicles. The *minimum* facilities to be provided include a substantial weatherproof shelter (which should have seats and be cleaned on a daily basis), a telephone booth (for emergencies) and lighting (essential for personal security in dark mornings and evenings). Public conveniences and tourist information should be considered at larger sites.

If drivers are to be encouraged to leave their cars at long-stay park-and-ride car parks they must be satisfied that the likelihood of finding them damaged upon their return is relatively low. Security for the vehicles is therefore critical. It is helped by the car park being well lit and supervised as well as being designed so that it is hard for vandals to gain unobtrusive entrance.

7. *Ample and continuing publicity must be given to the park-and-ride scheme.*

7.4 References

1. PIARC Committee on urban areas, 1995 Parking policies, *Routes/Roads*, 1995, Special 1, 57–133.
2. Ministry of Housing and Local Government, *Parking in town centres*. Planning Bulletin No. 7. London: HMSO, 1965.
3. Pickett, M.W., and Gray, S.M. *Informal park and ride behaviour in London*. Project Report 51. Crowthorne, Berks: Transport Research Laboratory, 1993.
4. Prideaux, J.D.C.A., Rail/road interchanges. *Highways and Transportation*, 1984, **31** (4), 12–16.
5. *Pedestrianisation guidelines*. London: The Institution of Highways and Transportation, 1989.
6. Pickett, M.W., Perrett, K.E., and Charlton, J.W., *Park and ride at Tyne and Wear metro stations – a summary report*. Crowthorne, Berks: Transport Research Laboratory, 1986.
7. Macpherson, R.D., Park and ride: progress and problems. *Proceedings of the Institution of Civil Engineers: Municipal Engineer*, 1992, **93** (1), 1–8.
8. *Parking principles*. Special Report 125. Washington, DC: Highway Research Board, 1971.

CHAPTER 8

Planning for pedestrians, cyclists and disabled people

G.R. Leake

8.1 Introduction

In the past transport planning has tended to concentrate on providing for the needs of vehicular movement, to the detriment of pedestrians, cyclists and, especially, disabled people. This has resulted in an imbalance in the provision of quality transport, an imbalance which is particularly serious in view of the importance of walking and the increased interest in cycling.

It is important to recognise the forces influencing the demand for provision of more and better pedestrian and cyclist facilities. Undoubtedly one important factor has been the increased awareness of the environmental problems created by the rapid national and worldwide growth in vehicle travel, but of equal importance has been the recognition by many people of the need for physical fitness and the role that walking and cycling can play in achieving this.

The demand for pedestrian and cycle facilities is influenced by a number of factors, of which some of the most important are:

1. *The influence of topography* – Cycling and pedestrian activity, particularly the former, tend to be at a higher level in flat areas than in hilly ones.
2. *The nature of the local community* – Cycling and walking are more likely to occur in a community that has a high proportion of young people.
3. *Car ownership* – The availability of the private car reduces the amount of walking and cycling, even for short journeys.
4. *Local land use activities* – Walking and cycling are primarily used for short distance trips. Consequently the distance between local origins and destinations (e.g. homes and school, homes and shops) is an important factor influencing the level of demand, particularly for the young and the elderly.
5. *Quality of provision* – If good quality pedestrian and cyclist facilities are provided, then the demand will tend to increase.
6. *Safety and security* – It is important that pedestrians and cyclists perceive the facilities to be safe and secure. For pedestrians this means freedom from conflict with motor vehicles, as well as a minimal threat from personal attack and the risk of tripping (particularly important for elderly persons and pregnant women) on uneven surfaces. For cyclists, there is also the security of the parked cycle at the journey destination.

8.2 Identifying the needs of pedestrians, cyclists and disabled people

Before deciding on the appropriate extent and standard of pedestrian, cyclist and disabled people's facilities, it is important to assess the potential demand. Here the possible methods of obtaining such estimates are examined. There are two steps in the process: obtaining reliable estimates of the existing demand and projecting this demand to a future design year. Most survey techniques are concerned with the first step.

8.2.1 Manual counts

Manual counts are concerned with counting the flow of pedestrians or cyclists through a junction, across a road, or along a road section/footway. Because it is important to determine conflicts with motor vehicles, vehicle counts (by type) are normally carried out at the same time.

If manual counts are to be useful, they need to satisfy the following:

1. The time period(s) in the day over which the counts are undertaken must coincide with the peak times of the activity of study (e.g. school trips).
2. The day(s) of the week and month(s) of the year when observations are made must be representative of the demand. School holidays, early closing, and special events should be avoided since they can result in non-typical conditions.
3. The survey locations need to be carefully selected in order to ensure that the total existing demand is observed.

The advantages of manual counts are: they are simple to set up and carry out; and they are flexible – changes in the agreed survey schedule can be introduced quickly in response to observed changes in the demand on the site. Their disadvantages are: they are labour intensive; only simple information is obtained, namely flows over given time periods (often 5–10 minutes); and no detailed information is obtained on problems encountered by pedestrians and cyclists, nor on their desired requirements.

8.2.2 Video surveys

Cameras are set up at the selected sites and video recordings taken of the pedestrian and/or cyclist movements, together with their interaction with vehicles where appropriate, during the selected observation periods. A suitable elevated vantage point for the camera is important.

Such surveys produce a permanent record of pedestrian and cyclist movements and their interaction with vehicles. In addition a record of delays and behaviour patterns (e.g. the reluctance of an elderly person to cross the road) is obtained. From such information, for example, pedestrians' crossing difficulties can be analysed.[1]

8.2.3 Attitude surveys

The two survey techniques covered briefly above obtain an estimate of what is happening now. What they do not do, and what they cannot do, is determine the circumstances under which increased walking or cycling might take place. This requires detailed questionnaire-based attitude surveys, often directed at particular target groups such as schoolchildren, shoppers, disabled people etc. Devising such questionnaires is a skilful task[2] and requires considerable expertise if complete, unbiased information is to be obtained.

Such surveys can be carried out at the home (including old people's homes), at destination points (such as schools, shopping areas, community centres), or during the journey. The first two alternatives are to be preferred since the interviews can be conducted in a leisurely manner indoors. In all cases it is necessary to obtain an unbiased sample of the travelling (or potential travelling) population.

Attitude surveys enable complete information to be obtained on why existing trips are made by the current mode of travel and not by others. In addition they can ascertain why certain types of trips or certain modes of travel are only made infrequently or not at all. They can also gather information on what new facilities, or improvements to existing facilities, need to be provided to divert trips to walking or cycling, or increase the amount of current pedestrian and cycling activity. Clearly the results of such an exercise will be an important input into the prediction part of the process.

8.2.4 Prediction of demand in the design year

Once an estimate of the *existing* demand has been obtained it is necessary to predict what the demand is likely to be at some future date. This is carried out in two stages.

The first stage is concerned with deciding on an appropriate target or design year. This will depend on the particular local circumstances and planning context, but for pedestrian and cyclist facility planning, a 10-year target figure beyond the end of the planning period is often considered to be appropriate.

The second stage involves predicting the probable level of change in walking and cycling demand to the design year. In considering this it should be stressed that this is not a precise process, since the future demand cannot be estimated using carefully researched and formulated mathematical expressions. All that can be hoped for is that the estimated flows will be sufficiently close to those which will finally occur so that the correct decision on the provision of facilities will have been taken.

A number of local factors will influence future changes in demand, and these local influences will need to be isolated, and their importance ascertained, using local planning knowledge and judgement, together with any relevant inputs from recently conducted attitude surveys. Some of the main factors to be considered are:

- planned and anticipated changes in land use developments over the design period, and hence possible changes in the number of generated trips being made which fall within the distance ranges normally covered by walking and cycling
- possible changes in the level and cost of public transport provision, together with constraints on car usage
- changes in the age structure and socio-economic status of the local residents and the effect of this on car ownership and car usage
- anticipated changes in local attitudes towards walking and cycling as viable/acceptable modes of travel.

Although, as has been said, prediction is a difficult and imprecise exercise, it is one that has to be attempted if an acceptable basis is to be established for justifying the need for any particular set of design proposals.

8.3 Identifying priorities of need

Since financial resources are inevitably limited, it is always necessary to put plans in implementation priority order. A number of important considerations have to be taken into account if the most effective proposals are to be identified and subsequently implemented. Among the most important of these are:

Safety

In planning any transport proposals safety must be of prime importance. Most developed countries have either specific or implied accident reduction targets at which to aim. When it is considered that in Great Britain approximately 36 per cent of all road fatalities and 23 per cent of all road accidents involve pedestrians or cyclists,[3] then it is clear that any proposals which can provide safer facilities for these two traveller groups will contribute significantly to the achievement of both national and local safety targets.

One way of prioritising is to identify locations (existing and future) where accident risk is high. To date most risk assessments have been concerned with establishing pedestrian risk when crossing the road,[4] where pedestrian risk was defined as:

$$\text{Pedestrian risk} = \frac{\text{Casualties per year}}{\text{Pedestrian flow per hour}} \times 10^2 \qquad (8.1)$$

Similar ratios can be established for cycling. From such simple measurements the worst safety locations for pedestrians and/or cyclists can be identified and prioritised.

Conflict

Accident risk, based on actual accidents, is not the only measure which needs to be taken into account. Luckily accidents are a relatively rare event at any particular location, but conflicts between competing travellers are much more common and often result in 'near misses'.

In the UK the Department of Transport (DoT) has set up a formal design procedure for establishing whether formal pedestrian road crossing facilities are necessary based on the volume of potential conflicts between pedestrians and vehicles.[5, 6, 7] The procedure is based on PV^2, where P = pedestrian flow (ped/h) across a 100 m length of road centred on the proposed crossing location, and V = number of vehicles on the road in both directions (veh/h).

The PV^2 value is the average over the four busiest hours of the day. A formal crossing is normally justified if the value of $PV^2 > 10^8$. The design (or warrant) diagram in use in Great Britain is shown in Fig. 8.1.

Satisfying policy objectives

The satisfying of any formal transport policy objectives must be an important factor when identifying priorities. For example, if an important local transport policy objective is to provide an extensive pedestrian route system aimed at increasing the amount of walking by elderly and disabled people, then those road crossing points in the vicinity of concentrations of such groups of people are likely to have a higher priority than those elsewhere.

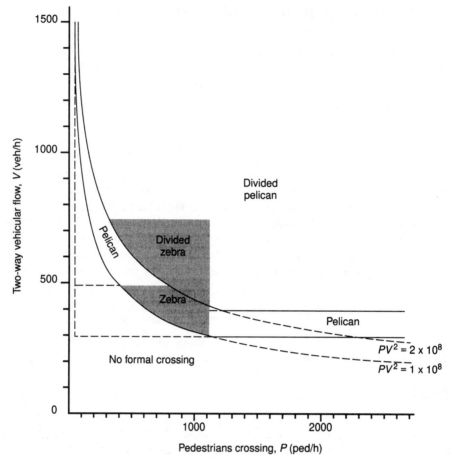

Fig. 8.1 Design warrant for deciding on appropriate formal road crossing facilities in Great Britain[5]

Cost effectiveness

Because of the competing demands for finance within the transport sector, all proposals given top priority on the basis of the above three criteria should be subjected to an appropriate cost-effectiveness analysis (see Chapter 4). This would be aimed at establishing the benefits to pedestrians and/or cyclists resulting from reduced delays, easier movement and increased safety, and to compare these benefits with possible increased delays incurred by vehicle occupants and the capital and maintenance costs of construction.

8.4 Pedestrian and cyclist characteristics and requirements influencing design

It is not the purpose of this chapter to describe and discuss the actual layout design of pedestrian and cyclist facilities. This is covered in Chapter 25. What is proposed is to outline some of the more important user characteristics and requirements which need to form part of such design layouts. For simplicity coverage will be under three headings –

pedestrian characteristics and requirements, needs of cyclists, and pedestrian areas. Some of the more important needs of elderly and disabled people are covered in Section 8.5.

8.4.1 Pedestrian characteristics and requirements

Walking speeds

An important design element, particularly when planning at-grade road crossings, is the need to provide sufficient crossing time to enable all pedestrians to complete the road crossing manoeuvre before traffic begins to move. This is an aspect which requires careful local study if the 'best' solution is to be provided, and is likely to be both time-of-day and area dependent. For example, during the morning and evening peak travel periods, most pedestrians will be physically fit and active, whereas in areas with many old peoples' homes, there will be significant numbers of elderly and disabled people with lower walking speeds.

Some research into road crossing speeds[8] has indicated an average value in the range 1.2 m/s to 1.35 m/s at busy crossings with a mix of pedestrian age groups. However, if crossings are less busy, then average walking speeds approximating to the free-flow walking speeds in pedestrian concourses of 1.6 m/s can be expected.[9] However, for disabled people a more appropriate value is 0.5 m/s if the needs of most disabled people are to be satisfied.[10]

Walking distances

Walking distance is an important design aspect, since the shorter the journey distance, the higher the probability that it will be made on foot. It has been found that over 60 per cent of all journeys under 1.5 km are made on foot, and that pedestrian journeys rarely exceed 3 km in length.[11] This means that if walking is to be encouraged, then the distance between origins (e.g. home) and destinations (e.g. shops) should preferably be less than 1 km.

The above relates to normal walking distances for active, fit people. The walking capability of elderly and disabled people is considerably lower. Such limitations in walking capability are important not only when walking along a footway, but also when moving about a pedestrian area. It has been argued that any design provision should be based on satisfying the needs of 80 per cent of the pedestrians within a particular disability group. If this is accepted then the recommended maximum walking distances without a rest are those set out in Table 8.1. The presence of steps and uphill gradients will reduce these distances.[10,12]

Table 8.1 Recommended maximum walking distance without a rest by impairment group[12]

Impairment group	Recommended distance limit without a rest (m)
Wheelchair users	150
Visually impaired	150
Stick users	50
Ambulatory without walking aid	100

Seating

The provision of seating is an important design feature, particularly in pedestrian areas. Published guidelines on the layout of pedestrian areas[13] have specified that:

> seating should be in the right place, functional, of robust design, aesthetically pleasing, and have a low maintenance requirement. Two types can be identified:
>
> (a) primary purpose-built seating, e.g. benches and seats
> (b) secondary informal seating e.g. steps, planter walls around flower beds etc.

Seating areas should not be affected by or hinder circulation, nor be subject to adverse microclimatic conditions. They should stimulate social interaction through their grouping ... They should allow for activities to be viewed, to enhance visual interest and be sited parallel to major pedestrian flows, especially in narrow street spaces.

Seating fulfils a distinct role for disabled people, since it provides needed resting areas. This means that the accessibility range of disabled people can be increased by ensuring that seating provision is provided within the maximum distances indicated in Table 8.1. Hence, if the requirements for ambulatory disabled pedestrians with sticks, for example, are to be satisfied then seating provision should be provided at 50 m intervals.

One further aspect which requires careful consideration is the type and amount of seating to be provided. It is important that this is based on the results of any local need surveys as well as on discussions with representatives of local disability groups. The latter are particularly important since, to satisfy the needs of disabled and elderly people, different types of seating ranging from conventional seating at a height 420–450 mm above walking surface level, to narrow perch seating 800 mm above surface level, will be needed.[14]

8.4.2 Needs of cyclists

Cycling has played an important part in the transport system in the past and no doubt will continue to do so. Therefore, it is important that the future needs of cyclists are identified and appropriate resources allocated.

The main objectives relating to the provision of new or improved cycle facilities can be summarised as follows:[5]

- to encourage increased cycling activity by providing facilities which give increased convenience, comfort and safety to cyclists
- to minimise direct conflict between cyclists and vehicles, especially at busy roads and junctions
- to ensure that where there are significant numbers of cyclists (e.g. in residential areas), traffic speeds are kept low.

There are a number of design principles influencing the location of a cycle route which need to be taken into consideration. The most important of these are:

- routes should be located to maximise demand
- routes should be as direct as possible to minimise travel distance
- segregated cycle routes should have frequent access points onto the local road system

- steep gradients should be avoided, where possible
- routes on high and open ground in windy, wet or icy conditions are unpopular with cyclists and should be avoided
- segregated routes are preferable, for safety reasons, because of the absence of motor vehicles.

If cycling activity levels are to be increased, then it is important not only to provide good facilities but also to ensure that as many journey destinations as possible are less than 5 km in length, the normal maximum length of cycle journeys.[11] Furthermore, it is essential that gradients are not excessive. Normally they should be no greater than 4 per cent. However, they may be greater than the above for short distances (e.g. where a cycle route rises out of a subway). Such gradients may be as high as 5 per cent but over distances no greater than 100 m, or 6.7 per cent for distances no greater than 30 m.[15]

Finally, cycle parking racks must be conveniently located, easy to use, plentiful in supply and secure from theft if cycling is to be encouraged.[16]

8.4.3 Pedestrian areas

Many areas (such as central and district shopping areas) and streets (such as residential streets) experience considerable pedestrian/vehicle conflict with its attendant accident risk, traffic noise, air pollution and visual intrusion. The introduction of pedestrian areas overcomes these problems by creating

> an environment which respects a human rather than a vehicular scale. However, to be successful, such schemes require satisfactory provision to be made for local and through traffic and for public transport and parking, including in particular the access needs of mobility handicapped people. Access for servicing of premises in the restricted area must also be provided ... [Pedestrianisation] options range from full pedestrianisation, with total vehicle exclusion, to partial ... pedestrianisation, with some or all vehicles excluded to differing degrees, at different times of the day, or on different days of the week.[5]

One potential adverse effect of pedestrianisation, particularly in the larger schemes, is the lengthening of the walking distance between car parks and public transport stops, which normally will be located outside or on the fringe of the pedestrian area, and the destinations within the area that people want to reach. This can create serious problems for elderly and disabled people and these need to be identified and appropriate solutions prepared, including seating provision.

The design of pedestrian areas[13] is a difficult and complex one if the diversity of all the user and trader demands are to be met satisfactorily.

8.5 Special needs of elderly and disabled people

Some of the special needs of elderly and disabled people relating to walking speeds, walking distances and seating provision have already been covered. However, these groups have a number of other special needs, the most important of which are discussed below.

8.5.1 Ramp gradients

Long steep ramps are very tiring and difficult to use, particularly for wheelchair users. Ideally a ramp gradient should not be greater than 5 per cent, and in no circumstances greater than 8 per cent.[17,18] To aid wheelchair users, as well as other elderly and disabled users, handrails should be provided on both sides, whose dimensions can be found in Fig. 8.2. Flat rest platforms should be provided at intervals of 10 m on the steeper ramps for wheelchair users, and should be at least 1350 mm long.

8.5.2 Steps

The recommended design of steps, together with the associated handrail which should be provided on both sides, is shown in Fig. 8.2.[14]

Blind and partially sighted people often have problems at steps. Consequently there should be an area of textured paving at both the top and bottom of a flight of steps to alert them to its presence. The top and bottom of a flight of steps should always be well lit.

8.5.3 Street furniture

Any piece of street furniture (including seating) is a potential obstruction and needs to be located away from pedestrians' main movement paths. To help visually impaired people, the colour of any street furniture should contrast with surrounding objects, particularly the walking surface and building frontages.

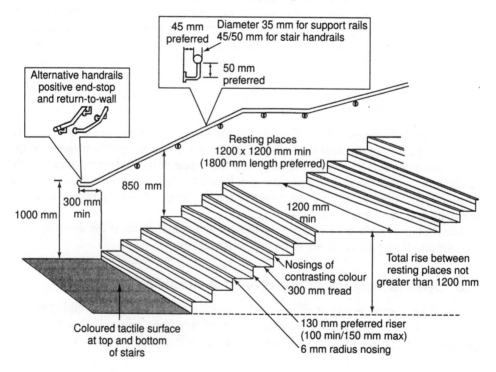

Fig. 8.2 Step and handrail dimensions[14]

Special design features can be provided to indicate the presence of street furniture. For example, features such as seating and telephone kiosks can be located in small groups, with the walking surface being a different colour and/or texture to the adjacent movement areas.[19]

A particular problem to blind and partially sighted people is caused by 'tapering obstructions' such as spaces below steps and ramps. Protection from such obstructions can be provided by a tap rail, raised kerbs or textured surfacing.

8.5.4 Walking surface quality

Many disabled people have severe problems when walking along poorly maintained footways and pedestrian areas. It is, therefore, important to establish a regular monitoring and maintenance programme to ensure that the quality of the walking surface is kept to a high standard. It is also important to ensure that the walking surface provides a good grip under wet weather conditions.

8.5.5 Information needs

It is essential that the particular information needs of elderly and disabled people are recognised and provided for. These are of two kinds: directional information in order to reach a specific destination; and information on when it is safe to cross the road.

Directional information is supplied by signing. However, since travel distances have to be minimised for elderly and disabled people and since potential obstructions to movement (such as kerb upstands) have to be avoided, the required route information has to be provided at frequent intervals and using signs which are clearly understood and easily seen. The directional information should focus on important destinations such as major shops, banks, libraries, car parks, bus stops and toilets, and be signed with a 'wheelchair' symbol to indicate that the route can be followed by all disabled people.

Research is continuing into the design of suitable embossed area maps and mobility aids to help blind people determine where they actually are and the direction in which they need to go to reach a particular destination.[20]

To aid blind and visually impaired people when crossing the road at signalised crossings, audible and tactile signals have been introduced to complement the tactile surfacing.[21,22] Two audible signals are available – one for staggered pelican crossings and a different one for single pelican crossings and traffic signals. Audible signals are activated when it is safe to cross.

The standard tactile signal is a small cone which protrudes from the underside of the standard push button box at signalised crossings, and rotates when the green pedestrian signal shows, thus allowing visually impaired pedestrians to 'feel' when it is safe to cross.

8.6 References

1. Hunt, J., and Abduljabbar, J., Crossing the road: a method of assessing pedestrian crossing difficulty. *Traffic Engineering and Control*, 1993, **34** (11), 526–532.
2. Oppenheim, A.N., *Questionnaire design and attitude measurement*. London: Heinemann, 1972.

3. Department of Transport, *Road accidents, Great Britain 1994 – The casualty report.* London: HMSO, 1995.
4. Grayson, G.B., Pedestrian risk in crossing roads: West London revisited. *Traffic Engineering and Control*, 1987, **28** (1), 27–30.
5. Institution of Highways and Transportation/Department of Transport., *Roads and traffic in urban areas.* London: HMSO, 1987.
6. *Pedestrian crossings: Pelican and zebra crossings.* TD 28/87. London: Department of Transport, 1987.
7. *Design considerations for pelican and zebra crossings.* TA 52/87. London: Department of Transport, 1987.
8. Wilson, D.G., and Grayson, G.B., *Age-related differences in the road crossing behaviour of adult pedestrians.* LR 933. Crowthorne, Berks: Transport Research Laboratory, 1980.
9. Daly, P.N., McGrath, F., and Annesley, T.J., Pedestrian speed/flow relationships for underground stations. *Traffic Engineering and Control*, 1991, **32** (2), 75–78.
10. Berrett, B., Leake, G.R., May, A.D., and Parry, T., *An ergonomic study of pedestrian areas for disabled people.* CR 184, Crowthorne, Berks: Transport Research Laboratory, 1990.
11. Mitchell, G.C.B., *Pedestrian and cycle journeys in English urban areas.* LR 497. Crowthorne, Berks: Transport Research Laboratory, 1973.
12. May, A.D., Leake, G.R., and Berrett, B., Provision for disabled people in pedestrian areas. *Highways and Transportation*, 1991, **38** (1), 12–18.
13. Institution of Highways and Transportation, *Pedestrianisation guidelines*, London: Institution of Highways and Transportation, 1989.
14. Institution of Highways and Transportation, *Guidelines – Reducing mobility handicaps.* London: Institution of Highways and Transportation, 1991.
15. Ketteridge, P., and Perkins, D., The Milton Keynes redways. *Highways and Transportation*, 1993, **40** (10), 28–31.
16. Institution of Highways and Transportation, *Providing for the cyclist – Guidelines.* London: Institution of Highways and Transportation, 1983.
17. Ministry of Transport & Public Works, *Manual – Traffic provisions for people with a handicap.* The Hague: Road Safety Directorate (DVV), 1986.
18. Lyon, R.R., Providing for the disabled highway user. *Municipal Engineer*, 1983, **110** (4), 120-125.
19. Lilley, A., Flexible brick paving: applications and design. *Highways and Transportation*, 1990, **37** (10), 15–19.
20. Gill, J., *A vision of technological research for visually disabled people.* London: The Engineering Council, 1993.
21. *Audible and tactile signals at pelican crossings.* TAL 4/91. London: Department of Transport, 1991.
22. *Audible and tactile signals at signal controlled junctions.* TAL 5/91. London: Department of Transport, 1991.

CHAPTER 9

Technologies for urban, inter-urban and rural passenger transport systems

G.R. Leake

9.1 Introduction

Although car ownership has risen significantly over the last few decades in developed countries, and is set to continue to rise, there will always be a large number of people who will not have access to a car and who will therefore be dependent on some form of public transport if their mobility needs are to be met. At the same time it is generally accepted that even if people do not have access to a car because they are too old (or too young), cannot afford to buy one, cannot drive because of illness, or simply because the car is being used by another member of the family, Government (whether national or local) has a duty to ensure that they have access to the fullest range of available activities.

Changes in our lifestyle and land use development patterns have increased our reliance on

> mobility to access opportunities for employment, shopping, education, health, leisure and recreation … Collectively, it means that there is an ever-growing demand for movement. Much of this increase is arising outside conventional peak periods, in areas previously unaffected by congestion or as a result of non-work trips. [In addition,] encouraging the use of public transport by all sectors of the community will help to ensure its availability to those who have no other choice.[1]

One further important aspect is the role of transport planning. This has been discussed already in detail, but there are some aspects relevant to public transport provision which need to be emphasised. Generally, movement is a means to an end, not an end in itself. In other words people move about in order to carry out pre-determined activities at the locations where those activities can take place (e.g. a shopping trip to a retail centre). As was pointed out by Buchanan there is a strong relationship between land use planning (or town planning) on the one hand, and transport planning on the other.[2] Thus transport planning cannot be considered in isolation, but undertaken within the wider perspective of land use planning.

The various travel modes constituting the transport system are interrelated.[3] Hence changing the level of service on one mode (e.g. providing improved facilities for the private car) has an impact upon the others. Thus one of the oft stated objectives of transport planning is to produce a plan which results in the *best* balance between *all* the available travel modes so that the movement demands of *all* sections of society can be met. An important question always to be answered is 'what is meant by the best?'

Any discussion about the need and role of public transport must take place against the above background. Since this will vary by type of journey, for convenience it is proposed to consider it under three headings: urban travel, inter-urban travel and rural travel.

9.2 Role of passenger transport systems in urban and non-urban areas

9.2.1 Urban travel

In urban areas there are five main passenger transport modes available: walking, cycle, motorcycle, private car, and some form of public transport. As discussed previously (Chapter 8) walking and cycling are for short distance journeys, especially the former. If motorcycle activity is recognised as catering for a relatively small proportion of trips, then the main competition between modes in urban areas lies between the private car and some form of public transport.

In looking at the role of urban public transport it is useful to consider the types of journey that take place. This can be done by considering an urban area to consist of three levels of trip end generation/attraction: high density (e.g. central area, large airport), medium density (e.g. high-density residential or industrial development) and low density (e.g. low-density suburban housing).

With this arrangement, six types of urban journey can be identified, as set out in Fig. 9.1.

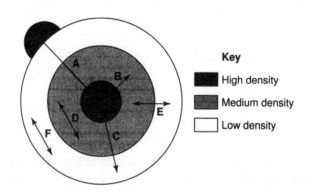

Fig. 9.1 Different types of urban journey

A (high–high)

This journey type has high concentrations of potential travellers at both ends of the journey. This means that the journey requirements can be met most satisfactorily by a high-capacity transport mode with good access/egress facilities at both ends. Some form of public transport satisfies these conditions.

B (medium–high)

This journey type has less concentrated journey densities at one end. A strong argument can still be made for some form of public transport, but there is a greater problem of picking up and dispersing people in the medium density areas.

C (low–high)

This type of journey typifies the 'journey to work' problem. At the high-density end, a high-capacity, low space-demand mode (such as some form of public transport) is required, while at the other a mode having good pick-up and distribution characteristics is needed, of which the private car is the best for those who have one. Opportunities for public transport fall into two heads: some form of bus operation, or park-and-ride, linked with a fast, reliable and segregated form of public transport.

D (medium–medium), E (low–medium), and F (low–low)

With the lower concentrations of journey origins and destinations, the private car is the most effective means of satisfying traveller requirements, especially for journey types E and F. However non-car travellers have to be provided with public transport, and this clearly creates level of service, as well as economic, problems.

The above scenario suggests that an effective and economic role for public transport is only feasible for journey types A, B and C, and that for journey types D, E and F it is unlikely to be particularly efficient even though it has to be provided. The above discussion also reinforces the role of land use planning, since it indicates that if public transport is to play an effective role then high- or medium-density developments are needed, and this is a land use planning activity.

So far the discussion has centred on a simplified and somewhat idealistic scenario, and one which has assumed that car owners would be prepared to forego using their car and transfer to public transport. However, since most people who have a car available will use it, transfer to public transport will only take place if there are perceived advantages to the individual. This is most likely to occur if travel by car has become difficult, costly and unreliable. Therefore, before proceeding further, it is necessary to examine the existing urban travel situation, particularly that occurring during the peak travel periods.

As discussed in Chapter 1, as car ownership has increased, more and longer journeys have been made by car and fewer by public transport (in the UK conventionally by bus). Reduced travel by public transport means reduced revenue to the operators, who then reduce the service frequency, or increase the fare, or both. This leads to the loss of more passengers and increased travel by car. More car travel means more road congestion which makes travel by bus even slower and more unreliable, thus further encouraging car travel. Thus the downward spiral of bus public transport usage continues (Fig. 1.5).

However, increasing car travel has created a number of serious detrimental consequences, of which the three most important are:

Traffic congestion

This is a phenomenon which is to be found in most urban areas and is increasingly occurring at times outside the normal commuter peak periods of travel. The effects of this expanding traffic congestion include not only lost economic output, but also increased noise and visual pollution, diversion of traffic onto unsuitable roads as drivers attempt to find alternative routes to avoid the congestion (often through residential areas), increased journey times and journey unreliability, and unacceptable driver stress and frustration. In addition this congestion, as already indicated, impinges adversely on the level of service that can be provided by bus public transport (but paradoxically can make rail-based public transport more attractive), and is impossible to eliminate, physically and economically, by providing the necessary additional road and parking capacity.

Energy consumption and air pollution

Cars consume a finite fossil fuel, petroleum, and although manufacturers are continually developing more efficient engines, the relative energy consumption of different travel modes favours public transport, as shown in Fig. 9.2. Greater use of public transport relative to the car will help to conserve this vital resource.

A second important consideration is the additional air pollution resulting from increased car travel. It has been 'estimated that 88 per cent of CO_2 emissions, 48 per cent of NOx and 37 per cent of volatile organic compounds emanate from road vehicles. Catalytic converters which help to reduce these emissions are less effective on short urban journeys and do nothing to reduce the CO_2 emissions which are contributing to global warming'.[1] Clearly this is an important issue which needs to be tackled urgently if the health of people, as well as the planet, is to be safeguarded.

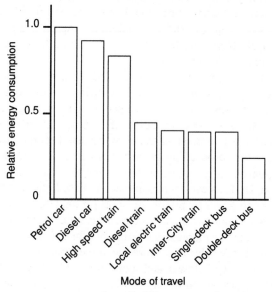

Fig. 9.2 Relative energy consumption of different travel modes (based on reference 1, amended)

Safety

With increasing car travel, there is a greater risk of accidents, particularly to vulnerable users of the road system such as motorcyclists, pedal cyclists and pedestrians (see Chapter 18). Greater use of public transport would have significant safety advantages.

It can be concluded that although car travel has brought many benefits to urban residents, it has already reached such a level that severe congestion extends beyond the traditional morning and evening peak periods, and is resulting in ever increasing and unreliable travel times. Thus it seems probable that if a satisfactory and reliable public transport service were to be introduced, for many journeys a transfer from car to public transport could be anticipated, and that this will become even more likely in the future as further increases in car ownership cause a further deterioration in the situation.

9.2.2 Inter-urban travel

In the UK long-distance or inter-urban travel is defined as journeys over 40 km in length.[4]

A reasonably comprehensive picture of the inter-urban travel market in Great Britain was obtained in the Long Distance Travel Survey conducted between 1973 and 1980. This showed that only a relatively small proportion of person journeys (3 per cent) were longer than 40 km and that this was as low as 1 per cent for shopping journeys, 8 per cent for work journeys, but as high as 28 per cent for holiday travel. Table 9.1 shows the number of long-distance person journeys by purpose and travel mode. It shows the predominance of rail and car travel for work journeys.

Table 9.1 Percentage of long-distance trips by main mode of transport and trip purpose, Great Britain 1979–80 (based on reference 4, amended)

	Trip purpose			
Mode	To and from work	In course of work	Non-business	Total
Rail	36	12	14	20
Bus/coach	2	2	8	7
Car	51	84	75	68
Air (domestic only)	–	2	–	–
Employer's bus etc.	11	–	–	3
Other modes	1	1	2	2

The overall picture outlined in Table 9.1 indicates that car travel predominated with 68 per cent of the total market, but with significant contributions from rail at 20 per cent and coach at 7 per cent. Except for work journeys, air travel was relatively unimportant, primarily because of the cost and the low distances (generally less than 350 km) between the major urban centres.

The two principal inter-urban public transport modes in the UK are rail and coach. Both modes normally have terminals in or adjacent to town/city centres and hence are readily accessible to most travellers. But rail, having its own segregated right of way, is immune from delays extraneous to its mode of operation, whereas coach is liable to be caught up in the increasingly occurring traffic delays. Although coach journeys are usually cheaper, they take longer and their journey times cannot be guaranteed. By a similar argument, car travel is also becoming more unreliable, as well as more stressful for the driver. Hence any improvements to the level of public transport service, coupled with any lowering of cost, will make it more attractive.

9.2.3 Rural travel

Rural areas are characterised by low-density development. Closures of rural rail branch lines, together with a reduction in the network coverage and service frequency of rural bus services, have created a situation where the possession of a car is considered to be essential, even though it can result in considerable financial hardship to families on low income. Indeed many families are obliged to buy two cars in order to enable the travel needs of all the family to be satisfied.

In rural areas the travel problems of elderly people, and those without a car, become acute because of the greater distances involved and the greater scattering of facilities such as shops and schools. However, as car ownership has increased, the scale of the problem has tended to decrease, but for those people unable to drive or without a car available for use at particular times the problem has worsened, as the difficulty of providing economic public transport services in a shrinking market has increased, resulting in a reduction in mobility and quality of life.

Attempts have been made in some rural areas to improve the standard of provision of public transport services by introducing unconventional transport such as postbuses. These will be considered later in this chapter.

9.3 Desired characteristics of public transport systems

As will have become clear from the previous discussion, the main competition to public transport for all three types of travel (urban, inter-urban, rural) is the private car. Hence if public transport usage is to increase, then it must have some operational characteristics which give it advantages over the car in the particular situation in which it is to be introduced or improved. In brief, the *desirable* characteristics can be summarised as:

Convenience

- The service needs to go to the destination(s) that travellers wish to reach, preferably without the need to interchange.
- The service frequency must be high enough to ensure that wait times are acceptably short.
- The service must be reliable, i.e. the scheduled arrival times and scheduled line-haul times must be consistently maintained.
- The door-to-door journey time must be comparable/competitive with that of the car.
- The public transport vehicle must be comfortable, with adequate seating for those who need it, and acceptable standing passenger densities during peak periods.
- The vehicle must be clean and easily accessible for all members of the travelling public, especially elderly and disabled people.
- Stations/stops should be well designed, have good waiting facilities, be protected from the weather, within easy walking distance for most potential users, and provided with feeder or park-and-ride facilities if necessary.
- Any interchanging should be undertaken without the necessity of having to change level (i.e. no steps) and should involve short distances.
- Pedestrian access routes to stations/stops should be attractive, well lit and well maintained.

Image

In the view of many people public transport, especially the bus, has an old-fashioned, antiquated, run-down image. This needs to be reversed, and improvements in the following aspects would go some way to doing this:

- seat comfort, the amount of leg room, and the noise level within the vehicle
- quality of ride and smoothness of acceleration and deceleration
- the design of the vehicle to give the impression that it is part of a modern, well-run service
- overall impression created by the design and upkeep of the stations/stops, which should be clean and free of damage and graffiti
- attitude and helpfulness of the staff.

Information

This is of vital importance if the system is to be perceived as being user friendly. There are several significant aspects:

- Details of service frequencies, times and fares (by route) should be readily available, clearly presented, and kept up-to-date.
- 'Real-time' information should be available at stations/stops, giving current information on actual running times and the time of arrival of the next vehicle.
- Details of any pre-booking arrangements should be clearly presented.

Security

- All travellers should feel safe when using public transport. This problem is accentuated by the fact that many stations, and all bus stops, are not staffed.
- Public transport facilities, including access routes, should be well lit and continually monitored by closed circuit TV to help reduce the risk of personal attack, and at the same time to induce the feeling of being safe in the mind of the traveller.

9.4 Urban, inter-urban and rural technologies

It is not the purpose in this chapter to determine what kind of public transport facility would be appropriate in a particular situation and what level of provision would be needed. This would be an output from an in-depth land use/transport study. What is proposed, however, is to outline the main features of some of the public transport technologies which currently play a part in providing a good transport system and at the same time help to alleviate the deteriorating environmental situation. A chart of those to be included is presented below.

Bus-based systems	*Track-based systems*	*Unconventional transport facilities*
Conventional bus	Light rail	Those which are applicable for use
Guided bus	Metros	in rural areas
	Conventional rail	
	Monorail	

9.4.1 Bus-based systems

Conventional bus

Although the particular design, internal layout and seating capacity of the conventional bus varies between make of bus, there are legal limits in the UK: maximum length of 12 m and maximum width of 2.5 m.

Bus vehicles typically take one of four forms. First there are the standard single-deckers of length 10–12 m and a seating capacity of around 50. Secondly there are the double-deckers which are typically about 10 m long and with a seating capacity of around 75. Thirdly there are the articulated single-decker vehicles, which are rare in the UK but more common on the continent with a length of around 16 m and a high proportion of standees to give a capacity of over 100 passengers. And finally there are the minibuses which have a length of up to 7 m and a seating capacity typically ranging from 16 to 20 passengers.

Normal bus service operation is too well known to require detailed description. Bus stops are located at appropriate intervals along the bus route at which passengers board and alight. Some shelter (with seating) may be provided, but many stops are fully exposed to the weather. Some limited timetable information may be provided at the stop. With bus stops at street level, passengers have to negotiate steps into and out of the bus. For disabled and elderly people, and people with pushchairs or luggage, this can create problems, especially as many entrances are narrow, as well as increasing the wait time for the bus at the stop. Fare collection by the driver also increases the wait time for the bus.

Since conventional bus services operate on-street they suffer from any congestion delays, thus leading to unreliable service timings, bunching of buses (a bus running behind schedule will have to pick up more passengers and hence will be stationary at the bus stop for a longer time; the following bus will have fewer passengers to pick up, and hence will 'catch up' on the leading bus), and hence a less attractive service to travellers.

In conventional bus operation the actual location of the bus stops is of importance. Immediately two conflicting considerations become apparent which need to be resolved: the bus operator wishes to locate them close to junctions, since pedestrian concentrations often occur at such locations; the highway authority wishes to locate them away from junctions because of their adverse effect on junction capacity and safety (obstruction of sight lines).

A number of simple guiding principles can be laid down governing the location of bus stops.

1. They should be located at/near places of pedestrian concentration. In particular they should be close to any designated homes for elderly people.
2. They should be positioned where they will not cause a safety hazard and where disruption to other traffic will be minimised.
3. They should be located on the *exit* side of a junction, preferably in a busbay. This has particular advantages if there is a right-turning bus route entering from a side road.
4. The positioning of the bus stop should not obstruct any accesses to adjacent properties.
5. Bus stops should be located preferably where busbays can be provided. However, in urban areas it will not always be possible to provide the standard dimensions (see Chapter 20) because of lack of space. In such cases there are usually advantages in

providing the best possible layout, adopting the principle that 'something is better than nothing'.

6. Wherever possible, footways adjacent to the bus stop should be sufficiently wide (3 m minimum) to prevent queuing passengers obstructing other pedestrians. Where a bus stop is associated with a formal pedestrian crossing (zebra, pelican), it should be located on the exit side of the crossing.

7. Often it will be advisable to prevent on-street parking in the vicinity of a bus stop. This not only ensures that passengers are visible to oncoming traffic, but also that buses can stop close to the kerb.

One important question on which guidance is also needed is the desirable spacing between bus stops. Clearly this will be affected by the type and arrangement of land uses within the 'catchment area' of the stop, the local topography, and the acceptable walking distance. It has been indicated[4] that the total door-to-door journey time is minimised with a bus stop spacing of around 550 m.

As mentioned earlier, conventional bus services use the same road space as other vehicles and hence are subject to congestion delays which reduce their level of service. Bus lanes, which can be with-flow or contra-flow, can be introduced to enable buses to bypass slow-moving or queuing traffic.[5] These are created when part of the available road width is allocated for the use of buses for all or part of the time (see Chapter 24).

Roads may be constructed for the exclusive use of buses (busways), and may be at ground level or elevated. The former are cheaper but have the disadvantage that they create a barrier to other movements (including pedestrians) which need to cross them, and since buses would normally be given signalled priority over other traffic at junctions this also results in some increase in traffic delay. These disadvantages do not occur with the more expensive elevated alternative.

Bus services on exclusive busways avoid the delays to services operating on the ordinary road network, and can be readily linked with residential, industrial and shopping areas with normal bus operation. Further details are given in Chapter 24.

To reduce busway costs and land take, a guided bus system may be preferred to conventional busways which rely on manual steering of the buses. This is taken up in the following section.

Guided bus

A guided bus

is a form of dual-mode system (also seen as an alternative to Light Rail systems) designed to enable a conventional bus to operate on both ordinary roads and special guideways. The system seeks to combine the advantages of buses, in offering direct services from suburban areas without interchange, with those of a rail vehicle, in providing an efficient, high capacity level of service along major radial corridors.[6]

Guided bus systems can be separated into two main groups – those with track guidance and those with electronic guidance. With track guidance the bus is mechanically controlled by direct contact with track guide rails. The simplest form (illustrated in Fig. 9.3a) relies on vertical guide rails to direct the front wheels of the bus. A typical track

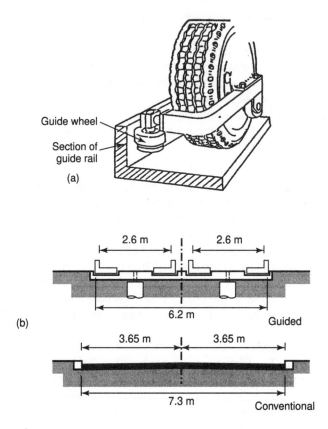

Guide wheel

Section of guide rail

(a)

(b)

2.6 m

2.6 m

6.2 m

Guided

3.65 m

3.65 m

7.3 m

Conventional

Fig. 9.3 (a) Mechanical guidance system on guided bus, and (b) minimum right-of-way for guided and conventional busway[6]

cross-section is shown in Fig. 9.3b. This shows that a guided system requires a two-way track which is 1.1 m narrower than a conventional busway. The vertical faces of the guide rails are 0.18 m high.

With electronic guidance the bus is steered electronically by a buried wire in the track. One advantage with such guidance systems is that the track is flush with the adjacent roadway, as compared with track guidance where there is a height difference of 0.18 m, and hence does not cause a physical barrier to any transverse movement (whether pedestrian or vehicular).

To date few guided bus systems have been implemented.[7,8,9] Hence little operational experience is available to help planners determine the advantages that a guided bus system might have over a conventional busway, or alternatively a light rail system, in a particular situation. Nevertheless it is important to attempt to determine the conditions under which guided bus systems could be worthy of in-depth consideration. This has been attempted by Read *et al.*[6] After commenting that 'in many situations guided bus will prima facie be preferred to a conventional busway', they proceed to examine the effect that a large number of stops, coupled with high passenger flows, would have on the efficiency and effectiveness of guided bus operation. They conclude that 'guided bus technology is better suited ... in a limited-access corridor ... . To perform best the system needs a separate, mostly uninterrupted right-of-way. It would not be effective in a busy residential or shopping street with frequent side roads, main road junctions, pedestrian crossings and bus stops.'

Three important, but restricted, situations were identified where a guided busway might be appropriate: former railway corridors; relatively straight road central medians, with few road intersections; and where new tunnel or elevated construction was being considered.

9.4.2 Track-based systems

There are four main types of tracked system: light rail transit (LRT), metropolitan railways (Metros), suburban railways and inter-city rail.

LRT can be defined as a system which 'uses a right of way which is physically segregated from other traffic over all or most of its length, but may have sections with controlled level crossings and may share some sections with road traffic'.[10] Metros always operate on segregated rights of way. They resemble, to some extent, the LRT systems but trains are longer and they carry higher passenger numbers. Because of this stations are longer and the whole system more costly than LRT. Metro systems (e.g. Paris Metro) may be rubber tyred.

Suburban railways typically form part of a national railway network. Station spacings are greater than for Metros and hence operating speeds are higher. At the top end of the service provision are the inter-city services. These operate at average journey speeds of 160–250 km/h and offer the highest level of service, comfort and reliability. It is not proposed to consider these high-level services in detail.

Some typical comparisons between LRT, Metros and suburban railways are outlined in Table 9.2. In terms of passenger carrying capacity the relative position of LRT, compared with conventional bus and other types of railway operations, is illustrated in Fig. 9.4.

Table 9.2 Comparison between LRT, Metro and suburban railway systems (based on reference 10, amended)

	LRT	Metro	Suburban rail
Operational characteristics			
Maximum speed (km/h)	70–80	80–100	80–130
Operating speed (km/h)	20–40	25–60	40–70
Reliability	High	Very high	Very high
Vehicle characteristics			
Vehicles/train	1–4	1–10	1–10
Vehicle length (m)	14–32	16–23	20–26
Passengers/vehicle	200	250	180
Other			
Station spacing (m)	300–800	500–2000	2000+
Average journey length	Short to medium	Medium to long	Long

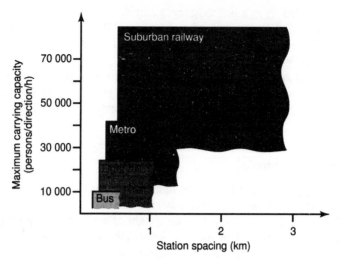

Fig. 9.4 Maximum pasenger capacity of different public transport systems[19]

Light rail

As indicated above LRT (light rail) may not be fully segregated from road traffic. However, if segregation is introduced, or priority given to LRT at road junctions, then speed and service reliability are increased.

LRT uses steel wheel on steel rail and operates on track at the standard 1435 mm gauge. Vehicles may be single or articulated and are electrically powered with the power normally being collected from an overhead wire at 750 V. Overhead power collection is required for all on-street working. Vehicles are normally manually driven, although automatic operation is possible if the system is fully segregated.

LRT vehicles have a wide range of design layouts. Typically they are between 25 m and 32 m in length (although a few designs are longer) and with a width between 2.3 m and 2.7 m.[11] Although vehicles can be linked together to form trains of 2–4 vehicles, normal operation, particularly with on-street running, is planned around single vehicles.

In recent years the big revolution in design has been the extensive introduction of low-floor vehicles. These have considerable access advantages for passengers when running on-street. In addition to these low-floor vehicles (floor level 160–420 mm above rail top), there are also middle-floor vehicles (floor level 440–530 mm) and high-floor vehicles (floor level 870 mm). Vehicle costs tend to increase with decrease in floor height. High-floor vehicles, in addition to being cheaper, are to be favoured when LRT is operating on the same track and using the same station platforms as suburban railways.

An LRT vehicle is shown in Fig. 9.5. Typically such vehicles have seating for 70–80 passengers, plus allowance for up to 130 standing passengers at peak periods.

Station spacing for LRT operation is a function of three parameters: land use developments within the station 'catchment area'; whether park-and-ride is to be an integral element in providing access to the station; and the required operational (commercial) LRT journey speed. The distance between stops generally lies between 250 m (exceptional) and 1 km. The greater the distance between stations, the higher the LRT journey

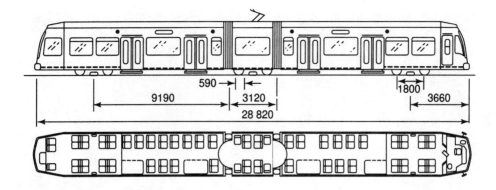

Fig. 9.5 Typical LRT vehicle layout[11]

speed but the lower the number of potential passengers within 'easy' walking distance. The relationship between maximum speed (usually 80 km/h), operating speed and station spacing is shown in Fig. 9.6. For a typical station spacing of around 500 m, the operational speed would be 30 km/h.

It is important that all passengers have easy station and vehicle access and egress. This not only makes travel by LRT more attractive, especially to elderly and disabled people and those with pushchairs and luggage, but also ensures minimum vehicle stop time at the station and hence higher service operational speeds. In order to achieve this, wide automatic doors (usually four in number) are provided on each vehicle, together with floor-level access off the station platform. The advantage of this for travellers in wheelchairs, in particular, will be obvious and is illustrated in Fig. 9.7(a).

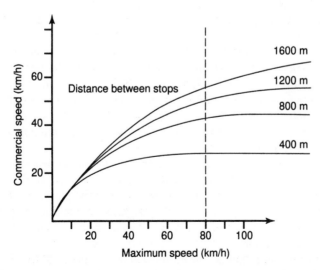

Fig. 9.6 Relationship between operating (commercial) speed, station (stop) spacing and maximum speed[19]

(b)

(a)

Fig. 9.7 (a) Floor level vehicle access, with (b) tactile paving defining the platform edge and door position[20]

Another important consideration, if use by blind and partially sighted people is to be encouraged, is to ensure that the edge of the platform is adequately defined and that they can position themselves on the platform opposite the doors. This can be accomplished by using two types of tactile paving, as shown in Fig. 9.7(b).

It is also important to provide well defined stations, particularly in those areas where on-street running is in operation (e.g. in a pedestrian area). These should provide adequate platform width to accommodate the anticipated peak passenger numbers, have platforms at vehicle floor level, be relatively unobtrusive, and be easily accessible to all potential passengers irrespective of their age and physical condition.

All platforms should be accessed by a low gradient ramp (not more than 5 per cent to allow easy wheelchair access), and possibly steps as well. One of the attractions of low-floor vehicles is that the platforms are also low with short ramp lengths, an important design feature for on-street working.

Access to station platforms on segregated rights-of-way or in pedestrian areas does not usually pose a problem.[12] However, if LRT is being provided in a central reservation, then passengers have to cross traffic in order to reach the platforms. In such cases, the pedestrian movement has to be controlled by signals. In addition the platforms will often be staggered, as shown in Fig. 9.8, in order to keep the station within the width limits of the central reservation.

The usual type of construction is conventional standard gauge (1435 mm) ballasted track for segregated off-street operations. However, for on-street running this is unsatisfactory, and some form of grooved rail is used. A typical example of such construction

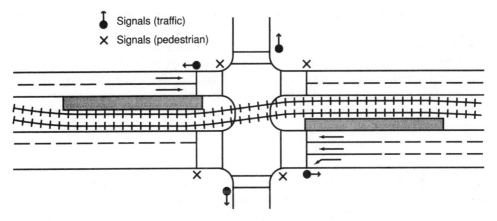

Fig. 9.8 Access to an LRT station with staggered platforms located in a central reservation

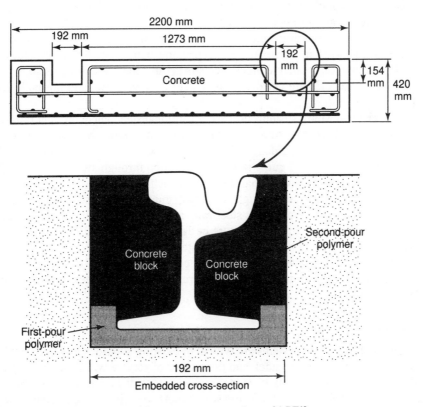

Fig. 9.9 Grooved rail track used for on-street running of LRT[13]

is shown in Fig. 9.9.[13] The rails are embedded in grooves in the continuous concrete track bed using a polymer material. This method of construction obviates the need for mechanical rail fastenings, acts as a shock absorber which reduces vehicle vibration and noise, and improves passenger comfort.

Any public utilities (e.g. gas, electricity) should be diverted from the line of the proposed route before construction of the track. This ensures that the LRT service will not be disrupted when public utility repairs have to be carried out.

One of the positive attractions of LRT is that it can cope with gradients of up to 6 per cent and negotiate radii as low as 25 m.[14] This means that, if necessary, the alignment can follow close to existing ground level (with considerable savings in capital cost) and be constructed in existing urban areas without the necessity of extensive (or even any) property demolition.

Metros

Metro systems have many elements in common with LRT and some of the design issues considered above when discussing LRT are equally applicable, such as access to stations and the relationship between station spacing and operational speed. However, there are also differences, and some of the more important of these are set out below.

- Unlike LRT, which normally operates as single vehicle units, Metro systems (e.g. the London Underground) operate in trains, often of six or more vehicles.
- Since Metro systems are completely segregated, the track can be constructed on concrete sleepers set in conventional deep stone ballast. In tunnel or elevated track sections, however, concrete block sleepers set in concrete will be used.
- The single vehicles are usually smaller than LRT vehicles, and normally have high floors.
- Although most Metro systems utilise conventional steel wheel on steel rail, some systems (e.g. the Paris Metro) have rubber tyres. With rubber tyre operation the support wheels do not provide guidance. This is done by small horizontal rubber-tyred wheels bearing against vertical guide rails. For operation through switches, steel wheels are provided as well as conventional steel rails which are raised to come into contact with the steel wheels. Thus, through switches, guidance and support is via the steel wheels. It should be noted that steel rails are provided throughout the alignment to provide vehicle support should a tyre deflate.[15] Rubber tyred trains result in slightly less noise and vibration than those with steel wheels.
- Metro stations will have many features in common with LRT stations located on segregated track. The two main differences are that they are longer because of the longer trains, and they will have high platforms.

Conventional rail

Conventional heavy rail systems are typical of existing suburban and inter-city services, although the type of rolling stock and the quality of service will differ between the two.

Conventional rail systems have steel wheel on steel rail, have heavy rolling stock vehicles, are to be found in trains of two or more vehicles, are completely segregated from road traffic, have high vehicle floor heights, have widely spaced stations which creates the need to provide access facilities (e.g. park-and-ride, feeder bus) and operate at high speeds (especially the inter-city services).

As has been indicated,[15] 'In recent years few urban areas have been able to contemplate introducing totally new heavy urban rail systems. The capital costs tend to put such systems out of financial reach of all but the most economically prosperous areas: the disruption to the urban fabric envisaged during construction can cause powerful objections on environmental grounds.' Thus it is clear that the future role of heavy rail will lie with suburban and inter-city services and that future changes will concentrate on reducing journey times and increasing traveller comfort by a combination of improvements to the track alignment and the vehicle stock.

If conventional heavy rail services are to be successful, then passengers have to be able to get easily and reliably to the more widely spaced stations. Since few potential passengers will be within walking distance, access has to be provided by feeder bus, and park-and-ride (including kiss and ride).

Feeder bus services, if they are to be satisfactory, must be reliable so that a traveller can guarantee arriving at the station in time to catch the train. The implementation of park-and-ride (including kiss and ride) can be an important element in improving the accessibility of widely spaced stations. Furthermore, as has been indicated,[16] 'greater use of public transport could be encouraged by the provision of park-and-ride facilities', and that park-and-ride is 'one means of easing interchange between modes and integrating private and public transport'.

However, if park-and-ride is to be successful (see also Chapter 7), then a number of key issues have to be addressed, of which some of the more important are:

- providing sufficient space for the estimated number of park-and-ride car parking bays and 'kiss and ride' waiting areas
- ensuring that congestion at entrances and exits does not occur, even with large numbers of travellers arriving or departing over short time periods
- providing facilities which are usable by elderly and disabled people
- charging policies for using the park-and-ride facilities which encourage, rather than discourage, use
- security.

Monorail

There is little evidence to suggest that monorail systems have inherent economic and operational advantages over wheeled systems. However, they do have novelty value, and hence a number of monorail applications have been in amusement parks. Monorail systems fall into two categories: *suspended systems*, where the vehicle/car is suspended from an overhead guideway; and *straddled systems*, where the vehicle/car straddles the guidebeam.

Suspended systems tend to be uneconomic, particularly in tunnel operation, and in addition produce greater potential visual intrusion, since the height of the vehicle and guideway approximates to 4 m.[15] Straddled systems have been introduced in some amusement parks and as airport links, where they operate at low speed. One example of a straddled system is the Maglev system introduced at Birmingham airport in the UK in 1984. This is a levitation system where the force between the magnets and the support rail enables an air gap of 15 mm to be maintained. A schematic diagram of the system is shown in Fig. 9.10. The vehicle is 6 m long, 2.25 m wide and can carry 40 people with luggage. It has a maximum speed of 50 km/h and an acceleration rate[17] of 0.8 m/s^2. In the event of a power failure, the vehicle de-levitates safely on to the track.

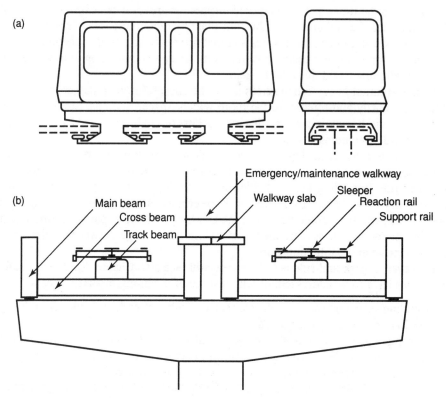

Fig. 9.10 Schematic diagram of (a) Maglev vehicle and (b) track support[17]

9.4.3 Unconventional transport facilities

When looking at the transport problems of rural areas, it was concluded that there were severe problems for people without access to a car. Recognising this, a number of alternative transport ideas have been introduced and tested in a number of rural areas where public transport was unable to satisfy local needs. These incorporated one or more of the following features:

- voluntary effort by local residents in organising and operating community transport services
- using small vehicles where demand is small and/or roads narrow
- arranging routes (and, in some cases, schedules) according to pre-booked demand
- carrying passengers on vehicles making regular journeys for other purposes
- linking remote areas to the main public transport network with feeder services.[18]

A wide range of unconventional services have been introduced, including:

- shared hire cars making published connections with existing services, with a pre-booking arrangement. Routes were not fixed and passengers were picked up, and returned to, their homes
- volunteer drivers carrying passengers in their own cars and accepting payment for the pre-booked journeys

- a 'community bus' using unpaid local volunteers for driving and managing the service, operating on a flexible route system picking up pre-booked passengers
- a car sharing service where passengers book journeys through a local organiser who keeps a list of when drivers participating in the scheme are available
- postbuses, where the carriage of passengers and the collection and delivery of mail are combined using a minibus instead of the conventional mail van.

Such services have been found to make a limited contribution to improving transport provision in rural areas, although the precise impact and most effective type of service varies from area to area. What has become clear, however, is that local volunteers are capable of organising and operating such services.

9.5 Final comment

It is now recognised that increased use of public transport, especially in urban areas where transport problems are at their most acute, results in significant transport movement and environmental benefits. Increasingly light rail systems are being seen as the most likely form of public transport which is capable of making a major contribution to solving the urban transport problem and which is affordable. At the same time it is clear that it can only be part of a larger package of measures, including traffic management.

For inter-city travel there will be a continuing effort to improve the speed, reliability and quality of service, and in rural areas, where deficiencies in transport availability can be identified, there will be continued experimentation to ascertain what unconventional services can be introduced locally to meet the identified deficiencies.

9.6 References

1. *Public transport: What's in it for me?* London: Institution of Civil Engineers, 1993.
2. Buchanan, C.D., *Traffic in towns*. London: HMSO, 1963.
3. Independent Commission on Transport, *Changing directions*. London: Coronet, 1974.
4. White, P., *Public transport: its planning, management and operation*. London: Hutchinson, 1986.
5. Institution of Highways and Transportation/Department of Transport, *Roads and traffic in urban areas*. London: HMSO, 1987.
6. Read, M.J., Allport, R.J., and Buchanan, P., The potential for guided busways. *Traffic Engineering and Control*, 1990, **31** (11), 580–587.
7. Heath, R.J., Adelaide's guided busway: the operating experience. *Proceedings of Seminar on Guided Bus Rapid Transit*. Adelaide: Sagric International, 1988.
8. Teubner, W., The extent, performance and future of the O-Bahn system in Essen. *Proceedings of Seminar on Guided Bus Rapid Transit*. Adelaide: Sagric International, 1988.
9. *Proceedings of Seminar on Guided Bus Rapid Transit*. Adelaide: Sagric International, 1988.
10. Jenkin, P., Urban railways: system choice. In *Urban railways and the civil engineer.* London: Thomas Telford, 1987.
11. All-low-floor cars dominate orders, developing metros. Sutton: *Railway Gazette International*, 1993.

12. Young, A.P., Greater Manchester's metrolink project. *Highways and Transportation*, 1989, **36** (11), 9–12.
13. Boak, J.G., South Yorkshire supertram: route and civil works. *Proceedings of Institution of Civil Engineers – Transport*, 1995, **111** (1), 24–32.
14. Ridley, T.M., Light rail – technology or way of life. *Proceedings of Institution of Civil Engineers – Transport*, 1992, **95** (2), 87–94.
15. Howard, G., Light, heavy or innovative? A review of current systems, In B.H. North (ed.), *Light transit systems*. London: Institution of Civil Engineers, 1990.
16. Niblett, R., and Palmer, D.J., Park-and-ride in London and the South East. *Highways and Transportation*, 1993, **40** (2), 4–10.
17. Sumnall, G., The Birmingham airport Maglev system. *Highways and Transportation*, 1987, **34** (1), 22–28.
18. Balcombe, R.J., *The rural transport experiments: A mid-term review*. SR 492. Crowthorne, Berks: Transport Research Laboratory, 1979.
19. Howard, D.F., The characteristics of light rail. *Highways and Transportation*, 1989, **36** (11), 6–8.
20. Russell, J.H.M., and Horton, R.J., Planning of the South Yorkshire supertram. *Proceedings of Institution of Civil Engineers – Transport*, 1995, **111** (1), 15–23.

CHAPTER 10

Planning for public transport

C.A. Nash

Earlier chapters have discussed how public transport may be an important element in any integrated transport plan. It has also been demonstrated that there is a variety of technologies available for the provision of public transport services.

Whereas in most countries provision of roads, with the exception of certain main roads, is seen largely as a responsibility of government, the arrangements regarding public transport vary much more widely. The spectrum ranges from being largely the concern of privately owned commercial operators with a minimum of regulation, through highly regulated private operators to monopoly provision by publicly owned companies.

In this chapter the appropriate roles of the different public transport modes are first considered, followed by discussion of the behaviour of a privately owned monopolist and of what might be appropriate for a government owned company. Reasons why a purely commercial approach to public transport operations may be inappropriate are discussed next and, finally, the way in which private ownership may be combined with public obligations is considered.

10.1 Appropriate public transport modes

As discussed in Chapter 9 there are several modes of public transport from which to choose.[1]

1. The bus is ubiquitous. It operates as a single vehicle, is usually diesel powered, can share road space with other traffic, and can combine the role of local feeder service with trunk express.
2. The train operates on segregated track which may be on the surface, underground or above ground. It can comprise a string of vehicles.
3. The tram operates on the street surface, usually with electric traction, and it can comprise several vehicles joined together.

However, the differences between these systems are blurred. For instance, in Essen, guided buses run on-street in the suburbs using diesel power, but run in tunnels in the centre under electric power. In Curitiba in Brazil, articulated buses carrying 180 passengers run on segregated busways. In many German cities, light rail systems operate in tunnels in the centre and on-street in the suburbs, while in other cities the position is reversed. Choice of the appropriate public transport mode to install in a particular location is therefore not easy; the spectrum of options available is wide. In general however, there are some key trade-offs to be made.

1. The operating costs of a public transport system may be minimised by operating high-capacity vehicles or trains at relatively low frequencies. If the volume of traffic is high enough, it may be worth installing a segregated fixed track system to permit longer trains than can be operated on-street.
2. The costs to the user include the time spent walking to, waiting for and travelling on the service. These are of varying importance according to the distance travelled. For short journeys, the convenience of a readily available service at high frequencies may be more important than the speed when travelling in the vehicle; for long journeys the reverse may be the case.
3. On-street systems may both delay and be delayed by other traffic. The greater the volumes both of public transport and of other traffic, the stronger the case for a segregated system.

Thus in practice a system which operates on-street is likely to have lower overhead costs, but to create and suffer from congestion. The more it is given priority over other traffic, the more likely it is to delay other vehicles. A segregated system will have a much higher initial cost but create lower congestion costs for other traffic. A system which can run high-capacity trains will be particularly valuable where volumes of public transport traffic are high, and/or where distances are relatively long. Where volumes are lower, or distances shorter, high-frequency low-capacity vehicles may be preferable. A system which is capable of frequent stops will be preferable where mean trip lengths are low, and walking time forms a high proportion of total travel time; where trips are longer, in-vehicle speed becomes more important.

10.2 Commercial services

For the moment consider a purely commercial operation, and also – to make life really easy – ignore the possibility that the operator has competitors who may respond to his or her actions. It is then possible to stipulate two simple rules which a commercial operator seeking to maximise profits will follow: (1) raise fares as long as this either increases revenue (which it will do if a 1 per cent increase leads to a less than 1 per cent reduction in demand) or reduces revenue by less than any cost saving as a result of a reduction in traffic (and *vice versa*); and (2) improve services as long as the increase in revenue is more than the increase in costs (and *vice versa*).

These rules obviously suggest a need to know how a change in price or service level affects the demand for the services, and how a change of service level or traffic volume affects costs. These relationships constitute two fundamental relationships in economics: the demand curve and the cost curve. These two relationships can be written algebraically.

Demand curve $$Q = Q(P,B) \tag{10.1}$$

where Q = volume of passengers, P = fare, B = level of service (e.g. bus miles run).

Cost curve $$C = C(B,Q) \tag{10.2}$$

where C = total cost, and B and Q are as defined previously.

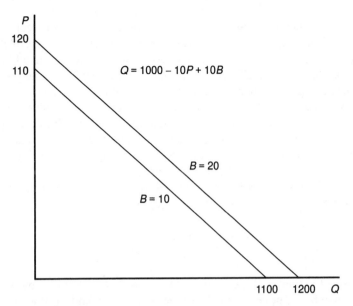

Fig. 10.1 The demand curve

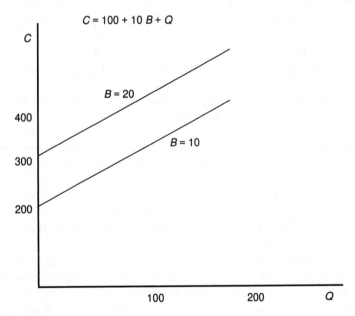

Fig. 10.2 The cost curve

Simple linear examples are given in Figs. 10.1 and 10.2, where the relationships between P and Q and C and Q are shown for alternative values of B.

Note that the concept of total cost used in this example is different from that of marginal cost used in Chapter 4. Total cost represents the entire cost of the operation including any fixed cost, whereas marginal cost is the extra cost of an additional unit of output, or more formally dC/dQ. For the example in Fig. 10.2, the marginal cost is 1.

A commercial operator then seeks to choose P and B to maximise the difference between its revenue and costs (i.e. $P.Q\,(P, B) - C\,(B, Q)$). (This is of course a gross simplification; in fact it is possible to have a whole range of fares (peak, off-peak, single, return, saver, child etc.), and it is necessary to know the effect of each of these on demand. Similarly, different passengers – peak versus off-peak – and different service improvements may have different effects on cost.)

The effect of fares on traffic is measured simply as the own price elasticity of demand. The own price elasticity of demand (e_D) is:

$$e_D = \frac{\%\Delta \text{ in demand}}{\%\Delta \text{ in price}} \tag{10.3}$$

Thus, if $e_D = -0.3$, then a 10 per cent price rise will reduce traffic by 3 per cent and increase revenue by 7 per cent (since revenue = price × quantity). If $e_D = -1$, then a 10 per cent price increase will reduce traffic by 10 per cent and leave revenue unchanged.

Thus it makes sense to charge higher fares in less elastic markets (i.e. markets where the own price elasticity of demand is lower in absolute terms). Generally, it is found that peak elasticities are lower than off-peak, since a high proportion of peak journeys are essential (journeys to work or school) and have fixed destinations (at least in the short run). A larger proportion of off-peak journeys are discretionary (e.g. social or leisure trips) or may change destination (e.g. a shopping trip by bus to the town centre may be replaced by a walk to the local shops).[2]

The price elasticity of demand may be measured in a number of ways. The most straightforward is from a time series of observations of the level of fares and the level of public transport patronage. However, it is important to allow for other factors which may be influencing demand (such as service levels, incomes and car ownership) using a multivariate statistical technique such as multiple regression analysis.[3]

One may introduce an analogous concept of a service level elasticity as the percentage change in traffic from a 1 per cent change in the bus miles run. This is a less useful concept, however, as there are numerous different ways of changing bus miles (changing frequencies, route density, times of day or days of week operated) and each is likely to have a different effect on demand.

It is of course likely that a commercial public transport service will be subject to competition. If there are close rivals, than a change of price or service level by one operator may lead to a change by the rival, thus having a further effect on the first operator's traffic and possibly making a further change worthwhile. For instance a rival may take the opportunity posed by a price increase to increase prices as well; alternatively the diversion of traffic caused by the price increase may lead the rival to increase his or her services. These responses are difficult to predict, but since they will in turn affect traffic and revenue for the first operator he or she will certainly need to consider them.

Even if there are no existing competitors, setting a price/service combination which leads to a high level of profits may still attract rivals in. This depends on how easy it is to set up in the service in question. Bus operators generally face the prospect of easy entry, particularly from existing bus companies. Even though governments are increasingly requiring rail operators to provide access to the infrastructure for rival operators, entry into rail operation remains very much more difficult because of the need to secure track access, rolling stock, maintenance facilities and trained staff in a much more specialised market than that of bus operation.

The result is that it may be thought that actual or potential competition is sufficient to prevent monopoly exploitation by the charging of excessively high fares or provision of poor services in the bus industry. This is less likely for rail, where public operation, or regulated private operation under a franchise arrangement, is the norm.

10.3 Subsidised services

The above discussion has concerned only services where the aim is to maximise profits. However, some bus services and most rail services continue to receive subsidies. To the extent that provision of these services is by commercial operators through competitive tendering (bus) or franchising (rail), operators will still follow the above rules to the extent that contractual arrangements with the provider of the subsidy permit. Where there is public ownership, the objectives may be quite different.

Both in bus tendering and in rail franchising, the provider of the subsidy is likely to specify minimum levels of service to be provided; both may also specify fares. In the former case, the operator can then apply the normal rules for commercial operation subject to meeting the service level constraint; in the latter, their commercial freedom is much more limited.

From the point of view of setting constraints on service levels and fares for subsidised services, and from the point of view of the provision of services by non-commercial publicly owned operators, it is the objectives of the provider of the subsidy or the service that are important. These may include:

- providing a minimum level of accessibility for all
- maximising the benefits to users from provision of a service
- obtaining benefits for non-users by means of relieving congestion and environmental damage caused by road traffic
- promotion of economic development.

A method of seeking to quantify these benefits and to compare them with the costs of the subsidy exists in the form of cost-benefit analysis (see Chapter 4).

Example

A simple example may help to illustrate the implications of the alternative objectives that public transport operators may pursue. Table 10.1 shows the optimal policies under a variety of objectives for a simple hypothetical but moderately realistic example. The exact equations used, which contain the variables specified above, are given in the Appendix to this chapter.

It will be seen (row 1) that in this case it is possible to make substantial profits by running limited services at high prices (although the likelihood that this would prompt

Table 10.1 Example of alternative objectives

Objective	Constraint	Bus miles (off-peak)	Bus miles (peak)	Fare (off-peak)	Fare (peak)	Passenger miles	Revenue	Cost	Profit	Net social benefit
Profit maximum	None	2.3	1.1	10.5	21.0	30	479	159	320	774
Service maximum	Break-even	20.5	1.1	10.5	21.0	34	525	523	2	502
Social welfare	Break-even	3.6	1.7	2.4	4.8	75	270	264	6	1135
Social welfare	None	3.9	1.8	0.5	1.0	92	69	291	−222	1159

Source: Model specified in Appendix 10.1.

Table 10.2 Typical public transport cost structures (%): bus[13] and rail[14]

Bus		Rail	
Variable (Crew, fuel, oil, tyres)	55	Train operation	25
		Vehicle maintenance:	
Semi-variable (Vehicle maintenance, depreciation)	25	Depreciation	20
		Terminals	15
		Track and signalling	25
Fixed (Garages, overheads)	20	Admin and general	15

another operator to enter the market is ignored in this example). If the aim were to provide the maximum level of bus miles possible, subject to at least covering costs with revenue, these profits would all be spent on providing additional cheap off-peak services (row 2). If, on the other hand, the aim were to provide maximum benefits to users as measured by a cost-benefit analysis (row 3), the surplus would be spent partly on lowering fares and partly on improving both peak and off-peak services. Finally, maximising the difference between benefits and costs without a budget constraint (row 4) leads to some further improvements in service levels and very low fares, but of course a subsidy is now needed.

10.4 Socially optimal pricing and service levels in public transport

Suppose that it is desired to find the socially optimal pricing policy for a public transport operator to follow. In Chapter 4 it was shown that the most efficient pricing policy was to price at marginal cost. Because public transport is a non-storable commodity subject to peaks in demand, the marginal cost in the peak is likely to be very much higher than in the off-peak. In the peak marginal cost can be assessed by considering an increment of capacity (e.g. an extra bus) and dividing the cost of providing and operating it (depreciation, interest, crew, maintenance, fuel, etc.) by the extra passenger miles provided.

In the off-peak one may effectively have a marginal cost of zero if load factors are low. Even if accommodation of extra traffic does require provision of extra services, the marginal cost will be low if there are spare buses and/or crews.

The above all applies to rail as well, but with an extra reason to suppose that marginal cost will be lower than average, since infrastructure costs may be largely fixed up to capacity (see the information on public transport cost structures provided in Table 10.2).

Suppose that it is necessary to raise more revenue than will be raised by marginal cost pricing. Governments may be reluctant to provide high levels of subsidy for public transport because of the damage raising the funds may do elsewhere in the economy, or because of a fear that high levels of subsidy lead to inefficiency (this fear should be minimised if the service is provided as a result of competitive tendering for the franchise to run the service). The best solution is to raise the price most where the price elasticity of demand is lowest, very much as a commercial operator would, but in this case only to the extent necessary to meet the budget constraint. As has been seen, generally the elasticity is lowest in the peak, when most people are obliged to make the journey in question (at least in the short run – in the long run this elasticity may be considerably higher as people adjust their homes and places of work).

The question now arises of what is the optimal level of service to provide. This may be most readily explained by using the concept of generalised cost. In the case of transport services 'price' in the demand equation is commonly replaced by 'generalised cost', which is a linear combination of price and quality of service attributes, e.g.:

$$G = P + vT \tag{10.4}$$

where G = generalised cost,
   T = journey time, and
   v = value of time.

For example, assume that for a particular journey the fare is 50p and the journey time 20 minutes. If the value of time is estimated to be 2p per minute, then the generalised cost of the journey is 90p.

More complex formulations may divide journey time into walking, waiting and in-vehicle time and add in other factors such as the expected value of time spent standing.

Thus demand is a function of generalised cost. Obviously some parts of generalised cost are borne by the transport users themselves (e.g. their own journey time), but an efficient outcome now will require price to equal the marginal cost borne by the operator plus any marginal congestion and other external costs borne by third parties. Measuring the marginal cost of public transport use necessitates taking into account the benefits an increase in output provides for existing passengers. Generally, as the volume of demand builds up, so the frequency of service improves. In other words, far from being an external cost there is an external benefit to other users. Although this is fairly obvious it is often referred to by transport economists as the 'Mohring effect',[4] as it was Mohring who first analysed its consequences for prices and subsidies. The benefit from this should be set against the marginal cost of providing the extra frequency, and may justify provision of a higher frequency than is required simply to provide adequate capacity.

There is a further point to be considered. This is the extent to which reductions of fares or improvements in services on public transport may divert passengers from car. The level of this diversion is determined by the cross-price elasticity of demand, which measures the per cent change in consumption of one good – such as motoring – brought about by a 1 per cent change in the price of another good – e.g. public transport. The cross-price elasticity of demand for car with respect to public transport (e_{CP}) is:

$$e_D = \frac{\%\Delta \text{ in demand for car travel}}{\%\Delta \text{ in demand for public transport}} \tag{10.5}$$

Thus, if $e_{CP} = 0.1$, then a 10 per cent reduction in public transport fares will lead to a 1 per cent reduction in car traffic.

If use of the car is underpriced relative to its marginal social cost, there will be a 'second best' argument for underpricing public transport as well. Table 10.3 shows the

Table 10.3 Benefits of subsidies to London Transport (pence per passenger mile: January 1982 prices)[5]

	Bus	Train	Overall
Average operating cost per passenger mile	17.1	15.1	16.1
Marginal operating cost per passenger mile	13.1	6.2	9.6
less			
Benefits to existing passengers from improved frequency	6.0	1.0	3.5
Benefits to other road users	2.0	4.0	3.0
Appropriate fare	5.1	1.2	3.1

results of a study in London which brings together all the arguments for subsidy used so far, and which found covering some 80 per cent of public transport costs by subsidy to be worthwhile.[5] This argument has been developed further in a computer model.[6]

In very congested conditions reducing public transport fares may yield a worthwhile benefit in terms of reduced congestion despite the fact that the cross-elasticity of demand between car and public transport is rather low. Moreover there is some evidence that public transport quality improvements may be more effective in attracting car users. The extent of this diversion should not be exaggerated, however; typically only 20 per cent or less of passengers attracted by improved public transport services will have come from car, the remainder diverting from other public transport, walking or making totally new journeys. In any event, the road space may quickly fill up again, if there is suppressed traffic as a result of congestion. This reinforces the need for public transport improvements to be seen as part of a package including traffic restraint.

In practice, it is the benefits of improved services to public transport users themselves that provide the strongest theoretical case for subsidy, although these are often politically less influential. Nevertheless, all this leads to a strong theoretical case for public transport subsidies in urban areas. As against that, as has already been mentioned, there are worries that raising the money to pay for public transport subsidies may be costly and that subsidies may 'leak' into inefficiency. One study suggested that up to 60 per cent of subsidies may be wasted in this way.[7] This is best guarded against by competitive tendering for the operation of services wherever this is feasible. In terms of their contribution to the solution of problems of congestion and environmental degradation caused by road traffic, public transport improvements will generally be most effective if combined with other action to restrain road traffic as part of an integrated strategy.

10.5 Public transport provision in practice

The above discussion gives some indication of the principles to be followed when planning public transport services. But there are many further practical issues to consider. For a comprehensive discussion, see reference 8; here they will only be briefly touched upon.

An important factor concerning the attractiveness of services to the general public is the ease with which they may be used. This extends to cover a variety of issues:

- how easy it is to find out about services
- how well the services mesh together into a network, and the ease of interchange at the points where they do
- how easy it is to understand the fare structure and to pay the right fare.

The logic so far might lead one to a set of services which had, for instance, a dense structure of high-frequency services in the peak, a less dense structure and lower frequencies between the peaks, further different service patterns in the evenings and on Saturdays and Sundays, fares which differ by route, origin and destination and time of day. Such a system might fulfil all the criteria set out above but still be unpopular and difficult to use.

Thus practicality limits the extent of the variation in patterns of routes and frequencies that it is sensible to operate. A 'clock-face' timetable whereby services operate at regular intervals for much of the day is easily understood. A fares system which is simple enough to be easily remembered is similarly popular. In the latter context, a simple flat fare which is the same whatever the journey being made is obviously the easiest to remember and will generally aid the speed of ticket sales. This is particularly important where tickets are being sold by the driver. But it does have severe disadvantages if the operator needs to cover a substantial proportion of costs from revenue and trip lengths vary a lot. For then the flat fare will have to be relatively high, and this will discourage the use of public transport for short trips. A better solution is likely to be a zonal one, in which the fare varies according to the number of zones the trip passes through. Zonal fares typically provide for free transfer between services within a certain time limit, and this is attractive, as it avoids penalising people whose trip requires them to change. It can also encourage transfer to rail-based systems with low marginal costs for the trunk part of the trip, with bus used as a feeder.

A further issue to consider is the encouragement of advance sales of multi-journey tickets. Again this is particularly beneficial where tickets are otherwise sold by the driver, in that it speeds boarding. Multi-journey tickets take a variety of forms, from the traditional season ticket, which is only valid between a particular pair of points, through tickets which are valid for a certain number of trips throughout the system to cards which are valid for any number of trips on all or a predefined part of the network. The latter type of ticket has become very popular in recent years; for instance in London the Travelcard is credited with a major contribution to the boom in public transport use in the second half of the 1980s.

All the above ticket types have advantages and disadvantages and these will have to be considered in the light of any particular situation. For instance, the operator of local services in a small town may opt for a single flat fare, perhaps higher in the peak than the off-peak. A larger town may have a simple zone system with a small number of different fares. A large city may have a more complex zonal system with a heavy emphasis on Travelcards.[9]

10.6 Ownership and regulation

As noted above, public transport operations may be in public or private ownership and may be subject to varying degrees of regulation. In most countries, public transport operations are either publicly owned or are subject to strict control on what levels of fares and services may be provided. In either case, usually only a single operator is allowed on a particular route, and frequently a single operator has control of a whole area. This was broadly the position in Britain until 1985, and is still the position in many countries. This may be referred to as a planned approach to the provision of public transport services.

In a planned approach, all providers of transport infrastructure and services are owned or controlled by the state; pricing decisions are governed by social cost-benefit considerations and are subject to political control; and investment and service level decisions are similarly based on social cost-benefit analysis. In principle, therefore, this should allow all decisions to be taken to give an optimal trade-off between the objectives specified. However, decision-taking may not always be competent and may give undue weight to short-term political advantage rather than long-term objectives; and the construction and operation of transport services may be inefficient, because the organisations involved lack strong incentives to achieve high-quality services at minimum cost.

Thus in recent years several countries, and in particular Great Britain, have moved towards a market approach to public transport provision.[10] In a market approach:

- infrastructure is provided and services operated by private sector organisations
- government control is reduced to a minimum, concentrating on safety and environmental regulation rather than on direct control of prices or services
- governments seek to ensure appropriate pricing policies through tax and subsidy mechanisms rather than direct control. Any subsidies are the result of competitive tendering.

This approach tends to emphasise economic efficiency, minimum cost and maximum scope for innovation. However, the resulting pattern of prices and services may fail to achieve either a truly economically efficient pattern of resource allocation (as many externalities are difficult to reflect adequately either in taxes or regulations), and certainly fail to achieve the other objectives of transport policy.

Obviously most countries lie somewhere between the two, although Britain was close to the first pattern in the 1970s and moved closer to the second in the 1980s. A good example of a mixed approach is Sweden, where:

- many transport operators are private sector, although mainline rail services remain a government owned monopoly and there are still some publicly owned bus companies
- provision of road and rail infrastructure is in the hands of government owned agencies, who work to comparable cost-benefit analysis criteria
- freight services are generally only regulated regarding safety and environmental issues
- most local passenger services are planned by regional or local authorities but provided by private or public operators as a result of competitive tendering
- taxes on road vehicles and charges for the use of the rail infrastructure are calculated by government to reflect the social costs of the flow in question.

Given that Britain has been in the lead in moves towards the deregulation and privatisation of public transport, it is worth commenting further on experience to date. In the freight sector this policy appears to have worked well, although there remains concern about the adequacy of environmental and safety regulation. In the passenger transport sector the policy has been much less successful. While deregulation and privatisation have secured an increase in bus miles operated at a very much lower cost, far from securing a diversion of passengers from private transport, bus patronage has actually dropped by 25 per cent since deregulation. It is widely believed that fragmentation of services between a host of operators, lack of information, connections and through ticketing and instability in the pattern of services have been significant causes of declining patronage.[11] Some evidence for this lies in the fact that in London, where the approach has been to introduce competitive tendering for a network of services planned by a single agency, patronage has risen rather than fallen.

Nevertheless, the British government is proceeding with the privatisation of the rail system as well.[12] It might be argued that the structure proposed for the rail industry in Britain, with minimum service levels defined as part of the franchise agreement for passenger services, requirements for the continued production of a national timetable and through ticketing and limited scope for competitive entry, at least in the first few years, should guard against the worst problems experienced in the bus sector. Unfortunately, however, the rail sector appears to have a wide range of new problems of its own. Of these, the complexity of the new regime, the number of contractual arrangements the operator has to enter into and the big increases in the prices charged for basic inputs such as infrastructure and rolling stock appear to be the most important. It may be that the freight sector, where existing BR operations are to be privatised outright and there are no constraints on competitive entry, may benefit more from privatisation; certainly a number of existing BR customers and other companies appear to have an interest in running their own trains, and the fact that most freight trains nowadays carry traffic for a single customer means that the problem of maintaining 'network benefits' is not such an issue for freight traffic. But even here there are high entry barriers in the form of the need to make a safety case, to acquire drivers and guards with appropriate route knowledge and to negotiate satisfactory slots on the infrastructure with Railtrack.

Many other countries are also moving towards the privatisation of rail systems. But in no other case is such a radical fragmentation of the existing rail operator proposed. In most cases, the intention is to split the existing company into an infrastructure company, a long distance passenger company, a short-distance passenger company and a freight company. In some cases, other organisations will be able to enter the market to provide profitable services, or to bid for the franchise to operate local services with a subsidy.

10.7 Conclusions

Rational planning of fares and services can take place with a variety of objectives. Commercial operators will of course be interested in profits, but franchising or tendering authorities may have a variety of other objectives. In all cases, however, sensible decision-taking relies on an understanding of the demand and cost characteristics of the services concerned.

There are many practical considerations which prevent the direct translation of theoretical results on appropriate pricing and service levels into practice. Fares systems and service patterns must be attractive to, and readily understood by, the public at large. This, plus the desire to use public transport in pursuit of the full range of transport policy objectives, means that most governments retain control over the fares and services offered. In bus systems this is perfectly compatible with private ownership of the operating companies, especially if competitive tendering is used to secure provision of services at minimum cost. In rail systems, the issues involved in private ownership are more complex, but a small number of urban light rail systems (such as that in Manchester) are operated on a long-term franchise, and a number of countries, including Great Britain, are planning to privatise their rail systems in their entirety.

10.8 References

1. Armstrong-Wright, A., *Urban transit systems: Guidelines for examining options.* Washington, DC: World Bank Technical Paper Number 52, 1986.
2. Goodwin, P.B., A review of new demand elasticities with special reference to short and long run effects of price changes. *Journal of Transport Economics and Policy*, **26** (2), 155–169.
3. Fowkes, A.S., and Nash, C.A. (eds), *Analysing demand for rail travel.* Aldershot: Avebury, 1991.
4. Mohring, H., Optimisation and scale economies in urban bus transportation. *American Economic Review*, 1972, **62**, 591–604.
5. Fairhurst, M., *Public transport subsidies and value for money.* London: London Transport, 1982.

6. S. Glaister, *Transport subsidy*. Newbury: Policy Journals, 1987.
7. *The demand for public transport: Report of the International Collaborative Study of the Factors Affecting Public Transport Patronage*. Crowthorne, Berks: Transport Research Laboratory, 1980.
8. White, P., *Public transport planning, management and operation*. London: Hutchinson, 1986.
9. Nash, C.A., *The economics of public transport*. London: Longman, 1982.
10. Gomez-Ibanez, J.A., and Meyer, J.R., *Going private – The international experience with transport privatisation*. Washington, DC: Brookings Institution, 1993.
11. Mackie, P.J., Preston, J.M., and Nash, C.A., Bus deregulation ten years on. *Transport Reviews*, 1995, **15** (3), 229–251.
12. Preston, J.M., and Nash, C.A., Franchising British Rail. In A. Harrison (ed.), *From hierarchy to contract*. Newbury: Policy Journals, 1993, 147–165.
13. Webster, F.V., *The importance of cost minimisation in public transport operations*. Laboratory Report SR766. Crowthorne, Berks: Transport Research Laboratory, 1982.
14. University of Leeds/BR, *A comparative study of European rail performance*. London: British Railways Board, 1979.

10.9 Appendix: Alternative objectives for public transport

This appendix provides the model which generates the results in Table 10.4 for the interest of those readers who know enough economics to follow it.

Suppose that a public transport operator faces the following demand and cost functions:

$$\text{Log } Q_P = 4.05 - 0.05 P_P - 0.3/B_P \tag{10.6}$$

$$\text{Log } Q_O = 4.05 - 0.1 P_O - 0.7/B_O \tag{10.7}$$

$$C = 80 B_P + 20 B_O + Q_P + 0.5 Q_O \tag{10.8}$$

where Q = passenger miles carried (thousands per week), P = fare per passenger mile (p), B = bus miles run (thousands per week), and the subscripts P and O refer to peak and off-peak operations respectively.

It will be noted that elasticities of demand with respect to fare and bus miles are given by:

$$P \frac{\delta \log Q}{\delta P} = \beta P \tag{10.9}$$

where $\beta = 0.05$ (peak) and 0.1 (off-peak), i.e. demand is price inelastic up to a fare of 20 (peak) and 10 (off-peak)

$$B \frac{\delta \log Q}{\delta B} = \frac{\gamma}{B} \tag{10.10}$$

where $\gamma = 0.3$ (peak) and 0.7 (off-peak).

Demand is quality inelastic down to a level of bus miles of 0.3 (peak) and 0.7 (off-peak). (Elasticities are, for simplicity, assumed to be independent. This is, however, undoubtedly unrealistic.)

CHAPTER 11

Freight transport planning – an introduction

C.A. Nash

A substantial part of the growth in road traffic in most countries consists of freight traffic. Although it is not concentrated on peak urban roads, freight traffic is still widely perceived as a problem. On motorways and trunk roads it takes a considerable amount of capacity, adding to the problem of providing for traffic growth. Off the trunk network, it is seen as an environmental and safety problem, contributing substantially to problems of noise, vibration, particulates, visual intrusion and fear. On the other hand, it is obviously recognised that freight vehicles play an essential role in distributing goods from factories, and in supplying shops. Any actions which make it more difficult or expensive to service such facilities in the area may have repercussions for jobs, and for prices in the shops (although freight transport costs are generally not a large part of the total costs of supplying most commodities, so any claims for dramatic repercussions need to be carefully examined).

In this chapter the factors behind the growth of road freight transport are first considered, and then possible ways of alleviating its effects are examined. Concentration will be on public policy issues rather than on planning methods used by the industry itself.

11.1 Trends in freight transport

In most countries, freight transport shows a tendency to grow at least in proportion with gross domestic product (a measure of the output of the economy). It may appear obvious why this is: as GDP grows so there are more goods being produced which require transport between factories, depots and shops.

In fact, in a modern economy, the relationship is not nearly this obvious. A substantial part of the output of a modern economy consists of services, such as healthcare, education, banking and insurance, which give rise to passenger transport rather than freight. Within manufacturing, there is a tendency for expansion to be concentrated on high-value products such as electronics, which lead to a low level of tonnes to be transported per unit of value of output. One might therefore expect that freight transport would grow very much more slowly than GDP.

On the other hand, there are some countervailing tendencies.[1] Market areas are expanding, as national and international barriers to trade decline; reductions in transport cost are of course an important part of this tendency. Production is becoming more specialised on any particular site, so that, for instance, inputs into the manufacture of motor vehicles in Britain may now be sourced all over Europe. Distribution systems are becoming national – or international – with stockholding concentrated at a very small number of locations. Just-in-time distribution systems are increasingly being used as a way of reducing the costs of stockholding, but these require that goods be delivered when they are needed, making necessary distribution systems that combine frequency of service with a very high reliability.

The outcome is that the demand for freight transport is still rising rapidly. Each tonne of goods is shipped several times during the course of its development from raw materials through intermediate goods to final products, and lengths of haul for each shipment are rising rapidly. The concentration on frequent small consignments of high-value goods also tends to favour road transport rather than rail or water; the latter both tend to specialise much more in low-value bulk commodities which do not require the same high quality of service.

11.2 Roads and economic growth

It is often argued that the needs of the road freight industry should be paramount when planning roads. This is in order to minimise costs of the road freight industry, and in that way to maximise the competitiveness of production in the area concerned.

A number of studies have found a statistical link between investment in roads and economic growth at an aggregate level.[2] However, there are several reasons why such a link could exist: for example, government spending on any project tends to promote economic growth through the standard 'multiplier' effect (it directly creates jobs and incomes, which in turn are spent on other goods and services, thus creating more jobs etc.); and it may be that rapid rates of economic growth lead to high investment in roads (both by creating the demand for them and providing the resources for such investment), i.e. that the direction of causation is the reverse of that postulated.

Nevertheless, road investment does undoubtedly reduce the costs of the road freight industry, and this may in turn make industries in locations so favoured more competitive and promote relocation to places where transport costs have fallen. Particularly in developing countries with a very poor existing road network these points may be very significant. On the other hand, in a country such as Britain, where the transport system is well developed and where transport costs are on average only some 8 per cent of the final delivered price of goods and services, the extent to which competitiveness may be stimulated by even major road building exercises is very limited. For evidence on this, see studies of the M62 motorway[3] and the Humber Bridge.[4] Also, the relative efficiency of road building and other forms of assistance to industry must be considered.

In some circumstances, road building may actually be counter-productive. For instance, where remote areas have retained local production and distribution facilities largely because of the time and cost of bringing in goods from outside, improved links to the outside world may actually lead to centralisation of facilities in order to exploit economies of scale. Only if the remote area has some particular advantage as a production location will improving its connections to the outside world stimulate production.

So far it has been argued that the overall effects of road improvements on economic performance are likely to be small. One important counter-argument to the discussion in this section is that transport costs are only a small part of total distribution costs.

Warehousing and stockholding costs usually make up more than half of the total costs of distributing products. By allowing for faster and more reliable deliveries, improved roads may permit economies in distribution networks by means of further concentration of warehousing and stockholding on a small number of locations. A recent study examined this hypothesis for a major brewery and a supermarket chain, and found some evidence that it is the case, but again the additional benefits appeared relatively small.[5]

In summary, then, the impact of road improvements and other transport projects on the freight sector must be carefully considered, as it will certainly effect the prices of goods and the relative competitiveness of alternative locations. The evidence is, however, that in a developed economy these effects are relatively small and should not lead to an automatic assumption that road planning should be heavily geared to what the freight industry would like to see regardless of other costs.

11.3 Policy issues

Historically, governments have intervened in the freight market for a range of reasons:[6]

- control monopoly power
- benefit specific regions or industries
- permit the exploitation of economies of scale on the railways
- relieve traffic congestion
- protect the environment
- reduce accidents
- save energy
- prevent unfair competition
- protect consumers and/or workers in the industry
- reduce wear and tear on the roads.

In the past, governments tended to regulate (or own outright) freight transport operators in part because they were seen as holding significant monopoly power. This was particularly true of rail operators, whose prices were usually controlled by government. It also applied to road haulage, where the number of vehicles allowed to operate for hire and reward was typically limited, and in many countries road haulage charges were also regulated. Over the past 20 years, however, most countries have dismantled this regulatory framework, and today the road freight transport industry is usually privately owned and highly competitive. In a number of countries, rail freight operations are also being privatised, or new private operators allowed access to the infrastructure. The rationale for controls has now shifted much more towards issues of congestion, safety and environmental protection.

The main weapons governments have are regulatory and control mechanisms, taxes and subsidies, and traffic management measures.

11.3.1 Construction and use regulations regarding vehicles

These specify the basic characteristics of vehicles that will be allowed to operate (e.g. maximum gross weight, maximum axleweight, height, length).

This is obviously important in terms of the costs of constructing and maintaining roads; for instance, heavier lorries require stronger bridges, and heavier axleweights require thicker pavements. Such regulations may also control emissions of noise and air pollutants.

11.3.2 Drivers' hours regulations

The regulations are designed to ensure that safety is not threatened by drivers who have taken inadequate rest.

11.3.3 Quality and/or quantity control on entry to the industry

There has been a shift away from controls on number of vehicles entering the market towards tougher quality controls and environmental regulation.[7] The intention has been to promote efficiency through greater competition. In practice, it is not clear that controls on the number of vehicles greatly restricted growth of the industry, partly because in most countries they only applied to professional road haulage. Customers could still acquire their own vehicles in order to move their goods by road.

11.3.4 Taxes and subsidies

The level of taxes on goods vehicles, and especially those with heavy axleweights which do most damage to the road pavement, has long been a controversial issue. The level and structure of road taxes is an important instrument for influencing both the overall level of road haulage costs, and hence the size of the market, and the relative costs of operating different types of vehicle. Since taxes typically comprise at most 20 per cent of the overall costs of road haulage, and the overall market is seen as very price inelastic, it may be that the latter effect is the more significant. Thus it is important to have a tax structure which gives adequate incentives for the use of vehicles which are less damaging to the road pavement and to the environment.

At the same time as penalising environmentally damaging modes of transport by the tax system, governments may wish to aid more environmentally friendly modes via grants. These may cover either capital or operating costs (or both). For instance, in Britain grants are available to cover both capital costs of equipment such as private sidings and rolling stock and charges for use of the infrastructure where rail offers significant environmental benefits compared with road.

11.3.5 Tariff controls

In the past, tariff controls have been used both to prevent exploitation of monopoly power (maximum tariffs) and to prevent excessive competition leading to instability in the industry (minimum tariffs). Both are now seen as reducing the effectiveness of prices as market signals, and governments now generally prefer to influence the level of road haulage rates through the tax system.

11.3.6 Traffic management measures concerning where and when heavy goods vehicles may be used

Proposals for controls on lorry use may take a number of forms, but in each case there is a trade-off to be made between the environmental, congestion and safety benefits that would result and the additional costs. For instance, British Department of Transport studies suggested that confining lorries to a national network of 5000 km (3100 miles) of road (except for access) would add 25 per cent to road haulage costs.[8] Lorry bans (except for access) often deal with local problems of short cuts, but some schemes cause major rerouting with substantial costs and benefits. Total bans above a certain weight may be by time – for instance the French and Swiss have a total ban on movement on Sundays; by place – for instance the restrictions on the size of vehicles that can be brought into cities; or by time and place – for instance restrictions in Paris on the hours at which heavy goods vehicles can be operated.

Restrictions may, however, lead to increases in distribution costs. A major study for London[9] examined the following options: (a) banning lorries above 7.5, 16 or 24 tonnes; (b) banning them from all roads within the M25 motorway, from all but some radials, and the North Circular Road; and (c) banning them at all times, at weekends only, or at nights only. The study found that a big problem in estimating costs and benefits related to predicting how shippers would react. They might, for example, switch to smaller vehicles for the entire journey. This would have little cost if 24-tonne vehicles were permitted, but 16- or 7.5-tonne restrictions would lead to large increases in both traffic and costs. It was not clear whether a larger number of smaller vehicles would be seen as environmentally preferable. There was also considerable controversy on the issue of how vehicle numbers would change. Would the smaller vehicles operate at higher load factors, so that the expansion would not be nearly as great as would be predicted simply from looking at their relative capacities? Moreover, for many goods it is the cubic capacity of the vehicle that is the constraint on how much the vehicle can take rather than the weight.

Also, it might be possible to tranship outside London from heavier vehicles or rail. It was found that trunk transhipment costs averaged £4 per tonne, but there could be countervailing advantages from the use of larger vehicles (or even rail) for trunk haul to minimise trunk haul costs while using smaller vehicles for local delivery. Demountable bodies or drawbar trailers might be used to facilitate this transfer. Finally, and this is the fear many local authorities have, haulage companies might relocate or close completely.

The Greater London Council finally implemented a night-time ban although enforcement (which passed to individual boroughs with the abolition of the GLC) was never very effective.

11.4 Potential for rail and water

Rail (and even more so water) freight is frequently seen as less environmentally damaging than road freight. It frequently offers better energy efficiency, the possibility of electric traction, lower noise levels and a better safety record. However, transferring freight from road to rail is not easy, given the trends in the freight market described above. For bulk commodities which may be carried throughout by rail from a siding at the point of origin to another siding at the destination, rail may be both environmentally attractive and cost effective if the volume to be transported is large. With smaller volumes it is not feasible to connect the origin and destination direct to the rail network, and in any case the costs of transporting small volumes in wagonloads that have to be shunted into full train-loads by combining them with traffic from a wide variety of points are often prohibitive. Thus such traffic usually has to be collected and delivered by road, and transhipped between rail and road at a depot. Again, this adds substantially to the costs, and can generally only be worthwhile where there is a relatively long distance on the trunk haul mode, so that low costs on the trunk haul offset the high costs at each end. Moreover, terminals need to be well located; otherwise use of rail or water may actually lead to an increase in the distance travelled by lorries in congested urban areas.

Thus the tendency now is for the use of rail for manufactured goods to rely on some form of intermodal service, using either containers or swap bodies (where the load carrying unit is readily transferred from road to rail) or piggyback (where the entire lorry semi-trailer – and sometimes the tractor as well) is transported on a railway wagon. The advantage of the latter is that no specialised equipment is needed; its disadvantage is an increase in the deadweight to be hauled when on rail or water, and a problem where (as throughout Britain) loading gauges are too restrictive to carry ordinary semi-trailers on truck without infrastructure alterations. Intermodal services are still generally only competitive with road over longer distances, and although their use in many countries within Europe and North America is growing rapidly, they have not achieved the market penetration that is desired by many governments in order to reduce the impact of long-distance road freight transport.[10] Within Europe, a key problem is achieving high-quality services for international movements involving a number of different railway companies; although European legislation permits new entry by international intermodal operators, who have a right of access to the rail infrastructure of all member countries, this has so far been little used.

In Britain, as in many countries, rail freight is required to operate without subsidy, but customers can apply for the so-called Section 8 grants (strictly now Section 139 grants under the 1993 Railways Act) towards the costs of facilities needed in order to be able to use rail or water transport and for a grant (under Section 137 of the same act) to cover track costs if use of rail or water transport rather than road confers environmental advantages. The case for the grant rests on the number of lorry miles removed from the road system, with a higher level of grant being payable if the miles are on environmentally sensitive roads.

11.5 Conclusions

Tackling the problems posed by the expansion of freight transport requires a number of measures. Some of these (for instance the levels of taxation on goods vehicles, or the support given to the national rail network) lie within the province of national government, but others, such as lorry routing and lorry bans, are very much a part of local transport planning. A whole variety of measures are possible, but need to be carefully weighed up to assess their cost effectiveness, remembering that with a competitive road haulage industry, any measures imposed which raise costs are likely to find their way very quickly through into the prices of goods in the shops.

11.6 References

1. McKinnon, A., *Physical distribution systems*. London: Routledge, 1989.
2. Aschauer, D.A., Is public expenditure productive? *Journal of Monetary Economics*, 1989, **23**, 177–200.
3. Dodgson, J.S., Motorway investment, industrial transport costs and sub-regional growth: A case study of the M62. *Regional Studies*, 1974, **8**, 91–95.
4. Mackie, P.J., and Simon, D., Do roads benefit industry? A case study of the Humber Bridge. *Journal of Transport Economics and Policy*, 1986, **20** (3), 377–384.
5. Mackie, P.J., and Tweddle, G., Measuring the benefits gained by industry from road network improvements – two case studies. *Proceedings of the PTRC Summer Annual Meeting: Seminar H,* 1992, 103–115.
6. Foster, C.D. (Chairman), *Road haulage operators' licensing. Report of the Independent Committee of Inquiry.* London: HMSO, 1978.
7. ECMT, *Deregulation of freight transport*, Paris: ECMT Round Table 84, 1991.
8. Bayliss, B.T., *Planning and control in the transport sector.* Aldershot: Gower, 1981.
9. Wood, D. (Chairman), *Report of the Committee of Enquiry into Lorry Routing in London.* London: Greater London Council, 1983.
10. Fowkes, A.S., Nash, C.A., and Tweddle, G., Investigating the market for inter-modal freight technologies. *Transportation Research*, 1991, **2** (4), 161–172.

PART II

Traffic surveys and accident investigations

CHAPTER 12

Issues in survey planning and design

P.W. Bonsall

Data are an essential input to the effective planning and design of transport systems, either directly by describing the current state of the system, or indirectly by allowing the calibration of models which yield insights into the processes at work in the system or help to predict how the system is likely to perform in the future with and without policy intervention. This chapter indicates how, with careful attention to the initial planning

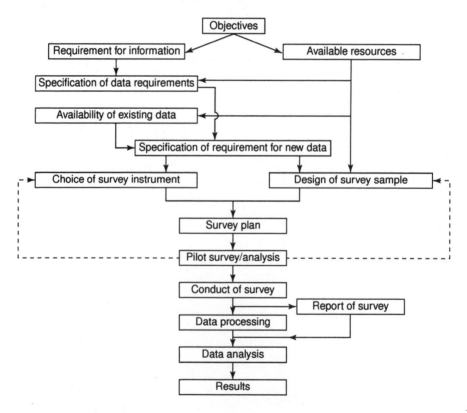

Fig. 12.1 Stages in the design and conduct of a survey

and design of a survey, it can be tailored to meet the specified task. Chapters 13, 14 and 15 will discuss the techniques involved in specific types of survey.

Figure 12.1 indicates the main stages in the design and conduct of a transport survey. It begins with a consideration of the objectives of the exercise which determine the requirement for information and the availability of resources.

12.1 Defining the data requirements

The requirement for information needs to be refined into a precise specification of data requirements in terms of the variable(s) to be studied and the hypotheses to be tested. The specification of the variables and of hypotheses at this early stage very usefully concentrates the mind on the real purpose of the survey and ensures that subsequent effort is properly targeted. The process may well involve interaction between the survey planner and the commissioning agency in order to ensure that the needs of the latter have been correctly interpreted. Thus, if an initial brief asked for a survey of speeds on such-and-such road, the survey planner will want to know whether the interest is in average speeds (rather than, for example, the number of cars exceeding a specified limit), speeds at a point (rather than speeds along a short stretch of road), speeds of all vehicles (rather than just those which are free-flowing) and so on.

The precise specification of an experimental hypothesis removes any ambiguity; for example the hypothesis might be that 'the mean speed of cars passing location x has decreased since the installation of new speed limit signs'. Specification of a hypothesis can be a useful discipline even when the survey is intended to be speculative or exploratory in nature; for example, a brief to study residents' attitudes to a proposed traffic calming scheme might be made more concrete by specifying hypotheses such as that 'most adults resident in area x are in favour of the proposals' or that 'the visual appearance of the proposed scheme is a serious concern to more than 10 per cent of residents in area x'.

Strict experimental method requires that each hypothesis should be accompanied by an equivalent null hypothesis whose validity is to be tested. Thus, if the hypothesis is that speeds have decreased, the null hypothesis might be that they have not decreased or alternatively that they have increased. The difference between these alternative null hypotheses is that the first is less restrictive, and therefore less likely to be rejected than the second. (A fuller description of the issues involved in specifying hypotheses and null hypotheses may be found in standard statistical textbooks but a useful discussion of their relevance to transport and traffic studies can be found in references 1 and 2).

12.2 Secondary sources

Once the data requirements have been specified it is important to consider whether they can be met by making use of existing data, thereby avoiding the need for a special survey. Use of existing data or 'secondary sources' can save time and money and is particularly desirable when a survey is likely to cause disruption and annoyance to the travelling public. (On the other hand, of course, there will be some circumstances where, even though secondary sources may be available, it is important for political reasons to be seen to be conducting a survey.) The three main sources of secondary data are published databases, previous local area surveys and data produced as a by-product of a control or management system. Each of these will be briefly considered while noting that more detailed information can be found in reference 3 or in occasional publications by bodies such as the Transport Statistics User Group.[4]

A wide range of data is available in *published tabulations* or in *commercially available databases*. Much of it is collected by governmental agencies as part of regular monitoring exercises and made available in summary form in publications such as reference 5. Such data are very useful for establishing background trends or contextual information on issues such as levels of car ownership, occurrence of accidents or seasonal trends in traffic flow, but they may be of little use for local area studies. Increasingly, however, it is becoming possible, at a price and subject to privacy restrictions, for analysts to access the disaggregate data or to request special tabulations to suit their particular purposes. Before doing this, of course, they should establish that the variable definitions are appropriate and the sample sizes adequate.

Detailed *local data* are likely to be available only if there has been some previous survey in the study area. This may have been part of a local monitoring programme or part of an ad hoc study. In either case the analyst will need to establish that the variable definitions and sample characteristics are appropriate and that the data have been archived in an accessible format. It is not unusual for a potentially useful database to remain untapped because of inadequate documentation or eccentric archiving conventions. Attempts to make use of data from previous ad hoc studies are particularly fraught in this respect, even when the data were collected by one's own organisation. Increasing use of Geographical Information Systems (GIS), as described in Chapter 13, should reduce this problem.

Recent years have seen the emergence not only of databases containing the results of national and local traffic monitoring programmes but also of *databases derived as a by-product of management and control systems*. Examples of these 'by-product' databases include traffic flow and congestion captured during the operation of fully automated urban traffic control systems, usage of car parks with computer controlled entry and exit barriers, flows of vehicles past toll points and flows of passengers through automatic ticket barriers within LRT or rail systems. Users of such data must satisfy themselves that the data are statistically robust and representative of the system they wish to study. Some of these data, particularly those relating to use of car parks, toll roads and other charged facilities, may be commercially sensitive whereas some may be available fairly freely; for instance, one can obtain real-time data on traffic flows and congestion on certain Californian freeways free of charge on the Internet.

12.2.1 Resource requirements

Assuming that, after careful consideration of all these secondary sources of data, the need for further data collection is established, the next stage is to produce a detailed specification of this requirement. The specification must take account of the resources available since they may seriously constrain what can be achieved in terms of accuracy or coverage. It is part of the survey planners' role to alert the commissioners of the study to any problems at an early stage so that they can decide to increase the budget, reduce the specification or abandon the whole exercise.

If there is no chance, within the budget allowed, of achieving the minimum required accuracy, it would be a waste of resources to proceed.

12.3 Choice of survey instrument

Once a decision to proceed has been made the survey planner must define the survey instrument. Definition of the survey instrument involves choosing the most appropriate technique – a choice which is likely to be heavily constrained by the resources provided and by the extent of in-house experience with the various techniques available. Chapters 13 to 15 give details of some of the main techniques, indicating their particular strengths and weaknesses.

It must be emphasised that while some techniques may be ideal for collection of one type of data it may sometimes be appropriate to use another technique which, while not so efficient for that item of data, may yield additional data at relatively little extra cost. This is, for example, the logic underlying the increased popularity of video-based surveys.

12.4 Design of sampling strategy

Design of the sampling strategy is intimately associated with the definition of the experimental hypotheses and of the variables of interest and involves definition of the sampling units, the target population, the sampling frame, the sampling method and the sample size.

The *sampling units* are the basic units whose characteristics or behaviour are to be logged. They might be individual 'actors' in the transport system (e.g. citizens, travellers, vehicles, or companies), individual locations (e.g. junctions, links, car parks, zones or geographical points), or individual test samples (e.g. a sample of air content, or a sample of sound levels).

Definition of the *target population* indicates which sample units are 'in' the survey. If the data requirements have been well specified the definition of the target population should be quite straightforward, for example 'cars using such-and-such stretch of road between 7 a.m. and 7 p.m. during the summer months', but some tightening up may be required to remove possible ambiguities (for instance, which are the 'summer months'?).

The *sampling frame* is a 'register' of the *target population* which defines all the sampling units within the target population and which provides the framework for the sampling process. This 'register' may exist in tangible form (e.g. as a list of residents of a particular zone or a list of registration plates of cars seen at a particular site) or may exist only in an abstract way (e.g. as a notional list of pedestrians crossing a particular stretch of road during the survey period).

12.4.1 Sampling methods

The choice of sampling method will depend on the objectives of the survey and perhaps on the survey technique being employed. Some commonly used methods are: true random, systematic, stratified and cluster sampling.

True random sampling ensures that each sampling unit has an equal chance of being selected. It is a theoretically attractive method, not least because it keeps the subsequent analysis simple, but since it involves allocating random numbers to each sampling unit to determine whether that unit is to be included, it is not really practical for surveys of 'live' events. The use of random sampling is therefore usually restricted to off-line surveys, such as household interviews, for which the sampling frame exists as a document.

Systematic ordered sampling involves the selection of every *n*th unit from the sampling frame (e.g. for a 20 per cent sample it might be every fifth unit). This concept is easily understood and implemented even by inexperienced field staff and is therefore widely applied in surveys of live events. The method is random in as much as that until the first unit is selected, all units have equal probability of being chosen, but it is not truly random and may produce biased results if the sequence of units has any significance – as for example in a queuing system where capacity is provided in discrete lumps (as might be the case in a saturated network with linked signals).

Stratified sampling involves division of the population into groups on the basis of some characteristic and applying a different sampling rate in each group. The method is usually applied when it is necessary to ensure adequate representation of a minority within the population (one might, for example, need to have a higher sample rate for motorcyclists in a speed survey if the speed of motorcyclists needs to be separately identified in the results). If used on the basis of a prominent but supposedly irrelevant characteristic (such as vehicle colour) it can provide a practical means of 'random' sampling – thus if approximately one in ten vehicles is blue and one wants a 10 per cent sample of vehicles one might instruct the field staff to survey all blue vehicles and no others.

Cluster sampling involves selecting groups of adjacent units (e.g. addresses in a given street or a group of vehicles following one another in a traffic stream). This technique usually results in increased survey efficiency, offers the possibility of studying interactions between adjacent units and can provide an 'enriched' sample in areas of particular interest (for instance, an opinion survey might be targeted on residents in politically marginal wards) but it obviously risks biasing the survey results.

12.4.2 Sample sizes

The final stage in the process of sample design is to determine the sample size or sampling rate. Using the common assumption that the data are normally distributed, the sample size (n) required to estimate the population mean $(\bar{x})$ for a large (effectively infinite) population is:

$$n = \frac{S^2}{se(\bar{x})^2} \qquad (12.1)$$

where S^2 = the sample variance (the best estimate of the true variance σ^2), and $se(\bar{x})$ = standard error of the population mean.

Where the population size (N) is finite, the equation becomes:

$$n = \frac{S^2 / se(\bar{x})^2}{1 + (S^2 / se(\bar{x})^2 / N)} \qquad (12.2)$$

The problem, of course, is that $se(\bar{x})$ will not be known before conducting the survey. The problem is solved by making best use of any information available and by taking a view on how much uncertainty can be tolerated in the result. Thus information from a previous survey or a pilot survey can be used to estimate S^2 and the properties of the normal distribution, together with one's attitude to certainty, can be used in lieu of $se(\bar{x})$. In a normal distribution a known proportion of the population will be found within a given number of standard deviations of the true mean(μ); for example, 95 per cent will be within $\mu \pm 1.96 \, \sigma$, while 90 per cent will be within $\mu \pm 1.64 \, \sigma$.

The above property can be expressed in general terms as:

$$se(\bar{x}) = p\mu/t \tag{12.3}$$

where p is the permitted error expressed as a percentage of the true mean (i.e. the smaller p becomes, the greater the precision achieved), and t is a value, derived from standard normal distribution tables, depending on the required degree of confidence in the result (i.e. the proportion of the population which is required to lie within the permitted range).

Thus if the permitted error is ± 10 per cent and it is desired to be 95 per cent confident that the true result lies in the range $\mu \pm 0.1\mu$ then: $se(\bar{x}) = 0.1\mu/1.96$, whereas if the permitted error is only ± 5 per cent with 90 per cent confidence, then $se(\bar{x}) = 0.05\mu/1.64$.

Substituting Equation 12.3 into Equations 12.1 and 12.2, the required sample size for a survey in an effectively infinite population is:

$$n = \frac{S^2}{(p\mu/t)^2} \tag{12.4}$$

while for a finite population it is

$$n = \frac{S^2/(p\mu/t)^2}{1+[S^2/(p\mu/t)^2/N]} \tag{12.5}$$

As S/μ is the coefficient of variation (cv), Equations 12.4 and 12.5 can also be expressed as:

$$n = \left(\frac{cvt}{p}\right)^2 \tag{12.6}$$

for an effectively infinite population, and

$$n = \frac{(cvt/p)^2}{1+[(cvt/p)^2/N]} \tag{12.7}$$

for a finite population.

These equations reflect the fact that the required sample size is positively correlated with the amount of noise in the data, the required precision of the result and the required confidence in that result.

The value of t in Equations 12.3, 12.4 and 12.5 depends on whether one is concerned with a 'one-tailed' or a 'two-tailed' test, i.e. whether the null hypothesis is defined to identify differences in one direction or two. (A null hypothesis phrased to say that x *is less than y* would require a one-tailed test, while a null hypothesis to say x *is not equal to y* would require a two-tailed test). Note that Equations 12.1 to 12.7 relate to situations where it is reasonable to assume that the data are normally distributed in the population and that the sample is unbiased; where this is not the case matters can be more complicated and beyond the scope of this book – interested readers should consult a specialist text on sampling method or (for a transport-oriented introduction) references 1 or 2.

Example 1

Suppose that it is desired to estimate the average daily flow of traffic past a site and to be 95 per cent confident that the answer obtained is within 10 per cent of the correct value. It is suspected that the flow varies quite markedly from day to day. For how many days will it be necessary to collect flow data?

It is first necessary to obtain an estimate of the daily variation. Suppose that, on the basis of data from similar sites elswhere, the coefficient of variation is expected to be 0.3. Given that the population (= days) is very large it is appropriate to use Equation 12.6 rather than Equation 12.7. The required number of days is thus $(0.3 \times 1.96/0.1)^2 = 34.6$ which is rounded to 35 days. If the daily variation had been much less (say $cv = 0.1$) the result would have been much lower: $(0.1 \times 1.96/0.1)^2 = 3.8$.

Alternatively, even with a cv of 0.3, the required number of days could be reduced by accepting a lower degree of confidence and/or a wider margin of error. Thus if an answer with 90 per cent confidence and a result within 20 per cent of the correct value had been deemed satisfactory, the requirement would have been $(0.3 \times 1.64/0.2)^2 = 6$.

Example 2

Suppose that it is desired to estimate the average household trip rate in a large city and to have 95 per cent confidence that the answer obtained is within 10 per cent of the true mean. How many households need to be surveyed?

If prior estimate of the coefficient of variation on trip rates is 0.5, and the number of households is assumed to be very great, then the answer is $(0.5 \times 1.96/0.1)^2 = 96$ households.

If instead of a large city the survey is to be carried out in a small village with only 100 households then the calculation should use Equation 12.7 to allow for the fact that the population is of finite size 100. The result is then $(0.5 \times 1.96/0.1)^2/\{1 + [(0.5 \times 1.96/0.1)^2/100]\} = 49$.

12.5 The survey plan

Having chosen the survey instrument and defined the sampling strategy, the survey itself can now be planned. The plan should comprise a detailed schedule of all the procedures and stages required in the implementation of the survey and the production of reports. Figure 12.2 shows an example of such a schedule. Note the inclusion of administrative details such as discussions with third parties such as the police, staff recruitment/training and a documentation schedule as well as more fundamental items such as a pilot survey and the survey itself.

A good survey organiser should seek to ensure the smooth running of the survey schedule by careful advance checking of all equipment and by using only reliable staff and suppliers. Inevitably, there will be unforeseen problems and upsets and so a certain amount of slack should ideally be built into the schedule to allow for this. In practice, however, the survey organiser will often have to balance the desirability of a generous schedule against the demands of the analysts for data to be made available as soon as possible.

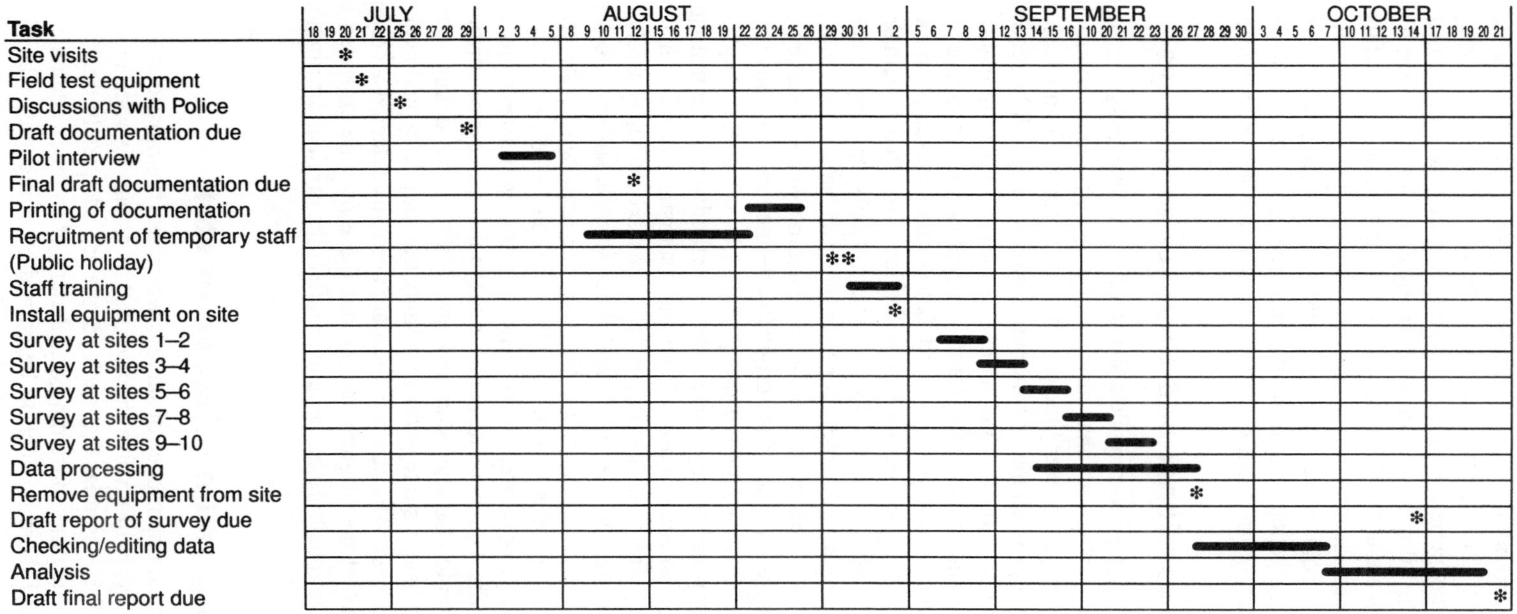

Task

Task	JULY 18 19 20 21 22 25 26 27 28 29	AUGUST 1 2 3 4 5 8 9 10 11 12 15 16 17 18 19 22 23 24 25 26 29 30 31	SEPTEMBER 1 2 5 6 7 8 9 12 13 14 15 16 19 20 21 22 23 26 27 28 29 30	OCTOBER 3 4 5 6 7 10 11 12 13 14 17 18 19 20 21
Site visits	*			
Field test equipment	*			
Discussions with Police	*			
Draft documentation due	*			
Pilot interview		▬		
Final draft documentation due		*		
Printing of documentation		▬		
Recruitment of temporary staff		▬ **		
(Public holiday)				
Staff training		▬ *		
Install equipment on site			*	
Survey at sites 1–2			▬	
Survey at sites 3–4			▬	
Survey at sites 5–6			▬	
Survey at sites 7–8			▬	
Survey at sites 9–10			▬	
Data processing			▬	
Remove equipment from site			*	
Draft report of survey due				*
Checking/editing data				▬
Analysis				▬
Draft final report due				*

Fig. 12.2 A typical survey schedule (source – reference 1)

A *pilot survey* provides an opportunity for checking that all procedures, documentation and instruments are adequate, and for fine tuning where necessary. If the survey team has previous experience with this type of survey and if no innovations have been introduced then the pilot may become a formality – unless of course it is being used to gather information about the variance within the population as part of the sample design process. If, however, the survey has innovative aspects the pilot survey becomes a crucial part of the whole exercise. It is at this stage that problems with the survey instrument, the documentation and the staff training should be revealed so that remedial action can be taken, or the main survey postponed or abandoned before it is too late.

12.6 Cross-sectional and time series surveys

Surveys may be designed to provide a snapshot picture of the system state at a particular point in time (a 'cross-sectional survey') or may be intended to provide a picture of the evolution of the system over time (a 'time series survey'), in which case a series of repeat surveys, or a programme of continuous monitoring, will be required.

If a survey is to be part of a time series it is clearly important that the definitions of the sampling units and the sampling frame are kept constant and that potential bias due to changes in survey procedures is avoided. Selection of the sample for a time series survey requires particular care because its definition will need to remain valid over the entire time period – possibly several years. There are two basic approaches to this issue: one involves drawing a new sample for each wave of the survey while maintaining the same sampling rules (this is termed a 'repeat cross-section survey') and the other involves repeated observation of the originally defined sample (this is termed a 'longitudinal survey' or, if the sample is a group of people, a 'panel survey').

Longitudinal or panel surveys have a theoretical advantage in that, since the sample remains constant, a potential source of unwanted statistical variation is removed and any differences detected from one survey period to the next can fairly safely be attributed to an underlying change. On the other hand these surveys suffer from a particular phenomenon known as 'sample attrition' whereby the units within the originally defined sample become ineligible or unavailable. For example, a household may decide that it no longer wishes to co-operate, it may move to a new address outside the study area or its members may die. Unless some action is taken this process of attrition will lead to a change in the sample characteristics and, eventually, to its demise. One approach to avoiding this is to have a strategy for refreshing the sample by adding new members to replace those who leave. The new members are normally selected according to the same criteria used to select the original sample but these may be amended if it is thought appropriate to try to rebalance the sample to reflect the original distribution of characteristics.

12.7 Training and motivation of staff

Despite the increased importance of surveys which use 'automatic' methods of data collection, most surveys will involve the use of staff as survey assistants, enumerators or interviewers and it is essential that such people are adequately trained and motivated. Training can be quite time-consuming, particularly if complex equipment or procedures are to be used, and there is much to be gained by using staff who already have the necessary experience even if they command a higher rate of pay. Similarly, if the task requires certain aptitudes such as keyboard skills or technical knowledge, this should be taken into account during recruitment.

Proper motivation of staff will be influenced by a number of factors including conditions of service and rates of pay but will most crucially depend on the design of duties and rotas which reduce the chances of staff becoming tired, bored, hungry or cold. More generally it requires a management style which convinces them of the value of the exercise and gives them confidence in its accuracy. Staff should understand the importance of their own contribution and may be motivated by the knowledge that their contribution can be identified in the data.

12.8 Administration

Even a well designed survey with a well defined purpose, an adequate sampling strategy and a well trained and motivated staff can fail if the administration is poor. Good administration involves preparation of a realistic and detailed schedule and maintenance of comprehensive records of all procedures and documentation involved in the design and conduct of the survey.

12.9 References

1. Taylor, M.A.P., Young, W., and Bonsall, P.W., *Understanding traffic systems*. Aldershot: Avebury, 1995.
2. Richardson, A.J., Ampt, E.S., and Meyburg, A.N., *Survey methods for transport planning*. Melbourne: Eucalyptus Press, 1995.
3. Bonsall, P.W., Transportation and traffic planning and engineering. In K. Mildren and P. Hicks (eds), *Information sources in engineering*. 3rd edition. London: Bowker-Saur, 1996.
4. *Transport Statistics Users' Group directory of data sources*. 2nd edition. London: TSUG, 1995.
5. Department of Transport, *Transport statistics Great Britain*. London: HMSO, London, 1994 edition.

CHAPTER 13

Observational traffic surveys

P.W. Bonsall and C.A. O'Flaherty

Surveys which do not require the active involvement of the public can be termed observational surveys with the 'observation' being effected automatically or by trained survey staff. The most common of the observational surveys are inventory and condition surveys, traffic flow surveys, speed, travel time or delay studies, parking surveys, origin–destination surveys, pedestrian and cyclist movement studies, and environmental impact studies.

13.1 Inventory and condition surveys

As indicated in Table 13.1, a comprehensive transport system inventory normally requires details of the networks, infrastructure and facilities associated with all modes of transport. Some of the required information can be obtained by site inspection but other information, such as ownership, usage restrictions, charging structures, public transport fleet and timetable information, will probably involve contacting owners or operators. It might be thought possible to obtain almost all the required information from files held by the responsible planning authority. In practice, however, the information is often spread between a variety of departments and organisations and it may be quicker to do a field survey than to try to locate all the requisite files. This state of affairs is likely to change with the advent of Geographical Information Systems (GIS) which allow this type of information to be stored and then easily accessed by location (e.g. 'list details of all infrastructure at location x') or according to some system characteristic (e.g. 'list locations of all pelican crossings'). Even so it will be necessary to conduct periodic field surveys to verify the information on file and to ascertain the current condition of the infrastructure. Even with the availability of software to predict the condition of the infrastructure – e.g. to predict pavement condition as a function of its construction and the amount of traffic of various types that has used it – it will still be necessary to conduct periodic field surveys to confirm these predictions.

An *inventory field survey* will usually begin by obtaining as much information as possible from pre-existing files, using this information to determine which locations should be visited and producing annotated maps and plans whose contents the surveyors are required to verify or update. Only when no previous information can be found will it be necessary to ask the field staff to compile a complete inventory *ab initio*. Where the prior information exists on a GIS or similar system it is becoming possible to dispense entirely with a hard copy plan and instead to send the surveyor into the field equipped with a portable computer containing all the relevant information. The most sophisticated of such systems can incorporate an automatic auto-location system such as GPS (Global Positioning Systems) so that the computer can tell the surveyors precisely where they

Table 13.1 Components of a comprehensive transport system

Highways
Network description (details of links and junctions, usage restrictions)
Fundamental structures (details of bridges, embankments, causeways etc.)
Pavement (construction and surfacing)
Drainage system
Lighting (type, location of lights, control system)
Pedestrian and cyclist crossing facilities (type and location)
Traffic monitoring systems (location and type of sensors)
Traffic control systems (location and type of signals etc.)
Toll systems (location and type of toll systems, charging structure)
Road signs and markings (location, type and wording)

On-street parking/loading facilities
Location and details of restrictions (e.g. start and end point of 'no waiting' area)
Location and arrangement of designated spaces or bays
Details of charging structure and enforcement method
Infrastructure (location and type of any meter, ticket machines etc.)

Off-street parking facilities
Location and capacity of spaces and of access/egress points
Type of structure (surface, underground, multi-storey)
Usage restrictions (category of user, type of vehicle)
Details of owner and operator
Method of operation (attendant, entry-barrier, exit-barrier, ticket machine etc.)
Details of charging structure

Public transport infrastructure
Fixed track systems (network details – number and gauge of tracks, segregated or shared right of way, points, power source, signals, detectors, crossing facilities, rolling stock, timetable of services)
Bus systems (bus lanes, busways, guide ways, power sources, turning facilities, laybys, bus detectors, special signals, vehicle fleet, timetable of services)
Passenger access and interchange facilities (location of bus stops, bus stations, rail or LRT stations, airports, ferry terminals etc.; facilities available)

Facilities for cyclists and pedestrians
Network details (location of designated cycleways, paths, walkways etc.; details of width, surfacing, lighting, markings etc.)

Canals and navigable rivers
Network details (location, width and depth of channels, locks, tidal restrictions, speed restrictions, fleet information, timetable of services)
Docking facilities for vessels carrying passengers, vehicles and freight

Freight interchange facilities
Location and description of facilities (e.g. which modes?)
Physical or other restrictions on use (size of vehicle, size or weight of container, type of goods)

Traveller facilities
Designated/recommended routes (e.g. HGV routes, scenic routes, diversion routes)
Location and description of rest/recuperation/refuelling/accommodation facilities
Location and details of sources of traveller information

are and what infrastructure should be visible and any new data that are input can be automatically referenced to the current location.

The most common forms of *condition survey* are those relating to the road surface, lighting and traffic signs. These surveys are usually conducted on a routine basis according to some carefully planned schedule and it has become standard practice to use hand-held computers to guide the surveyors along prescribed routes and to prompt them to record the required items of data. Many of the required observations can be made by eye by a trained surveyor but some require specialist equipment to quantify aspects such as deformation or breakdown of the pavement, deterioration in the luminance of lighting or of reflectivity of signs and so forth.

Recent years have seen increasing use of semi-automatic survey equipment such as videos, lasers and light meters which, if housed in a special survey vehicle driven at 'normal' speed along a designated route, can provide a permanent record of key parameters which can be examined by skilled staff back in the laboratory with a view to determining which sites or locations require more detailed investigation.

13.2 Vehicle flow surveys

Information on the flow of vehicles past a given point in a specified time period provides a key input to decisions on the planning, design and operation of transport systems. The information can be gathered automatically or by survey staff and will normally be disaggregated to indicate the number of vehicles of different types in the traffic stream. Disaggregation by vehicle occupancy may also be achieved if the count is made by survey staff rather than by automatic equipment. Depending on the purpose of the survey, the count may relate to traffic passing along a specified link or may relate to traffic making a particular turning movement at an intersection.

13.2.1 When to survey

It is usual to require estimates of the flow for 'typical' days, hours or peak hours in the year and so it is normal to process the survey data to give an estimate of the annual average daily (AADF), hourly (AAHF) or peak hour (PHF) flows for the site. If data are collected at a site on a continuous basis for a year or more then these quantities can be calculated precisely in arrears, but if the count is of limited duration it becomes necessary to make estimates based on what is known about seasonal flow patterns at other sites.

Table 13.2 shows factors devised by the Department of Transport for estimating AADF from short period counts at various times of year – note that the expansion factors are different for different types of road and that each has an associated coefficient of variation reflecting the uncertainty inherent in the process. Although AADF and other parameters can be estimated from very short counts, a longer count will always give a better estimate. It is recognised that flows are more stable in some months than in others and that, in the UK at any rate, the most stable periods are generally April/May and September/October, but even in these 'neutral' months the flow can be distorted by the irregular occurrence of public holidays such as Easter, variability in weather conditions and local factors such as the dates of school terms. For these reasons it is always wise to treat flow estimates derived from short period counts with considerable caution.

Table 13.2 Factors for converting short period counts to 24-hour AADF

(a) Expansion factors (and associated average coefficients of variation) for converting short period counts to 16-hour flows

Length of count	Period of count	Urban/Commuter	Non-recreational low flow	Rural long-distance (Mon–Thurs)	Recreational
2 hour	15.00–17.00	6.41 (9.7%)	6.15 (17.3%)	6.19 (8.8%)	5.62 (13.2%)
	16.00–18.00	5.48	5.50	5.70	5.41 (9.2%)
4 hour	14.00–18.00	3.18 (6.2%)	3.05 (10.2%)	3.08 (6.0%)	2.88 (9.2%)
	15.00–19.00	3.13	3.10	3.19	3.06
6 hour	13.00–19.00	2.25 (6.2%)	2.18 (7.2%)	2.22 (4.6%)	2.09 (7.7%)
	14.00–20.00	2.28	2.23	2.33	2.20
8 hour	12.00–20.00	1.79 (4.3%)	1.73 (6.0%)	1.78 (3.9%)	1.66 (6.7%)
	13.00–21.00	1.84	1.81	1.90	1.78
10 hour	10.00–20.00	1.48 (3.9%)	1.41 (4.0%)	1.41 (4.0%)	1.34 (6.0%)
	11.00–21.00	1.51	1.50	1.50	1.40

(b) Road type M-factors (and associated coefficients of variation) for converting 16-hour flows to 24-hour AADF

Month	Main/Urban	Main/Inter-urban	Recreational/Inter-urban
April	1.016 (6.5%)	1.115 (8%)	1.271 (12%)
May	0.989 (6.5%)	1.071 (8%)	1.140 (10%)
September	1.005 (6.5%)	1.016 (8%)	0.062 (11%)
October	1.000 (6.5%)	1.068 (8%)	1.142 (12%)

Source: DoT Traffic Appraisal Manual Supplement August 1991

13.2.2 Manual data collection methods

The traditional method of collecting traffic flow data was to station enumerators beside the road and require them to record the number of vehicles passing in the designated direction(s). The record might be kept on one or more tally counters (perhaps with separate counters for different directions of movement or different classes of vehicle) or on prepared forms with provision for different categories of vehicle. Depending on the design of the form, the surveyor might record each vehicle using the five-bar gate counting

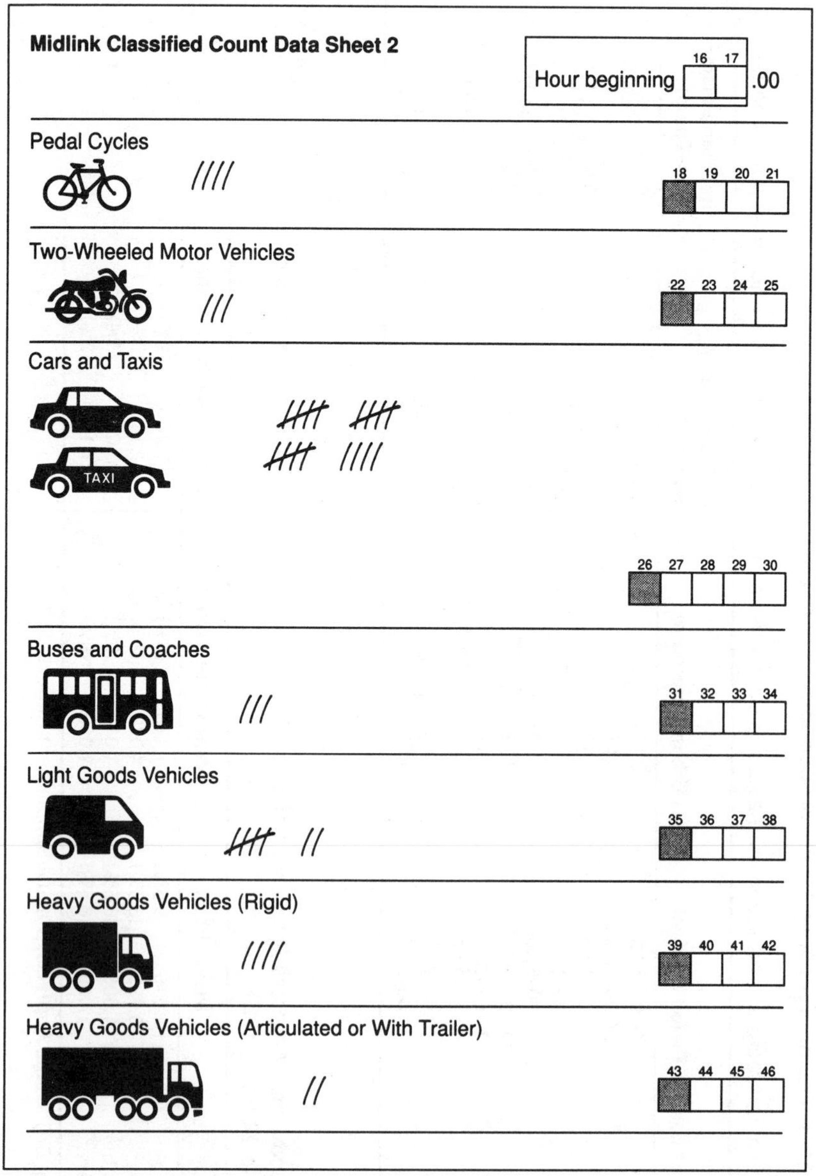

Midlink Classified Count Data Sheet 2

Hour beginning [16][17] .00

Pedal Cycles — //// — 18 19 20 21

Two-Wheeled Motor Vehicles — /// — 22 23 24 25

Cars and Taxis (TAXI) — 𝍸 𝍸 𝍸 //// — 26 27 28 29 30

Buses and Coaches — /// — 31 32 33 34

Light Goods Vehicles — 𝍸 // — 35 36 37 38

Heavy Goods Vehicles (Rigid) — //// — 39 40 41 42

Heavy Goods Vehicles (Articulated or With Trailer) — // — 43 44 45 46

Fig. 13.1 A typical 'five-bar gate' turning movement traffic survey form (reduced from an A4 original)

system (see Fig. 13.1) or by scoring out the next in a sequence of numbers. At the end of the survey the totals would then be transferred to a summary sheet. Since about 1980 some survey teams have been replacing the tally counters and survey sheets by hand-held data loggers which have an inbuilt clock to indicate the start and end of the survey period and which can have their keys designated for different classes of vehicle or different streams of traffic. Use of these devices is therefore particularly advantageous when the survey requires simultaneous monitoring of several streams of traffic and separate recording of different categories of vehicle (see reference 1 for further details).

Manual recording of traffic flow by survey staff is still common for one-off or ad hoc short period counts and by organisations who cannot justify the investment in more expensive equipment, but such surveys are now becoming the exception. It is now much more usual to use one or other of the automatic counting methods.

13.2.3 'Automatic' data collection methods

Automatic counters comprise one or more sensors and a recording device. There are many different types of sensor and the choice between them will depend on site conditions. The most popular types include: (a) *pneumatic tubes* of about 25 mm O.D. that are fixed across the surface of the carriageway and are momentarily 'squashed' as each wheel passes over them, causing a pulse of air to be sent to an airswitch and hence to the recording device; (b) various forms of *noisy cable* that are fixed across the surface of the carriageway and contain concentric cores which, by a tribo-electric or piezo-electric effect, generate or vary an electric current when squashed by a vehicle's wheel; (c) *inductive loops* buried 25–50 mm below the surface of the carriageway and energised with a low voltage, or *magnetic field detectors* mounted in small housings on the surface of the carriageway, which can detect the electromagnetic disturbance caused by the proximity of large metal objects such as vehicles; (d) the interruption by a vehicle of *photo-electric beams* transmitted across the carriageway can be recorded, or alternatively the Doppler effect can be used to detect moving objects causing a disturbance in the reflected signal from a beam transmitted along the carriageway; and (e) computer analysis of the pixels making up a *video* image can detect the presence of moving vehicles.

Of these, the inductive loop is particularly suited to permanent installation because, although expensive and disruptive to install in the first place, it has a long life with minimal maintenance; it is therefore popular at sites where a continuous count is required. Many authorities are now adopting the practice of installing loops in new or resurfaced roads before opening them to traffic and then connecting them up as and when they are required.

The pneumatic tube is easier and less disruptive to install than loops but, being surface mounted, is more prone to general wear and tear and more specific damage by heavy vehicles or vandals. Pneumatic tubes are therefore usually restricted to counts of limited duration – typically less than two weeks, depending on traffic flow and site conditions.

The most popular of the photo-electric beam systems is an infra-red beam transmitted by light-emitting diodes (LEDs) to photo-transistor receivers. The transmitters and receivers may be located on opposite sides of the carriageway or, if a reflector is used on the opposite side, adjacent to each other. Alternatively, if the flow in separate lanes is to be differentiated, the beam can be transmitted vertically downwards from an LED mounted over the carriageway and reflected back up to a receiver by a reflector on the carriageway surface. Photo-electric beam equipment can be moved from site to site but can operate successfully at a fixed site under a range of climatic conditions as discussed in reference 2.

Modern recording devices are able to accept data from several sensors simultaneously and, using appropriate logic, can deduce parameters such as the direction, speed, and spacing and headway of vehicles from an array of sensors across the carriageway. These multi-sensor sites may use a range of different types of sensor or may take advantage of the fact that different configurations of inductive loop sensors can provide data on vehicle type and speed. This information, or a simple count if that is all that is required, can be stored in solid state memory in the roadside unit pending a visit by survey staff or can be sent by telemetry to a central site. Use of telemetry is becoming the norm for the counting sites which make up the permanent core of national vehicle flow monitoring systems.

The relative advantages and disadvantages of the different methods of counting vehicles are reviewed in more detail in reference 3.

The accuracy of vehicle flow data derived from automatic counters depends crucially on favourable site conditions and correct installation of the equipment; tubes, loops and cables require a firm and even road surface, and all sensors perform best if the traffic flow is relatively smooth, without erratic vehicle trajectories and sudden changes in speed and without any overtaking within the sensor's field. Care needs to be taken to ensure sensors are correctly installed – for example cross-carriageway sensors must be securely fixed at right angles to the flow of traffic. Care should also be taken to ensure that equipment is adequately secured against vandalism and theft. Once correctly sited and installed the equipment should be tested and its sensitivity adjusted to detect all required vehicles (for example the sensitivity may need to be increased if bicycles are to be detected) while minimising extraneous noise. Accurate 'tuning' of inductive loops is a particularly skilled task because as the sensitivity is increased so too is the possibility of unwanted 'side firing' by vehicles in adjacent lanes.

Periodic site visits may be necessary to check that sensors are still in good condition. Common problems include surface mounted sensors which have come loose or been dislodged, pneumatic tubes which have been cut or suffered from moisture ingression, and photo-electric reflectors which have become obscured. If the sensor is connected by telemetry to the central office, a fault on site may show up as an anomaly in the readings and site visits may not be necessary until there is evidence to suggest that a problem has developed.

13.3 Vehicle weight surveys

Vehicle mass data are critical to the structural design and management of highways and bridges, and are used regularly in transport planning and economic studies. Historically these data were collected in surveys which involved diverting samples of commercial vehicles into roadside weigh stations equipped with portable or fixed measurement devices (*capacitive mats* and *weigh bridges* respectively). These gave very accurate mass results and drivers could be interviewed for supplementary information. However, the operational costs were high both for the surveyors and vehicle operators, and suitable survey sites could be difficult to find and expensive to establish. More importantly, however, the widespread usage of CB radios in the freight industry in recent times means that advice regarding the locations of survey sites is easily disseminated and the data can be biased by avoidance.

The trend is now toward using *weigh-in-motion* systems not easily detected by drivers. Typically these involve strain gauges attached to bridges or box culverts or piezo-electric cables set into the carriageway. In the case of piezo-electric sensors an electrical charge is generated in the cable as the vehicle wheel passes over it, the charge being proportional to the pressure exerted. In either case the data are read directly into computer storage. These systems are often integrated with other sensors which can provide data on vehicle numbers, axle loads, classifications and speeds. Weigh-in-motion systems have the advantage that representative samples of vehicle mass are easily obtained and analysed, and time-related relationships can be established. Further, the axle masses obtained are the actual dynamic forces applied and thus may be more useful than static masses for design and monitoring purposes.[4]

13.4 Spot speed surveys

Spot speeds are the instantaneous speeds of vehicles at the observation site. They are commonly required in accident analyses, and used to assess the need for, and impact of, traffic management and control measures/devices, e.g. speed limits, signal settings, road markings, speed-change lanes, no passing zones, traffic signs, and pedestrian crossings. Spot speed surveys are carried out as inconspicuously as possible to minimise the influence of the observer(s) and equipment upon the vehicles surveyed.

13.4.1 The sampling process

Manual spot speed data collection is normally based on randomly sampling individual vehicle speeds over short time periods. If the target population is free-flowing vehicles, the survey should be carried out during off-peak travel periods, when the speeds of isolated vehicles and bunch leaders are easily measured. If the target population comprises all vehicles, care must be taken to ensure that the speeds of successive vehicles in the sample can be measured; if this cannot be achieved, for example if there is too much bunching, sampling bias can be minimised by measuring the speed of the second vehicle to arrive after completing the previous observation.

The minimum number of vehicles to be sampled in order to achieve a given degree of accuracy can be determined from

$$n = \frac{s^2 k^2 (2 + u^2)}{2e^2} \tag{13.1}$$

where n = minimum number of measured speeds; s = estimated sample standard deviation; e = allowable error in the speed estimate; k = a constant corresponding to the number of standard deviations associated with the desired confidence level (e.g. 1 for 68.27 per cent, 2 for 95.45 per cent, 3 for 99.73 per cent), and u = a normal statistical deviate corresponding to the percentile being estimated, e.g. 0.00 for the mean or 50 percentile speed, and 1.04 for the 15th or 85th percentile speed, assuming the speed distribution is normal. For example, the minimum sample size required to say with 95.45 per cent confidence that the 85th percentile speed of measured vehicles lies within ±1.5 km/h of the true target population speed is given by

$$n = \frac{s^2 (2)^2 (2 + 1.04^2)}{2(1.5)^2} = 2.739s^2 \qquad (13.2)$$

The estimate of the standard deviation is often available from previous studies on similar roads. If a reliable estimate is not available, the s should be obtained from a pilot survey (which can also be used to check on the adequacy of the sampling method, efficiency of organisation, etc.).

13.4.2 Measurement methods

The three methods most commonly used to collect spot speed data are (1) electronically measuring the time taken by a vehicle to cross two parallel detectors located closely together, on or in or over the carriageway, e.g. two pneumatic tubes, cables, inductive loops, or infra-red beams; (2) using video or closed circuit television (CCTV) to measure the distances travelled by vehicles during a fixed time-interval or the times taken to cover a fixed distance; and (3) manual measurement using a radar speedmeter.

With *parallel detector methods* it is important to ensure that the detectors are precisely parallel and far enough apart to allow accurate readings (see reference 5 for a discussion of the errors associated with different spacings). Pairs of pneumatic tubes or cables are suitable for temporary surveys but, due to the visibility of the tubes and cables to drivers, they may not always provide unbiased data. When choosing between cables and tubes for this task it should be noted that the cables provide a cleaner signal but are more expensive. Inductive loops are expensive and disruptive to install but are less visible to the drivers and are very suitable for use at permanent survey sites. The equipment involved in using pairs of photo-electric beams can, if appropriately located, be invisible to drivers and this technology is very popular for automatic speed surveys in some countries.

Video and *CCTV* are increasingly popular as an 'automatic' source of spot speed data. Their particular advantage, which they inherit from the now obsolete time-lapse photography using 16 mm film cameras, is that, if the images are stored on videotape or digitally, the raw data can be revisited to provide information on other aspects of the scene such as vehicle classifications, volumes, headways and overtaking manoeuvres. A convenient way of extracting speeds from video images is to record the distance travelled by all vehicles between a particular pair of frames and then to estimate the *space-mean speed* ($\bar{x}_s$) as:

$$\bar{x} = \frac{\sum l_j}{nt} \qquad (13.3)$$

where l_j is the distance travelled by the jth vehicle, n is the number of vehicles common to consecutive frames, and t is the time elapsed between frames. Alternatively, particularly if the observed traffic is slow moving, one can determine the average time (measured by the number of frames taken by a vehicle to cover a specific distance e.g. between two points marked on the kerb). This produces a *time-mean speed* ($\bar{x}_t$) defined as

$$\bar{x} = \frac{dn}{\sum t_j} \qquad\qquad (13.4)$$

where d is the specified distance, t_j is the time taken by the jth vehicle and n is the number of vehicles sampled. The time-mean speed is, of course, the version of spot speed derived from pairs of detectors.

The time-mean and space-mean spot speeds will always be different from each other except in the unlikely event that all vehicles are travelling at the same speed. In practice this means that when conducting before-and-after speed surveys the same measurement method should be used to ensure that any speed differences are true reflections of the change in road conditions.

Manual surveys are usually conducted using radar speedmeters. These are convenient to use and can produce accurate data but particular care needs to be taken to avoid the surveyors being visible to drivers because radar is associated with police enforcement and so causes drivers to slow down. It should also be noted that some vehicles carry, legally or otherwise, equipment which can detect radar and so the survey may be known about even if the surveyors themselves are inconspicuous. This can be a particular problem when the target population of vehicles includes a high proportion of commercial vehicles which are in contact with one another by CB radio.

Radar speedmeters work by directing a continuous beam of high-frequency microwaves onto a target vehicle which bounces them back to the transceiver at a slightly different frequency. This change in frequency (the Doppler effect) is directly proportional to the speed of the vehicle irrespective of whether it is approaching or moving away from a stationary speedmeter.

The operating range of a radar speedmeter can be as high as 2–3 km in open country, but a typical range is about 500 m. There is a limit to the closeness of a vehicle to a speedmeter as it must be in the transmitted beam for about two seconds if a reading is to be obtained. Radar speedmeters are used most effectively when vehicles are moving freely and unlikely to overtake or mask each other, e.g. on single-lane roads, and on two-lane and multi-lane roads with traffic volumes less than 500 veh/h and 1000 veh/h, respectively. Speeds observed by these speedmeters are typically slightly less than the true speeds, with the relative error decreasing as speeds increase. This results from the fact that most of the equipment rounds the observed speed down to the nearest whole number, and from a consistent angle-of-incidence error; for example, if the angle between the beam and the vehicle path is 15 degrees (typical practice) and the true speed of the vehicle is X_t km/h, the observed speedmeter reading will be $X_t \cos 15$ (which is consistently less than X_t).

13.5 Journey speed, travel time and delay surveys

Typically journey speed, travel time and delay surveys are carried out over road-network sections to: identify the causes and locations of traffic congestion; measure the before-and-after impacts of road and traffic management improvements; provide inputs for transport planning/trip assignment/route diversion models; and facilitate the economic analysis of alternative transport improvement proposals.

Unlike spot speed surveys, these surveys provide information about speeds over a significant stretch of road or an entire route. *Journey speed* is defined as the distance travelled divided by the total *travel time*. This travel time includes the running time, i.e. when the vehicle is in motion, and stopped time, i.e. when the vehicle is at rest at traffic signals, in traffic congestion, etc. The total travel time also contains the *travel-time delay*, i.e. the difference between the actual travel time and the travel time measured under non-congested traffic conditions.

The survey data can be collected by observers located at strategic positions along the designated route ('stationary observers') or by vehicle-borne observers ('moving observers'). In each case there are several techniques available.

13.5.1 Stationary observer methods

The most popular of the stationary observer techniques is the licence plate survey whereby the time taken by each vehicle to travel between two points is deduced from the difference between the times at which its licence plate was recorded passing the two locations. In its simplest form this survey involves observers with synchronised watches and a data sheet onto which they write the licence number and precise time of each passing vehicle. The task is simplified if the watches have digital readouts. If the traffic flow is too high to record the complete licence number of every vehicle an agreed observation of the number (e.g. first 4 digits) may be used and an agreed sample of the flow (e.g. odd numbered vehicles only) can be taken. A single observer can accurately record about 90 per cent of plates at vehicle flows up to about 600 veh/h. Higher flow rates (up to 1100 veh/h) can be accommodated if two observers are available at each site, i.e. one to read out the numbers while the other writes them down, or if the single observer speaks the data into an audio tape recorder. The problem with the audio tape method is that significant time and effort is subsequently required to transcribe the tapes. Automatic transcription machines involving speech recognition may eventually solve this problem but have not done so yet.[6]

If the target traffic flow is less than 500 veh/h (or can be reduced to this by an appropriate sampling strategy) then the most efficient method of collecting the licence plate data is to have the observers type them directly into a portable computer or datalogger which will automatically provide an accurate time label for each vehicle and which can transfer the data directly to an analysis computer without any need for transcription.[1]

Video has for some years been used as a medium for collecting licence plates but it has suffered from the fact that there are relatively few sites at which a good camera angle can be found (the traffic has to be well spaced and an end-on view is desirable – such as can be had from gantries or bridges over motorways) and the transcription process has been expensive, time consuming and error prone. However, the recent development of software which can locate and read licence plates from a digital image has overcome the transcription problems and it is now economically feasible to set up continuous, fully automatic, video-based licence plate surveys.[7]

The effectiveness of license place surveys as a source of travel time data is, of course, dependent on there being a good number of records 'matched' between the two survey sites. (Depending on the variability of travel times it is generally thought desirable[8] to obtain at least 50 matches before attempting to calculate a mean.) The probability of obtaining matched records depends of course on the volume of traffic at each site and the amount of traffic common to both sites. The method is therefore most effective where there is relatively little traffic joining or leaving the route between the two sites.

Where the amount of joining and leaving traffic is negligible or non-existent (as on a motorway between intersections) a simplified survey method known as the *input-output method* can be used to estimate the mean travel time ($\bar{t}$). The method typically involves designating a cohort of 50 or so vehicles between a pair of marker vehicles (which might be driven by survey staff or, preferably, a pair of conspicuous vehicles whose identity can be radioed, or phoned, from the upstream site to the downstream site) and then recording the clock time at which each one passes the upstream site and the clock time at which each one passes the downstream site and then applying the following formula:

$$\bar{t} = \frac{\sum cd_j}{n} - \frac{\sum cu_i}{n} \qquad (13.5)$$

where cu_i is the clock time for the ith vehicle at the upstream site and cd_j is the clock time for the ith vehicle at the downstream site. Note that this method allows vehicles to overtake one another within the cohort but is dependent on no vehicle overtaking, or being overtaken by, either of the marker vehicles.

The input-output method can also be used to estimate queuing time in a single lane queue but a related technique known as the *queue analysis method* is more generally applicable for estimating mean delay time in a queue. It simply involves recording the number of queuing vehicles at periodic intervals and keeping a note of the number of vehicles which get through to the end of the queue. The average delay ($\bar{d}$) is then given by the following formula:

$$\bar{d} = t\frac{\sum q_i}{n} \Bigg/ f \qquad (13.6)$$

where t is the length of the survey period, q_i is the number of vehicles queuing in the ith interval, n is the number of intervals and f is the total throughput of vehicles during the survey period. It is important, when using this method, to avoid specifying an interval between queue counts which is a factor or multiple of the cycle time of any adjacent traffic signals since this would introduce a bias into the queue length estimate.

The above methods can provide useful estimates of journey time or queuing time but they cannot reveal anything about the location or cause of delays along the route. This information can only be obtained by using one of the moving observer methods described below or by positioning an observer in a high building or mounting a video in a high building or perhaps on a telescopic mast, such that the progress of individual vehicles can be monitored, noting their arrival time at predetermined points and the nature of any delay en route. This *path-trace* method is very labour intensive and must be very carefully operated to avoid bias[3] but, nevertheless, it can be an effective method of collecting speed and delay data on relatively short stretches of city streets.

13.5.2 Moving observer methods

These methods involve vehicle-borne observers measuring the time taken for their vehicle to complete a specified journey. The simplest and most popular method is the *floating car method* whereby the driver of the test vehicle is instructed to drive so as to safely pass as many vehicles as pass the test vehicle over the length of the route. If this is achieved, the test vehicle is said to be 'floating' in the stream of traffic and its measured travel time can be taken as an estimate of that for the stream as a whole.

On urban roads with significant levels of platooning from traffic signals, safety considerations may not make it possible to achieve a perfect 'float'. In such cases, the driver is instructed to travel at the legal speed limit except when impeded by the traffic conditions; this is called the *maximum-car method*. More accurate results are obtained by gathering extra data with which to 'correct' for failure to achieve a perfect float.[9] The *corrected moving observer method* involves driving the test vehicle over the study route at a safe and comfortable speed while another vehicle travels in the opposite direction. The following equations are then used to calculate journey time ($\bar{t}$):

$$\bar{t} = t_w - \frac{y(t_a + t_w)}{x + y} \tag{13.7}$$

and

$$\bar{v} = l / \bar{t} \tag{13.8}$$

where t_w = measured journey time for the specified direction, t_a = measured journey time in the opposite direction, x = number of vehicles in the stream travelling in the specified direction which are met by the observer travelling in the opposite direction, y = number of vehicles that overtake the observer minus the number overtaken while travelling in the specified direction, l = route length, and $\bar{v}$ = average space-mean speed in the specified direction. While 12–16 runs in each direction are normally sufficient to give reasonably accurate results, the heavier the traffic the less the number of runs required. Whatever the number of runs, the values of x, y, t_a and t_w used in Equation 13.7 are the averages of the individual values obtained. If the mean speed of a certain class of vehicle is required the procedure is the same except that the only vehicles taken into account are the designated type. Resources are saved by using one car instead of two but this risks introducing sampling errors associated with changing traffic environments.

Moving observer methods can be used to record not only the extent but also the causes of travel time delay. For example, as the test vehicle moves along the route an observer can log the stopping and starting times, and locations of delays caused by events such as parking/unparking vehicles, pedestrians, conflicting vehicles at uncontrolled intersections, or queues at signals. These can then be analysed to determine the most important causes of delay or travel time variability.

13.6 Origin–destination cordon and screenline surveys

A useful picture of local traffic movements can be built up by defining one or more cordon(s) and screenline(s) in a local area, noting the licence plates of vehicles crossing these cordons or screenlines and then using special software to match the records so as to produce a matrix of flows. The licence plates can be recorded by one of the methods described in the previous section, except that there is no need to keep so precise a record of the time at which each vehicle passes the station. Again, if the flow at some sites is too high to allow the licence plates of all vehicles to be recorded, it is wise to agree in advance what particular subset of the total vehicle population is to be surveyed.

The method is, of course, dependent on accurate records having been kept and on sufficient data being collected to quantify the flow in the least popular cells in the matrix. A misrecorded plate cannot, of course, be matched and, given the likelihood of a certain amount of misrecording, some of the software allows 'near misses', such as those which might be due to recording an R for a B or a C for an L, to be matched. This can improve the number of matches but must not be overused lest it result in spurious matches which would distort the matrix.[1,10] Spurious matches may also be caused if licence plates have not been recorded in full. For example, it can be seen that A123 BCD and E123 FGH could be spuriously matched if only the digits had been recorded. Yet another potential source of error in the matrix would be a failure to allow for the normal travel time between survey sites. However, sophisticated matching software is now available that will 'realise' that the first few records of vehicles leaving the study area at the beginning of the survey period, and the last few entering it at the end of the survey period, should be ignored rather than regarded as representing internal-to-external and external-to-internal flows respectively.

The matrix resulting from the matching process can be factored up to a complete matrix by using the Furness procedure (see Chapter 5) to match it to total flow counts taken for each direction at each site.

13.7 Parking use surveys

Information about the amount, location and duration of parking is clearly essential to the formulation of parking policy and to proper management of the parking stock. An important preliminary to any parking use survey is the careful definition of the survey area and of the best time to conduct the study. The survey area may need to be more extensive than first envisaged because drivers may be walking considerable distances from their parking space to their final destination. The timing of parking surveys is dependent on objectives of the surveys, the characteristics of the survey area, and the variation in usage likely to be experienced throughout the week. In most British towns, seasonal demand for parking is at an 'average' level in September and October. Days immediately preceding or following holidays, special shopping days, or days when shopping hours differ from the usual are not normally selected; however, there may well be instances when it is appropriate to carry out a special study on, say, a market day or, in the case of a holiday resort, at special times of the year. Where the survey area includes fringe residential areas, care should be taken to ensure that the results can also be considered separately from the remainder of the data.

The simplest form of parking use survey is the *accumulation survey* (sometimes known as a *concentration survey*). This survey, which provides information regarding the total parking accumulation within the survey area at any given time, can be obtained by making a tour around the area and noting the number of vehicles parked in each part of the area or by drawing a cordon around the area and counting the number of vehicles entering or leaving the cordon (the accumulation of vehicles inside the cordon at any given moment can be deduced from the difference between the numbers entering and leaving). The cordon count is usually started early in the morning when the numbers of vehicles already within the cordon are most easily estimated, and continued through the day. The total parking accumulation at any given time (usually every hour) is then the previous accumulation plus the vehicles entering the cordoned area minus the number of vehicles exiting the area less an allowance for the number of vehicles estimated to be circulating within the survey area.

While the above accumulation survey provides information on the total number of parked vehicles, it provides no data regarding their duration of stay. To obtain this information it is necessary to carry out a *duration survey*. The simplest form of duration survey is equivalent to the input-output survey described in Section 13.5.1 and uses average entry times and average exit times to produce an average duration of stay. If more detailed information is required on the distribution of durations of stay then it is necessary to identify individual vehicles by noting their licence plates. The most straightforward method of doing this is the *sentry survey* whereby the surveyors are positioned where they can record, using either pencil and paper, audio tape recorders or hand-held computers, the licence plates, and entry and exit times, of all vehicles entering or leaving the parking area.[11] Matching software can then be used to deduce the length of stay of each vehicle in the same way as was done to deduce journey times. This technique is very efficient for studying large car parks with a limited number of entry/exit points or for area studies when the area can be defined by a cordon with a limited number of entry/exit points.

If parking use data are required for individual on-street spaces or for small car parks, the sentry method is not efficient and the preferred method is the *parking beat survey*. This method requires surveyors to patrol a predefined beat at a fixed interval and to record the registration plates of vehicles which are parked when the surveyor reaches them. The duration of stay of each vehicle at a given location can then be estimated by multiplying the number of occasions on which their presence was noted at that location by the beat interval. The method is, of course, not completely accurate and can produce unrealistically low estimates of usage and unrealistically high estimates of average duration if the beat interval is too long to pick up the more transient parkers.

The traditional version of the parking beat survey, based on pencil and paper records, required the surveyor to check his or her records on each visit to ascertain whether a given car was 'newly arrived' or 'still there' – the licence plates of newly arrived vehicles were recorded, while for each 'still there' vehicle a tick was added to show that it was present for another interval. This method was well suited to semi-manual analysis (the number of ticks for each vehicle indicated the number of visits for which it was present) but the method was slow and error prone – particularly when one surveyor passed records to another at the change of shift. More recent practice involves keying in *all* licence plates to a portable computer on each visit and then allowing specialist software to do all the necessary matching and analyses.

Parking beat surveys may be made more detailed by requiring the surveyors to note not only the licence numbers but also the type of space, the vehicle type, whether the vehicle is legally parked, whether it is displaying a disabled badge or other permit and so on. Collection of such data enables separate estimates to be made of the parking behaviour of different categories of user and the usage of different categories of space.

At the increasing number of sites where a fee is payable, or access/egress are controlled for some other reason, it may be possible to obtain data on parking usage, and perhaps on durations of stay, direct from the car park operators. Ticket receipts may record the numbers of parkers and the amount of parking time purchased. Where the payment is made on exit, data on precise durations of stay may be available. Factors are required to correct for underestimation of parking volume due to failure by some drivers to buy tickets, or of some attendants to issue them, and to correct for over-purchase of time at pay-on-entry car parks, on-street meters, or other pre-purchase systems. The necessary factors may be estimated by comparing estimates derived from analysis of ticket receipts with data from a more conventional survey carried out inconspicuously so as not to affect the behaviour of parkers or attendants. Further details of this source of parking data are contained in a recent review.[12]

Data from any of these sources can be processed to produce estimates of a number of useful quantities including: (a) *space occupancy*, i.e. the proportion of spaces occupied at any time (this is the concentration divided by the total number of spaces); (b) *turnover*, i.e. the average number of vehicles using each space throughout the survey (this is the total number of vehicles observed divided by the total number of spaces); (c) *turnover rate*, i.e. the turnover divided by the number of hours over which the survey was carried out; (d) *average duration*, which is the summation of all vehicle-hours of parking divided by the total number of vehicles observed; and (e) the *proportion of overtime parkers*, i.e. the number of vehicles with durations in excess of the legal parking time-limit divided by the total number of vehicles parked in legal spaces.

13.8 Surveys of pedestrians, cyclists and public transport use

The planning and design of pedestrian and/or cycle facilities and of public transport services should obviously reflect their anticipated usage. Surveys to provide information about current usage are often carried out at intersections, at mid-block crossings, along pathways, or at public transport stopping places. Parameters measured in observational pedestrian and cycle surveys typically include: volume and classification, (local) origin–destination, travel time and delay, conflict, path route, speed, arrival time and queue behaviour. Most surveys are carried out manually with observers recording the data on paper, audiotape or portable computer. Counts may be disaggregated according to the age and sex of the pedestrian or cyclist and by any obvious handicap or encumberment they are suffering. Counts may be derived automatically in the case of cyclists or, even, using photo-electric beams, of pedestrians if they are moving through a well defined passage (e.g. a turnstile) with little opportunity for overtaking.

Local origin–destination information, travel times, speeds, routes, and delays and their causes may be examined by observers stationed at high vantage points, as described in Section 13.5.1. Video recording is particularly useful, despite the high costs of analysis, because it provides a complete picture of events which can be analysed and reconstructed without danger of missing anything.

Estimates of the usage of particular public transport vehicles or services can be made by estimating vehicle occupancy from the roadside or, more accurately, by counting boarders and alighters at each stop. Depending on the vehicle capacity and stop frequency this may be best achieved by putting survey staff at each stop or on board each vehicle. Photobeam technology is available[13] to count automatically the number of boarders and alighters but it is as yet too expensive to be warranted on more than a small sample of the fleet. Depending on the ticketing system in force it may be possible to derive key statistics about passenger flows from analysis of receipts but the growing use of flat rate tariffs and of prepaid tickets makes this increasingly difficult. (On the other hand the introduction of smartcard-based tickets which might be read on exit as well as on entry to the vehicle may herald a very rich source of data.) A major source of data which is already available is the on-board ticket interview survey whereby passengers are asked to show their tickets and provide data about their journey – this however is a participation survey rather than an observation survey and so is more appropriately dealt with in Chapter 14.

13.9 Environmental impact surveys

The impact of some environmental detriments, e.g. noise and air pollution, can be measured quantitatively. Others, e.g. severance and visual intrusion, involve qualitative inducement. The quantifiable aspects are dealt with in this section while leaving the qualitative issues to Section 14.6.

13.9.1 Noise impact surveys

Noise is unwanted sound. Sound is the sensation produced in the ear as a result of fluctuations in air pressure. The basic pressure, zero decibel (dB), is the level of the weakest sound at 1000 Hz which is just audible to a person with good hearing at an extremely quiet location. The decibel scale is logarithmic, so that an increase of one dB indicates a tenfold increase in the sound pressure level. Sound energy is emitted in different frequencies or pitches. Thus the decibel by itself is limited as a measurement of sound as it does not take into account the ear's decreasing response to low frequencies. Consequently, sound meters are usually fitted with special attachments which allow sound to be studied in different frequency ranges. One such attachment, the A-filter, gives a greater emphasis to the medium and high frequencies to which the human ear is most sensitive. The A-filter is used in traffic noise studies, and its measurements are noted as decibels weighted on the logarithmic scale, dB(A).

In practice the ambient noise level at the roadside will vary continuously. Consequently, a number of measures are used to represent noise levels; for instance, the L_1 level, often referred to as the peak level, is the sound level exceeded 1 per cent of the time, the L_{10} level is the sound level exceeded 10 per cent of the time, L_{50} is the median sound level, and L_{eq} is a sound level which is representative of all sound levels throughout the period of measurement, e.g. a full day. Most usually, however, noise levels are expressed in terms of the indices L_{10} (1 h) dB(A), which is the level that is exceeded 10 per cent of the time over a measurement period of one hour, or $L_{10}(18$ h) dB(A) which is the arithmetic average of the $L_{10}(1$ h) values for each of the 18 one-hour periods between 6 a.m. and 12 midnight. Social surveys have shown that people's annoyance with noise correlates well with an $L_{10}(18$ h) level of about 68 dB(A) and hence it is this measure which receives greatest attention in noise surveys.

Most noise measures consider the entire traffic stream rather than individual vehicles as the noise source. Thus the source of traffic noise is normally taken to be on a line 3.5 m in from the nearside carriageway edge (excluding bus laybys, hard shoulders and hard strips) and 0.5 m above the carriageway. Measuring equipment used in surveys typically comprises a microphone with windshield and preamplifier, measurement amplifier, attenuators, magnetic tape recorder (for subsequent analysis), meter for visual display of noise readings, and a calibration noise source. The measurement (reception) point for the microphone is normally at a height of 1.2 m and at least 4 m from the nearside edge of the carriageway but within the boundaries of the site being evaluated. In the case of a survey involving residential buildings the measurements are usually taken 1 m from the building facade facing the roadway.

Weather conditions, especially wind speed and direction, are critically important when taking noise measurements.[14] Thus, the microphone should be located so that (a) the wind direction gives a component from the nearest part of the road toward the measurement point that exceeds the component parallel to the road, (b) the average wind speed midway between the road and the measurement point does not exceed 7.2 km/h toward the microphone, and (c) the wind speed at the microphone does not exceed 36 km/h in any direction. In all cases a windshield should be used on the microphone and measurements taken only when the peaks of wind noise at the microphone are at least 10 dB(A) below the measured L_{10}. Unless there are special reasons the noise measurement is taken only when the road surface is dry.

The sampling time required to obtain a valid L_{10} measurement can normally be obtained from the following equation:

$$t = \left(\frac{4000}{q} + \frac{120}{r} \right)$$
(13.9)

where t = minimum sampling time, minutes, q = total vehicle flow, veh/h, and r = registration rate, samples/minute. This equation is applicable provided that $q \geq 100$ veh/h and $r > 5$ samples/minute; also t should never be less than 5 minutes in any one hour. If $q < 100$ veh/h, then r should be at least one sample per second and measurements taken for the full hour.

13.9.2 Air quality surveys

The main air pollutants emitted by motor vehicles are carbon monoxide, nitrogen oxides, oxides of sulphur, hydrocarbons, and lead and other particulate matter and fibres. As these are dissimilar in their chemical properties, each must be measured in a different way.

Vehicle emission surveys are carried out either in the laboratory or on the road. Laboratory testing, which involves simulating the acceleration, deceleration and speed profiles of typical vehicles in the traffic stream (e.g. floating cars) and capturing the exhaust emissions for evaluation, has the advantage that the test conditions are strictly controlled and more easily replicated. Its disadvantage is that its correlation with traffic stream conditions may be suspect. On-road testing is carried out using instrumented vehicles travelling in the traffic stream, and measuring their fuel consumptions. Emission rates are then correlated with fuel consumption rates to estimate actual emissions.

Where an estimate of air quality at a particular site is required detectors can be placed at the site and readings taken. Many cities now have permanent air pollution stations located throughout their metropolitan areas. However, these area-wide field measurement systems measure pollution data from all sources and not just traffic pollutants. Thus, when monitoring the effects of various transport schemes on air pollution, it is usually necessary to take before-and-after measurements at a number of road-side reference points.

In response to the growing interest in air quality issues a number of manufacturers have begun to produce equipment for monitoring specified pollutants. Interesting recent developments include equipment for instantaneous analysis of gases such that they can be attributed to individual passing vehicles, and air quality monitoring equipment which is combined with traffic counting/classifying equipment to produce a combined record of traffic flow and composition along with levels of key pollutants.

13.9.3 Visual impact surveys

Estimation of the visual impact of the infrastructure associated with a new scheme obviously and quite properly involves qualitative/aesthetic judgement (see Section 14.6). It is, however, possible to estimate the impact in quantitative terms as the proportion of the pre-existing view which has been, or is likely to be, obscured by the infrastructure. Such estimates can be made from representative locations (e.g. at intervals along an existing road or from the front entrances of existing premises) by analysis of photographs or, if the scheme is not yet in place, of photomontages.

The method involves taking a photograph of the existing view and measuring the proportion of the previous view which is obscured by the infrastructure. If the infrastructure is already in place this is a simple procedure; if it is not then it is first necessary to create a photomontage of what the view would be like after completion of the scheme. Weighting systems can be used to give greater importance to obscuration of the psychologically dominant central field of view and less to the peripheral view.

13.9.4 Estimation of severance

A quantitative estimate of the extent to which a new scheme has disrupted, or might disrupt, a given community cannot tell the whole story because it cannot fully capture the attitudinal dimension, but it does nevertheless provide a useful indicator of community severance. The recommended method of estimating severance involves quantifying the pre-existing flows, by all modes of transport including walking, that are to be affected by the scheme and, for each affected movement, estimating the extra inconvenience (normally measured in additional journey time or distance) that is caused by the scheme.

The basic data requirement is therefore a field survey of movements across a screen-line prior to commencement of work on the scheme. This might be done manually or using video or automatic counters depending on the nature or volume of the flows.

13.10 References

1. Bonsall, P.W., Ghahri-Saremi, F., Tight, M.R., and Marler, N.W., The performance of hand held data-capture devices in traffic and transport sector surveys. *Traffic Engineering and Control*, 1988, **29** (1), 10–19.
2. Garner, J.E., Lee, C.E., and Hwuang, L., Photoelectric sensors for counting and classifying vehicles. *Transportation Research Record 1311*, 1991, 79–84.
3. Taylor, M.A.P., Young, W., and Bonsall, P.W., *Understanding traffic systems*. Aldershot: Avebury, 1995.
4. *Traffic data collection and analysis: Methods and procedures*. NCHRP Synthesis of Highway Practice 130. Washington, DC: Transportation Research Board, 1986.

5. Bennett, C.R., and Dunn, R.C.M., Error considerations in undertaking vehicle speed surveys. *Proceedings of the 16th Australian Road Research Board*. 1992, Part 7, 182–196.

6. Dew, A.M., and Bonsall, P.W., Evaluation of automatic transcription of audio tape for registration tape surveys. In E.S. Ampt, A.J. Richardson, and A.H. Meyburg (eds), *Selected readings in transport survey methodology*. Melbourne: Eucalyptus Press, 1992, 147–61.

7. Gibson, T.G., Vehicle identification by automatic licence plate reading. *Traffic Technology International '95*. Dorking: UK and International Press, 1995.

8. Robertson, H.D., Hummer, J.E., and Nelson, D.C., *Manual of transportation Engineering Studies*. Englewood Cliffs, NJ: Prentice Hall, 1994.

9. Wardrop, J.G., and Charlesworth, G., A method of estimating speed and flow of traffic from a moving vehicle. *Proceedings of the Institution of Civil Engineers*, Pt II, 1954, **3**, 158–71.

10. Hauer, E., Correction of licence plate surveys for spurious matches. *Transportation Research A*, 1979, **13A** (1), 71–8.

11. Dean, G., *Parking surveys – A comparison of data collection techniques*. Research Report 222. Crowthorne, Berks: Transport Research Laboratory, 1989.

12. Bonsall, P.W., The changing face of parking related data collection and analysis: The role of new technologies. *Transportation*, 1991, **18**, 83–106.

13. Stopher, P.R., The design and execution of on-board bus surveys: Some case studies. In E.S. Ampt, A.J. Richardson, and W. Brog (eds), *New survey methods in transport*. Utrecht: VNU Science Press, 1985.

14. Department of Transport/Welsh Office, *Calculation of road traffic noise*. London: HMSO, 1988.

CHAPTER 14

Participatory transport surveys

P.W. Bonsall and C.A. O'Flaherty

Surveys which require the active involvement of those being surveyed in some form of interview, questionnaire or discussion may be termed 'participatory surveys'. These surveys can supplement observational surveys by providing much more detail on individual travel patterns and by providing information on the characteristics of the travellers, their attitudes towards travel and their reasons for making one kind of trip rather than another. A major transportation study may require information on all these aspects and may use a combination of discussion groups, interviews and questionnaires to obtain it. This chapter covers only the principal issues involved in organising such surveys and the reader is advised to consult more specialist sources for further details (see for example references 1 and 2).

14.1 Group discussion

Discussion groups or *focus groups* are primarily used at the early stages of a study to obtain an insight into the factors or issues which feature most strongly in the minds of residents and users or potential users of a transport facility. The discussion group will often be made up of community representatives and be 'moderated' by a skilled interviewer who leads the discussants to express their views on the central topic freely without introducing any bias into the discussion. Analysis of the points raised, often by means of careful analysis of a taped transcript, can help decision-makers to take account of public opinion or, more usually, can provide the basis for a more formal questionnaire or interview. (See reference 3 for further details of the use of discussion groups and see section 14.6 for further detail on analysis methods.)

The discussion/interviews may be conducted at people's homes or workplaces or in halls or, more briefly, at the roadside or on board a public transport vehicle. Each of these possibilities is suited to a particular style and content of interview.

14.2 Household interview surveys

The *household interview* has long been a major source of information on the characteristics, behaviour and attitudes of travellers. The first stage in the process is clearly to select a sample of households using the principles indicated in Chapter 12. Most household interview surveys are intended to give a representative snapshot picture at a particular point in time. The most usual sampling method is therefore a *random sample* of addresses in the designated study area. A *clustered random sample* focusing on particular neighbourhoods in the area may be justified to reduce survey costs or if those neighbourhoods merit particularly detailed treatment by virtue of some local factor (e.g. a major impending investment in the area). Similarly, a *stratified random sample* may be justified if it is thought necessary or desirable to enrich the database with respect to some characteristic such as household tenure, income profile or car ownership level. The address lists which form the sampling frame may be derived from lists of local electors, local taxpayers or registered customers of the major utilities – the choice will depend on the completeness and reliability of the list (e.g. whether it contains hotels and institutions), its cost, convenience and availability.

14.2.1 Conventional home interview surveys

The US Bureau of Public Roads began to issue advice on the conduct of household interview surveys in the early 1950s[4] and recommended sampling rates as high as 20 per cent in small towns and 4 per cent in cities of over one million inhabitants. These targets are rarely met and are widely criticised as unrealistically expensive. Standard practice now is instead to calculate the sample size which will achieve the desired precision for key indicators at the required level of confidence (see Section 12.4).

Once a sample of addresses has been selected the survey itself can begin. Individual householders should first be contacted by mail indicating the purpose of the survey and that an interviewer will call on them shortly. Publicity about the survey in the local media may also be appropriate. The interviewer should then visit the household, keeping a record of abortive calls, and arrange a convenient time for the main interview. It may of course be possible to conduct it on the first visit, but given that it is often necessary to have several members of the household present, this will not always be feasible. It is important to persevere in arranging the interview even if this is a time-consuming process, because the exclusion of households from the sample on the grounds that they are rarely at home could obviously bias the data.

The main interview at the heart of a traditional household interview survey will be conducted with at least one responsible adult from the household and will normally collect information about the household composition, income and car ownership, the age, sex, employment status and driving licence tenure of all household members and the frequency with which certain types of trip are made. Precise details of travel patterns may also be collected at this time but it is more usual to collect these by leaving travel diaries with each household member for a few days or for as long as a week. If a diary approach is adopted the initial interview will end by arranging a date and time for the completed diaries to be collected. On this follow-up visit the diaries will be checked for obvious omissions or inconsistencies so that necessary corrections can be made. If the survey requires questions to be asked on the householders' attitudes towards the local transport system, these should be asked on the final visit.

Data collected during household interviews are usually recorded on specially printed forms which are then checked prior to entering the data into a computer. Some savings in transcription cost can be achieved by taking portable computers into the field and entering the data directly. However, despite the advantages of automatic question selection and logic checks on answers, the current generation of survey staff generally prefer the more traditional methods.

14.2.2 The Kontiv survey method

The classic household interview as described above is very expensive and time consuming and each interview may well involve five or more hours of interviewer time in
arranging visits, abortive journeys, conducting the interviews and making repeat visits.
An alternative approach, pioneered in Germany as the 'Kontiv' method and now in
widespread use,[5] manages to reduce this cost by replacing the face-to-face interviews by
reliance on carefully designed self-completion mailback forms, carefully written
instructions on their use and frequent follow-up and/or phone calls to encourage
response. Only as a last resort does an interviewer visit the household. The results from
this technique are impressive and, by continuing politely to contact households until
they do respond, surveyors manage to avoid what might otherwise be a serious problem
of non-response bias.

14.2.3 Telephone surveys

Another potential means of increasing the efficiency of household surveys is to conduct
all or part of the process by telephone. This will obviously avoid time spent in travelling
to the interviewee's home but can yield further savings if the interviewers are equipped
with CATI (computer assisted telephone interview) technology.[6] A complete CATI system might include automatic dialling of next household, prompts to the interviewer to
ask the next question, automatic 'skipping' and 'branching' within the list of questions
depending on the answer to the previous question, immediate logic checks on answers
provided, dynamic adjustment of the interview sample to correct for any emerging biases
and automatic alerting of a supervisor to provide assistance to staff who seem to be log-
ging data at an unusually slow rate or with an unusually high proportion of outliers or
errors.

A previous objection to telephone interviewing was that the resulting sample would
be biased because of low phone ownership. This is no longer a problem in industrialised
countries but other problems are emerging to limit the use of this technique. First, many
phone owners are now ex-directory and so would not be included in a survey which used
the phone directory as the sampling frame; and, second, many phone owners now have
call screening facilities which they can use to avoid being 'at home' to callers whom
they suspect may be part of a telephone sales organisation. Random digit dialling is only
a partial answer to the ex-directory problem because selection of business or fax lines
will represent an inefficiency and because some households have more than one phone
line – so invalidating the use of random phone numbers as an unbiased source of household addresses.

14.2.4 The household activity travel simulator (HATS) approach

At the other extreme from the Kontiv approach and a reliance on telephone interviewing is the interactive household interview exemplified by the HATS (Household Activity
Travel Simulator) approach.[3] This style of interview involves a relatively in-depth discussion between household members and the interviewer around the question of
possible modification of the household's travel patterns in response to some change in
the land use transport system or household resources. The original HATS approach
facilitated this discussion with a kit comprising a map of the area and coloured blocks

in a grid to indicate each household member's activity schedule. More recent variations on the technique have used portable computers to display the activity schedules and to suggest new ones.[7]

14.3 Trip end surveys

For some survey purposes the household may not be the appropriate sampling unit and it may be more appropriate to survey people at their trip end – be it their place of work, a shopping centre or leisure facility. Interviews at such locations are typically shorter than the classic household interview and will usually concentrate on the journey to or from the interview location and the individual travellers' reasons for choosing that destination. Such questions can be put in a relatively brief interview or by a self-completion questionnaire. The surveys are relatively cheap to organise but do, of course, require the permission of the site owner.

A quite specialist variation on the standard trip end survey is the *hall survey*, so called because it is normally conducted in a hall adjacent to a shopping centre or other facility. Potential interviewees are approached on-street and asked to come into the hall to take part in the survey – a cup of tea or other refreshment and perhaps a token payment is often offered as an inducement. These surveys are based on standard market research practice[8] and generally involve stated preference techniques, ranking, rating, or other methods of gathering attitudinal data (see Section 14.6 for further details).

14.4 En-route surveys

14.4.1 Roadside interview surveys

The most efficient way of collecting information about people making a particular journey will often be to intercept them en route. The classic example is the roadside interview survey in which motorists are stopped at a screenline or cordon and asked a number of questions about their current journey. Such surveys can obviously be disruptive to the traffic flow and are not normally permitted unless there is a convenient layby or a spare lane into which the interview traffic can be diverted.

It is not normally possible to interview every motorist and so a sampling strategy is defined. If a constant sampling rate (of, say, $1:n$) is required throughout the survey period then the driver of every nth vehicle will be interviewed or, for greater efficiency, a cluster of c vehicles will be taken from each group of $c \times n$ passing vehicles. Use of a constant sampling rate is, however, quite inefficient because, in order to have the fraction small enough to avoid the build-up of queues at peak flow times, the survey staff will be underemployed at low flow times. The practical solution to this is to take a cluster of c vehicles out of the stream, interview them, release them and immediately take another cluster of c vehicles. This procedure produces a variable sampling rate (higher at times of low flow); however, this can be corrected for by using a continuous count of the total flow passing the site in each hour as the basis for differential weighting of the interviews in each hour.

The interview must be kept brief in order to avoid too much disruption to the traffic or too much annoyance to the motorists. The classic interview comprises only four questions: Where have you come from? What were you doing there? Where are you going? What will you do there? Additional questions may be asked about the trip frequency and the upstream or downstream route. The interviewer will normally also make a visual assessment of vehicle type and occupancy. Where the vehicle is a freight vehicle additional questions may be asked about the goods being carried and the vehicle owner/operator.

14.4.2 Questionnaire surveys

If more questions are required, or if it is not convenient to ask even these few questions, the roadside interview may be replaced or supplemented by a *self-completion mailback questionnaire* handed out to each driver as they pass the site. Questions included on a self-completion survey must be kept very simple and straightforward and this may, in practice, limit the extent to which this technique can be used to obtain detailed information about individual stages of a trip or about its context. A much more serious problem associated with self-completion surveys, however, is the low response rates typically achieved when there is reliance on voluntary mailback.

Response rates vary depending on the nature and complexity of the questions and the type of people being surveyed but rates as low as 30–40 per cent are not unusual. Low response rates become a problem if they are associated with an unquantified base, which makes the responding sample unrepresentative of the target population. There is, for example, some evidence to suggest that response rates may vary with sex, journey frequency and journey length.[9] If the bias can be identified and quantified, for example by using interviews to achieve a 100 per cent sample at some stage during the survey and comparing the characteristics of this group with the mailback group, then it may be corrected by differential weighting. Most usually, however, it is not possible to make an accurate estimate of the potential bias and so the problem remains.

The use of incentives, such as prize draws based on completed questionnaires, may increase the response rate, but does not eliminate the bias. It may indeed introduce its own bias, due to the different susceptibility of different types of people to such incentives, and risks introducing incorrect data by encouraging response by people who are motivated by the prospect of the prize draw rather than by a desire to cooperate with the survey.

If it is thought that response bias can be overcome or is not likely to be serious, it may be possible to dispense with the interview site altogether and instead hand out the forms at a natural stop-line, such as at traffic signals, thus avoiding all disruption to traffic. This latter procedure is known as a *stop-line survey* and may be the only feasible method of conducting questionnaires in congested urban areas. It is important when conducting stop-line surveys to ensure that the sample of traffic likely to be stopped at the location is properly representative of the target population – bearing in mind factors such as that the phasing of linked traffic signals may result in an under-representation of some flows among the vehicles stopped at a given set of lights.

All the roadside interview methods carry an associated road safety hazard to the survey staff and all due precautions must be taken to minimise the risk – such as, for example, the obligatory wearing of fluorescent jackets and, in the case of stop-line surveys, the appointment of a member of the team to alert the surveyors to an imminent change in the traffic signals.

14.4.3 Licence plate surveys

If none of the above methods of roadside survey are feasible in a given situation then an alternative means of obtaining a sample of drivers from the flow of traffic may be to use a licence plate sample survey. This involves recording vehicle licence numbers (manually or by video) and then sending questionnaires to the registered owners of these vehicles through the vehicle licensing authorities. This technique is not available in some countries due to privacy legislation and must, in any event, be used with care. It should not, for example, be relied on to yield a representative sample of drivers, because the database is likely to refer to vehicle owners (or registered keepers) rather than drivers. Also, due to the inevitable delays involved in contacting the drivers, it must be expected that many of the factors surrounding a specific journey will have been forgotten by drivers by the time they receive the questionnaires.

14.5 Public transport user surveys

Just as the roadside interview is the classic means of questioning car users about their current journeys, so the public transport passenger interview survey is the standard method of collecting information about public transport journeys. The simplest form of interview involves surveyors riding on board a sample of public transport vehicles and asking a simple set of questions of each passenger. The questions will typically include boarding point, alighting point and fare paid (all of which may sometimes be available from the ticket without needing to question the passenger) as well as trip origin, trip destination and journey purpose. Additional questions on trip frequency and access modes may also be asked. Such surveys are now often undertaken on a regular basis by specialist survey staff equipped with portable computers to record the answers and perhaps bar code readers to extract ticket details.[10]

A more extended interview, gathering more detailed data about the tripmakers, their reasons for making the journey and their attitudes to the service provided, may be feasible where the journey is relatively long – as in the case of inter-city train journeys – and thus allows time for the survey staff to ask more detailed questions. Self-completion surveys may be more feasible among public transport users than among car drivers because, particularly if the journey is relatively long, there may be an immediate opportunity for the subject to complete the forms and for the survey staff to collect them again before the journey is over, thus avoiding the problems of low response expected in a mailback survey.

14.6 Attitudinal surveys

A number of the surveys outlined in the previous sections will also involve the collection of attitudinal data in addition to more prosaic facts about the subjects' journeys or travel patterns. There are many different methods of collecting attitudinal data and it is perhaps worth outlining some of them here.

Protocol analysis and *keyword analysis* are techniques whereby free-format discussions are recorded and their content is then analysed to reveal the apparent concerns and preoccupations of the discussant. One might, for example, analyse a transcript of a conversation to find out the relative frequency of use of positive and negative adjectives to describe a particular transport facility. A training in relevant aspects of psychology would clearly be a prerequisite to any serious attempt to undertake this type of analysis.

Ranking exercises of various types can be used to discover the relative importance that the subject puts on different aspects of a transport service, the relative levels of satisfaction that they have with these different aspects or their overall preference for different services. The ranking can be achieved by asking them to prepare a preference ordered list or, since this is not always easily achieved, by a series of pairwise comparisons (Do you prefer A or B? ... Do you prefer A or C? ...). These ranking exercises may relate to transport services or facilities which already exist or may relate to as yet unavailable options.

A technique known as *Stated Preference analysis*[11] (SP analysis) has become very popular as a means of quantifying underlying preferences which can then be used to predict preference for as yet unavailable options (e.g. new LRT services) or for substantially redesigned or repriced options (e.g. an executive train service with more comfortable seating at a higher price). The technique involves asking subjects to indicate their preference between pairs of hypothetical options each with specified attributes (e.g. a price of x, a journey time of y and a frequency of z). The components of the options presented in each pair are carefully specified such that, by examining the pattern of preferences stated by the subject, his or her underlying rationale for preferring one option over another can be deduced even if it was not obvious to the subject.

SP analysis was developed in the 1980s and 1990s and has now become a key element of many transport studies. SP exercises are usually conducted by face-to-face interviews, perhaps with the exercise set up on a computer such that the answers to initial questions can be used to help generate the most appropriate and experimentally efficient set of follow-up questions. Considerable success has, however, been had with self-completion SP exercises.

Rating and *scaling* exercises involve asking the subjects to indicate not simply their order of preference but rather the strength of their feelings about some aspect of a service. The rating or scaling task can be approached in a variety of ways. For example, the subjects may be invited to gauge their satisfaction with some aspect such as price, comfort, convenience or cleanliness using a scale from 1 (very good) to 5 (very poor), they may be invited to indicate how important they consider such and such an attribute to be when making their real-life choices or they may be asked to indicate their overall satisfaction score for different services or facilities. Another possibility, known as the *transfer price technique*, seeks to quantify the subjects' strength of preference by asking how much the price of a currently preferred option would have to rise before some other option became the preferred one. A full description of the alternative techniques is beyond the scope of this book but interested readers will find useful advice elsewhere.[1, 2, 12, 13, 14]

Simulators and *mock-ups* can be used as a source of information on attitudes and underlying preferences. Examples include the use, in environmental impact studies, of *photomontage*, computer enhanced *video montage* and, most recently, *virtual reality* as a means of exploring attitudes to potential visual impact. These techniques are used to show subjects how a particular scene might look after completion of a new scheme and they are then asked to indicate their attitudes or preferences using techniques such as those described in the previous paragraphs. Photographs, video and virtual reality techniques can also be used, along with physical mock-ups and models, to obtain user reaction to proposed redesigns of facilities such as railway coaches, platforms and airport concourses.

The increasing power of cheap portable computers has enabled the development of very sophisticated data collection tools known as *travel simulators*. Simulator subjects are invited to make journeys to given destinations using travel facilities represented in the simulator. As they make the journeys they will be given audio and visual feedback through computer-generated sound and images or sequences of digitised images taken from the equivalent journeys in the real world. The simulator can alter the conditions they meet, delaying them in queues or making them wait for a bus to arrive and so on. The software can record the travellers' behaviour in the simulated environment and subsequent analysis can reveal the extent to which this behaviour was affected by deliberate stimuli such as road signs, radio messages, timetable information, parking charges and so on, thus providing data more efficiently than is possible when relying on observation of real-world conditions where the stimuli are not under the experimenters' control. Results from using this type of simulator to explore drivers' responses to variable message signs suggest that the technique can produce very reliable results.[15]

14.7 References

1. Richardson, A.J., Ampt, E.S., and Meyburg, A.H., *Survey methods for transport planning*. Melbourne: Eucalyptus Press, 1995.
2. Moser, C.A., and Kalton, G., *Survey methods in social investigation* (2nd edition). London: Heinemann Educational Books, 1979.
3. Jones, P.M., Interactive travel survey methods: the state-of-the-art, In E.S. Ampt, A.J. Richardson, and W. Brog (eds), *New survey methods in transport*. Utrecht: VNU Science Press, 1985.
4. Bureau of Public Roads, *Conducting a home-interview origin and destination survey; Procedure manual 2B*. Washington, DC: US Public Administration Service, 1954.
5. Brog, W., Erl, E., Meyburg, A.H., and Wermuth, M.J., Development of survey instruments suitable for determining non home activity patterns. *Transportation Research Record 944*, 1983, 1–12.
6. Morton-Williams, J., *Interviewer approaches*. Aldershot: Social and Community Planning Research, Dartmouth Publishing Company, 1993.
7. Jones, P.M., Bradley, M., and Ampt, E.S., Forecasting household response to policy measures using computerised activity-based stated preference techniques. In IATBR (ed.), *Travel Behaviour Research*. Aldershot: Avebury, 1989.
8. Gordon, W., and Langmaid, R., *Qualitative market research: A practitioners' and a buyers' guide*. Aldershot: Gower, 1988.
9. Bonsall, P.W., and McKimm, J., Non response bias in roadside mailback surveys. *Traffic Engineering and Control*, 1993, **34** (12), 582–593.
10. Hamilton, T.D., Public transport surveys and the role of the microcomputer. *Proceedings of the PTRC 12th Summer Annual Meeting*. London: PTRC, 1984.
11. Bates, J.J. (guest editor), Stated preference methods in transport research, *Journal of Transport Economics and Policy*, 1988, **22** (1), 1–137.
12. Grigg, A.O., *A review of techniques for scaling subjective judgements*. Supplementary Report SR 379. Crowthorne, Berks: Transport Research Laboratory, 1978.

13. Tischer, M.L., Attitude measurement: psychometric modelling. In P.R. Stopher, A.H. Meyburg, and W. Brog (eds), *New horizons in travel behaviour research.* Lexington Books, 1981.
14. Bonsall, P.W., Transfer price data – its definition, collection and use. In E.S. Ampt, A.J. Richardson, and W. Brog (eds), *New survey methods in transport.* Utrecht: VNU Science Press, 1985.
15. Bonsall, P.W., Clarke, R., Firmin, P., and Palmer, I., VLADIMIR and TRAVSIM – powerful aids for route choice research. *Proceedings of the PTRC European Transport Forum.* London: PTRC, 1994.

CHAPTER 15

Accident prevention, investigation and reduction

C.A. O'Flaherty

15.1 Traffic accident terminology[1]

A *traffic accident* on a public road may involve a single road vehicle (e.g. a vehicle which skids and overturns), or it may involve a vehicle in a collision (e.g. between a vehicle and one or more vehicles, a pedestrian, an animal, and/or fixed object). In this context a public road includes footways, road vehicles can be motorised vehicles (including mopeds, motorscooters, motorcycles, and invalid tricycles) or pedal cycles, and the accident may involve an injury to a person (fatal, serious, or slight) or damage only to property.

A *fatal traffic accident* in Britain is one which involves a person who dies as a result of an injury sustained in the accident (usually within 30 days); it excludes traffic incidents involving confirmed suicides. A *serious accident* involves a person who is detained in hospital as an in-patient, or who suffers any of the following injuries: fractures, concussion, internal injuries, crushings, severe cuts and lacerations, or severe general shock, that require medical treatment. A *slight accident* is one involving a person who is only slightly injured, e.g. a person who sustains a sprain, bruise or cut, which is not judged (by the police) to be severe, or slight shock requiring only roadside attention. A *damage-only accident* does not involve people who sustain personal injuries.

It can be expected that all fatal traffic accidents and most serious accidents are reported to the police and, hence, appear in 'official' traffic accident statistics. Official statistics usually underestimate slight accidents and damage-only accidents.

15.2 Accident prevention

Accident prevention[2] is generally considered to be concerned with the application of safety principles to new road improvement or traffic management schemes that are initiated to satisfy traffic or environmental demands, and are thus not justified on the basis of accident savings *only*. As such work on accident prevention usually involves the carrying out of safety checks on designs for these schemes to ensure that no problem feature is introduced, and to identify whether any safety measures need to be added to lower their accident potential.

For example, if an intersection is being redesigned to improve traffic flow, the best safety principles of layout should obviously be applied. However, in addition, accident prevention checks should be carried out to ensure that carriageway surface changes, new road markings, or additional road furniture do not introduce hazards, and that any proposed speed control measures or pedestrian facilities enhance safety.

When accident prevention checking is applied to new road or traffic management schemes it is now usually termed a *safety audit*.[2, 3] Roadworks also require special monitoring, including frequent on-site accident prevention checks, to ensure that they are not the cause of accidents while a road or traffic management scheme is being implemented.

15.3 Accident investigation and reduction

In direct contrast with accident prevention, the term 'accident reduction' is normally associated with remedial proposals that result from direct accident investigations on existing roads and which are mainly justified by savings in the number and severity of accidents. The schemes/measures can usually be classified as low-cost engineering proposals, publicity and the education/training of various groups of road user (see Chapter 18), and/or advice regarding the enforcement of traffic laws (see Chapter 18).

There are four main investigative approaches used to develop accident reduction programmes. In Britain these are described[4] as: single site schemes; mass action programmes; route action programmes; and area action programmes. All involve four major planning steps: (a) data collection, storage and retrieval, (b) identification of hazardous locations for further study, (c) diagnosis of the accident problem(s), and (d) the final selection of sites to be included in the remedial implementation programme (see Fig. 15.1).

15.3.1 Single site approach

This, which is probably the most commonly used approach, involves the identification of 'blackspot' locations on the basis of the number of accidents clustered at single locales within a given period of time (usually three years). Blackspot locations are typically individual intersections, short lengths (300–500 m) of roadway, or small areas (100–200 m squares).

The total number of accidents (i.e. the reaction level) required to enable a site to be listed for subsequent investigation is usually determined using numerical or statistical techniques. A large accident reduction (typically 33 per cent) and a high first-year economic rate-of-return (typically not less than 50 per cent) should be expected[2] from blackspots that are included in a final remedial action programme for single sites.

15.3.2 Mass action approach

This, the second most commonly used approach, involves searching for sites that are clearly associated with a particular predominant type of accident for which there is a well-proven engineering remedy. Once identified these sites can normally be economically treated en masse, for instance by the resurfacing or surface dressing of skid sites. Table 15.1 lists some common accident situations and well-tried remedial actions.

Mass action programmes should be expected[2] to achieve an average accident reduction of 15 per cent at treated sites and a first-year economic rate-of-return of not less than 40 per cent. They are particularly applicable to publicity and road user training schemes and as an aid to more effective deployment of police traffic personnel.[4]

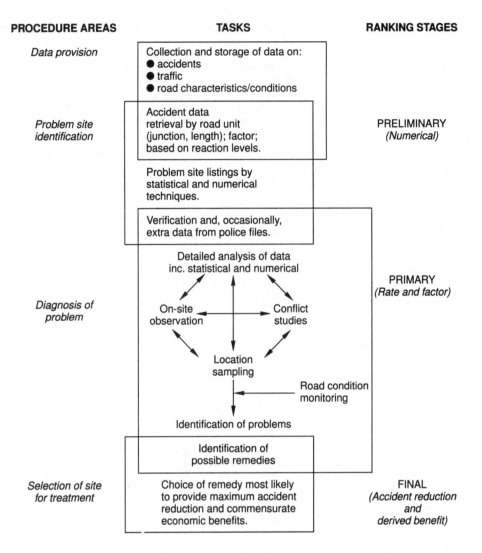

PROCEDURE AREAS

Data provision

Problem site identification

Diagnosis of problem

Selection of site for treatment

TASKS

Collection and storage of data on:
● accidents
● traffic
● road characteristics/conditions

Accident data retrieval by road unit (junction, length); factor; based on reaction levels.

Problem site listings by statistical and numerical techniques.

Verification and, occasionally, extra data from police files.

Detailed analysis of data inc. statistical and numerical

On-site observation   Conflict studies

Location sampling

Road condition monitoring

Identification of problems

Identification of possible remedies

Choice of remedy most likely to provide maximum accident reduction and commensurate economic benefits.

RANKING STAGES

PRELIMINARY
(Numerical)

PRIMARY
(Rate and factor)

FINAL
(Accident reduction and derived benefit)

Fig. 15.1 Outline of procedure for identifying, diagnosing and selecting hazardous road sites for remedial treatment[2]

15.3.3 Route action approach

With this approach the main 'blacksites' along a particular road or class of road, i.e. sites with more accidents than the norm for that type of road and level of traffic usage, are identified from an analysis of traffic data collected over a recent period (usually 1–3 years). Usually the search process involves dividing the road(s) into section lengths of 0.5–1.5 km and, typically, selecting for listing those sections with accident levels of 1–2 standard deviations above the norm or, alternatively, using a statistical test for a predetermined level of significance above the norm (see reference 4).

Table 15.1 Proven remedies for some common accident situations (adapted from reference 4)

Situation	Remedies
Wet road	
Skidding	Restore micro/macro texture with use of a high polished stone-value aggregate in surface dressing or a conventional layer of bituminous surfacing, or by grooving surfacing
Splash obscuring visibility	Restore macro-texture
Poor delineation	Contrast texture of markings with carriageway surface
Darkness – lit road	
Poor/uneven surface luminance	Match surface texture with lighting installation
Inadequate illuminance	Renew lighting installation
Darkness – unlit road	
Poor delineation	Install edge markings or improve reflectorisation of lane/edge markings; install cats'-eyes
Roadside obstacles	
Hazard by presence	Install safety fences or guardrails; remove obstacle; install frangible columns (rural, non-footpath locations only)
Cross-reserve or running off road with steep side-slopes or drops	Install safety fences or guardrails
At urban intersections	
Turning traffic problems	Install channelisation; install roundabout or traffic signals; add exclusive turning phase to signals; install traffic islands (physical or ghost)
Overshoot from minor road	Install physical traffic islands
Pedestrian problems	Have traffic signal with pedestrian phase
Sight restriction (at bend)	Remove or relocate obstruction
Between urban intersections	
Pedestrian problems	Install light-controlled crossing, subway or bridge; install guardrails; prohibit parking; restrict access and/or introduce traffic-calming measures (residential areas)
Parked vehicle problems	Restrict parking
Bicycle problems	Delineate special cycle lanes
Excessive speeding	Improve enforcement; introduce speed-reducing humps
At rural intersections	
Turning traffic	Install channelisation; install roundabout or traffic signals; add extra lanes (deceleration/ acceleration); install traffic islands (physical or ghost); prohibit turns
Overshoot from minor road	Install physical traffic islands
Overshoot at roundabout on main road (excess speed)	Add chevrons and warning signs; add yellow bar markings (on dual carriageways)
At rural intersections with bends	
Collisions while waiting	Install physical traffic islands; widen and dual
Sight restriction	Relocate minor access; remove vegetation; realign oblique-angled intersection; adjust minor road profile
Between rural intersections	
Overtaking problems	Install double-line markings
Loss of control at bends	Correct superelevation; install advisory speed signs
Heavy commercial vehicle problems on hills	Add extra lane uphill; add run-off provision downhill
Opposing vehicle problems on crest of hill	Add double-line markings and deflection arrows
Excessive speed	Improve enforcement

In contrast with mass action programmes, remedial actions implemented as a result of route action programmes vary with the accident situation encountered. Blacksites treated should achieve average accident reductions of 15 per cent and first-year economic rates-of-return of not less than 40 per cent.[2]

15.3.4 Area action programmes

In urban areas in particular a significant proportion of accidents are sparsely scattered and do not lend themselves to selection for treatment by the three previously described methods. In this case the distribution of accidents throughout the urban area is searched over a recent one-, two- or three-year period in order to identify discrete areas, e.g. neighbourhoods or 1 km Ordnance Survey grid squares, having accidents per unit area or per unit of population above a predetermined level.

Traffic calming measures that are aimed at reducing traffic movement on local access roads within the identified areas are often the remedial outcomes of area action programmes. Appropriate objectives set for area action plans might be to achieve an accident reduction of 10 per cent and a first-year economic rate-of-return of 10–25 per cent within each area addressed.

15.3.5 The planning process

The database

Fundamental to any ongoing accident study programme, whatever the approach used, is the establishment of a databank of accident records for the road authority's area of responsibility. In Britain the national Stats 19 police reporting system for accidents is the core database upon which most local authority data systems are built. Key data included in a Stats 19 form for every reported accident are:

1. a basic accident description, e.g. its severity, the number of casualties and vehicles involved, contributory factors (if collected), and the time, day, date and location at which it occurred
2. road features, e.g. class and identification number, carriageway type or markings, speed limit, and intersection type and control
3. environmental features, e.g. weather and light conditions, surface condition, any carriageway hazards, or any special conditions
4. vehicle features, e.g. vehicle type, manoeuvres, movements and location, skidding, and hit object(s)
5. driver features, e.g. age, sex, breath test results, and whether a hit-and-run was involved
6. casualty details, e.g. sex, age and severity of injury, and pedestrian location, movement and direction, and any school pupil involvement.

Additional data regarding each accident that are normally also included in the local authority's database include the exact road location of each accident and such site characteristics as road geometry, surface type and texture, adjacent land use, speed limits, physical aids, e.g. lighting, signs and markings, and permanent extraneous features, e.g. advertisement hoardings and noticeboards, posts, poles, tree trunks and guardrails, and street furniture such as bus shelters or telephone boxes.

Traffic data included in the database normally relate to the time and location of each accident, e.g. traffic flow and composition by class of road user, pedestrian flows across and along roads and, as appropriate, vehicle and pedestrian delay information and queue lengths.

Nowadays most road authority data systems are computerised, and many authorities' computers directly link the sources of data for accidents, roads and traffic. With the aid of the computer large quantities of data are easily and economically stored and manipulated, and accident ranking lists, tables, graphs and automatic plots can be output on demand (see reference 4 for examples).

Very many local authorities also continue to maintain pictorial representations of accident clusters, be they at locations such as intersections, bends, along stretches of road, or within geographic areas. Often these involve manual plots on transparencies to overlay 1:2500 O.S. area maps for urban areas and 1:10 000 O.S. strip maps for rural areas. Blackspot wall maps, using pins with differently coloured heads to signify different severities of accident, are particularly useful as a public relations and education tool when housed in local road safety centres patronised by the general public. An historical record of these maps is easily maintained by photographic means.

Site identification

This phase of the investigation is concerned with the sifting of the accident data to obtain a preliminary ranked list of sites that are most likely to be susceptible to engineering treatment. There is no uniformly accepted procedure for doing this, and so the tendency has been to use fairly simple numeric criteria to select initially the sites, roads or areas for further investigation. For example, the reaction level might be twelve injury accidents per defined spot (or per road length, or per defined urban area) that occurred within the last three years, and any sites with more than that number require further investigation. The sites identified in this preliminary analysis may then be listed in numerical order of accident occurrence, from the highest (most dangerous) to the lowest.

Although the numeric criterion method of ranking sites is very simple, using it on its own leads to priority of subsequent examination being biased toward sites on major traffic routes, simply because they have large numbers of accidents. For this reason, many local authorities prefer to base their rankings on an accident rate which relates accident occurrence to a measure of exposure, e.g. injury accidents per 10^6 vehicles passing through a blackspot intersection (or per 10^6 vehicle-kilometres for a blacksite section of road, or per 10^3 population for an urban neighbourhood). When accidents per 10^6 vehicles is used as the measure, it should lead to the selection of intersections on low-trafficked minor roads even though they have relatively few accidents. Overall, the use of an exposure rate is intended to give greater assurance to the investigators that the more the rate is above the norm for, say, the level of traffic flow through that type of intersection, the more likely it is that the site is truly a high-risk blackspot.

A simple method of identifying hazardous locations that has been commonly used in the United States is to assign a rating number to each accident on the basis that accident severity is the most important consideration. Thus, for example, 9 points might be assigned to each fatal accident, 1 point to each serious-injury accident, 0.1 points to each slight-injury accident, and 0.01 for each damage-only accident, and the site with the greatest number of points would then be regarded as the most dangerous. (The ratings used in this example are roughly based on the relative costs of fatal, serious, slight and damage-only accidents in Britain at this time.)

Some investigators also divide their priority list into 'easy' and 'hard' sites. In this context, easy sites are those that appear to have consistent accident patterns, and, where remedial action requires no land purchase and can probably be easily implemented at low cost within a reasonable time-frame. A hard site is one with many accidents but where each appears to involve a different cause, so that further extensive and detailed investigation is likely to be required before any remedial action can be implemented.

Diagnosis

The next stage in the planning process is the detailed analysis of the accident situation at each site. The objective is to diagnose the main cause(s) of each accident and develop a remedial proposal that will obviate the common dominant causes.

Detailed statistical analyses are often carried out at this stage to confirm that the sites are indeed high-risk and to lay the basis for subsequent 'after' studies which may be carried out to test and confirm the beneficial effect of the eventual treatment. These analyses can be complex (see, for example, reference 5) and it is advisable to get expert statistical advice if the investigator has doubts about the adequacy of the statistical testing that is proposed.

As part of the engineering analysis collision and condition diagrams are very often prepared for each site to assist in the determination of the dominant accident cause(s), particularly if physical features are believed to be influencing the accident situation. Figure 15.2 is a hypothetical *collision diagram* drawn for an intersection (usually developed on a 1:500 scale plan) to illustrate how and when each accident happened, based on the official reports. Also shown are the conditions prevailing at the site at the time of the accident; colour coding can be used if further information needs to be added. A *condition diagram* is a scaled drawing or photograph illustrating the physical and environmental conditions at and about the accident site, e.g. the geometric features affecting the site, and the locations of all road signs and markings, pedestrian crossings, traffic signals, bus stops, parking spaces, sight obstructions, driveways, and fronting land uses. Usually the analysis of the collision and condition diagrams is supplemented by an on-site visit, preferably during the hours when the greatest number of accidents are known to occur. The field visit may simply involve taking skid-resistance measurements, or it may require the use of a detailed check-list to determine the adequacy of various physical design features. In some instances it may involve carrying out a comprehensive on-the-spot conflict study.

It is now internationally agreed that a *traffic conflict* [6] is an observable situation in which two or more road users approach each other in space and time to such an extent that a collision is imminent if their movements remain unchanged. In other words, traffic conflicts are events where there is a possibility of an accident, but a collision does not occur because at least one of the involved parties takes an evasive action. Conflicts occur more often than accidents so that a supplementary on-site conflict study, while expensive, may be able to identify unsafe driving and/or accident-generating manoeuvres at hard-to-diagnose high-risk accident locations.

In the case of a mass action approach, it should be possible at this stage to group together accident data for sites of similar physical characteristics and/or accident features to obtain sufficient data for further assessments to be carried out. This process,

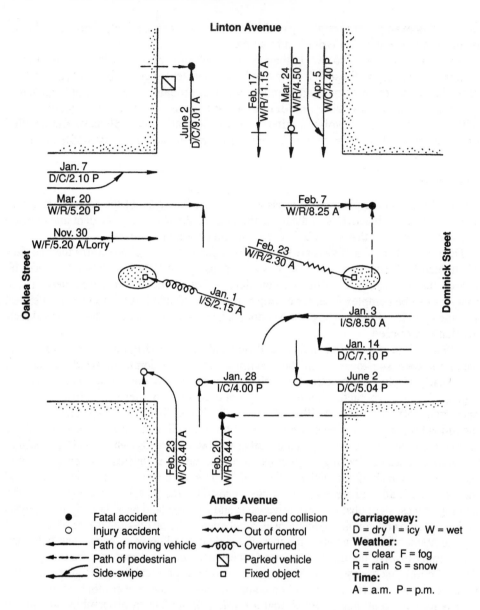

Fig. 15.2 Example of a hypothetical collision diagram at an intersection

known as *location sampling*, is fundamental to the production of a mass action accident-reduction programme.

Whatever the approach adopted the end effect of the detective work carried out to date is a diagnosis of the main cause(s) of the accidents at each site under investigation, and remedial proposals for their reduction (see Table 15.1). References 4 and 7 contain excellent examples of remedial solutions and comprehensive check-lists that can be used to help determine them.

Final selection

Before implementing a remedial proposal it is usual to test that it is economically justified. In some instance, two or more solutions of different cost are proposed for a given site and a further evaluation is required in order to determine which is the more viable. In many instances the local authority's budget for the year will be unable to meet the implementation costs of all the proposed remedies for all the sites on the preliminary list of hazardous locations, and it is necessary to develop a new priority list for implementation.

Various economic methods are used to resolve these and other issues of choice. In Britain a determination that is often used to resolve for this purpose is the first-year economic rate-of-return (*FYRR*):

$$FYRR, \% = \frac{\text{Net benefits in the first year after completion of scheme}}{\text{Total capital cost of scheme}} \times 100 \quad (15.1)$$

The first-year rate-of-return evaluation is appropriately used to compare proposals that have similar scheme lives and similar profiles of cost and benefit throughout their lives. Then the scheme with the highest *FYRR*-value can be given the highest priority.

If, however, the schemes are significantly different, the comparisons are better made on the basis of the net present values of the schemes as determined over their anticipated lives (see reference 4).

15.4 References

1. *Road accidents Great Britain 1993: The casualty report.* London: HMSO, 1994.
2. *Highway safety: Guidelines for accident reduction and prevention* (2nd edition). London: The Institution of Highways and Transportation, 1986.
3. Road safety audits, Part 2: HD 19/94 and Part 3: HA 43/94. In *The design manual for roads and bridges*, Vol. 5, Section 2. London: HMSO, undated.
4. *Accident investigation manual* (2nd edition), Vols 1 and 2. London: The Department of Transport, 1986.
5. Wright, C.C., Abbess, C.R., and Jarrett, D.F., Estimating the regression-to-mean effect associated with road accident blackspot treatment: Towards a more realistic approach. *Accident Analysis and Prevention*, 1988, **20** (3), 199–214.
6. *Highway safety: Guidelines for the traffic conflict technique.* London: The Institution of Highways and Transportation, 1987.
7. *Hazardous road locations: Identification and countermeasures.* Paris: OECD, September 1976.

PART III

Design for capacity and safety

CHAPTER 16

Introduction to traffic flow theory

A.D. May

16.1 Introduction

Road space is a scarce resource, and traffic engineers need to ensure that roads are able to accommodate as much traffic as possible, subject to safety and environmental constraints. In other words, we need to maximise the capacity of the road. But how is capacity measured, and what influences it? These questions lie at the core of traffic flow theory, a science which has attracted both theoretical and empirical analysis. In this chapter there is space only to introduce the concepts, the principal parameters and some relationships which enable us to estimate capacity. Further developments of these concepts are given in reference 1.

16.2 The principal parameters

16.2.1 The range of conditions

Every driver will have experienced the range of traffic conditions which can be experienced on the same length of road, and which are illustrated in Fig. 16.1. When traffic is light, the road is relatively empty (Fig. 16.1a) and drivers are free to choose their own speeds. As traffic increases (Figs 16.1b, c, d) drivers are more constrained by other vehicles, and less able to overtake; they are thus less able to choose their own speeds, and average speeds fall. As traffic levels increase further, traffic forms into platoons of slow-moving vehicles (Fig. 16.1e) which may stop and start. Finally (Fig. 16.1f) traffic levels become so great that queues form, and traffic may be at a standstill for considerable periods. These six conditions are referred to in the US *Highway Capacity Manual*[2] as Levels of Service A to F (see also Chapter 17).

Which of them, though, represents the capacity of the road? Clearly Fig. 16.1f has the highest concentration, but is not a desirable state of affairs.

16.2.2 Measures of quantity

There are in practice two ways in which the number of vehicles can be counted on a road. One is illustrated by Fig. 16.1. One can photograph a length of road, x, count the

Fig. 16.1 Levels of service

number of vehicles, n_x, in one lane of the road at a point in time, and derive a rate per unit distance. This measure is called the *concentration* of traffic (sometimes referred to as *density*) and is denoted by the parameter k (veh/m). Thus

$$k = \frac{n_x}{x} \qquad (16.1)$$

The second approach is to stand at the side of the road for a period of time, t, and count the number of vehicles, n_t, passing that point in one lane in that period and derive a rate per unit time. This measure is called the *flow* of traffic (sometimes referred to as *volume*) and is denoted by the parameter q (veh/s). Thus

$$q = \frac{n_t}{t} \qquad (16.2)$$

Since one is usually concerned to get as many vehicles through a road as possible in a given time, it is q rather than k which one wishes to maximise, and capacity is then described as the maximum value of q, q_m. However, it is not immediately clear which of Figs 16.1a–f represents this condition.

16.2.3 Measures of separation

While concentration and flow indicate how much traffic a road is handling, the traffic engineer is also interested in the separation between vehicles, which may affect safety (vehicles travelling close together may be more likely to crash) and the ease with which pedestrians and vehicles can cross the traffic stream. Once again there are two ways of measuring separation between vehicles in a lane.

The first involves measurement from a photograph, conventionally measuring the distance from rear bumper of the lead vehicle to rear bumper of the following vehicle at a point in time. This is called the *space headway* (or *spacing*), s. If all the space headways in the distance x over which concentration has been measured are added, then

$$\sum_1^{n_x} s_i = x \qquad (16.3)$$

Thus concentration is the inverse of the average space headway:

$$k = \frac{n_x}{x} = \frac{n_x}{\sum_1^{n_x} s_i} = \frac{1}{\bar{s}} \qquad (16.4)$$

The second approach involves recording the time between the passage of one rear bumper and the next past a given point. This is the *time headway* (or *headway*), h. If all the time headways in the time period, t, over which flow has been measured are added, then

$$\sum_1^{n_t} h_i = t \qquad (16.5)$$

Thus flow is the inverse of the average time headway h:

$$q = \frac{n_t}{t} = \frac{n_t}{\sum_1^{n_t} h_i} = \frac{1}{\bar{h}} \qquad (16.6)$$

16.2.4 Measures of quality

While traffic engineers are concerned to achieve flows approaching capacity, at least at busy times, individual drivers will be most concerned about the quality of their journeys. Their main concern will be with the speed at which they can travel. Individual speeds (in m/s or km/h) are easy enough to measure, but traffic engineers will be more concerned with average speeds. Here again there are two ways of measuring speed, and the distinction between them often causes confusion. The difference between the two methods can be illustrated by a simple example.

Three vehicles are recorded taking 6 s, 8 s and 10 s to cover 100 m. Method 1 would calculate the average time, 8 s, and divide this into the distance. This gives an average speed of 100/8 or 12.5 m/s. Method 2 would calculate the individual speeds: 16.7, 12.5 and 10.0 m/s and average them, as 39.2/3 or 13.1 m/s. Both are correct; they simply represent different ways of averaging. Clearly it is important to be consistent in the averaging method which is used. The two speeds can be related to the two different measures of quantity and separation. For example, suppose that a stream of traffic includes vehicles travelling at n different speeds, $u_1 \ldots u_n$, and that two different approaches are adopted to determine the overall average speed.

The first involves photographing a length of road on which there are k vehicles per unit length. Let the concentration of set i of vehicles with speed u_i be defined as k_i. The average speed of the k vehicles is referred to as the *space mean speed*, $\bar{u}_s$, since the average is of all vehicles in a given space (at one point in time). Then $\bar{u}_s$ is given by

$$\bar{u}_s = \frac{\sum_1^n k_i u_i}{k} \qquad (16.7)$$

The second involves recording all vehicles passing a point in a given time, i.e. q per unit time. Let the flow of set i of vehicles with speed u_i be defined as q_i. The average speed of the q vehicles is referred to as the *time mean speed*, $\bar{u}_t$, since the average is of all vehicles passing in a given time. Then $\bar{u}_t$ is given by

$$\bar{u}_t = \frac{\sum_1^n q_i u_i}{q} \qquad (16.8)$$

However, only one of these two speed averages is related to the parameters q and k.

16.3 The fundamental relationship

16.3.1 The relationship between concentration, flow and speed

For any given stable traffic condition, the three parameters k, q and u are directly related. This can be seen from the simple example in Fig. 16.2, which considers a kilometre of road, on which all vehicles are travelling at the same speed. The parameters K, Q and U are concentration in veh/km, flow in veh/h and speed in km/h respectively. By definition there are K vehicles, each with speed U, in the kilometre of road at any instant. If the flow is recorded at the end of the road, Q vehicles will pass per hour. The vehicle at the start of the kilometre will take $1/U$ hours to reach the end at a speed of U km/h. It will then be the Kth vehicle to pass the end of the road, and this will happen in K/Q hours. Thus

$$\frac{1}{U} = \frac{K}{Q} \qquad (16.9)$$

or

$$Q = KU$$

This is dimensionally correct, i.e. veh/hour = (veh/km)(km/hour). The same expression using metres and seconds is

$$q = ku \qquad (16.10)$$

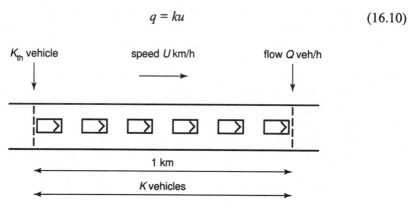

Fig. 16.2 A simple derivation of the relationship between concentration, flow and speed

16.3.2 The shape of the fundamental relationship

Although Equation 16.10 is important, it is still necessary to know the values of two of the parameters in order to calculate the third. In practice, on a given road, a given concentration is likely to give rise to a certain value of flow and a certain value of speed, subject always to the variations in driving conditions and driver behaviour. This relationship can best be seen by considering the three parameters a pair at a time, as in Fig. 16.3. In each of them there are two limiting conditions, which can be thought of by reference to Figs 16.1a and f.

In Fig. 16.1a flow is very low (approaching zero), and so is concentration; speed, which is likely to be at its highest, is referred to as free-flow speed, u_f. In Fig. 16.1f, speed is zero and so, since traffic is not moving, is flow; however, concentration is at its highest, and is referred to as jam concentration, k_j.

Figures 16.3a and 16.3b illustrate that as speed falls from u_f to zero, and as concentration rises from zero to k_j, flow first increases and later falls again to zero. It seems reasonable to expect that it only rises to one maximum, and this can be thought of as the capacity, q_m. As flow increases towards capacity it is relatively easy to understand what is happening; as Figs 16.1a–e show, vehicles increasingly disrupt one another, reducing the individual driver's ability to choose his or her own speed or overtake others. Speed thus falls and concentration increases. Beyond capacity it is less easy to explain what is happening. In practice such conditions are caused by queues from downstream conditions, perhaps a junction, or an accident, or even a gradient or tight curve whose capacity is slightly lower. The queue leads to increased concentration; speeds fall further, and vehicles which cannot flow past the point join the queue as it stretches upstream.

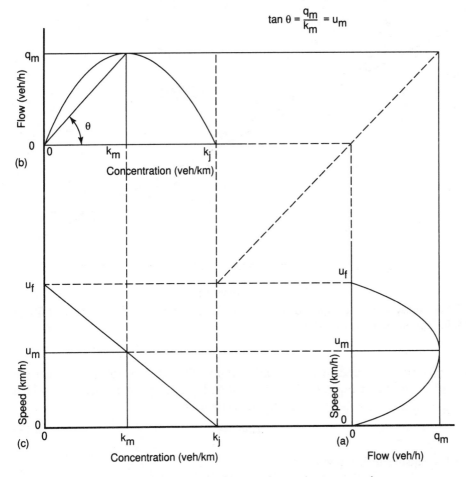

$$\tan \theta = \frac{q_m}{k_m} = u_m$$

Fig. 16.3 Flow–concentration, speed–flow, and speed–concentration curves

16.3.3 Empirical relationships

Several analysts have attempted to fit relationships to observed data. This is most easily done with concentration and speed (Fig. 16.3c), since the relationship is monotonic. However, care needs to be taken in manipulating data to fit such a relationship.[3]

Figure 16.4 presents a data set extensively used for such analysis, and collected in the United States.[4] It appears to suggest a linear relationship between speed and concentration, and this was first suggested by Greenshields[5] in 1934:

$$u = a + bk \qquad (16.11)$$

Using the two limiting conditions it can be shown that this becomes

$$u = u_f \left(\frac{1 - k}{k_j} \right) \qquad (16.12)$$

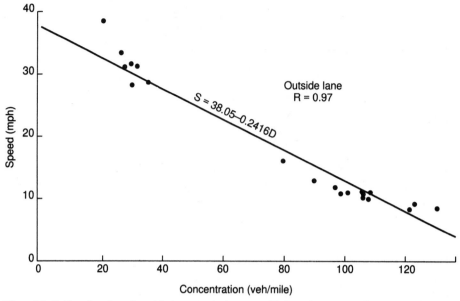

Fig. 16.4 Study showing high correlation coefficient between field data and linear model for speed vs. concentration

Other analysts have noted that the relationship in Fig. 16.4 is not quite linear (as suggested by the fit of Equation 16.12) but slightly concave. One suggestion which accounts for this is Greenberg's[6] logarithmic relationship:

$$u = a \log_e bk \qquad (16.13)$$

in which k_j equates to $1/b$, giving

$$u = \frac{a \log_e k}{k_j} \qquad (16.14)$$

but unfortunately u_f has a value of infinity. A second suggestion is Underwood's exponential relationship:[7]

$$u = ae^{-bk} \qquad (16.15)$$

in which u_f equates to a, giving

$$u = u_f e^{-bk} \qquad (16.16)$$

but k_j has a value of infinity.

Each of these relationships has been supported by theoretical analysis, primarily based on an analogy with fluid flow[8] and the interpretation of car-following behaviour.[9] However, it is important to stress that traffic flow is not a wholly scientific phenomenon, but one which depends on the vagaries of driver behaviour. Indeed, some authors have suggested that there is no reason why conditions above and below capacity should be part of the same relationship, since they arise in different ways.[10, 11]

16.3.4 Estimation of capacity

Subject to these caveats, it is possible to use any of Equations 16.12, 16.14 or 16.16 to provide a theoretical estimate of the capacity of a length of road from a pair of observed data points. This is most easily seen with Greenshields' relationship, Equation 16.12. Fig. 16.3b shows the relationship between flow and concentration which, combining Equations 16.10 and 16.12, is:

$$q = ku_f \left(\frac{1-k}{k_j} \right) \tag{16.17}$$

The maximum value of q can be determined by differentiation:

$$\frac{dq}{dk} = u_f - \frac{2u_f k}{k_j} \tag{16.18}$$

which, when equated to zero, gives:

$$k_m = \frac{k_j}{2} \tag{16.19}$$

and, from Equations 16.12 and 16.16:

$$u_m = \frac{u_f}{2} \tag{16.20}$$

$$q_m = \frac{k_j u_f}{4} \tag{16.21}$$

16.3.5 The appropriate value of average speed

Before applying this approach to estimating capacity, however, one needs to know which average of speed (from Section 16.2.4) to use. This can be seen immediately by reference to Equations 16.7 and 16.8, in conjunction with Equation 16.10. Equation 16.7 can be further simplified to give

$$\bar{u}_s = \frac{\sum\limits_{1}^{n} q_i}{k} = \frac{q}{k} \tag{16.22}$$

Equation 16.8 cannot be further simplified in this way. Thus it is *space mean speed* which is related to flow and concentration in the fundamental relationship.

16.3.6 A worked example

Two sets of vehicles are timed over a kilometre, and the flows are also recorded. In the first set, four vehicles take 52, 56, 63 and 69 seconds when the flow is 1500 veh/h. In the second, four vehicles take 70, 74, 77 and 79 seconds when the flow is 1920 veh/h. These vehicles have been timed over space, so the space mean speeds can be calculated from the average times for each set. These give:

	Set 1	**Set 2**
$\bar{u}_s$	60 km/h	48 km/h
q	1500 veh/h	1920 veh/h
k	25 veh/km	40 veh/km

Fitting these two pairs of values of u and k into Equation 16.12 (clearly a substantial approximation) gives u_f = 80 km/h and k_j = 100 veh/km. Incorporating these into Equation 16.21 gives q_m = 2000 veh/h. This is broadly represented by conditions in Fig. 16.1e, in which vehicles keep moving at a slow but uniform speed.

16.3.7 A comment

Care is needed in using these empirical relationships in this way, since there are no scientific laws explaining the behaviour of traffic. Moreover, in practice vehicles are not homogeneous, and practical analysis uses the concept of passenger car units (pcus) to describe the effect of vehicles of different sizes. Typical values of capacity (the highest flows recorded on a section of road) are in the range 1500–2500 pcu/h, and for jam concentration 150–200 pcu/km. Free-flow speeds are inevitably more variable, and depend on drivers' perceptions of the safety and acceptable maximum speed on the road. It is thus unwise to design for the capacity as calculated in Equation 16.21. British practice uses the concept of *practical capacity* and US practice *levels of service* (see Fig. 16.1), both of which are described in Chapter 17.

16.4 References

1. Gerlough, D.L., and Huber, M.J., *Traffic flow theory: a monograph*. Special Report 165. Washington, DC: Transportation Research Board, 1975.
2. Transportation Research Board, *Highway capacity manual*. Special Report 209. Washington, DC: US Transportation Research Board, 1994.
3. Duncan, N.C., A further look at speed/flow/concentration. *Traffic Engineering and Control*, 1979, **20** (10), 482–3.
4. Huber, M.J., Effect of temporary bridge on parkway performance. *Highway Research Board Bulletin 167*. Washington, DC: Highway Research Board, 1957.
5. Greenshields, B.D., A study of traffic capacity. *Proceedings of 14th Annual Meeting of the Highway Research Board*. Washington, DC: Highway Research Board, 1934.
6. Greenberg, H., Analysis of traffic flow. *Operations Research*, 1959, **7** (1).
7. Underwood, R.T., Speed, volume and density relationships. In *Quality and theory of traffic flow*. Yale: Bureau of Highway Traffic, 1961.
8. Lighthill, M.J., and Whitham, G.B., On kinematic waves II : A theory of traffic flow on long crowded roads. *Proceedings of Royal Society*, 1955, **A229** (1178).
9. Herman, R., Montroll, E.W., Potts, R.B., and Rothery, R.W., Traffic dynamics: studies in car following. *Operations Research* 1959, **7** (1).
10. Edie, L.C., Car following and steady state theory for non-congested traffic. *Operations Research*, 1961, **9** (1).
11. Hall, F.L., and Montgomery, F.O., The investigation of an alternative interpretation of the speed–flow relationship on UK motorways. *Traffic Engineering and Control*, 1993, **34** (9).

CHAPTER 17

Road capacity and design-standard approaches to road design

C.A. O'Flaherty

17.1 Capacity definitions

The capacity of a road can be described simply as the extent to which it is able to provide for traffic movement under given circumstances. This description allows for three different usages of the word 'capacity'.

The *economic capacity* of a road is a term used to describe the smallest of all traffic volumes which needs to be attained so that a road project is justified by cost-benefit analysis.[1] In practice, the traffic volume required to justify economically a given quality of road at one location may be significantly different from that at another, depending upon the many factors which affect the economic equation, such as the cost of land or the extent of the congestion experienced on the existing roadway(s) which the new facility is intended to supplement/replace. Most road proposals are now subjected to economic analysis before final decisions are made in respect of their implementation, to ensure that their economic capacities are at least achieved.

The *environmental capacity* of a road is a term that is most usually applied to road improvements that affect historic or residential areas. It has been described[2] as the capacity of a street or area to accommodate moving and stationary vehicles having regard to the need to maintain the (chosen) environmental standards. In other words, environmental capacity is the upper limit of traffic volume that is permitted on the road(s) in question without exceeding desired minimum standards relating to, say, noise, air pollution, pedestrian and cyclist safety and amenity, and visual intrusion. The environmental standards acceptable in a given locale can vary considerably depending upon, for example, the time of day or the type of street frontage. While there are no firm guidelines regarding environmental capacity, it has been suggested that the maximum hourly volumes that can normally be tolerated on access or local distributor roads are 300–600 veh/h.

The *traffic capacity* of a road has been defined[3] as the maximum hourly rate at which vehicles can reasonably be expected to traverse a point or uniform section of a road or lane during a given time period under the prevailing roadway, traffic and control conditions. Use of a maximum rate of flow in this definition recognises that there can be substantial variations in the flows experienced within an hour, especially under near-congested conditions, and it allows a rate of stable traffic flow that can be repeatedly achieved during peak periods, but which is measured over a shorter period of time (usually 15 minutes), to be extrapolated to an equivalent hourly rate.

Roadway conditions relate to the effect that the geometric features of the road have upon traffic capacity, e.g. road type and its development environment, width of lanes, shoulder width and lateral clearance, design speed, and horizontal and vertical alignment. Traffic conditions relate to the impacts of the numbers of different types of vehicle in the traffic stream(s), and their lane or directional distribution(s). Control conditions refer mainly to the effect of traffic control features and devices at intersections, which cause vehicles to stop or slow down irrespective of how much traffic exists. Ideal road/traffic/control conditions assume good weather, good pavement and cross-section conditions, users familiar with the facility, and no incidents impeding traffic flow, so that further improvement will not achieve any increase in capacity.

In practice, two categories of traffic capacity can be discussed: those determined for interrupted and uninterrupted flow conditions.

Uninterrupted flow conditions apply to traffic operations on road links *between* intersections, i.e. where intersectional flows do not interfere with continuous traffic movement. Interrupted flow conditions apply to at-grade intersections where the effects of intersecting traffic flows and associated control devices predominate. The following discussions regarding the *Highway Capacity Manual* and British design-standard approaches to determining the traffic capacities of roads are only concerned with design capacities for uninterrupted traffic flow conditions on road links between intersections.

17.2 The *Highway Capacity Manual* approach

First published in 1950, the U.S. *Highway Capacity Manual*[3] provides an assembly of analysis techniques to be used by traffic engineers to determine the capacity rate of traffic flow to use in road design, to ensure that a specified quality of service is provided to the motorists using a new/improved facility. The procedures can also be used to assess the quality of service provided on an existing road, thereby indicating whether or not it needs upgrading.

While the actual capacity values given in the *Highway Capacity Manual* are generally considered to be directly applicable only in the United States, the level-of-service approach espoused in this document – which is based on extensive research on American roads over many decades – is widely used throughout the world, and adapted to suit local conditions in most countries. It is not, however, directly used in road design in Britain.

17.2.1 The level-of-service concept

The *Highway Capacity Manual* approach is basically concerned with the quality of service provided by a roadway at a given rate of traffic flow per lane or per carriageway *as perceived by the driver of the vehicle*. Six levels of service are defined for each road type considered. These are designated as levels of service A through to F, with level A representing free-flow, low-volume, high-speed, comfortable operating conditions while level F represents forced-flow, stop-start, uncomfortable conditions (see Fig. 16.1). The levels recommended by the American Association of State Highways and Transportation Officials for use when designing various types of road in differing locales are given in Table 17.1.

Table 17.1 Recommended levels of service for use with various types of road[8]

Road type	Rural			Urban and suburban
	Level	Rolling	Mountainous	
Freeway	B	B	C	C
Arterial	B	B	C	C
Collector	C	C	D	D
Local	D	D	D	D

The following discussion provides some qualitative detail in relation to the application of the level-of-service concept to traffic operations on two types of road, multi-lane freeways (motorways) and two-lane single carriageway roads.

Freeways

Known as *motorways* in Britain, freeways are the highest form of dual carriageway roads, with two or more lanes for the exclusive use of traffic in each direction and full control of access and egress. Full control of access and egress means that through traffic on the mainline has priority of movement, at-grade crossings and private driveway connections are prohibited, and movements onto or off the mainline are only possible by way of specific roads (called ramps in America and slip roads in Britain) which are designed to ensure that high-speed merging and diverging manoeuvres can take place with minimal disruption to the mainline's through traffic. As a consequence, operating conditions on freeways are mainly influenced by the interactions between vehicles within the traffic stream, and between vehicles and the geometric characteristics of the freeway. Environmental conditions such as weather and the condition of the carriageway surfacing (e.g. whether there are potholes), and the occurrence of traffic incidents (e.g. accidents or vehicle breakdown) also affect operating conditions.

When vehicles are operating under *level of service A* conditions on freeways, they are almost completely unimpeded in their ability to manoeuvre within the traffic stream, and the effects of minor incidents or breakdowns are easily absorbed without standing queues being formed. These operating conditions afford the driver a high level of physical and psychological comfort. At *level B* operating conditions vehicles are still able to operate under reasonably free-flow conditions, and traffic manoeuvres are only slightly restricted.

Level of service C provides for stable operations at speeds that are still at or near the free-flow speed. However, flows are at the stage where freedom of manoeuvre within the traffic stream is noticeably restricted and lane changes require additional care and vigilance by the driver. Minor traffic incidents may still be absorbed but queues can be expected to form behind any significant blockage. These operating conditions cause the driver to experience a noticeable increase in tension due to the additional vigilance required for safe motoring. *Level D* is that at which speeds begin to decline slightly with increasing flows, and small increases in flow within this range cause substantial deterioration in density. Freedom to manoeuvre is more noticeably limited, and minor incidents cause substantial queuing because the traffic stream has little space to absorb disruptions. Drivers experience reduced physical and psychological comfort levels.

Level of service E's lower boundary describes driving conditions at capacity that are extremely unstable because there are virtually no useable gaps in the traffic stream. A vehicle entering the mainline, or changing lanes on the mainline, can cause following vehicles to give way to admit the vehicle; this can establish a disruption wave which works its way through the upstream traffic flow. At capacity any traffic incident causes a serious breakdown of flow and major queuing occurs. Freedom to manoeuvre within the traffic stream is extremely limited, and the driver's physical and psychological comfort is extremely poor.

Level F describes forced or breakdown flow. It is usually associated with mainline locations where the number of vehicles arriving is greater than the number that can traverse the locations, i.e. when the ratio of the arrival flow rate to the capacity exceeds 1. Level F operations are typically observed where traffic incidents cause a temporary reduction in the capacity of a short segment of freeway, or where recurring points of congestion exist, e.g. at merges or weaving areas.

Two-lane two-way roads

By definition these roads have one lane for use by traffic in each direction, and overtaking of slower vehicles requires the use of the opposing lane when sight distance and gaps in the opposing traffic stream permit. Two-lane roads compose the great majority of most national road systems.

Two-lane roads perform a variety of traffic functions. For example, efficient mobility is the main function of major two-lane roads that are used to link large traffic generators or as inter-city routes; thus consistent high-speed operations and infrequent overtaking delays are normally desired for these roads. High-volume two-lane roads are sometimes used as connectors between two major dual carriageway roads, and motorist expectations in relation to quality of service are also usually high for these roads. In the case of two-lane roads in scenic and recreational areas, a safe roadway is required but high-speed operation is not expected or desired. Two-lane roads in rural areas need to provide all-weather accessibility, often for fairly low traffic volumes; in these locales cost-effective safe access is the dominant policy consideration and high speed is of lesser importance.

When vehicles are operating under level of service A conditions on two-lane two-way roads, motorists are able to drive at their desired speed, overtaking demand is well below passing capacity, and almost no platoons of three or more vehicles are observed. Level of service B characterises the region of traffic flow where the overtaking demand needed to maintain desired speeds becomes significant. As volumes increase and approach the lower boundary of level B the overtaking demand approximately equals the passing capacity, and the number of platoons forming in the traffic stream begins to increase dramatically.

Traffic volumes are so high at level of service C that significant reductions in overtaking capacity occur and these are associated with noticeable increases in platoon formation and platoon size. Although traffic flow is still stable, it is more susceptible to congestion caused by turning traffic and slow-moving vehicles.

Unstable traffic flow is approached within level of service D and, at high volumes, the opposing traffic streams essentially begin to operate separately. Overtaking demand is very high while passing capacity approaches zero. Average platoon sizes of 5–10 vehicles are common. The proportion of no-passing zones usually has little influence on overtaking. Turning vehicles and/or roadside distractions are causes of major shock-waves in the traffic stream.

Overtaking is virtually impossible under level of service E conditions, and platooning is intense when slow vehicles or other interruptions are encountered. Operating conditions at the lower limit of level E, i.e. at capacity, are unstable and difficult to predict. As operating conditions approach that of level E perturbations in the traffic stream often cause a quick transition to level of service F. As with other road types, on two-lane roads, level F is associated with heavily congested flow and traffic demand seeking to exceed capacity.

17.2.2 Level-of-service measures of effectiveness

Objective level-of-service measures for three differing road types are summarised in Table 17.2.

Note that for freeways and for multi-lane single/dual carriageway rural and suburban roads that are not/only partially access-controlled the concentration (density) of traffic is assumed to be the prime measure of level of service. (Concentration, which is the inverse function of the linear spacing between following vehicles, is expressed in passenger cars per mile per lane in the United States, and in passenger cars per kilometre per lane in Britain.) As the concentration of vehicles increases, a driver's freedom to choose his or her speed becomes increasingly more constrained. However, freedom to manoeuvre within the traffic stream and proximity to other vehicles are equally noticeable concerns, and these are related to the concentration of the traffic stream. For any given level of service, the maximum allowable density on a freeway is somewhat lower than for the corresponding service level on a multi-lane road; as might be expected, this reflects the higher quality of service that drivers expect on freeways.

For two-lane roads 'per cent time delay' is used as the main surrogate for driver comfort, and (space) mean speed is a secondary consideration. Per cent time delay is defined as the average percentage of the total travel time that all motorists are delayed in platoons, unable to pass, while traversing a given section of road. Also, drivers are defined as being delayed when they are travelling behind a platoon leader at speeds less than their desired speed and at headways of less than five seconds; thus if 20 per cent of the vehicles have time headways of less than five seconds, the per cent time delay is assumed to be 20 per cent.

The mean speed and flow rate associated with the busier boundary of each level of service for each type of road, under ideal conditions, are also given in Table 17.2. Ideal conditions include: (a) 12 ft (3.65 m) lane widths; (b) 6 ft (1.82 m) clearance between the edge of the travel lanes and the nearest obstructions or objects at the roadside or in the central reservation; (c) 70 mile/h (112 km/h) design speed for freeways and multi-lane roads and 60 mile/h (96 km/h) for two-lane roads; (d) level terrain, and (e) all passenger cars in the traffic stream. In the case of two-way two-lane roads it also assumes a 50/50 directional split of traffic and no 'no-passing zones' on the road section under consideration.

Table 17.2 Summary of level-of-service characteristics of major U.S. roads, as described in the *Highway Capacity Manual*

Level of service	Freeways*	Multi-lane rural and suburban roads*	Two-lane roads*
A	Free flow. Max. concentration = 10 pcu/mile/lane. (Mean speed = free-flow speed and is dependent upon design speed or imposed speed limit. Max. v/c at 70 mile/h = 32% and 30% of ideal capacity for 4-lane and 6+lane freeways, respectively, under ideal conditions. Max. v/c is lower for lower mean speed)	Free flow. Max. concentration = 12 pcu/mile/lane. (Mean speed = 60 mile/h and max. v/c at this speed = 33% of ideal capacity, under ideal conditions)	Free flow. Most overtakings are easily made and % time delays never exceed 30%. (Mean speed ≥ 58 mile/h, and max. v/c = 15% of ideal 2-way capacity, under ideal conditions)
B	Reasonably free flow. Max. concentration = 16 pcu/mile/lane. (Mean speed is dependent upon design speed or speed limit. Max. v/c at 70 mile/h = 51% and 49% of ideal capacity for 4-lane and 6+lane freeways, respectively, under ideal conditions. Max. v/c is lower for lower mean speed)	Reasonably free flow. Max. concentration = 20 pcu/mile/lane. (Mean speed = 60 mile/h and max. v/c at this speed = 55% of ideal capacity, under ideal conditions)	Overtakings are fairly easily made but % time delays may be up to 45%. (Mean speed ≥ 55 mile/h and max. v/c = 27% of the ideal 2-way capacity, under ideal conditions)
C	Max. concentration = 24 pcu/mile/lane. (Mean speeds still at or near free-flow speeds, and max. v/c is at 68.5 mile/h and = 75% and 71.5% of ideal capacity for 4-lane and 6+lane freeways, respectively, under ideal conditions. Max. v/c is lower for lower mean speed)	Stable flow. Max. concentration = 28 pcu/mile/lane. (Mean speed = 59 mile/h, and max. v/c at this speed = 75% of ideal capacity, under ideal conditions)	Flow still stable. Overtakings are difficult and % time delays may reach 60%. (Mean speed ≥ 52 mile/h, and max. v/c = 43% of ideal capacity, under ideal conditions)
D	Concentration begins to deteriorate more quickly with increasing flow. Max. concentration = 32 pcu/mile/lane. (Mean speeds begin to decline slightly with increasing flows and max. v/c is at 63 mile/h and = 92% and 88% of ideal capacity for 4-lane and 6+lane freeways, respectively, under ideal conditions. Max v/c is lower for lower mean speed)	Approaching unstable flow. Max. concentration = 34 pcu/mile/lane. (Mean speed = 57 mile/h, and max. v/c at this speed = 88% of ideal capacity, under ideal conditions)	Approaching unstable flow. Overtakings are most difficult and % time delays are as high as 75%. (Mean speed ≥ 50 mile/h and max. v/c = 64% of ideal capacity, under ideal conditions)
E	Max. concentration = 36.7 and 39.7 pcu/mile/lane for 4-lane and 6+lane freeways, respectively. v/c at lower LoS	Max. concentration = 40 pcu/mile/lane. (Mean speed = 55 mile/h, and max. v/c at this speed = 100% of ideal	% time delays are more than 75%. (Mean speed ≥ 45 mile/h and max. v/c at this speed

Table 17.2 continued

boundary = 100% of ideal capacity, under ideal conditions, and speeds = 60 and 58 mile/h for 4-lane and 6+ lane freeways with free-flow speed of 70 mile/h, respectively)	capacity under ideal conditions)	= 100% of ideal capacity, under ideal conditions)
F	Highly unstable and variable flow	Highly unstable and variable flow	Heavily congested flow

*Ideal capacities for 4-lane and 6–8-lane freeways are 2200 and 2300 pcu/h/lane, respectively, under ideal conditions. Ideal capacity for multi-lane rural and suburban roads is 2200 pcu/h/lane. Ideal capacity (2-way flow) for 2-lane roads is 2800 pcu/h

In most capacity analyses, however, the prevailing operating conditions are not ideal; thus the ideal capacity for each road type (and the level-of-service capacity flows derived from each ideal capacity) must be adjusted to match the prevailing conditions. The manner in which this is done is illustrated by Equation 17.1, which takes an ideal capacity of 2800 pcu/h for a two-lane two-way road and adjusts it to reflect a volume to capacity ratio (v/c) appropriate for the desired level of service, directional distributions other than 50/50, lane width restrictions and narrow shoulders, and some heavy vehicles in the traffic streams.

$$SF_i = 2800 \times \left(\frac{v}{c}\right)_i \times f_d \times f_w \times f_{HV} \qquad (17.1)$$

where SF_i = total service flow rate in both directions for the prevailing road and traffic conditions, operating at level of service i (veh/h); $(v/c)_i$ = the ratio of flow rate to ideal capacity for level of service i, obtained from Table 17.3; f_d = an adjustment factor for directional distribution of traffic, obtained from Table 17.4; f_w = an adjustment factor for narrow lanes and restricted shoulder width, obtained from Table 17.5; and f_{HV} = an adjustment factor for the presence of heavy vehicles in the traffic stream, obtained from

$$f_{HV} = \frac{1}{\left[1 + P_T(E_T - 1) + P_R(E_R - 1) + P_B(E_B - 1)\right]} \qquad (17.2)$$

where P_T, P_R and P_B are the proportions of heavy commercial vehicles, recreation vehicles, and buses in the traffic stream, respectively, with each expressed as a decimal, and E_T, E_R and E_B are the passenger car unit equivalents for heavy commercial vehicles, recreation vehicles, and buses, respectively, obtained from Table 17.6.

Full details of the adjustment factors to be used with different types of road operating under varying prevailing conditions are available in the *Highway Capacity Manual*.[3]

17.2.3 Final comment on the level-of-service concept

The level-of-service concept marks a major contribution to traffic engineering and to road design. As well as being a useful design and operations classification tool for traffic engineers, it is a concept that is easily explained to and understood by the public – who ultimately pay for and use the roads.

Table 17.3 Level-of-service criteria for two-lane roads[3]

			Ratio of flow rate to ideal capacity of 2800 pcu/h (v/c) in both directions																				
			Level terrain							Rolling terrain							Mountainous terrain						
				% no passing zones							% no passing zones							% no passing zones					
LoS	% time delay	Ave.* speed	0	20	40	60	80	100	Ave.* speed	0	20	40	60	80	100	Ave.* speed	0	20	40	60	80	100
A	≤ 30	≥ 58	0.15	0.12	0.09	0.07	0.05	0.04	≥ 57	0.15	0.10	0.07	0.05	0.04	0.03	≥ 56	0.14	0.09	0.07	0.04	0.02	0.01
B	≤ 45	≥ 55	0.27	0.24	0.21	0.19	0.17	0.16	≥ 54	0.26	0.23	0.19	0.17	0.15	0.13	≥ 54	0.25	0.20	0.16	0.13	0.12	0.10
C	≤ 60	≥ 52	0.43	0.39	0.36	0.34	0.33	0.32	≥ 51	0.42	0.39	0.35	0.32	0.30	0.28	≥ 49	0.39	0.33	0.28	0.23	0.20	0.16
D	≤ 75	≥ 50	0.64	0.62	0.60	0.59	0.58	0.57	≥ 49	0.62	0.57	0.52	0.48	0.46	0.43	≥ 45	0.58	0.50	0.45	0.40	0.37	0.33
E	> 75	≥ 45	1.00	1.00	1.00	1.00	1.00	1.00	≥ 40	0.97	0.94	0.92	0.91	0.90	0.90	≥ 35	0.91	0.87	0.84	0.82	0.80	0.78
F	100	< 45	–	–	–	–	–	–	< 40	–	–	–	–	–	–	< 35	–	–	–	–	–	–

* Space mean speed of all vehicles (mile/h) for roads with design speed ≥ 60 mile/h; for roads with lower design speeds, reduce speed by 4 mile/h for each 10 mile/h reduction in design speed below 60 mile/h; assumes that speed is not restricted to lower values by regulation

Table 17.4 Adjustment factors for directional distribution of traffic on two-lane roads[3]

Directional distribution	100/0	90/10	80/20	70/30	60/40	50/50
Adjustment factor, f_d	0.71	0.75	0.83	0.89	0.94	1.00

Table 17.5 Adjustment factors for lane and shoulder widths on two-lane roads[3]

Useable shoulder width[a] (ft)	12 ft lanes LoS A–D	12 ft lanes LoS[b] E	11 ft lanes LoS A–D	11 ft lanes LoS[b] E	10 ft lanes LoS A–D	10 ft lanes LoS[b] E	9 ft lanes LoS A–D	9 ft lanes LoS[b] E
≥ 6	1.00	1.00	0.93	0.94	0.84	0.87	0.70	0.76
4	0.92	0.97	0.85	0.92	0.77	0.85	0.65	0.74
2	0.81	0.93	0.75	0.88	0.68	0.81	0.57	0.70
0	0.70	0.88	0.65	0.82	0.58	0.75	0.49	0.66

[a] Where shoulder width is different on each side of the roadway, use the average shoulder width. [b] Factor applies for all speeds less than 45 mile/h

Table 17.6 Passenger car unit equivalents for use on two-lane roads[3]

Vehicle type	Level of service	Type of terrain Level	Type of terrain Rolling	Type of terrain Mountainous
Heavy commercial vehicles, E_T	A	2.0	4.0	7.0
	B and C	2.2	5.0	10.0
	D and E	2.0	5.0	12.0
Recreation vehicles, E_R	A	2.2	3.2	5.0
	B and C	2.5	3.9	5.2
	D and E	1.6	3.3	5.2
Buses, E_B	A	1.8	3.0	5.7
	B and C	2.0	3.4	6.0
	D and E	1.6	2.9	6.5

In summary, the level-of-service concept accepts that (a) there is a maximum (ideal) capacity for a given 'ideal' lane or road section, (b) this maximum capacity value is reduced if the prevailing road, environmental and/or traffic conditions are less than ideal, (c) this maximum capacity describes vehicle operating conditions that are unstable and uncomfortable for drivers, (d) for each maximum capacity value there are a number of vehicle density or flow-rate descriptors that can be used to classify the operating conditions experienced by motorists (i.e. the level of service) on a given road section at a given time, and (e) there is a vehicle concentration (for freeways and multi-lane roads) or 'per cent time delay' (for two-lane two-way roads) that describes the capacity at the lower bound of each level of service range. When designing a new road, a design level of service is usually chosen (e.g. Table 17.1), and this is compared with the design-hour volume to provide guidance for the selection of an appropriate road cross-section. In the United States the design-hour volume most commonly selected for new roads in non-built-up areas is that for the 30th highest hour in the design year, adjusted for the peak 15-minute flow-rate within that hour.

Characteristic curves showing the highest hourly volumes, expressed as proportions of the annual average daily traffic (AADT) and hourly traffic (AAHT), are given in Fig. 17.1. Note that for non-urban roads in particular the rate of decrease changes rapidly and the 'knee' of each curve (i.e. where the steep slope associated with high peak hours changes to a more gentle lower peak hour relationship) lies between the 20th and 50th highest hours. On the basis of this form of relationship, it has been traditional American

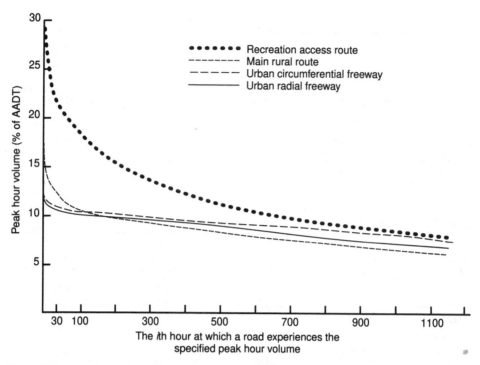

Fig. 17.1 Peak hour traffic volumes on various types of road in the central United States, shown in descending order from the highest (adapted from reference 3)

practice to assume that it would be uneconomic to attempt to cater for the traffic volumes experienced during the relatively few hours above the 30th highest hour on most rural roads. The case is not so strong, however, for urban roads with their more repetitive traffic flow patterns, and traffic volumes experienced during the 10th to 20th highest hours (adjusted for the peak 15 minutes within the hour selected) are often proposed as design-hour volumes for those roads in the United States. (The consistency of this relationship was established in the eastern United States as far back as 1947; see reference 4, p.86.)

17.3 The British design-standard approach

Unlike most other countries Britain has not adapted the *Highway Capacity Manual*'s level-of-service approach for road design. In this country a relatively independent approach has been developed which is based upon empirical British research studies related to different discrete aspects of road operation and analysis. These studies have resulted in the establishment of 'practical capacity' design standards for use in rural and urban road design, which form the basis for subsequent economic and environmental analysis before the final decision is taken as to which road layout to use.

17.3.1 Rural road design flows

Table 17.7 shows the traffic flow values currently used in the design of non-urban roads in Britain. In relation to this table it is important to appreciate that the recommended

Table 17.7 Design flow levels recommended for rural road assessment purposes in Great Britain[6]

[a]Road class (COBA classification)	24-h AADT flow (veh)	Edge treatment	Access treatment	Intersection options relating to flow	
				Minor road intersection	Major road intersection
Normal 7.3 m single carriageway, S2	[b]Up to 13 000	1m hardstrips	Restriction of access. Turning movements concentrated. Clearway at top of range	Simple intersections or ghost islands	Ghost islands, single lane dualling, or roundabouts
Wide 10 m single carriageway, WS2	10 000 to 18 000	As S2	As S2	Ghost islands, single lane dualling, or roundabouts	As S2
Dual 2-lane all-purpose carriageways, D2AP	11 000 to 30 000	As S2	As S2	Priority intersections. No other gaps in the central reservation	Generally at-grade roundabouts
	[c]30 000 to 46 000	As S2	Restriction of access severely enforced and left turns only. Clearway	No gaps in the central reservation	Generally grade-separation
Dual 3-lane all-purpose carriageways, D3AP	40 000 and above	As S2	Restriction of access severely enforced and left turns only. Clearway	No gaps in the central reservation	Generally grade-separation
Dual 2-lane motorway, D2M	[c]28 000 to 54 000	Motorway standards	Motorway regulations	None	Grade-separation
Dual 3-lane motorway, D3M	[c]50 000 to 79 000	As D2M	As D2M	As D2M	As D2M
Dual 4-lane motorway, D4M	77 000 and above	As D2M	As D2M	As D2M	As D2M

[a]The detailed dimensions of each of these carriageway layouts are given in reference 11 (see also Chapter 18). [b] The upper limit of flow range assumes a maximum diverting flow of about 2000 veh/day during maintenance works. [c] The upper limit of flow range assumes a maximum diverting flow of about 10 000 veh/day during maintenance works.

road classes do not guarantee any particular level of service for the traffic flows using them; rather they are simply guides to the road layouts that are likely to be economically and operationally acceptable in most instances, and which are therefore justified for subsequent analysis. These are a number of points about Table 17.7 that are worth emphasising.

1. The flow ranges are the 24-hour annual average daily traffic (AADT) flows for highway links in the design year (i.e. the 15th year after opening). They have no significance other than providing convenient starting points for assessment, and are not intended to indicate the ultimate flow (i.e. the maximum capacity) which a road can carry.[5] They result from economic and operational analyses which reflect the costs and benefits of different carriageway widths both during normal operating conditions and during maintenance, and assist designers by providing a starting point for scheme assessment of those road widths within whose range either or both of the high and low traffic forecasts fall.
2. The lower bound of a flow range is determined from an economic assessment; it indicates the lowest flow at which a given carriageway width is likely to be economically preferred to a lesser width. The upper bound of each flow range is based not only on economic assessment but also on an operational assessment of the acceptable diversion flow to other roads during maintenance works.[6] Table 17.8(a) relates the two-way carriageway flow (AADT) in the 15th year after opening to the maximum

Table 17.8 Maintenance arrangements relating to the derivation of the upper limits of the recommended traffic flows[5]

(a) Permitted diverting traffic

Road class	Diverting flow (veh/day):				
	500	2000	5000	10 000	12 000
	2-way carriageway flow (AADT × 10³ veh):				
S2	10	13	18	23	25
WS2	26	32	38	–	–
D2AP	27	32	38	46	49
D3AP	51	58	67	78	81
D2M	27	33	42	54	59
D3M	52	60	67	79	82
D4M	65	74	86	101	104

(b) Traffic management measures

Road class	Width available without works (m)	Traffic management during maintenance
S2	9.3	Shuttle working
WS2	12.0	2-way working
D2AP	2 × 9.3	1 + 1 contraflow on secondary carriageway
D3AP	2 × 13.0	2 + 2 contraflow on secondary carriageway
D2M	2 × 11.3	2 + 1 contraflow on secondary carriageway
D3M	2 × 14.3	2 + 2 contraflow on secondary carriageway
D4M	2 × 17.9	2 lanes open on primary carriageway
		3 + 2 contraflow on secondary carriageway

permitted diverting flow during maintenance; this table should not be regarded as applying directly to an individual scheme but rather as a guide to the maximum daily two-way diverting flow permitted in the 30-year (evaluation) life of the road – for instance, if the maximum diversion permitted from an S2 road at any time in the 30-year period is 2000 veh/day, then the carriageway flow should not exceed 13 000 ADT in the 15th year after opening. Table 17.8(b) shows the assumed traffic management arrangements during the maintenance works; in this table the 'primary carriageway' is the one being maintained, while the other is the 'secondary carriageway'.

3. The 15th year after the opening of the road to traffic is taken as the design year. This is considered to be a reasonable time to look ahead, particularly when the additional planning time is taken into account; it recognises that the further one looks into the future the more uncertain traffic predictions become. It also follows from the discounting technique used by COBA whereby the present value of benefits diminish the further into the future that the benefits accrue. The 15th year after opening is also used for other analysis purposes, e.g. noise calculations and the geometric design of intersections.

4. The unit of traffic flow used for design purposes is the 24-hour AADT, which is the total annual traffic planned for the road divided by 365. (Appropriate peak-hour flows for the main types of road can be estimated by applying the peak-hour flow factors in Table 20.3.)

5. Design flows are expressed in vehicles, not equivalent passenger car units. Reasons given for not using pcus are:[7] (a) the pcu-value of a heavy vehicle is different under different road and traffic conditions, so that the main attraction of its usage – its simplicity – is lost; (b) there is very little experimental backing for the conventional all-purpose pcu-values; and (c) even if traffic flows are expressed in pcus the average traffic speed is still a function of the proportion of heavy vehicles, so that the use of pcu-factors does not fully express the effect of traffic composition.

6. As traffic flows increase a higher standard of rural road is generally recommended. However, for a motorway to be considered it must be intended to connect it with the national motorway network, or form part of a new motorway at least 20 km long, as well as having the high-growth traffic forecast exceed 28 000 AADT in the 15th year after opening.

7. General rural road layout features, including suggested intersection types, are also given in Table 17.7 for the various design flows.

17.3.2 Urban road design flows

Department of Transport flow recommendations to be used in selecting two-way and one-way urban road layouts are given in Tables 17.9 and 17.10, respectively. The urban roads in these tables are defined as roads in built-up areas which are two- to six-lane single carriageways with speed limits of 40 mile/h (64 km/h) or less, and dual carriageway roads (including motorways) with speed limits of 60 mile/h (96 km/h) or less. As with the rural road data, the urban road design flows specifically apply to roads functioning as traffic links and are independent of intersection capacities.

All the design flows allow for traffic compositions containing up to 15 per cent heavy (>1.52 t unladen mass) commercial vehicles, and no adjustments are required for lesser commercial vehicle contents. However, the peak hourly flows in the design year should

Table 17.9 Recommended design flows for two-way urban roads[6]

	Peak flows (veh/h) for carriageways of width (m)											
	Single carriageways									Dual carriageways		
	2-lane*					4-lane†			6-lane†	2-lane†		3-lane†
Road type	6.1	6.75	7.3	9	10	12.3	13.5	14.6	18	6.75	7.3	11
A Urban motorway, no frontage access, grade-separation at intersections											3600	5700
B All-purpose road, no frontage access, no standing vehicles, negligible cross-traffic			2000		3000	2550	2800	3050		2950‡	3200‡	4800‡
C All-purpose road, frontage development, side roads, pedestrian crossings, bus stops, waiting restrictions throughout the day, loading restrictions at peak hours	1100	1400	1700	2200	2500	1700	1900	2100	2700			

* Total for both directors of flow; 60/40 directional split can be assumed; † for one direction of flow; ‡ includes division by line of refuges as well as central reservation; effective carriageway width excluding refuge width is used

Table 17.10 Recommended design flows for one-way urban roads[6]

Road type	Peak flows (veh/h) for carriageways of width (m)					
	6.1	6.75	7.3	9	10	11
All-purpose road, no frontage access, no standing vehicles, negligible cross-traffic		2950	3200			4800
All-purpose road, frontage development, side roads, pedestrian crossings, bus stops, waiting restrictions throughout the day, loading restrictions at peak hours	1800	2000	2200	2850	3250	3550

be reduced by the amounts given in Table 17.11 when the proportions of heavy commercial vehicles exceed 15 per cent.

Table 17.11 Total reductions in flow level (veh/h) to be applied to Tables 17.9 and 17.10 for heavy commercial vehicle contents in excess of 15 per cent[6]

Heavy vehicle content (%)	Motorway and dual carriageway all-purpose roads	Single carriageway roads	
		≥ 10 m wide	<10 m wide
	per lane	per carriageway	per carriageway
15–20	100	150	100
20–25	150	225	150

Parking is often permitted, or not prohibited, on wide single carriageway roads. The design flows for such links can be also obtained from Table 17.9 using the effective carriageway widths derived after application of the data in Table 17.12. Note also that this latter table shows that the effect of parking on traffic flow is not directly proportional to the number of vehicles parked, i.e. small numbers of parked cars have a relatively large impact on effective width and, therefore, on traffic flow.

Table 17.12 Effect of parked vehicles upon useful carriageway width (based on data from reference 9, p.211)

No. of parked vehicles per km (both sides added together)	Effective loss of carriageway width (m)
3	0.9
6	1.3
30	2.2
60	2.7
125	3.2
310	3.8

Unlike the data for rural roads the urban road design flows in Tables 17.9 and 17.10 are given in vehicles per hour rather than in terms of AADT. Peak-hour flow demand is defined as the highest flow for any specific hour of the week averaged over any consecutive 13 weeks during the busiest period of the year. In Britain the busiest three-month period is from June to August and, while local variations can occur, the weekday peak hour is normally 5–6 p.m., Friday. While new urban links are designed for maximum hourly flows in the design year, i.e. the 15th year after opening, these design flows are not necessarily determined by measuring current peak-hour flows and projecting them forward on the basis of unrestrained growth; rather these are usually adjusted to reflect the influence of restraint measures adopted by the governmental authority as part of the overall transport planning for the area.

Note also that design flows are not given for urban roads of lesser traffic-handling calibre than all-purpose type C roads, as their physical characteristics are expected to be more prohibitive to the flow of traffic. Also, environmental considerations may well predominate over traffic ones when determining acceptable traffic flow levels on some of these lower-order roads.

All-purpose roads are normally designed for standard carriageway widths. However, the reason that design flows for non-standard widths are given in Tables 17.9 and 17.10 is that special circumstances, in older urban areas in particular, often give rise to an unavoidable need for a non-standard carriageway.

In reality, one-way road schemes often use all-purpose roads that are already in existence so that the terminal intersection is often the prime determinant of the design flow. Similarly, for many existing two-way all-purpose roads their maximum capacities are more likely to be determined by the capacity of their terminal junctions than by the characteristics of the links themselves.

17.3.3 Final comment on the British design-standard approach

The procedure involved in using the design-standard approach to select an appropriate road layout is most easily summarised on a step-by-step basis for a rural road:

Step 1. Forecast the high- and low-growth AADT flows for the 15th year after opening, based on the original assignments to the road section under consideration. (Table 17.13 summarises the current national forecasts of road traffic and vehicles.)

Step 2. Select for local assessment those carriageway widths within whose Table 17.7 flow range either or both of the forecasts fall. For example, if the forecast flow range is 16 000 to 20 000 AADT, Table 17.7 suggests WS2 and D2AP roads for 16 000 AADT and a D2AP road for 20 000 AADT. Hence, WS2 and D2AP roads should be considered for further assessment.

Step 3. Consider whether there are any local circumstances which suggest that additional road types should be considered, e.g. unusually high or low costs, severe environmental impacts, operational changes, and/or anticipated major network changes during the evaluation period.

Step 4. Carry out detailed economic and environmental assessments to determine the optimal road width.

Overall, it can be seen that the above approach to selecting a road layout is significantly different from the *Highway Capacity Manual*'s approach whereby a given layout with a specified level-of-service capacity is matched with the predicted design-hour flow for the design year. The British approach of using a range of traffic forecasts for the 15th year after opening to identify the approximate carriageway requirements, and then to use assessment criteria to test alternatives consistent with this range over the assumed life of the scheme is, in fact, similar to that used when selecting a preferred route.

Table 17.13 Low and high forecasts of road traffic and vehicles in Great Britain for years 1995 to 2025.[10] (Note: the index for 1993 is taken as 100)

	Year						
	1995	2000	2005	2010	2015	2020	2025
Vehicle kilometres							
Cars and taxis	104	114	123	133	142	151	160
	107	123	138	153	167	180	193
Goods vehicles >3.5t	103	110	119	127	137	147	158
	105	118	134	151	171	194	219
Light goods vehicles	104	114	125	137	150	165	181
	106	124	145	169	197	230	268
Buses and coaches	100	100	100	100	100	100	100
	100	100	100	100	100	100	100
All motor traffic (except 2-wheelers)	104	113	123	132	142	151	161
	106	122	138	153	168	183	198
Car ownership							
Cars per person	104	112	119	126	132	138	144
	105	117	128	137	145	152	158
No. of cars	104	113	121	129	136	142	149
	106	119	130	140	149	157	164
Tonne-kilometres of road freight carried by goods vehicles >3.5 t	104	114	125	137	150	165	181
	106	124	145	169	197	230	268

17.4 References

1. OECD Scientific Expert Group, *Traffic capacity of major routes*. Paris: Organisation for Economic Co-operation and Development, 1983.
2. Buchanan, C.D., *et al.*, *Traffic in towns*. London: HMSO, 1963.
3. *The highway capacity manual*. Special Report 209. Washington, DC: Transportation Research Board, 1994.
4. Matson, T.M., Smith, W.S., and Hurd, F.W., *Traffic engineering*. London: McGraw Hill, 1955.
5. *Traffic flows and carriageway width assessment for rural roads*. Advice Note TA 46/85. London: Department of Transport, 1985.
6. *Traffic flows and carriageway width assessment*. Departmental Standard TD 20/85. London: Department of Transport, 1985.
7. Duncan, N.C., *Rural speed/flow relations*. Report LR 651. Crowthorne, Berks: Transport Research Laboratory, 1974.
8. *Policy on geometric design of highways and streets*. Washington, DC: American Association of State Highway and Transportation Officials, 1990.
9. Transport Research Laboratory, *Research on road traffic*. London: HMSO, 1965.
10. Department of Transport, *Transport statistics Great Britain*. 1994 edition. London: HMSO, 1994.
11. *Standard highway details*. London: HMSO, 1987.

CHAPTER 18

Road accidents

C.A. O'Flaherty

Worldwide, deaths and injuries from road accidents have reached epidemic proportions. Present indications are that about 0.5 million people are killed and 15 million injured on the world's roads every year. Next to circulatory diseases (including heart attacks and strokes) and cancer, road accidents are probably the third major cause of death in the developed world. Collisions on roads are now the main cause of death for young people in the age group 15–25 years.

The purpose of this chapter is to describe some trends in respect of road accidents, their main causes and characteristics, and some means by which they may be alleviated.

18.1 International comparisons

Absolute comparisons of fatality and (particularly) casualty rates in different countries must be treated with care as they contain results arising from differing traffic compositions, variations in the proportions of travel occurring in built-up areas, differences in quality of street lighting, road standards, vehicle legislation, etc. Accident reporting procedures also vary considerably between countries. Road fatalities are the simplest to compare but these are also complicated by problems of definition: in Britain a death is due to a road accident if it occurs within 30 days; in Italy, seven days; in France, six days; in Greece, three days; and in Japan, Spain and Portugal, one day.

The seminal work[1] on comparing fatality rates, which was based on data from 20 developed countries for the year 1938, gave the equation:

$$\text{fatalities/vehicle} = 0.0003 \ (\text{vehicles/population})^{-2/3} \tag{18.1}$$

Although this empirical relationship has been criticised for its simplicity, it suggests that the number of fatalities per vehicle is not a constant but decreases as the number of registered vehicles per person increases in a given country. In other words, road users become more careful and governmental authorities adopt more road safety measures as the roads are perceived to become more dangerous.

More recent work[2] draws a parallel with findings from studies of industrial processes, where the rate of failures (e.g. defective goods manufactured) normally declines exponentially over time. In this context road accidents and casualties are the failures of the national road transport system, and the rate of failure (expressed as accident, fatality, and casualty rates per billion vehicle-kilometres) can be expected to decline exponentially as a given country learns to cope with motorised transport. The decline is then the result of a combination of national factors, such as improved standards of vehicle design, and the increasing personal experience of modern road conditions.

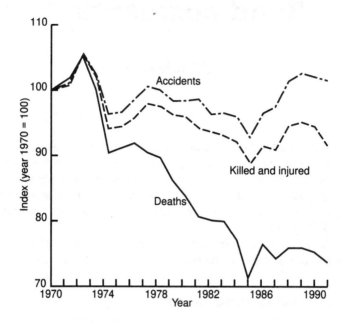

Fig. 18.1 European road safety trends[23]

In terms of absolute numbers nearly all countries saw increasing road casualties in the post-World War II period; this continued until the early 1970s, after which there were decreases, especially in the number of fatalities. Figure 18.1 shows overall road safety trends in 19 European countries (excluding countries with economies in transition) in the period 1970–91. Table 18.1 compares road fatality rates for some selected countries.

Table 18.1 International comparison of road accident fatality rates in 1992[13]

Country	Vehicles per 10^3 population	Deaths per 10^5 population	Deaths per 10^5 vehicles	Car user deaths per 10^9 car-kms	Pedestrian deaths per 10^5 population
Britain	441	7.5	17	6	2.4
Germany	574	13	23	14	2.2
Ireland (Rep.)	317	12	37	8	3.2
Netherlands	466	8	18	7	1.0
USA	*763	15	*20	9	2.2

* Excludes mopeds and moped users

The following points were highlighted by an OECD analysis[3] of international statistics.

1. Even though night-time traffic flows are much lower, approximately one third to one half of road accidents occur at night. Casualty rates and severities are much higher

at night than during the day. Alcohol is a major cause of accidents at night. Young drivers are statistically over-represented in night accidents.

2. Between one third and one half of all fatal accidents to adults involve drivers with measurable alcohol and/or drug presence.
3. The total number of accidents in fog is small (1–3 per cent); however, they often involve multiple collisions and, on motorways, increased severity. Wet weather typically accounts for 15–25 per cent of all accidents; lack of visibility and of skid resistance are equal causes of the increase in accident rates. Accidents under snow conditions are usually less severe than those under other adverse weather conditions.
4. Some 10–20 per cent of road fatalities are victims of collisions with natural or artificial obstacles alongside the road.
5. Children and elderly people often represent the majority of casualties on residential streets. Accident locations on residential streets are scattered widely, with few clusters occurring.
6. Increases in traffic speed give rise to increased accident occurrence and, other things being equal, a reduction in mean traffic speed gives rise to a reduced accident rate and lower accident severity.
7. Heavy commercial vehicles (HCVs) pose low accident risks for their occupants but are relatively high accident risks for other road users. Fatality rates for car/HCV accidents are 6 to 80 times higher than those for car/car accidents, depending on the direction of impact.
8. Young drivers ($\leq$ 25 years) have three to four times the average accident rate. Lack of driving experience may be a contributing factor in 10–15 per cent of all car accidents.
9. Pedestrians have a higher accident risk per kilometre travelled than most other road users. Pedestrian accidents occur mainly in urban areas. Children and elderly pedestrians have higher accident risks per capita than adults in the 'between' group.
10. The average elderly (65+) accident victim is about three times as likely to die from injury as a victim from the 25–64 age group. Elderly drivers are relatively less involved in single-vehicle accidents (a category in which the young are over-represented); however, the elderly are relatively more involved in collision accidents, especially at intersections, when changing direction or when joining a traffic stream.

18.2 Accident trends in Great Britain

Figure 18.2 summarises the scale and extent of the road accident and casualty problems in Britain since 1949. In 1993 Britain's population was approximately 56.4 million, there were 24.826 million vehicles, and travel by motor vehicle amounted to 4.1×10^{11} veh-km.[4] 3814 people were killed, and some 45 000 seriously injured and 257 200 slightly injured in 1993; this number of fatalities is the lowest since records began in 1926 and compares well with 7985 deaths (the peak) in 1966. While a significant part of the containment in fatal and serious accidents can be attributed to the relative decline in cycling and walking as transport modes, as well as to higher standards of medical care, much is undoubtedly due to the application (through engineering, education and enforcement) of the lessons learnt from accident research.

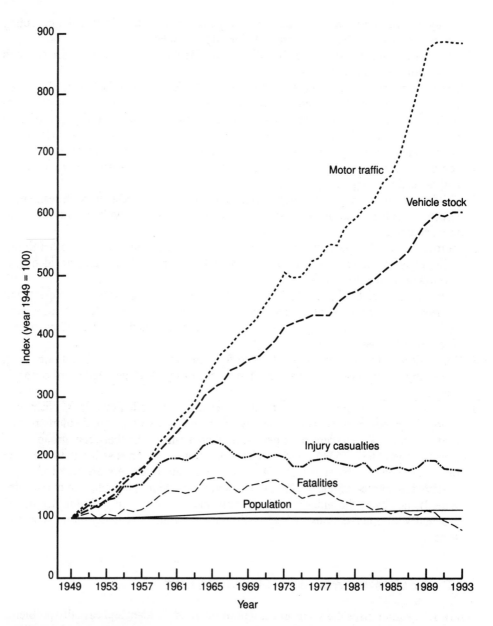

Fig. 18.2 Indices of population, vehicles registered, traffic, and reported fatalities and injury casualties in Britain[13]

Numerous accident studies carried out in Britain have generally confirmed the OECD trends described previously. Some other trends/statistics of relevance in Britain are shown below.

1. The risk of an accident is greatest on built-up roads (i.e. with speed limits ≤ 64 km/h) and least on motorways (see Table 18.2). About three quarters of all accidents occur

Table 18.2 Accident rates and numbers on various types of roads in Britain in 1993 and average for 1981–85[13]

Accidents	Motorways	Non-built-up*		Built-up*	
		A roads	Other roads	A roads	Other roads
1993					
Rate per 10⁸ veh-km	11	28	44	95	87
Numbers	6863	33 053	20 412	74 774	93 763
1981–85					
Rate per 10⁸ veh-km	13	40	52	132	134
Numbers	4246	31 952	21 681	87 538	103 761

*Excluding motorways

on built-up roads; however, about half of the fatal accidents occur on non-built-up roads (mainly) and motorways. Note that only 3–4 per cent of all fatal accidents occur on motorways, which cater for some 15 per cent of all motor vehicle travel.

2. On rural single carriageway roads:[5,6] (a) one third of all accidents involve one vehicle only; (b) single-vehicle accidents are more likely than others to be associated with B and C roads, night, the youngest drivers, and with 'going ahead on a bend'; (c) about one fifth of all accidents involve a single vehicle leaving the carriageway while going ahead between intersections; (d) high-performance cars are disproportionately involved in non-intersection accidents, in single-vehicle accidents, and in overtaking accidents; (e) accidents at (i.e. within 0–20 m of) intersections account for more than one third of the accidents; (f) about 80 per cent of accidents at intersections occur at unsignalised layouts; (g) the most frequent type of accidents at T-intersections are those involving a right-turning vehicle from the minor road (27 per cent), a right turn from the major road (22 per cent) and rear shunts (20 per cent); (h) the most frequent type of accidents within 20–100 m of a T-intersection are single-vehicled (42 per cent), head-on (26 per cent) and rear shunts (20 per cent); and (i) the annual frequency, A, of accidents at T-intersections is given by

$$A = 0.24 \, (QP)^{0.49} \tag{18.2}$$

where QP is the product of major and minor road inflows in units of thousands of vehicles per day.

3. Cross-over intersections on rural roads are more dangerous than T-junctions.
4. About one half of all annual personal-injury road accidents occur at urban intersections, and about one half of these occur at major-minor intersections.
5. The percentage of all accidents involving skidding is much higher on wet roads than on dry ones. Motorcycles are 10 times more likely to be involved in personal-injury skidding accidents than cars.
6. Provided that dual carriageway intersection accidents are excluded, the overall accident rate on modern dual carriageways with minimal frontage development is similar to that for motorways; however, the proportion of fatal and serious accidents is much higher on all-purpose dual carriageways. Modern dual carriageways have lower accident rates than old ones.

Table 18.3 provides some data regarding the proportions of different categories of road user killed and injured on the roads in 1993. Table 18.4 compares, by age group, the average numbers and percentages of those killed and seriously injured on the roads in the 1981–85 period with those in 1993.

Table 18.3 Road user casualties in Britain, 1993

Type	Killed (%)*	All (%)*	Type	Killed (%)*	All (%)*
Pedestrians	32.5	15.7	Bus/coach drivers and passengers	0.9	3.0
			Light goods vehicles drivers and passengers	2.4	2.4
Cyclists	4.9	7.9			
2-wheeled motor vehicle riders (incl. passengers)	11.2	8.2	Heavy goods vehicles (<3.5t) drivers and passengers	1.5	1.1
Car drivers and passengers	46.1	61.3	Total number killed		= 3814
			Total number of casualties		= 306 020

* Percentages do not add to 100 as total numbers include other road users and type not reported

Table 18.4 Persons killed or injured on Britain's roads, by age group

Age group	Average in 1981–85 period No.	%	In 1993 No.	%	Change from 1981–85 period to 1991 %
Killed					
0–15	563	10.1	306	8.1	−45.6
16–59	3549	63.6	2382	62.7	−32.9
60+	1470	26.3	1111	29.2	−24.4
All casualties					
0–15	50 333	15.8	42 590	14.2	−15.4
16–59	233 996	73.6	224 767	74.8	−3.9
60+	33 540	10.6	33 027	11.0	−1.5

18.2.1 Policy

In 1987 the Government set a target to lower road casualties by one third by the year 2000, compared with the 1981–85 average. On average there were 321 912 injured (all severities), 74 532 seriously injured, and 5598 killed in each year of the 1981–85 period. By 1993 the total number of casualties and the total killed had declined by 5 and 32 per cent, respectively. Over this time the casualty rate dropped by approximately 32 per cent, from 108 to 74 per 10^8 veh-km, while the vehicular travel increased by 39 per cent.

With this policy priority is given to lowering the casualties amongst the most vulnerable road users – children, pedestrians, cyclists, motorcyclists, and the elderly – particularly in urban areas where the great majority of road accidents occur.

18.2.2 Car accidents

Car (including taxi) drivers and their passengers account for the greatest number of persons killed and injured on the roads. In each year of the period 1981–85 they accounted for, on average, 2198 fatalities and 143 941 casualties (all severities); by 1993 the total number of car travellers killed had declined by 24 per cent, but the total number of casualties rose by 30 per cent.

The 1980s were a decade of considerable expansion in car usage. Between 1981 and 1985 cars and taxis accounted for 234.49×10^9 veh-km of travel each year, on average; in 1993 the figure had risen by over 43 per cent. On an exposure-to-risk basis the average annual 1982–85 car casualty rate of 62 per 10^8 veh-km dropped by 9.5 per cent by 1993, even though the total number of casualties increased.

Accident analyses (see reference 7) indicate that the likelihood of a car driver being involved in a road accident is mainly dependent on his or her exposure to risk (i.e. total distance travelled each year), age *and* driving experience (measured as the number of years since passing the driving test). The injury-accident involvement rates of 17 to 20-year-old male car drivers per kilometre driven is about seven times that of drivers aged about 50; after the age of 65 injury-accident involvement rates begin to rise again. Female driver injury-accident involvements show a similar pattern although young females exhibit only half the rate of their male counterparts, but involvement rates for older female drivers rise to a rate nearly twice that for men of similar age.

18.2.3 Pedestrian accidents

Next to car drivers and passengers, pedestrians form the second largest group of road casualties. In 1993 pedestrians reported as involved in road accidents amounted to *ca* 16 per cent of all casualties; this compares with *ca* 20 per cent throughout the 1970s and 19 per cent (average) during 1981–85. The pedestrian proportion of all fatalities also declined from 39 per cent (*ca* 3000) in the late 1960s/early 1970s to 33 per cent (1863 average) during 1981–85; the proportion has remained essentially constant since then, although the number of pedestrians killed declined to 1241 by 1993.

Children ($\leq$15 years) comprised 13 per cent (165) and 38 per cent (18 249) of pedestrian fatalities and casualties, respectively, in 1993. Whereas over 90 per cent of pedestrian accidents occur on built-up roads, those involving children tend to be scattered across the road network (vis-à-vis those involving adults which tend to be more clustered). More child accidents occur during the week between 8–9 a.m and 3–7 p.m. then at other times, coinciding with regular trips to/from school.[8] Almost 25 per cent more child pedestrian casualties occur in summer than in winter (due at least partially to a greater exposure to risk in the longer summer evenings). For both boys and girls casualty rates rise to a peak at age 12; however, boys are more at risk than girls as they tend to play more often in the streets, especially between the ages of six and ten years.

Various studies have also shown that there is a general tendency for child pedestrian fatality rates to increase with town size, and for casualty rates to be highest in the inner areas of conurbations. Also, there are two to four times fewer accidents in areas with modern residential road layouts than in older ones. Higher child casualty rates are also reported amongst low income groups (probably because they tend to live in areas of higher residential density) and ethnic minorities (possibly related to their travel patterns and lesser access to safety education).

An adult pedestrian involved in a road accident is 3.5 times more likely to die than a child pedestrian involved in an accident. The greater recuperative powers of children vis-à-vis adults (especially the elderly) are undoubtedly an influence here.

A Transport Research Laboratory analysis of pedestrian accident data in 1980 gave the information in Table 18.5 regarding the relative blameworthiness of drivers and pedestrians. Another more recent analysis of adult pedestrian fatalities[9] concluded that 38 per cent had been drinking prior to the accident, and that 77 per cent of these had blood alcohol concentrations in excess of 80 mg/100 ml (the legal limit for driving in Britain).

Fatal accidents to drinking pedestrians are most common between 10 p.m. and midnight. Seasonal variations in the numbers of drinking pedestrian fatalities are broadly similar to those for adult pedestrians as a whole, both peaking in December and having lowest levels in mid-summer. The great majority of drinking fatalities are males. Although the fatality rate for retired pedestrians involved in accidents is very high, the incidence of heavy drinking in this group is low. The drinking problem is particularly acute amongst those in their early 20s and middle 40s from lower socio-economic groups.

Table 18.5 Who are at fault in pedestrian accidents

Degree of involvement	Drivers (%)	Pedestrians (%)
Primarily at fault	41	65
Partially at fault	19	14
No blame allotted	40	21

18.2.4 Two-wheeled motor vehicle accidents

After climbing during the 1970s, two-wheeled motor vehicle (motorcycles, scooters and mopeds) numbers and usage declined throughout the 1980s. While the fatality and total casualty numbers for two-wheeled motor vehicle riders declined by 57 per cent (to 427) and 62 per cent (to 25 066), respectively, between 1981–85 and 1993, these road users are still at the greatest risk. In 1993 two-wheeled motor vehicles accounted for only 2.6 per cent (650 000) of the motor vehicle fleet and 1 per cent of the vehicle-kilometres travelled; however, they also contributed 8.2 per cent of the fatalities and 11.2 per cent of all casualties. Their casualty rate in 1993 was 598 per 10^8 veh-km, the highest for all motorised vehicles.

Within the two-wheeled motor vehicle category of motor vehicle casualties, 78 per cent involved motorcycle riders and their passengers.

The conclusions of various analyses of accidents in Britain involving motorcycles (see reference 10) follow.

1. The likelihood of being involved in an accident is dependent predominantly on a rider's age, experience and exposure, and accident frequency decreases rapidly with increasing age and riding experience (with the rate of decrease reducing with increasing age and experience) while it increases with increasing mileage ridden (with the rate of increase reducing with increasing travel).
2. Accident frequency increases from riding predominantly in built-up areas and from riding in the winter.
3. About two thirds of motorcycle accidents occur at intersections.

4. Up to age 28 motorcyclists tend to ride larger machines as they grow older, but after this age the size of the machine appears to be of less importance and it tends to fall progressively.
5. The carriage of passengers increases with engine capacity.
6. Casualty and accident-involvement rates per kilometre travelled are highest for 51–125 cc machines (probably due to the number of riders who are learners) but they fall gradually with increasing engine capacity.
7. The fatality rate per 10^8 veh-km is half the average for the smallest machines, rises to twice the average for 251–500 cc machines, and is 40 per cent above average for machines over 500 cc.
8. The fatality rate is highest on non-built-up roads whereas the all-casualty rate on non-built-up roads is less than half that on built-up roads.
9. Most motorcyclist fatalities result from head injuries.
10. The risk of death or serious injury due to head injury alone is much less for those wearing a safety helmet than for those without, and full-face helmets reduce the chance of head injury as compared with open-faced ones.

18.2.5 Bicycle accidents

Bicycle usage and bicycle accident trends vary considerably from country to country. In Britain bicycle usage and the number of reported cyclist casualties and fatalities fell steadily from post-World War II until the mid-1970s. Following the 1973–74 energy crisis, bicycle usage and cyclist casualties and fatalities rose until peaking in 1982–84, and then declined again. Between 1981–85 and 1993 the *reported* number of fatalities and total casualties declined by 40 per cent (to 186) and 15 per cent (to 24 068), respectively, while cycle traffic was estimated to decline by 26 per cent. (Note: pedal cyclist traffic only accounts for about 2 per cent of all vehicle-kilometres travelled on Britain's roads.) The official statistics for cyclist travel and casualties suggest that the casualty rate increased from 463 (average) to 533 per 10^8 veh-km over the same period.

Many bicycle accidents are not reported; one study concluded[11] that the number of pedal cycle accidents per year could be 3.3 times that reported in the national statistics.

Most bicycle accidents occur at intersections on minor roads in built-up areas in daylight; conventional roundabouts appear to be particularly hazardous sites. Many bicycle accidents involving children do not involve a motor vehicle; playing, doing tricks, travelling too fast, and general loss of control are major causes of accidents involving children under thirteen. Older cyclists are more likely to have an accident involving a motor vehicle. A significant change in recent years has been a shift in risk from child cyclists, who are cycling less, to adult cyclists, who are cycling more. Head injuries are the major cause of death to cyclists involved in road accidents.

18.2.6 Public service vehicle accidents

Accident and casualty rates for public service vehicles showed continuing declines throughout the 1980s. Deregulation of buses and coaches in 1985 does not appear to have increased accident rates.[12] A limited national study suggests that small buses have substantially lower accident rates than larger ones.

18.3 Accident costs

Accidents represent a significant cost to the national community. The total cost-benefit value of road accidents in 1993 was estimated to be £10 780 million, of which £8740 million was attributed to personal-injury accidents; damage-only accidents accounted for the rest.[13] Table 18.6 shows the average costs per injury accident and per casualty.

Table 18.6 Average costs of accidents, by road type, and of road casualties in Britain in 1993

Accident type	Costs (£) per accident on roads that are:			Costs (£), all roads, per	
	Built-up	Non-built-up	Motorways	Accident	Casualty
Fatal	805 090	908 910	990 880	863 370	744 057
Serious	95 600	114 330	122 850	101 990	84 262
Slight	9380	11 520	13 570	9970	6540
All injury	30 200	61 650	51 800	38 200	27 160
Damage only	940	1380	1330	1000	–

Note that the average cost per casualty for each type of severity is lower than the average cost per accident. This is because accident costs include elements that are not casualty specific (such as police and insurance administration, and property damage) and because there is normally more than one casualty involved in each accident.

Table 18.6 also shows that the average cost of an injury accident is lowest on built-up roads. Accidents on built-up roads have, on average, fewer casualties per accident and a lower proportion of fatal and serious casualties per accident vis-à-vis the other (higher speed) roads.

18.4 Reducing the accident toll

18.4.1 Factors contributing to road accidents

The main factors contributing to accidents as determined in a major on-the-spot analysis of 2042 accidents[14] are summarised in Fig. 18.3. Note that these factors are distributed such that the human element contributed to nearly 95 per cent of all accidents, road factors to 28 per cent, and vehicle factors to 8.5 per cent. The road user was the sole contributor in 65 per cent of the accidents; in contrast the road and vehicle factors were usually linked with a road user factor.

The principal *road user factors* are:

- perceptual errors, e.g. driver or pedestrian looks but fails to see, distraction or lack of attention, or misjudgement of speed or distance
- lack of skill, e.g. inexperience, lack of judgement, or wrong action or decision
- manner of execution, e.g. deficiency in actions (such as too fast, improper overtaking, failure to look, following too closely, wrong path), or deficiency in behaviour (such as irresponsible or reckless, frustrated, aggressive)
- impairment, e.g. alcohol, fatigue, drugs, illness, emotional stress.

Road environment factors which present difficulty to the driver include

- adverse road design, e.g. unsuitable layout and intersection design, or poor visibility due to layout

- adverse environment, e.g. slippery road, flooded surface, lack of maintenance, weather conditions
- inadequate road furniture or markings, e.g. insufficient and/or unclear road signs and carriageway markings, poor street lighting
- unexpected obstructions, e.g. roadworks or parked vehicles.

Vehicle factors contributing to accidents are mainly those resulting from a lack of regular maintenance by the user of the vehicle, e.g. defective brakes and tyres.

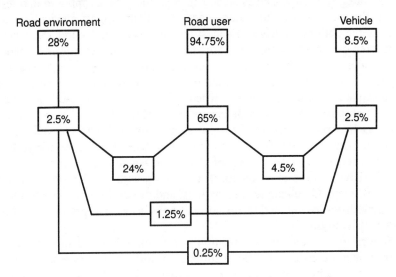

Fig. 18.3 Contributions to road accidents[14]

From Fig. 18.3 it can be deduced that the greatest potential for reducing accidents lies in influencing road users to act more responsibly and in designing new roads/amending old roads so as to be safer. The tools for doing this are education (accompanied, where appropriate, by enforcement) and engineering.

18.4.2 Role of education in road safety

There is no standard definition of 'education' as used in relation to road safety. Generally, the term is used to refer to a diverse range of activities, from early childhood road training to initial driver training (where a person is taught the basic repertoire of perceptual, motor and informational skills necessary to operate a vehicle safely) to attitude-changing programmes (which are intended to inform adult road users of the need to change existing dangerous practices, e.g. drinking and driving, speeding, not wearing seat-belts, etc.).

In Britain the road safety educational tasks tend to be shared between those who have a statutory responsibility, such as the Department of Transport and local authorities, and those with a professional interest in the area, such as the Royal Society for the Prevention of Accidents, driving schools, health educationalists, the Motorcycle Association and the police.

Teaching safety

With very young children the emphasis is usually placed on public information programmes aimed at educating parents to impose on children the habit-forming discipline
of keeping off the carriageway unless absolutely necessary and to 'stop, look, listen'
when crossing the road. Once children are in school there is the opportunity for formal
road safety education directly aimed at the child. For the five- to nine-year-olds this may
range from a consolidation of basic road behaviour (including training them on how to
develop safe strategies for themselves on how to cross the road, as opposed to rule-based
training[8]), to the development of outdoor activities that support curriculum activity
while enabling the children to develop pedestrian and cyclist skills under supervision.

Older children can be encouraged to participate in interesting exercises on pedestrian
and cyclist accident causes and effects, in discussions on the bicycle as a means of transport (as against a toy to play tricks on) and on the value of being conspicuous and of
wearing protective clothing and headgear when cycling, as well as practical cycle training covering both basic control and defensive riding skills. How to carry out difficult
manoeuvres, such as right turns and the negotiation of roundabouts, is a skill that needs
to be taught to youngsters in their early to mid-teens. Safety education for young people in their later teens may involve tests of vision and reaction, observation practicums
of road user behaviour (and misbehaviour), interviews with accident victims and vulnerable persons, visits to vehicle testing stations, and discussions on common accident
situations, drink/drug-driving, and the use of seat-belts, as well as practical motorcycle-
riding and car-driving training.

The introduction of road safety education into schools has had mixed success. In
most schools it is not considered to be of sufficient priority to justify a separate course,
and is mostly covered by visiting speakers (e.g. road safety officers), during assembly,
and incidentally in normal lessons. The limited use made by teachers of road safety curriculum materials, their limited preparation to teach road safety education, and the lack
of a well-developed organisation structure for road safety education within primary and
middle schools, have been identified[15] as factors requiring urgent attention if road safety
is to be improved for the 5–13 age group.

Analyses of driver training courses, as distinct from road safety education courses, in
American high schools have concluded that there is no evidence that they had any extra
effect upon subsequent accident records. (Note also that studies of the effectiveness of
commercial driving schools have concluded that they have greater success in preparing
people to pass a driving test than private individuals, but there is no evidence that they
have any special value in road safety terms over informal training methods.)

Safety programmes for the elderly

As people grow old their driving skills tend to deteriorate as a result of the degradation
of functional and cognitive skills associated with aging. Night vision, especially glare
resistance and recovery time, worsens with old age; reflexes also slow and older drivers
have slower reaction times in complex driving situations. Older drivers are more likely
to misunderstand signs; they also tend to drive more slowly which, paradoxically, can
be more dangerous in certain situations. The hearing, attention span, walking speed and
agility of older pedestrians also decrease with age. When accidents happen, the elderly
are more likely to be severely injured because of their physical vulnerability.

Public information programmes aimed at elderly drivers and pedestrians, which advise them on how to cope with their disabilities and with difficult road situations, have considerable potential to reduce accidents involving older persons.

Drinking-and-driving campaigns

Alcohol is a drug which depresses the central nervous system. As a result people's perceptions are blunted, coordination is impaired, reaction times are slowed, accurate judgements of speed and distance are lessened, and the propensity to take risks is increased.

Upon intake alcohol is absorbed into the bloodstream and then distributed through the fluids and tissues of the body; however, the level of intoxication attained by any individual then depends upon the concentration of alcohol in the brain and central nervous system, and this is closely related to the blood alcohol concentration. The blood concentration at any given time depends upon the rate at which the alcohol is absorbed into the bloodstream and the rate at which it is burnt up by the body or excreted in the urine, breath or perspiration. The rate of absorption depends upon each individual's constitution, the time since the last meal, the food consumed, and how quickly the alcohol is taken. Alcohol is eliminated from the body at the more or less standard hourly rate of 15 mg per 100 ml of blood.

Typical minimum amounts that will result in an average male adult attaining a certain alcohol concentration in the blood are given in Table 18.7; these data are based on tables published by the British Medical Association. Driving skills are reduced at blood alcohol concentrations (BAC) above 30 mg/100 ml; they are seriously affected above 80 mg/100 ml, i.e. the legal limit for driving in Britain (which corresponds to a breath alcohol concentration of 35 mg/100 ml). Clumsiness and impaired judgement follow at 100 mg/100 ml, and at 200 mg/100 ml motor control and behaviour are seriously affected.

Table 18.7 Minimum intake of alcohol needed to raise the blood alcohol level in a typical 70 kg male to a given concentration

BAC (mg/100 ml)	Beer or stout (pints)	Single whiskys
20	0.5	1
40	1.0	2
55	1.5	3
75	2.0	4
100	3.0	6
150	4.0	8

A major study (at Grand Rapids, Michigan, in 1962/63) found that drinking drivers at 80 mg/100 ml BAC are twice as likely as non-drinking drivers to be involved in an accident, and that the risk rises sharply to 10 times at 150 mg/100 ml and to 20 times at 200 mg/100 ml. Subsequently it was determined that with inexperienced and infrequent drinkers the sharp risk increase occurs at much lower levels than for more experienced drinking drivers. A more recent British study of the risks involved in drink-driving gave the results shown in Fig. 18.4.

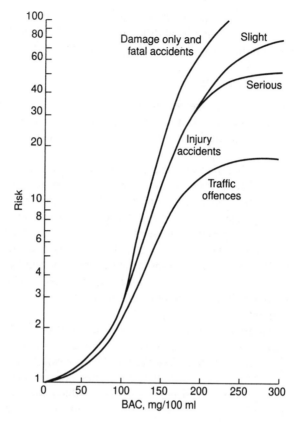

Fig. 18.4 Effect of blood alcohol concentration levels on accident probability[24]

It might also be noted here that, excluding alcohol, the evidence linking drugs which act upon the central nervous system (CNS drugs) with road accidents is limited at this time, although there is no doubt that driving skills can be adversely affected by such drugs. One British study of road accident fatalities[16] concluded that the overall incidence of CNS drugs (including medicinal ones) amongst all fatalities was 7.4 per cent; for car drivers and motorcycle riders it was 6.7 and 8.0 per cent, respectively. The greatest incidence of medicinal CNS drugs was in road users over 60 years. Drugs of abuse, notably cannabis, were most common amongst young and middle-aged male drivers and motorcycle riders; in 40 per cent of the fatalities which involved CNS drugs, cannabis was used with alcohol.

Public information campaigns which have targeted the drink-driving problem have been run in Britain since 1964. On the whole these have focused on making the public aware of the dangers associated with drinking and driving, and on making it socially unacceptable to drink and drive. While the success of these campaigns is reflected in subsequent reductions in accident numbers and rates, it might also be noted that surveys carried out in 1979 and 1991 indicated that the numbers of people who admitted driving after drinking in the previous week declined from 51 to 29 per cent over that period.

Seat-belt campaigns

The majority of injuries to car occupants involved in road accidents are the result of (a) frontal impacts by cars with fixed roadside obstacles, or with the front, sides or rear ends of other vehicles, and (b) occupants continuing to move forward at very nearly the full speed of the car before impact and striking (usually) the windscreen or instrument panel. The primary function of seat-belts is therefore to restrain the car occupants and prevent or reduce injury to heads and bodies during an accident; a secondary function is to hold occupants in their seats, especially during overturning accidents. Recognition of their value in achieving these functions resulted in the fitting of seat-belts being made compulsory in the front seats of new cars in Britain in 1965 (1967 for vans), and many publicity campaigns were subsequently mounted to educate motorists as to why seat-belts should be worn.

The law now requires drivers and their passengers to wear seat-belts in moving vehicles. However, the evidence would suggest that there is considerable value in continuing these campaigns, and especially in targeting rear seat occupants about the benefits of wearing seat-belts.

Campaigns against speeding

Speed control/reduction is another area where public education is being usefully employed to change attitudes. This application is particularly timely as speed is increasingly being identified as a major factor in accident causation. Public opinion is currently being encouraged to shift to become more critical of speeding and close following – especially when children are about.

A traffic stream consists of many drivers who individually choose their speeds and headways in relation to other vehicles on the road. Figure 18.5 illustrates that the factors which influence each individual's speed choice are complex, as are the relationships between speed, speed limits, and accidents.

In general, between 22 and 32 per cent of accidents in Britain have excessive speed as a contributory factor; i.e. some 75 000 injury accidents and about 1000 road deaths each year involve excessive speed.[13] The higher the average speed on a given road the more likely it is that an accident will have severe consequences. There is evidence[17] to the effect that drivers who drive much faster or much slower than the general traffic stream are more likely to be involved in accidents.

Changing the speed limit tends to result in a smaller change in mean traffic speed in the same direction; as a guide, the change in the mean traffic speed is roughly a quarter that of the posted difference. It is well established that the imposition of a speed limit or the lowering of an existing limit is usually associated with significant reductions in road accidents, and vice versa. The data available at this time also suggest that the percentage change in accidents is directly proportional to the absolute change in the posted speed limit, and that a greater decrease in the fatality rate can be achieved by reducing speed limits in urban areas in Britain than in rural areas. There appears to be a strong relationship between actual traffic speeds and accidents:[17] for every 1 mile/h (1.6 km/h) decrease in the mean traffic speed casualties fall by about 5 per cent and fatalities by 7 per cent.

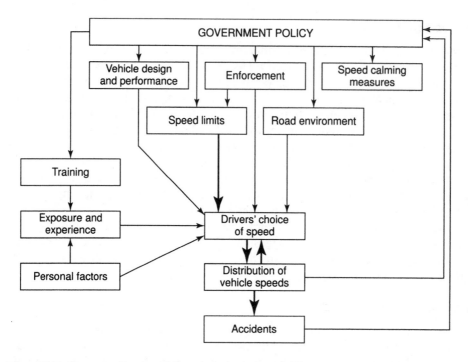

Fig. 18.5 Factors affecting drivers' choice of speed[17]

18.4.3 Importance of enforcement to road safety

Unfortunately, patterns of social behaviour are difficult to change quickly by education alone, and so change in relation to road safety is often accelerated by the use of complementary enforcement strategies. Thus education programmes aimed at attitude-changing often have a secondary objective, that of persuading the target audiences to accept the enforcement of related (possibly unpopular) legislation.

A major review of the literature on traffic law enforcement[18] concluded that the success of enforcement is dependent on its ability to create a meaningful deterrent threat to road users and, to achieve this, the emphasis must be on increasing enforcement activity to ensure that the perceived apprehension risk is high. Classic examples of the effective use of enforcement in conjunction with public education are in relation to drink-driving and seat-belt-wearing legislation.

Effect of enforcement of drink-driving legislation

A major step in respect of changing attitudes toward the drinking-driver problem in Britain occurred with the passing of the Road Safety Act of 1967 which made it an offence for anyone to drive a vehicle with a blood alcohol concentration of above 80 mg/100 ml. This act received wide publicity and its immediate effect was to bring about the biggest reduction in road casualties (−11 per cent) and fatalities (−15% per cent) known to have occurred anywhere following drinking-and-driving legislation. However, not too long after the implementation of this legislation its dynamic beneficial effects began to wear off, and nearly a decade later a governmental inquiry that looked into the reasons why concluded that it was not the law that was inadequate or the penalties too weak, but that the enforcement was insufficient and drivers no longer expected to get caught.

As a consequence of the inquiry the Transport Act of 1981 allowed (from May 1983) for the admittance into evidence by the police of alcohol levels established by breath test; the limit was set at 35 mg/100 ml BrAC, which is equivalent to 80 mg/100 ml BAC. Further, drivers who were convicted twice within 10 years of exceeding the legal limit by more than 2½ times (i.e. 200 mg/100 ml BAC) lost their licence indefinitely. These changes in the implementation of the law had both an immediate and a continuing effect on drink-driving: the number of casualties in drink-drive accidents fell by approximately one half between 1979 and 1988 as a consequence.[19] Between 1982 and 1992 the number of drivers breath-tested after an accident increased from just over 40 000 to nearly 110 000 per annum; despite this almost threefold increase in breath testing, the absolute number of positive tests fell by nearly 40 per cent over this decade. In 1984 about one in four road deaths were drink-related; in 1993 this proportion was estimated[20] at between one in six and one in seven.

Evidence elsewhere in the world is that the use of sustained and highly intensive breath-testing operations that are carried out at random (rather than being discretionary, as in Britain), are highly visible, and receive wide publicity, is one of the most effective means of deterring drink-driving.[18]

Effect of seat-belt legislation

The 1981 Transport Act also made it compulsory for drivers and front-seat passengers of cars and vans (from January 1983) to wear seat-belts. As a consequence the wearing rate by those persons rose from about 24 per cent (drivers only) in 1978, and 36–40 per cent in 1982, to 95 per cent in April 1983; over the ensuing six years it fell marginally to 93–94 per cent.[21] The wearing rate was lower for males than for females, and lower (at *ca* 19 per cent) for taxi drivers. Overall it was estimated that the seat-belt legislation brought about savings of about 7000 fatal or serious casualties (including about 500 fatalities) and 13 000 slight casualties in its first year.

In September 1989 it became compulsory for a child up to 13 years of age to be restrained when travelling in the rear of a car fitted with seat-belts or other child restraints; the child wearing rate rose to 80 per cent in such cars following the legislation. In October 1989, however, only 10 per cent of adult rear-seat passengers wore seat-belts, and the wearing rate tended to decrease with increasing age. In July 1991 seat-belt wearing by rear-seat adult passengers became law in cars where belts are fitted and available.

The review of literature[18] suggests that the simplest and most effective seat-belt wearing enforcement strategy is one which ensures that checks are adopted as a standard operational procedure when carrying out other forms of traffic-policing activity which require the roadside stopping of motorists. These checks should be supported by high levels of publicity.

Effect of motorcycle legislation

The 1981 Transport Act also included measures to reduce the high motorcyclist accident rate: the riding test was extended, the duration of a provisional licence was limited to two years, and the maximum engine size of motorcycles that learners could ride was reduced to 125 cc. The immediate effect of these changes was a reduction of over 50 per cent in the number of motorcyclists passing the riding test and an associated increase in their age; the amount of motorcycling also fell (possibly due to other factors as well). The numbers of motorcycling accidents and casualties fell as a consequence: casualties fell by 7010 in the first year (2430 of whom would have been killed or seriously injured). The engine-size restriction was estimated to reduce casualties amongst learner riders by about one quarter, independent of any reduction from the declining number of motorcyclists.[22]

Enforcing speed-limit legislation

The first demonstration of the effect of a speed limit in Britain was in 1935 when reductions in fatal (−15 per cent) and all accident severities (−6 per cent) were reported following the introduction of the 30 mile/h (48 km/h) limit in all urban areas. A compulsory 50 mile/h (80 km/h) limit was imposed and enforced on all roads not subject to a lower limit during the 1973–74 oil crisis. While there were no reductions on accident rates on roads subject to lower limits, there were statistically significant reductions in the rates on motorways (−40 per cent) and on all-purpose roads with speed limits normally ≥ 80 km/h (−21.5 per cent), while the emergency legislation was in force.

It is well established that highly visible stationary and mobile enforcement procedures are very effective in ensuring that drivers keep within speed limits, particularly when it is perceived that they are supported by unmarked (non-visible) mobile operations which increase the unpredictability of when, where and how enforcement will be encountered. The 1991 Road Traffic Act now enables radar-operated automatic camera technology to be used to enforce speed limits in Britain and it can be expected that these cameras will be increasingly deployed by the police to enhance the deterrent effect of speed enforcement.[17] Maximum safety benefits (and community acceptance) are gained if these cameras are targeted at accident locations where speed is known to be a causal factor.

Given the propensity for accidents at intersections in urban areas, there is considerable potential for the use of automated cameras to reduce the level of deliberate red light running at high accident-risk signalised intersections. To maximise both the deterrence and accident-reduction effectiveness of red light cameras consideration needs to be given to (a) the use of warning signs at intersection approaches and the use of highly visible hardware installations, (b) the rotation of several cameras through a large number of treated intersections, (c) visible deployment of the camera flash unit when cameras are not installed at treated sites, and (d) the use of high levels of supporting publicity.[18]

The thrust of governmental educational campaigns in relation to speeding is to make it socially unacceptable to speed.[20] Particular attention is being paid to influencing driver speeds in residential areas, with the aim of reducing the pedestrian (especially child) casualty toll.

18.4.3 Engineering and road safety

The term 'engineering' can be generally used to refer to technological changes to the vehicle or to the roadway and its environs in order to improve road safety.

Considerable changes (many now accepted as commonplace) have been deliberately made to motor vehicles, since the 1970s in particular, to make motoring safer. These include:

- automatic transmissions which make driving less complicated
- systems which ensure better driver visibility, e.g. window defrosting/defogging, more and better wipers, tinted windows, better headlights, exterior mirrors adjustable from inside the vehicle, anti-glare interior mirrors, fewer driver 'blind spots' in the vehicle
- anti-lock braking systems which considerably reduce the risk of skidding
- tyres which give improved wet weather traction
- better vehicle signalling systems, e.g. front and rear turn-indicators, braking lights
- greater discernibility, through lighting, of heavy commercial vehicles
- injury attenuation systems, e.g. seat-belts, collapsible steering wheels/columns, airbags, non-lethal 'safety' windscreens, neck restraints.

In-vehicle safety engineering is possibly the most important factor affecting motor car sales today. Still of major concern is the injury outcome of collisions between larger and smaller vehicles. Accidents between heavy commercials and cars are normally associated with serious injury to the car occupants. Collisions between large and small cars usually result in greater damage to the smaller vehicles and a greater likelihood of the driver of the smaller car being injured (see Table 18.8).

Table 18.8 Percentage of drivers injured, by size of car, when involved in a two-car injury accident: 1989–1992[4]

Car size*	Fatal or serious injury (%)	All (%)
Small	9	71
Small/medium	7	63
Medium	6	57
Large	4	45

* Small cars are typically 3.56–3.81 m long, small/medium 3.94–4.19 m, medium 4.32–4.57 m, and large >4.57 m.

It is quite likely that an important factor influencing further reductions in the numbers and severities of road accidents in the future will be the development of new technologies associated with in-vehicle information systems. As well as providing navigational advice and real-time information regarding road conditions, these systems may be able to alert the driver regarding unsafe vehicle conditions, e.g. too-close proximity to other vehicles in the traffic stream or the presence of an overtaking vehicle in driver 'blind spots', so that avoidance manoeuvres (manual or automatic) can be carried out in sufficient time to prevent an accident.

Road and traffic engineering developments which have contributed significantly to controlling and reducing the accident rate include:

- greater control over accesses to roadways (see Fig. 18.6 which shows that the total accident rate increases linearly with the number of commercial accesses permitted on a highway)
- improved geometric design, especially in respect of intersections, curves, hills, and the flanks of the roadway (see Chapters 19 and 20)
- improved lighting and carriageway delineation which helps drivers carry out the driving task in the dark and in wet weather (see Chapters 23 and 28)
- improved road signing which provides the driver with notice of intersections well in advance of the actual turn-offs (see Chapter 28)

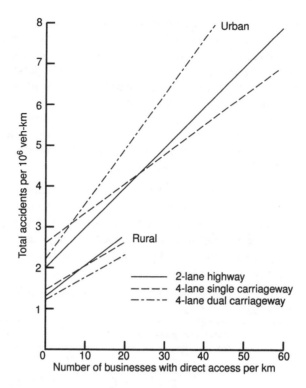

Fig. 18.6 Total accident rate on non-interstate highways in the USA, by number of businesses per kilometre (reported in reference 25)

- implementation of educational and enforcement processes to control drink-driving and speeding
- implementation of physical methods of traffic control to prevent speeding in, for example, residential areas (see Chapter 25)
- separation of pedestrians and cyclists from motor vehicles at dangerous locations (see Chapter 25)
- separation of moving vehicles by time and space at complex intersections (see Chapters 20 and 26)
- improved processes for identifying and correcting accident blackspots (see Chapter 15).

18.5 References

1. Smeed, R.J., Some statistical aspects of road safety research. *Journal of the Royal Statistical Society*, 1949, Series A, **112**(1), 1–23.
2. Broughton, J., Long-term trends in road accident casualties. In *Road accidents in Great Britain 1993*. London: HMSO, 1994, 27–32.
3. *OECD Road safety research: A synthesis*. Paris: OECD, 1986.
4. *Transport statistics Great Britain*. London: HMSO, 1994.
5. Taylor, M.C. and Barker, J.K., *Injury accidents on rural single-carriageway roads – An analysis of stats 19 data*. Research Report 365. Crowthorne, Berks: Transport Research Laboratory, 1992.

6. Pickering, D., Hall, R.D., and Grimmer, M., *Accidents at rural T-junctions*. Research Report 65. Crowthorne, Berks: Transport Research Laboratory, 1986.

7. Maycock, G., Lockwood, C.R., and Lester, J.F., *The accident liability of car drivers*. Research Report 315. Crowthorne, Berks: Transport Research Laboratory, 1991.

8. O'Reilly, D., Child pedestrian safety in Great Britain. In *Road accidents in Great Britain 1993*. London: HMSO, 1994, 33–40.

9. Everest, J.T., *The involvement of alcohol in fatal accidents to adult pedestrians*. Research Report 343. Crowthorne, Berks: Transport Research Laboratory, 1992.

10. Taylor, M.C., and Lockwood, C.R., *Factors affecting the accident liability of motorcyclists — A multivariate analysis of survey data*. Research Report 270. Crowthorne, Berks: Transport Research Laboratory, 1990.

11. Mills, P.J., *Pedal cycle accidents — A hospital based study*. Research Report 220. Crowthorne, Berks: Transport Research Laboratory, 1989.

12. Astrop, A., Balcombe, R.J., and Finch, D.J., *Bus safety and maintenance following deregulation*. Research Report 337. Crowthorne, Berks: Transport Research Laboratory, 1991.

13. *Road accidents in Great Britain 1993*. London: HMSO, 1994.

14. Sabey, B.E., and Taylor, H., *The known risks we run: The highway*. Report SR567. Crowthorne, Berks: Transport Research Laboratory, 1980.

15. Spear, M.G., Singh, A., and Downing, C., *The current state of road safety education in primary and middle schools*. Research Report 101. Crowthorne, Berks: Transport Research Laboratory, 1987.

16. Everest, J.T., Tunbridge, R.J., and Widdop, B., *The incidence of drugs in road accident fatalities*. Research Report 202. Crowthorne, Berks: Transport Research Laboratory, 1989.

17. Finch, D.J., Kompfner, P., Lockwood, C.R., and Maycock, G., *Speed, speed limits and accidents*. Project Report 58. Crowthorne, Berks: Transport Research Laboratory, 1994.

18. Zaal, D., *Traffic law enforcement; A review of the literature*. Report No. 53. Leidschendam, The Netherlands: The Institute for Road Safety Research (SWOV), 1994.

19. Broughton, J., *Trends in drink/driving revealed by recent road accident data*. Research Report 266. Crowthorne, Berks: Transport Research Laboratory, 1990.

20. Hooke, S., Road safety publicity campaigns. In *Road accidents in Great Britain 1993*. London: HMSO, 1994, 41–5.

21. Broughton, J., *Restraint use by car occupants, 1982–1989*. Research Report 289. Crowthorne, Berks: Transport Research Laboratory, 1990.

22. Broughton, J., *The effect on motorcycling of the 1981 Transport Act*. Research Report 106. Crowthorne, Berks: Transport Research Laboratory, 1987.

23. Trends in the European transport sector 1970–1991. *Routes/Roads*, 1993, **279**, 21–23.

24. Broughton, J., and Stark, D.C., *The effect of the 1983 changes to the law relating to drink/driving*. Report 89. Crowthorne, Berks: Transport Research Laboratory, 1991.

25. *Access to highways — Safety implications*. Advice Note TA4/80. London: Department of Transport, 1980.

CHAPTER 19

Geometric design of streets and highways

C.A. O'Flaherty

Geometric design relates the layout of the road in the terrain with the requirements of the driver and the vehicle. Main features considered in the geometric design of roads are horizontal and vertical curvature and the visible features of the road cross-section. Good geometric design ensures that adequate levels of safety and comfort are provided to drivers for vehicle manoeuvres at the design speed, that the road is designed uniformly and economically, and that it blends harmoniously with the landscape.

19.1 Design speed

A fundamental consideration in the design of a road section is the design speed to be used. Notwithstanding the apparent simplicity of the term 'design speed' there is still argument over its exact definition. In practice, however, it simply means the speed value used as a guide by road designers when determining radii, sight distances, superelevations, and transition lengths.

In most design guides, the selection of design speeds for road sections of a particular classification is primarily influenced by the nature of the terrain (e.g. whether level, rolling, or mountainous) and by motorists' expectations in relation to the free speed at which it is safe to drive (in rural areas) or legal to drive (in urban areas). The 85th percentile free speed is generally regarded as the most appropriate choice for design speed; to design for the 99th percentile speed would be excessively expensive while use of a 50th percentile value would be unsafe for the fastest drivers.[1]

19.1.1 Rural roads

The speeds at which motorists travel along road sections are generally individual compromises between that which will minimise the journey time and the perceived risk associated with high speed. On straight sections the risk is seen to be small and speeds are essentially determined by driver preference and vehicle capability, while speeds on road sections in difficult terrain are governed by the perceived risk associated with each section's geometric features (especially horizontal curvature) and by the volume and composition of traffic. In practice, designers obviously cannot take direct account of actual speed expectations on proposed alignment sections for *new roads or major improvements* and they rely instead on typical speed distribution data previously obtained on similar existing roads. An important relationship so derived in Britain is that on rural single and dual carriageways and on rural motorways both the 99th to 85th and 85th to 50th percentile speed ratios are approximately constant at 2%.

In order to derive geometric design standards for Britain's non-urban roads the free speeds of some 620 000 light vehicles were measured on inter-urban road sections of various quality, and the following relationship (shown graphically in Fig. 19.1) was developed:

$$V_{L50} = 110 - A_c - L_c \qquad (19.1)$$

where V_{L50} = mean free operation speed in wet conditions (km/h), L_c = a layout constraint obtained from the standard verge-width rows in Table 19.1; and A_c = an alignment constraint. The alignment constraints for single and dual carriageway roads are derived from Equations 19.2 and 19.3, respectively:

$$A_c = 12 - \frac{VISI}{60} + \frac{2B}{45} \qquad (19.2)$$

$$A_c = 6.6 + \frac{B}{10} \qquad (19.3)$$

where B = bendiness of the road section (deg/km), obtained by summing all changes in direction in the section and dividing by the section length; and $VISI$ = harmonic mean visibility (m) derived from $Log_{10} VISI = 2.46 + VW/25 - B/400$, where VW is the average verge width (averaged for both sides of the road) and B is the bendiness.

The following should be noted in relation to Fig. 19.1:

1. the right-hand abscissa gives the equivalent 85th percentile speed after applying the 2% -factor to the 50th percentile wet speed
2. wet speeds are used because limiting design is based upon the tyre/road surface adhesion in the wet
3. the mean and median speeds are assumed equal
4. the 85th percentile speed is taken as the derived design speed
5. for practical purposes the 85th percentile design speeds are grouped into four bands, 120, 100, 85 and 70 km/h, each with upper and lower ranges (A and B)
6. use of a particular 85th percentile design speed should result in the situation where 99 per cent of the vehicles using that road section travel in or below the speed band above the design speed selected
7. the alignment constraint (and hence the design speed) is not dependent on the radius of curvature of any individual curve but on the total number of degrees turned through per kilometre, i.e. the bendiness
8. the layout constraint reflects the effects of those features which describe road type, i.e. carriageway and verge widths and the frequency of accesses (including laybys) and intersections
9. the effects of both upgrade (H_r) and downgrade (H_f) hilliness are excluded from the initial calculation of design speed and subsequent adjustments are applied specifically at hills; that is, speeds, in kilometres per hour, reduce on uphill sections of single carriageways by $H_r/6$ and increase on downhill sections of dual carriageways by $H_f/4$.

(Note: The hilliness (m/km) of a road section is the sum of all height changes, i.e. rises and falls, divided by the section length; it is composed of the upgrade hilliness, H_r (m/km) and the downgrade hilliness, H_f (m/km).)

Table 19.1 Determining the layout constraint L_c (km/h) for rural roads[1, 4]

Road type	S2				WS2		D2AP		D3AP	D2M	D3M
Carriageway width (excluding metre strips)											
							Dual		Dual	Dual 7.3 m and hard shoulder	Dual 11 m and hard shoulder
		6 m		7.3 m	10 m		7.3 m		11 m		
Degree of access and junctions (both sides)											
	H	M	M	L	M	L	M	L	L	L	L
Std. verge width	29	26	23	21	19	17	10	9	6	4	0
1.5 m verge	31	28	25	23							
0.5 m verge	33	30									

S2 = 2-lane single carriageway, WS2 = 2-lane wide single carriageway, D2AP and D3AP = 2-lane and 3-lane all-purpose dual carriageways, respectively, and D2M and D3M = 2-lane and 3-lane motorway dual carriageways, respectively; and L, M and H are a Low (2–5), Medium (6–8) and High (9–12) number of access points per km. No research data are available for 4-lane single carriageway roads (S4) and their design speeds should be estimated assuming a normal D2AP with an L_c of 15–13 km/h.

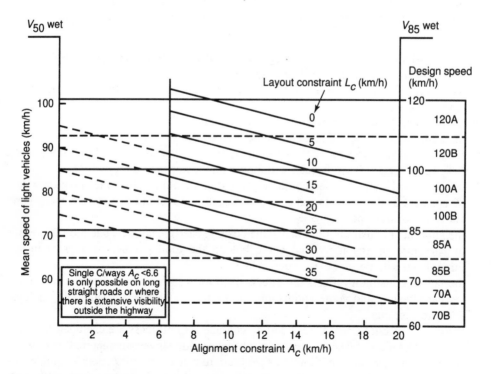

Fig. 19.1 Chart used to derive design speeds for rural roads in Britain[4]

Table 19.2 Geometric design standards currently in use in Britain[4]

Standard	Design speed (km/h)						V²/R
	120	100	85	70	60	50	
Stopping sight distance (m)							
A1 Desirable minimum	295	215	160	120	90	70	
A2 One step below A1	215	160	120	90	70	50	
Horizontal curvature (m)							
B1 Minimum R^* without elimination of adverse camber and transitions	2880	2040	1440	1020	720	510	5
B2 Minimum R^* with e = 2.5%	2040	1440	1020	720	510	360	7.07
B3 Minimum R^* with e = 3.5%	1440	1020	720	510	360	255	10
B4 Desirable minimum R with e = 5%	1020	720	510	360	255	180	14.14
B5 One step below desirable minimum R with e = 7%	720	510	360	255	180	127	20
B6 Two steps below desirable minimum R with e = 7%	510	360	255	180	127	90	28.28
Vertical curvature							
C1 Desirable minimum* crest K-value	182	100	55	30	17	10	
C2 Absolute minimum crest K-value	100	55	30	17	10	6.5	
C3 Absolute minimum sag K-value	37	26	20	20	13	9	
Overtaking sight distance (m)							
D1 Full overtaking sight distance (FOSD)	*	580	490	410	345	290	
D2 FOSD overtaking crest K-value	*	400	285	200	142	100	

*Not recommended for single carriageways

Determining the design speed(s) for a new road or a major improvement is an iterative process which involves dividing the proposed road into a series of sections, each at least 2 km long, and as a first iteration, designing an overall trial alignment using an assumed 'typical' design speed. This enables a design speed to be derived for each section using Fig. 19.1 and A_c and L_c values associated with the trial alignment. Each section's design speed is then compared with the assumed value and appropriate locations and standards used for the trial alignment are examined in detail to determine, say, whether they should be varied to obtain economic or environmental savings or, perhaps, whether they should be upgraded to meet standards associated with a higher derived design speed (see Table 19.2). Note that if a derived design speed falls within category B of a design-speed band, consideration is often given to testing whether the correct design speed estimate might actually be in a lower band. (This may also be done if it falls within category A, but is less likely and less desirable.) The iterative process is continued and trial alignment sections are revised until the designer is satisfied that the final design is geometrically consistent and cost effective. Note that with this procedure the final design speeds for various sections of the full road may be different; preferably, however, the design speeds for two adjacent sections should not differ by more than 10 km/h.

19.1.2 Urban roads

Research data are not available to model speeds for roads in urban areas in the same way as for rural roads. However, due to the many disruptions to flow (e.g. from intersections) experienced in urban areas, free speeds on urban roads are normally less than on rural roads, and drivers are more willing to adjust their desired speeds to the imposed speed limits. Consequently, the design speed for a new road or a major road improvement in an urban area is usually selected after taking into account the likely speed limit regime for the road.

Design speeds most usually used on urban roads in Britain are 60B, 70A, 85A and 100A km/h for roads with speed limits of 48, 64, 80 and 96 km/h, respectively. Use of these design speeds normally allows a small margin in each instance for speeds that may be in excess of the speed limit. If, however, the alignment is being improved on an existing road and speed measurements confirm that the actual 85th percentile speed is greater than the speed limit then, for safety purposes, the design speed selected should be higher than that which might be suggested by the speed limit.

The design speed for a primary distributor road that is at-grade is never less than 70A km/h; if grade-separated the design speed is normally 100A km/h.

19.2 Sight distance requirements

Sight distance – the length of carriageway visible to a driver in both the horizontal and vertical planes – is the most important feature in the safe and efficient operation of a highway. From a construction aspect the standards set for minimum sight distances have a major bearing on the cost of a road.

Stopping and overtaking sight distances are of special interest to road designers. If safety is to be built into all roads, then enough sight distance must be available to drivers in each lane of a single or dual carriageway road to enable them to stop before striking an object on the carriageway; this is the safe stopping sight distance (*SSD*). If efficiency is to be built into a two-lane single carriageway road, then sufficient sight distance must be available to drivers to enable them to complete normal overtakings in the face of oncoming vehicles; this is the full overtaking sight distance (*FOSD*).

19.2.1 Stopping sight distance

Calculating the minimum distance required to stop a vehicle before it hits a stationary or slow-moving object on the carriageway involves establishing values for speed, perception-reaction time, braking distance, and eye and object heights. These values can vary from country to country, which explains why different design distances are in use in many countries. The vehicle *speed* used in safe stopping sight distance (*SSD*) calculations is normally the design speed of the road section under consideration.

Perception time is the time which elapses between the instant the driver sees the hazard and the realisation that brake action is required. The reaction time is the time taken by the driver to actuate the brake pedal, after realising the need to brake, until the brakes start to take effect. Field measurements indicate that combined *perception-reaction times* typically vary from 0.5 s in difficult terrain, where drivers are more alert, to 1.5 s under normal road conditions. For safe and comfortable design, a combined time of 2 s is used in Britain. Expressed in formula form, the distance travelled by a vehicle during the perception-reaction time is given by $0.278tV$ where V = the initial speed (km/h), and t = the perception-reaction time.

The *braking distance*, i.e. the distance needed by a vehicle to decelerate to a stop on a level road after the brakes have been applied, is given by

$$d = \frac{v^2}{2w} = \frac{v^2}{2fg} = \frac{V^2}{254f} \tag{19.4}$$

where d = braking distance (m), w = rate of deceleration (m/s^2), g = acceleration of freefall = 9.81 m/s^2, v = initial speed (m/s), V = initial speed (km/h), and f = longitudinal coefficient of friction developed between the tyre and the carriageway surface.

The choice of the f-value to use in *SSD*-design is complicated. Major factors which affect the friction are:

1. carriageway conditions – wet roads are normally assumed as water acts as a lubricant between the carriageway and tyres
2. tyre quality – a well-patterned tread provides good escape channels for bulk water and a radial ply increases the tyre–road contact area
3. speed – the higher the vehicle speed the less is the contact time available to expel water from between the tyre and the carriageway
4. carriageway macrotexture and microtexture – a rough macrotexture helps with the removal of bulk water and the maintenance of skidding resistance at high speeds, while the harsh microtextures of surfacing materials add to skid resistance as they puncture and disperse the thin film of water remaining after the removal of the bulk water by the tyre tread and the carriageway macrotexture.

If motorist comfort is a major criterion, an f-value greater than 0.5 should never be used in normal road design as persons who are not wearing seat-belts will slide from their seats with decelerations in excess of about 0.5 g (with possible consequential injury or loss of vehicle control). Research has shown that an *f*-value of 0.25 can be achieved by worn tyres at high speeds on wet roads, as well as providing a very comfortable rate of deceleration, and hence this is normally used in road design in Britain.

Comfortable (rounded) *SSD*-values derived from a total driver perception-reaction time of 2 s and a rate of deceleration of 0.25 g, as used in road design in Britain for various design speeds, are shown in row A1 in Table 19.2. The values shown at A2 assume the same perception-reaction time and a deceleration rate of 0.375 g; this deceleration rate can be achieved in wet conditions on normally-textured surfaces without loss of control.[1] Note that the sight distance at a given design speed in row A2 is the same as the desirable minimum value for the next lower design speed, i.e. it is equivalent to the relaxation of one design speed step.

Note also that the *SSD*-values given in Table 19.2 relate to motor cars and light commercial vehicles on straight, level sections of road. Longer stopping distances should be provided prior to sharp curves, however, particularly after steep downhill grades. No adjustments are normally made for heavy commercial vehicles; while these vehicles require longer braking distances than cars, they have higher eye heights and thus have compensatory earlier warning of hazards ahead. Generally, also, the free speeds of most commercial vehicles are lower than those of cars.

The *eye and object heights* used should ensure that there is an envelope of clear visibility which enables drivers of low cars to see low objects on the carriageway, and drivers of high vehicles to see portions of other vehicles, even though bridge soffits at sag curves and overhanging tree branches may be in the way. The lower and upper eye heights used in Britain are 1.05 m and 2 m, respectively, as 95 per cent of car driver eye heights are above 1.05 m and 2 m is a typical eye height for the driver of a large commercial vehicle. The lower and upper object heights used are 0.26 m and 2 m, respectively; the 0.26 m height ensures that the tail lights of other vehicles can be seen and the upper bound of 2 m ensures that at least a part of the vehicle ahead can always be seen.

19.2.2 Overtaking sight distance

As shown in Fig. 19.2 there are four components of the minimum distance required for safe passing on two-lane single carriageway roads. The dimension d_1 represents the perception-reaction time taken or distance travelled by a vehicle while its driver decides if it is safe to pass the vehicle in front; d_2 is the overtaking time taken or distance travelled by the overtaking vehicle in carrying out the actual passing manoeuvre; d_4 is the closing time taken or distance travelled by the opposing vehicle during the time d_1 plus d_2, and d_3 is the safety time or distance between the overtaking vehicle and the opposing vehicle at the instant the overtaking vehicle returns to its correct lane.

There are significant differences in the full overtaking sight distance (*FOSD*) standards used in various countries; these mainly arise from differing assumptions regarding the component times or distances, assumed speeds, and driver behaviour. The *FOSD*-values used on two-lane British roads assume that:

1. the overtaking driver commences the manoeuvre two design speed steps below the design speed of the road section while the opposing vehicle is travelling at the design speed

2. 85 per cent of overtaking manoeuvres take less than 10 s
3. the safety distance $d_3 = 0.2d_4$
4. there is an envelope of clear visibility in both the horizontal and vertical planes between points 1.05 m and 2.05 m above the carriageway over the full overtaking sight distance.

The rounded data shown at row D1 in Table 19.2, which are those recommended for road design purposes in Britain, were derived from the formula

$$FOSD \text{ (m)} = T\left[\frac{1}{2}\left(\frac{v}{2^{\frac{1}{4}} \times 2^{\frac{1}{4}}} + v\right)\right] + 0.2vT + vT = 2.05vT \qquad (19.5)$$

where T = time taken to complete an overtake (s), and v = design speed (m/s). They provide the visibility for a reasonable degree of safe overtaking for 85 per cent of traffic, though not in peak hours and when heavy traffic prevails.[1]

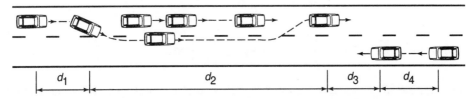

Fig. 19.2 Components of full overtaking sight distance requirements on two-lane roads

With *FOSD* no special provision is provided in Table 19.2 for commercial vehicles, even though they take longer than cars to overtake. Rather it is again assumed that their greater eye height compensates at least partially for any additional overtaking length that might be required, by enabling them to judge sooner and better whether a gap is suitable for overtaking.

Full overtaking sight distances are most easily implemented on relatively straight and level sections of road in rural locales; land use and cost reasons usually prohibit their design into most urban roads.

Where possible, overtaking sight distances used in design should be longer than the minimum specified in Table 19.2. If the environmental and/or topographic conditions are such that *FOSD* cannot be implemented in the road design and faster vehicles are constrained to follow slower ones over a particular rural road section, then (a) that should be made apparent to drivers with the aid of markings/signs, and (b) that section should be followed by one which allows for overtaking as otherwise the faster drivers will become increasingly frustrated and perhaps try to pass at increased risk on more dangerous alignments.

19.3 Horizontal alignment design

Horizontal alignment, which comprises a series of intersecting tangents and circular curves (with or without transition curves) is a most important feature affecting the safety, efficiency and cost of a road. Several studies have indicated that highway curves exhibit higher accident rates than tangent sections, and that the accident rate increases as the radius of curvature is decreased.[2] Generally, vehicle operating speeds decrease as the overall horizontal curvature increases; thus road user costs are affected by the bendiness of a road. Also speeds on individual curves are lowered as their radii are reduced, while construction costs usually increase as radii increase (especially in hilly terrain).

19.3.1 Curvature and centrifugal force

As a vehicle of mass M at speed v moves about a carriageway curve of radius R that is at an angle α with the horizontal it is subject to a reactive force $= Mv^2/R$, called the *centrifugal force*, which causes it to slide outward, away from the centre of curvature. Figure 19.3 illustrates the forces acting on the vehicle as it moves about the curve. As the forces are in equilibrium

$$\left(Mv^2/R\right)\cos\alpha = Mg\sin\alpha + P$$

$$= Mg\sin\alpha + \mu\left[Mg\cos\alpha + \left(Mv^2/R\right)\sin\alpha\right]$$

and, therefore,
$$\frac{v^2}{gR} = \tan\alpha + \mu + \left(\frac{\mu v^2}{gR}\right)\tan\alpha \qquad (19.6)$$

where M, N, P, R, g, v and α are as in Fig. 19.3, and μ = coefficient of lateral friction (or side friction factor). The quantity $(\mu v^2/gR)\tan\alpha$ is so small that it can be neglected. If $\tan\alpha$ is expressed in terms of the crossfall slope (or *superelevation*), e, then

$$\frac{v^2}{R} = \frac{(e+\mu)}{g} \qquad (19.7)$$

where v^2/R is the *centrifrugal acceleration*. If v m/s is replaced by V km/h and $g = 9.81\text{m/s}^2$, then Equation 19.7 becomes

$$\frac{V^2}{127R} = e + \mu \qquad (19.8)$$

Equation 19.8 is known as the *minimum radius equation*. If design speed and superelevation are known, it enables the minimum radius to be determined for an acceptable

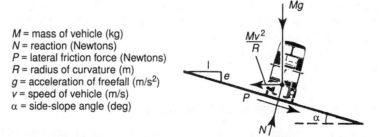

M = mass of vehicle (kg)
N = reaction (Newtons)
P = lateral friction force (Newtons)
R = radius of curvature (m)
g = acceleration of freefall (m/s^2)
v = speed of vehicle (m/s)
α = side-slope angle (deg)

Fig. 19.3 Forces acting on a motor vehicle moving about a horizontal curve

level of side friction (and of driver comfort). It should be noted that the friction factor used in Equations 19.7 and 19.8 is not the same as the longitudinal coefficient of friction discussed in relation to stopping sight distances and measured by machines such as SCRIM;[3] at this time no relationship has been established between these two friction measurements.

In most countries the maximum superelevation used in design is below 10 per cent; in Britain it is usually limited to a maximum of 7 per cent, however, with most curve designs being based on 5 per cent or less. This results from the recognition that a stationary or slow-moving vehicle tends to slide toward the inside of a curve in icy conditions (when the lateral coefficient of friction is about 0.1) if the superelevation is more than 10 per cent.

British design practice is based on the fundamental assumption that at absolute minimum radius the 99th percentile vehicle should not experience more than the maximum level of centrifugal acceleration acceptable for comfort and safety.[1] This was established at about 0.22 g some 70 years ago and has not been changed since. Thus, if

$$\frac{v^2_{99}}{R} = 0.22g$$

then the total centrifugal acceleration experienced at the 85th percentile speed, i.e. at the design speed, should not exceed

$$\frac{v^2_{85}}{R} = \frac{0.22g}{2^{\frac{1}{2}}} = 0.156g \tag{19.9}$$

since the 99th percentile speed is approximately equal to the 85th percentile speed multiplied by $2^{\frac{1}{4}}$. If a crossfall of 0.07 is applied, then from Equation 19.8 the crossfall's contribution to counteracting the total centrifugal acceleration at the design speed is 45 per cent, with the balance of 0.086 g (55 per cent) being met by the lateral friction. If the crossfall contribution is kept constant at 45 per cent then, for any 85th percentile design speed and radius greater than the absolute minimum, the design superelevation (per cent) is given by

$$e = \left(\frac{0.45v^2}{gR}\right)100 = \left(\frac{1}{2.82}\right)\left(\frac{V^2}{R}\right) \tag{19.10}$$

However, $V^2/2.82R$ is approximately equal to $V^2/2(2^{\frac{1}{2}})R$, and it is in accordance with this latter relationship that the V^2/R and R values shown at rows B1–5 in Table 19.2 were calculated. Note that the desirable minimum radius (with e_{max} = 5 per cent) calculation at row B4 assumes that the total centrifugal acceleration at the 85th percentile design speed is 0.11 g, i.e. only half the threshold of driver discomfort, with the same proportions being taken by superelevation (45 per cent) and by friction (55 per cent). As a guide, it is recommended that the normal 2.5 per cent adverse camber be replaced by a favourable crossfall (e = 2.5 per cent) when V^2/R exceeds 5, to ensure that the maximum net lateral acceleration which must be resisted by the tyre–road surface at the design speed does not exceed 0.064 g. Note also that the V^2/R and R values shown at row B6 are based on a superelevation of 10 per cent; since, however, e_{max} = 7 per cent is recommended for design purposes, it means that a higher net lateral acceleration, of (theoretically) 0.12 g, has to be met by the tyre–carriageway interaction for vehicles negotiating the curve at the design speed. The limiting radii given in row B6 are only permitted at very difficult sites and when the 85th percentile speed is expected to lie in the lower half of the design speed band, i.e. for category B design speeds.

19.3.2 Sight distances

As noted previously, difficulties in providing the required safe stopping and full overtaking sight distances are most commonly encountered in urban road design where the alignment constraints are such that the desired visibilities can only be achieved at considerable financial and environmental costs. In rural areas diverse obstructions at the side of the road, e.g. buildings, bridge supports, slopes of cuttings, solid fences, or uncut grass on or adjacent to verges, can hinder visibility. In both urban and rural areas safety fences in the central reservation between dual carriageways can hinder the achievement of the minimum stopping distance in the inside lane because of the low object height.

Figure 19.4 (a) illustrates the situation where the required sight distance lies wholly within the length of the curve and ACB is assumed equal to the required sight distance S. The minimum offset clearance M desired between the centre line and any lateral obstruction can be approximated by considering the vehicle track to be along the chords AC and CB. Then by geometry $R^2 = x^2 + (R - M)^2$ and $x^2 = (S/2)^2 - M^2$. Therefore

$$M = \frac{S^2}{8R} \qquad (19.11)$$

Figure 19.4(b) illustrates the situation where S is greater than the available length of curve L and overlaps onto the tangents for a distance l on either side. By geometry $(S/2)^2 = x^2 + M^2$, and $x^2 = d^2 - (R - M)^2$, and $d^2 = [(S - L)/2]^2 + R^2$. Therefore

$$M = \frac{L(2S - L)}{8R} \qquad (19.12)$$

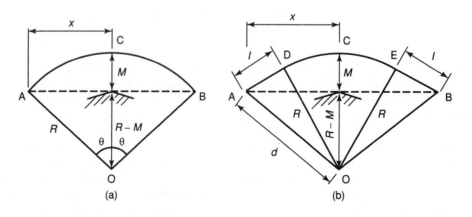

Fig. 19.4 Sight distance offset considerations on a horizontal curve

19.3.3 Transition curves

In practice transition curves are included in curve design, between curved and straight sections of road, if the curve radius to be used at a given design speed is less than that for a wholly circular curve without superelevation, e.g. for radii less than those at row B1 in Table 19.2. Their main purpose is to provide a driver travelling at high speed with an easy-to-follow path so that the radial acceleration increases and decreases gradually as the vehicle enters and leaves the circular curve. Superelevation, or the elimination of adverse camber, is also progressively applied on or within the length of the transition curve from the arc end; if the transition length defined by the design speed is less than that required by the superelevation turnover, a longer transition is usually applied to meet superelevation needs. Transition curves also facilitate widening on curves and improve the appearance of a road.

The essential feature of a well-designed transition curve is that its radius of curvature decreases gradually from infinity at the tangent–transition curve intersection to the radius of the circular curve at the transition curve–circular curve intersection. The spiral (or clothoid), lemniscate, and cubic parabola curves are most commonly used for transition purposes; however, in practice, there is relatively little difference in outcome between the three. The spiral curve is recommended for use in road design in Britain.[4]

Figure 19.5 shows a circular curve joined to two tangents by spiral transition curves. The dotted lines illustrate the circular curve as it would appear if the transition curves were omitted. TS and ST are the tangent–spiral intersection points and SC and CS are the junctions of the spiral and circular curves. The distance p by which either end of the circular curve is shifted inward from the tangent is the dimension K to P′C′; this is called the *shift*. Note that the offset from point K on the tangent to the transition spiral is very nearly $p/2$, and the line from K to P′C′ approximately bisects the spiral.

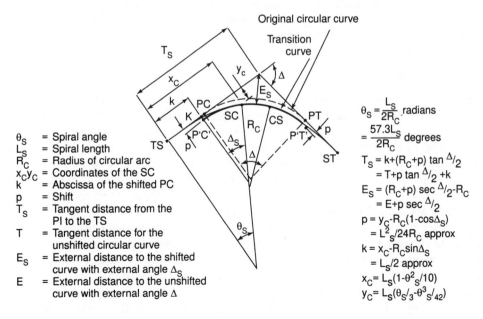

$\theta_S = \dfrac{L_S}{2R_C}$ radians

$\quad = \dfrac{57.3L_S}{2R_C}$ degrees

$T_S = k+(R_C+p)\tan\Delta/2$

$\quad = T+p\tan\Delta/2 +k$

$E_S = (R_C+p)\sec\Delta/2-R_C$

$\quad = E+p\sec\Delta/2$

$p = y_C-R_C(1-\cos\Delta_S)$

$\quad = L_S^2/24R_C$ approx

$k = x_C-R_C\sin\Delta_S$

$\quad = L_S/2$ approx

$x_C = L_S(1-\theta_S^2/10)$

$y_C = L_S(\theta_S/3-\theta_S^3/42)$

θ_S = Spiral angle
L_S = Spiral length
R_C = Radius of circular arc
$x_C y_C$ = Coordinates of the SC
k = Abscissa of the shifted PC
p = Shift
T_S = Tangent distance from the PI to the TS
T = Tangent distance for the unshifted circular curve
E_S = External distance to the shifted curve with external angle Δ_S
E = External distance to the unshifted curve with external angle Δ

Fig. 19.5 Basic properties of the spiral transition curve

Also shown in Fig. 19.5 are some of the more important formulae affecting the use of spiral curves in conjunction with circular ones; details regarding their derivation, and how spiral curves are set out on the ground, can be found in most surveying textbooks. As a vehicle passes along the transition curve in Fig. 19.5, the radial acceleration changes from zero at TS to v^2/R_c at SC, the time taken to travel from TS to SC is L_s/v, and the rate of change of radial acceleration is $(v^2/R_c)/(L_s/v)$. Then

$$C = \frac{\left(\dfrac{v^2}{R_c}\right)}{\left(\dfrac{L_s}{v}\right)} = \frac{v^3}{R_c L_s}$$ (19.13)

and

$$L_s = \frac{v^3}{CR_c} = \frac{V^3}{3.6^3 CR_c}$$ (19.14)

where C = rate of change of radial acceleration (m/s³), v = design speed (m/s), V = design speed (km/h), R_c = radius of circular curve (m), and L_s = transition length (m).

In railroad practice a value of $C = 0.3$ m/s³ is generally accepted as the value at which the increase in radial acceleration will be unnoticed by a passenger sitting on a railway coach over a bogie. Conditions are not the same for motorists, however, as the car or lorry driver, unlike the railway passenger, can make travel path corrections. Hence the tendency in many countries has been to use higher C-values as they result in shorter transition lengths. British practice is that C should not normally exceed 0.3 m/s³, although in difficult cases, e.g. in urban areas, it is often increased up to 0.6 m/s³. Also, on curves that are substandard for the design speed, the transition length is normally limited so that the shift of the circular curve, p, does not exceed 1 m; thus $p = (L_s^2/24R) < 1$ m, and $L_{s(max)} < (24R)^{\frac{1}{2}}$. For aesthetic reasons a minimum transition length of $L_{s(min)} = R/9$ is used in transition curve design.

19.3.4 Curve widening

The steering geometries of heavy commercial vehicles are based on the Ackerman principle whereby a vehicle turns about a point approximately in line with its rear axle, so that when a long vehicle is turning the rear wheels follow a path of shorter radius than that traced by the front wheels. The swept path developed has the effect of increasing the effective carriageway width required if the same clearance is to be maintained between opposing commercial vehicles travelling about curves as when they are on tangent sections of single carriageways, or between adjacent commercial vehicles travelling about curves in the same direction on dual carriageways.

To determine the extra carriageway width required in any particular situation, it is necessary to choose an appropriate design vehicle for the curve in question. Formulae have been developed[5] which enable the extra widths to be calculated for various vehicle types.

General British practice for widening on curves is summarised in Table 19.3. These widenings are applied to carriageways with substandard widths, and to low radius curves of standard width with design speeds below 70 km/h (usually in urban areas). No extra widening is applied to two-lane roads wider than 7.9 m. The extra width is applied

Table 19.3 British practice for widening on road curves

Radius (m)	Increase per lane width of carriageway (m)	Application
90–150	0.3	To standard width carriageways: 7.3 m (2 lanes), 11 m (3 lanes), and 14.6 m (4 lanes)
	0.6	To carriageways less than the standard widths, subject to max. widths of 7.9 m (2 lanes), 11.9 m (3 lanes) and 15.8 m (4 lanes)
150–300	0.5 ⎫	To carriageways of less than the standard widths, subject to the max. widths not being greater than the standard widths
300–400	0.3 ⎭	

uniformly over the entire length of a transition curve so that there is zero extra width at the beginning of transition and the full extra width at the end. When improving existing curves the widening is generally made on the inside of curves.

19.4 Vertical alignment design

Vertical alignment design refers to the design of the tangents and curves which compose the profile of the road. Its primary aim is to ensure that a continuously unfolding ribbon of road is presented to motorists so that their anticipation of directional change and future action is instantaneous and correct.

19.4.1 Gradients

Although the use of steep tangent gradients in hilly terrain generally results in lower road construction and environmental costs, it also adds to road user costs through delays and extra fuel and accidents. The extra user costs are most noticeable when traffic is heavy and commercial vehicles form a large proportion of the traffic stream. On uphill gradients in hilly terrain extra accidents often result from frustrated drivers of faster vehicles overtaking where normally they might not do so. Speeds normally increase on downhill gradients and if drivers have to brake to, say, traverse a curve at the bottom of the hill and the road is icy or wet serious accidents can result from vehicles skidding out of control.

Gradients of up to about 7 per cent have relatively little effect on the speeds of passenger cars. However, the speeds of commercial vehicles are considerably reduced on long hills with gradients in excess of about 2 per cent. When the uphill distances are short, gradients of 5 or 6 per cent may have little detrimental effect on commercial vehicle speeds; drivers usually accelerate upon entering an uphill section and the extra momentum can overcome the effect of the gradient over these shorter distances.

The *desirable maximum gradients* currently used in British road design are 3, 4 and 6 per cent for motorways, dual carriageways and single carriageways, respectively. In hilly terrain gradients of up to 8 per cent may be used on all-purpose single and dual carriageways, especially when traffic volumes are at the lower end of the design range. Gradients of 4 per cent are never exceeded on motorways in Britain.

A *minimum gradient* of about 0.5 per cent is needed to ensure the effective drainage of carriageways with kerbs. On flat urban roads drainage paths are usually provided by false kerbside drainage channels that are caused to rise and fall with minimum gradients of 0.5 per cent, rather than by artificially sloping the road. False channels can be avoided on flat rural roads by using over-edge lateral drainage to surface channels or filter drains.

Climbing lanes

The maximum gradient is not in itself a complete design control, and an extra climbing lane is often provided on long uphill climbs. The addition of a climbing lane is normally considered when the combination of hill severity and traffic volumes and composition is such that the operational benefits achieved are greater than the additional cost of providing the extra lane.

It is British practice[4] to provide an extra lane on hills with tangent gradients (as defined in Fig. 19.6) greater than 2 per cent and longer than 500 m, where it can be economically or environmentally justified. A measure of the economic justification required can be gained from Fig. 19.6 which shows, for various hill heights risen and different commercial vehicle contents of the traffic flow, the design year traffic flows which must normally be forecast for a two-lane road before an extra climbing lane is justified. (Similar justification data are available for non-motorway dual carriageways.[4]) The data assume that the terrain is relatively easy and there are no especially high costs or adverse environmental impacts associated with the climbing lane provision. Climbing lanes on motorways in Britain are rare occurrences and require site-specific evaluation.

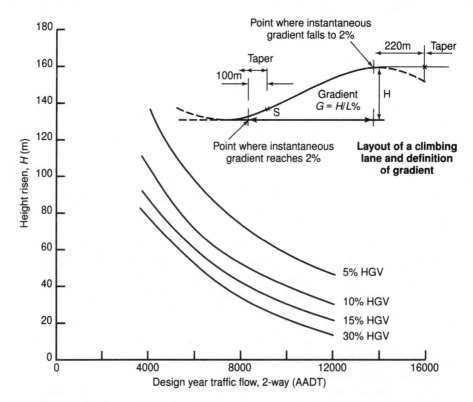

Fig. 19.6 Traffic flow criteria for climbing lanes on single carriageway roads[4]

Where a hill occurs which has varying gradients such that, for example, an easy gradient follows one justifying a climbing lane, an extra lane may be justified on both sections as the speeds of the heavy commercial vehicles will be very slow at the start of the second section. Local dualling is preferable to an extra climbing lane on a three-lane road; sometimes, however, the need for dualling can be avoided by applying road markings so that the two existing uphill lanes are given priority by means of offset double white lines.

Where a climbing lane is started depends upon commercial vehicle speeds as they approach the hill, and the lower the speeds the closer to the bottom of the hill it is initiated. As indicated in the layout diagram within Fig. 19.6 the full width of the extra lane is normally carried well beyond the crest of the hill (by 220 m in Britain). Tapers of 1:30 (single carriageways) or 1:45 (dual carriageways) should be used at both ends of the climbing lane.

19.4.2 Vertical curves

A vertical curve provides a smooth transition between successive tangent gradients in the road profile. When the algebraic difference of the two gradients is positive the curve is called a crest or summit curve; when the difference is negative it is called a sag or valley curve.

As a motorist traverses a vertical curve a radial force acts on the vehicle and tries to force it away from the centre of curvature and this may give the driver a feeling of discomfort; in extreme cases, as at hump-backed bridges, vehicles travelling at high speed may leave the carriageway. The discomfort experienced by the motorist is minimised by restricting the gradients (which reduces the force) and by using a type and length of vertical curve which allows the radial force to be experienced gradually and uniformly. Sight distance requirements are also aided by the use of long vertical curves on both crest and sag curves.

Circular, elliptical and parabolic curves can all be satisfactorily used in vertical curve design. In practice, however, the tendency is to use the simple parabola, primarily because of the ease with which it can be laid out as well as enabling a comfortable transition from one gradient to another. Vertical curves of this type are not necessary when the total grade change from one tangent to the other does not exceed 0.5 per cent.

With the simple parabola in Fig. 19.7 overleaf, the vertical offsets from the tangent at various points along the curve are used to lay it out. The offset at any given point with coordinates x, y is given by

$$y = -\left[\frac{(q-p)}{2L}\right]x^2 = \frac{-Ax^2}{2L} \qquad (19.15)$$

where A = algebraic difference between two gradients with grades p and q and L = length of the curve between the tangent points PC and PT. (Note that negative answers are measured downwards from the tangents for all summit curves; positive answers are measured upwards for sag curves.) The offset at the intersection point PI is given by

$$y = e = -(q-p)\frac{L}{8} \qquad (19.16)$$

The coordinates of the location of the highest (or lowest) point on the curve, which are often calculated to ensure that minimum sight distance (or drainage) requirements are met, are given by

$$x = \frac{Lp}{(p-q)} \tag{19.17}$$

and $$y = \frac{Lp^2}{2(p-q)} \tag{19.18}$$

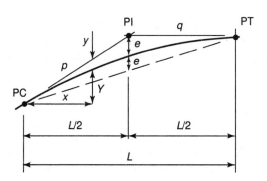

Fig. 19.7 A simple symmetrical parabolic curve

Figure 19.8 shows in exaggerated form the six different types of vertical curve. The algebraic signs given to the grades are important; the plus sign is used for ascending gradients measured from the point of curvature, and the minus sign for descending ones. Note that A, the algebraic difference in grades, can be either positive or negative, and that crest curves always have positive A-values and negative y-coordinates while sag curves always have negative A-values and positive y-coordinates.

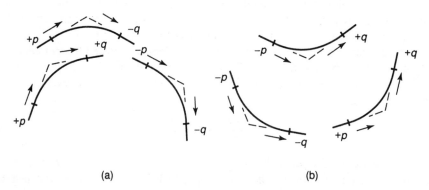

(a)   (b)

Fig. 19.8 Typical vertical curves: (a) crest curves and (b) sag curves

Crest curves

The *sight distance requirements for safety* are critical to the design of a crest curve. Thus when calculating the minimum lengths of crest curves, there are two design conditions that have to be considered (see Fig. 19.9): (a) where the required sight distance S is contained within the length of the vertical curve L and (b) where the required sight distance

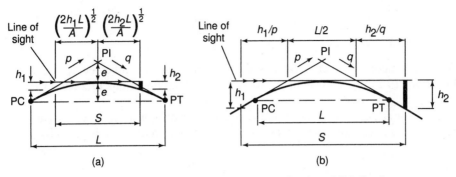

Fig. 19.9 Sight distances over crest curves when (a) $S \le L$ and (b) $S > L$

overlaps onto the tangent sections on either side of the vertical curve. Thus for $S \le L$

$$L_{min} = \frac{AS^2}{\left[(2h_1)^{\frac{1}{2}} + (2h_2)^{\frac{1}{2}}\right]^2}$$

(19.19)

and for $S > L$

$$L_{min} = 2S - \frac{2\left(h_1^{\frac{1}{2}} + h_2^{\frac{1}{2}}\right)^2}{A}$$

(19.20)

If $h_1 = 1.05$ m and $h_2 = 0.26$ m, and A = difference in grades expressed in decimal form, then the above equations refer to the safe stopping sight distance (*SSD*) and become, for $SSD \le L$

$$L_{min} = \frac{AS^2}{471}$$

(19.21)

and for $SSD > L$
$$L_{min} = 2S - \frac{471}{A}$$

(19.22)

If $h_1 = h_2 = 1.05$ m then the equations refer to full overtaking sight distance (*FOSD*), and then, for $FOSD \le L$

$$L_{min} = \frac{AS^2}{840}$$

(19.23)

and for $FOSD > L$
$$L_{min} = 2S - \frac{840}{A}$$

(19.24)

The decision as to which condition should be used at a given site can be made by solving either of the equations

$$e = \frac{(q-p)L}{8}$$

(19.25)

or

$$e = \frac{(q-p)S}{8}$$ (19.26)

depending upon whether L or S is the known value. In either case, if $e > h_1$ then Equations 19.21 or 19.23 should be used; if $e < h_1$ then Equations 19.22 and 19.24 where $S > L$ should be used.

If *motorist comfort* is taken as the main criterion governing vertical curve design then, for *both* crest and sag curves, the minimum length of curve is given by

$$L = \frac{V^2 A}{13a} = \frac{V^2 A}{390}$$ (19.27)

where L = curve length (m), V = vehicle speed (km/h), A = algebraic difference in grades expressed in decimal form, and a = vertical radial acceleration (m/s^2) and usually assumed = 0.3 m/s^2 for comfortable design for design speeds above 70 km/h.

Sag curves

When a road passes beneath an overpass, the driver's line of sight may be obstructed by the edge of the bridge. Then the minimum length of sag curve which meets minimum *stopping sight distance* requirements is given by

$$L_{min} = \frac{S^2 A}{\left[8D - \frac{8(h_1 + h_2)}{2} \right]}$$ when $SSD \le L$ (19.28)

and

$$L_{min} = 2S - \frac{\left[8D - \frac{8(h_1 + h_2)}{2} \right]}{A}$$ when $SSD > L$ (19.29)

If the eye height h_1 = 1.05 m and the object height h_2 = 0.26 m Equations 19.28 and 19.29 become

$$L_{min} = \frac{S^2 A}{(800D - 524)}$$ when $SSD \le L$ (19.30)

and

$$L_{min} = 2S - \frac{(800D - 524)}{A}$$ when $SSD > L$ (19.31)

where L = minimum length of the sag curve (m), S = minimum stopping sight distance (m), A = algebraic difference in grades expressed in decimal form, and D = vertical clearance (ideally taken as 5.7 m) to the critical edge of the overbridge.

In the above equations, the critical edge is assumed to be directly over the point of intersection of the tangents. In practice, both equations can be considered valid provided that the critical edge is not more than about 60 m from the point of intersection.

All of the above equations are based on visibility during daylight. At night the visibility is determined by the distance illuminated by a vehicle's headlights, and the minimum sag curve length is given by

$$L = \frac{S^2 A}{200(h_3 + S\tan\alpha)} \qquad \text{when } S \le L \qquad (19.32)$$

and

$$L = 2S - \frac{200(h_3 + S\tan\alpha)}{A} \qquad \text{when } S > L \qquad (19.33)$$

where h_3 = headlight height (usually 0.6 m above the carriageway), α = angle of upward divergence of the light beam (usually 1 degree), and L, A and S are as defined previously. Care has to be taken in using these formulae as: (a) the equations are very sensitive to the assumption of a 1 degree upward divergence of the light beam; (b) they erroneously assume that headlamps can illuminate an object on the carriageway at long distances, and they ignore the fact that many vehicles are driven on dipped lights; and (c) the effect of headlamps is reduced on horizontal curves.

British practice

General practice in Britain in respect of determining the lengths of vertical curve L (m) makes use of the simplified equation $L = KA$ where A = algebraic difference in grades (per cent) using the absolute value (always positive), and K = a rounded design speed related constant (m/percentage difference in grade) obtained from Table 19.2 for the appropriate design speed. If full overtaking is included in the design of single carriageways then the *FOSD* crest K-values at row D2 are used. Desirable minimum crest K-values (row C1) are used on dual carriageways (including motorways) while the absolute minimum crest K-values (row C2) provide for safe stopping on single carriageways for design speeds at or greater than 60 km/h; the absolute minimum crest K-value (row C2) at 50 km/h for single carriageways is based on the comfort criterion. The absolute minimum sag K-values (row C3) for design speeds at or above 85 km/h are based on comfort while speeds at 70 km/h or lower ensure that headlamps illuminate the carriageway for at least the (longer) absolute minimum safe stopping sight distances on unlit roads; these sag values are applicable to both single and dual carriageways.

Often minimum curve lengths greater than those given by the K-values in Table 19.2 are used in practice. For example, if a vertical curve length is started within a horizontal curve drivers can be faced with an optical illusion which may lead to an accident; in such instances the vertical curve length is normally increased so that the tangent points of both the vertical and horizontal curves coincide. (If the minimum vertical curve length is longer than the horizontal curve length, then the opposite should apply.)

19.5 Cross-section elements

By cross-section elements are meant those features of a roadway which form its effective width. There are two types of elements, basic and ancillary. The basic features of major interest are the width of carriageway (including the number of lanes), central reservation (or median strip), shoulders (including verges), laybys, camber of the carriageway, and the side-slopes of cuttings and embankments. Ancillary elements include safety fences and crash attenuation devices, anti-dazzle screens, and noise barriers. Figure 19.10 shows some simplified cross-sections containing the main basic elements.

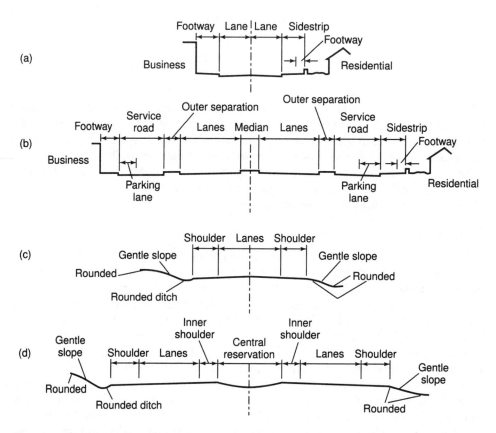

Fig. 19.10 Simplified road cross-sections: (a) 2-lane street, (b) urban motorway, (c) 2- or 3-lane rural highway, and (d) rural motorway

19.5.1 Carriageway width

Factors which influence the width of a carriageway are:

- the design volume, i.e. the greater the traffic volume the wider the carriageway and, normally, the greater the number of lanes
- vehicle dimensions, i.e. heavy commercial vehicles require wider carriageways to ensure adequate clearances when passing each other
- the design speeds, i.e. vehicles travelling at high speed, especially commercial vehicles, require wider carriageways to ensure safe clearances between passing vehicles
- the road classification, i.e. the higher the road classification the greater the level of service (and width of carriageway) expected.

Internationally, it is generally accepted that lane widths should normally be at least 3.5 m, although narrower lanes are often used for economic or environmental reasons on both rural and urban roads. However, research has established[6] that the accident rates for vehicles running off the road, and for head-on and sideswipe accidents between vehicles travelling in opposite directions, decrease with increasing lane width up to 3.65 m on two-lane two-way rural roads. Lanes on British roads are most commonly 3.65 m.

Table 19.4 summarises general practice in relation to carriageway widths used in road design in Britain.

Table 19.4 General practice with regard to carriageway widths in Britain

Road type	Description of carriageway(s)	Carriageway width (m)
Urban		
Primary	Dual 4-lane	14.60
distributor	Single 4-lane (undivided), with central refuges	14.60
	Single 4-lane, no refuges	13.50
	Dual 3-lane	11.00
	Dual 2-lane (normal)	7.30
District	Single or dual 2-lane (normal)	7.30
distributor	Dual 2-lane (proportion of heavy commercial vehicles is low)	6.75
Local	Single 2-lane in industrial areas	7.30
distributor	Single 2-lane in central business districts	6.75
	Single 2-lane in residential areas used by heavy vehicles (min)	6.10
Access	Single 2-lane, in residential areas (min)	5.50
roads*	2-lane back or service roads, used occasionally by heavy vehicles	5.00
	2-lane back roads in residential districts, for cars only (min)	4.00
Rural	Single 2-lane (min)	5.50
	Motorway slip road	6.00
	Single or dual 2-lane (normal)	7.30
	Single 2-lane (wide)	10.00
	Dual 3-lane	11.00
	Dual 4-lane	14.60

*In industrial and central business districts local distributor widths are used

New single-lane single carriageway roads are rarely constructed today; even though traffic volumes are light, considerations of safety and ease of traffic movement normally justify two lanes. Two-way two-lane single carriageway roads constitute the predominant type now constructed. A three-lane single carriageway road is theoretically justified when the design volume exceeds that for a two-lane road but is not sufficient to justify a four-lane road. However, safety considerations dictate that single carriageway roads marked with three lanes (other than for a climbing lane situation) should only be constructed in rural terrain where it is practical to provide near-continuous passing sight distances; it is far preferable to build a wide (say 10 m) single carriageway and mark it for two-lane operation. Uncontrolled two-way three-lane roads should never be built in urban or suburban locations because of their high accident potential.

Two or more lanes in each direction are needed for vehicles to overtake on lanes not used by opposing traffic. Normally these roads have dual carriageways separated by a central reservation or median strip. Motorways are dual carriageway roads with full control of access; all intersections are grade-separated and there are no crossings at-grade. Internationally, few countries currently have motorways wider than dual four-lanes, and most are dual two- and three-lanes. Britain is currently implementing a major programme of widening motorways to dual four-lanes. Motorways with dual four-, five- and six-lane carriageways are not uncommon in the major conurbations in the United States. Many motorways in urban areas are flanked by collector/distributor roads with frontage access, thereby adding to the total number of lanes within the highway corridor.

Particular circumstances often give rise to the use of non-standard lane widths on non-motorways in urban areas. For example, the width of the nearside lane(s) of multi-lane single or dual carriageways is often increased (sometimes at the expense of other lanes) to improve conditions for cyclists and allow more space for commercial vehicles. Lane widths as narrow as 3.25 m also have been used on multi-lane roads to help provide an extra lane when the width available for the carriageway is limited. Provided kerb parking is restricted, carriageway widths as low as 6.1 m can operate satisfactorily as two-way local distributor routes.

Note also that in rural areas in Britain the standard edge treatment on normal two-lane single carriageways consists of a 1 m strip of the same construction as the carriageway, on either side, and delineated by a solid white line. Thus the full width of the surfaced road pavement is actually 9.30 m.

19.5.2 Central reservation

Dual carriageways are normally divided by a central reservation or median strip. Central reservations have a number of uses, including: separating high-speed opposing traffic, thereby lessening the chances of head-on collisions; providing opportunities for erring vehicles to recover when they inadvertently leave the carriageway; providing a safe waiting place for pedestrians crossing a high-speed dual carriageway; controlling right-turning (in Britain) vehicles wishing to enter/leave premises with access to the highway; and providing space for road furniture such as road signs and street lighting. Gaps provided in central reservations can also be used to safely store vehicles from minor roads or driveways while they wait to turn right (in Britain). Deceleration and storage lanes are easily and safely provided in central reservations for right-turning vehicles wishing to leave the high-speed through lanes. Wide dish-shaped (shallow) central reservations also simplify solutions to drainage and snow clearance problems.

Research has shown that more than 90 per cent of the vehicles that inadvertently leave rural carriageways deviate less than 15 m from the carriageway edge while 80 per cent deviate less than 10 m. Thus, ideally, rural central reservations should be at least 15 m wide, and trees and fixed road furniture located within 10 m of the edge of the carriageway should be protected with guardrails or by other appropriate measures. A wide central reservation need not be a constant width; rather the width can be varied to blend into the topography. Where possible shrubs (but never trees) should be grown within the central reservation; as well as being aesthetically pleasing these also reduce headlight glare and help to dissipate the energies of out-of-control vehicles.

Dual carriageway roads in urban areas generally have narrower median strips because of the high value of land and lower vehicle speeds. Ideally they should be about 9 m, and preferably not less than 5.5 m, wide. The surfacing of the central reservation should be in direct contrast to the carriageways being separated, and be clearly visible to the motorist in wet as well as dry weather. Grass is a suitable surfacing for widths greater than about 2 m; below 2 m grass can be difficult to maintain and cut, and use is often made of raised medians with contrasting bituminous or concrete surfacings. Preferably also the kerbs of raised medians should be studded with reflector buttons or painted with reflectorised paint to emphasise the contrast.

Central reservations on high-speed heavily trafficked rural roads in the United States are typically 15 m to 30 m wide. In Europe they tend to be much narrower (say 4–10 m) and to be used with safety barriers. Those used in Britain are normally 4.5 m wide, and include a crash barrier. In urban areas they can be as narrow as 1 m, but 3 m is preferred so that a crossing pedestrian pushing a pram or a wheelchair has space to wait in safety.

19.5.3 Shoulders

A shoulder is that surfaced clear portion of the roadway cross-section immediately adjacent to the carriageway edge. Shoulders serve a number of purposes – for example, they provide: a refuge for vehicles forced to make emergency stops; a recovery space for vehicles that inadvertently leave the carriageway, or deliberately do so during emergency evasive manoeuvres; temporary extra traffic lanes during road maintenance or carriageway reconstruction; assistance in achieving desired horizontal sight distances; and structural support to the road pavement (usually by extending all or part of the roadbase through the shoulder width). Research has established[6] that the accident rates for vehicles running off the road, and for head-on and sideswipe accidents between vehicles travelling in opposite directions, decrease with increasing shoulder width up to 3 m on two-lane two-way rural roads. Also, without shoulders, obstructions adjacent to the carriageway edge and/or vehicles parked on the carriageway cause a reduction in the effective width available to moving traffic, thereby reducing road capacity.

About 3.35 m width of hard shoulder is needed by a lorry to enable a tyre to be changed without danger to the operator. Light vehicles require less clearance space and as these account for most stoppages, a shoulder width of 3 m is generally (internationally) considered adequate for most high-speed high-volume roads. However, because shoulders add significantly to the cost of a road, the widths actually built vary considerably. Thus, for example, the general practice in Britain is to use 3.3 m wide shoulders on motorways but, as noted previously, only 1 m wide hardstrips are used on the nearsides of other dual carriageway and single carriageway rural roads. These hardstrips are then usually flanked by grass verges about 2.5 m wide.

The 1 m hardstrips are also used on new (non-motorway) urban roads, although the use of kerbed edges is probably more common in built-up areas. Kerbs (which are typically about 100 mm high) assist road drainage as well as protecting adjacent areas from intrusion by vehicles. If footpaths (typically at least 2 m wide) flank new roadways, they are normally separated from traffic by 3 m wide raised verges; these verges, which are hard-surfaced or grassed so as to be unattractive for walking, may also be used to contain services. These verges are reduced in width where hardstrips or pedestrian guardrails are also provided.

While internationally there is general acceptance that wide nearside shoulders are desirable on all high-speed high-volume roads, there is no consensus as to the provision of off-side shoulders, i.e. flanking the central reservation. Although they offer the same advantages as nearside shoulders, they have the major disadvantage that their usage encourages deceleration (and acceleration) on the fast inner lanes. British practice is to provide 0.7 m wide (offside) hardstrips on all dual carriageways, except dual three- or four-lane motorways where no hardstanding extra width is provided next to the central reservation.

Shoulders should be flush with the carriageway and able to support vehicles under all weather conditions, so that there are no skidding or rutting problems when driving on their surfaces. Research has established[6] that greater accident rates are exhibited on unstabilised shoulders (including loose gravel, crushed stone, raw earth, or grass) than on stabilised (e.g. tar plus gravel) or paved (e.g. bituminous or concrete) shoulders. Shoulder surfacings should also appear distinctly different from the carriageway surface as otherwise motorists may be inclined to use them as traffic lanes. Thus bituminous surfacings with either an added pigment material or different coloured stone chippings are in common usage on shoulders.

19.5.4 Laybys and bus bays

When economic considerations do not favour the construction of shoulders on rural roads, *laybys* should be provided instead, at spacings that are appropriate to the traffic volume. Thus, for well trafficked and lightly trafficked single carriageways, it is British practice to provide 2.5 m and 3 m wide by 30 m long laybys at 1.5 km and 5.8 km intervals, respectively, on either side of the carriageway, while 3 m wide by 100 m long laybys are provided at approximately 1 km intervals on each side of dual carriageways. Laybys should be located at sites with good visibility and provided with tapered hardstrips at either end to assist in the safe deceleration and acceleration of vehicles using them.

In Britain laybys (3 m by 30 m, plus 16 m end tapers) are also provided at approximately 1.5 km intervals on each side of all-purpose urban primary distributors without hard shoulders. Laybys and bus bays are sometimes combined on these roads; typically these are 2.75 m to 3.25 m wide by at least 45 m long (excluding 20 m end tapers). Laybys are rarely provided on district and local distributor roads in urban areas.

Full *bus bays* (3.25 m by at least 12 m, plus 20 m end tapers) may be provided at bus stops in urban areas; however, the appropriateness of this provision is dependent on the traffic volumes on the road in question (see Chapter 24). Bus stops on non-urban roads without hard shoulders should always be located in full bus bays.

19.5.5 Camber

The term 'camber' is used to describe the convexity of the carriageway cross-section. Its purpose is to drain surface water from the road and avoid ponding in surface deformations on the carriageway.

Surfaced two-lane roads have either parabolic or circular cross-sections to ensure that the swaying of heavy commercial vehicles is kept to a minimum as they cross and re-cross the crown of a normal road during an overtaking manoeuvre. Common practice in the case of wide single carriageways with three or four lanes is to use a curved crown section for the central lane(s) and to have a tangent section on each of the outer lanes; the use of the flatter tangents avoids having undesirably steep cross-slopes in the outer lanes.

Ideally each carriageway of a dual carriageway road should be cambered, as it ensures that the difference between low and high points in each cross-section is minimised and that the run-off water is equally dispersed in both lateral directions. Also, it allows changes from normal to superelevated sections to be easily made. However, this solution is expensive because it requires underground drainage facilities on both sides of each carriageway.

An alternative approach for dual carriageways is to have a one-way cross-slope on each carriageway draining toward the central reservation. This solution is economical in that it minimises the drainage facilities needed to remove the run-off. However, it has the disadvantage that all of the water must pass over the high-speed inner lanes, and this can result in dangerous splashing of windscreens, as well as accentuating the risk of skidding in heavy rainfalls.

When each carriageway of a dual carriageway is sloped so as to drain away from the central reservation to the nearside, the greatest water accumulations pass across the slower, more heavily trafficked, outer lanes. However, some savings are effected in drainage structures, and the treatment of intersecting roadways is easier.

British practice is that the crossfall or camber should average 2.5 per cent from the centre of single carriageways, and from the central reservation edge of each carriageway of dual carriageways, to the outer drainage channels. At intersections other than round-abouts the cross-section of each major carriageway is retained across the junction, and the minor road cross-section is graded into the channel line of the major road.

19.5.6 Side-slopes

Soil mechanics analysis enables the accurate determination of the maximum slopes at which earth embankments or cuts can safely stand. However, these maximum values are not always used, especially on low embankments not protected by safety fences.

Figure 19.11 shows the three regions of a side-slope that are important from a safety aspect when designing and constructing low embankments.

The *hinge-point* at the top of the slope contributes to the loss of steering control as an erring vehicle tends to become airborne when crossing this point, particularly if the encroachment conditions (angle and speed) and embankment drop-off are severe. If the wheels are turned while airborne and dig into the side-slope upon landing, vehicle rollover is more likely. The *front slope* region is important in that an erring driver's natural instinct is to reduce speed and attempt a recovery manoeuvre before impacting the ground at the bottom of the slope – and there is a front-slope steepness at which a vehicle will roll during this recovery manoeuvre. If the *toe of the slope* is relatively close to the carriageway, then the likelihood of a vehicle reaching the ground (or bottom of the ditch) is high, and the sharper the transition at ground level the greater the danger for the vehicle.

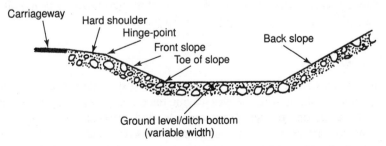

Fig. 19.11 The side-slope regions of interest in roadside safety design

Research has shown[7] that vehicle rollover need not be a problem at the hinge-point of an embankment with a slope of 2 horizontal to 1 vertical (or flatter), at encroachment angles up to 25 degrees and speeds up to 130 km/h. Also, rounding the hinge-point reduces the tendency of a vehicle to become airborne, helps the driver to maintain control, and lessens the chance of a rollover. The gentler the side-slope (preferably 3 to 1 or flatter), and the firmer it is and the higher its coefficient of friction (i.e. it should not be composed of soft material that will rut easily), the greater the chance of a driver making a successful recovery.

Less severe vehicle damage will result if the toe of the slope is rounded to allow a more gradual transition to the natural ground. If the side-slope forms part of a wide ditch, a trapezoidal configuration represents the safest cross-section for a vehicle to intrude upon at high speed.

While reasons of economy encourage the construction of steep side-slopes on embankments and cuttings, ease of maintenance is promoted by gentler slopes. The flatter the slopes the easier it is to grow grass and the less the erosion. Grading equipment used for maintenance can work most efficiently on slopes of 3 to 1 or flatter. Also, desired horizontal sight distances at curves in cuttings are more easily achieved by the use of flatter side-slopes.

19.5.7 Anti-glare screens

A driver's vision at night, particularly on unlit roads, is often affected by the dazzle from oncoming vehicle headlights. In heavy traffic, the result can be that a driver is subjected to dangerous 'glare-out'. On heavily travelled dual carriageways, therefore, anti-glare screens are sometimes erected on narrow central reservations (often in conjunction with safety fences) to alleviate this problem.

An anti-glare screen aims to cut out the light from oncoming vehicles that is directed toward the driver at oblique angles (up to 15–20 degrees). To be effective in screening the lights from all types of vehicle (including heavy commercials) the screen must reach to at least 1.73 m above the carriageway. Environmentally, it is also desirable that the screen not be very noticeable, and that as much open vision as possible be maintained in a sideways (perpendicular) direction.

Hedges of *rosa multiflora japonica* have been planted in wider central reservations in some countries to serve as anti-glare screens. Manufactured screens are more commonly used (typically of expanded metal mesh). An anti-dazzle screen using vertical plastic vanes of a dull green colour, mounted at 0.8 m centres above a central reservation safety fence, has been developed by the Transport Research Laboratory.[8]

When installing screens care must be taken to ensure that they do not impair minimum stopping sight distance requirements at horizontal curves.

19.5.8 Safety fences

Roadside safety fences form a longitudinal protective system designed to reduce the severity of accidents resulting from vehicles leaving the carriageway. They can be divided into two main groups: edge barriers (or guardrails) and crash barriers. *Edge barriers* tend to be used alongside roads on embankments with steep side-slopes and/or non-traversable ditches, and along verges where rigid obstacles, e.g. trees or rock outcroppings, are continuously close to the edge of the carriageway. Edge barriers are also used to deflect erring vehicles away from individual hazards, e.g. bridge abutments. In urban areas guardrails are also used to protect pedestrians and separate them from moving vehicles. *Crash barriers* are located within narrow central reservations and are primarily intended to prevent errant vehicles from crossing into the opposing traffic stream.

Longitudinal safety fences are themselves safety hazards. Hence their design and usage are based on the premise that their interaction with vehicles either results in the avoidance of accidents or reduces their severity.

The majority of longitudinal safety fences available commercially can be divided into three main types: steel beam, rigid concrete, and flexible cable.

The tensioned corrugated mild *steel beam* is probably the most widely used safety fence, both in Britain and worldwide. Such beams are normally about 300 mm deep and formed so that the traffic face has a central trough of at least 75 mm deep. Typically a beam is mounted at a height of 760 mm above and parallel to the edge of the adjacent carriageway; it is attached by shear bolts to mild steel posts, and tensioned between anchorages sunk in the ground. The type used in Britain is able to redirect a 1.5 t car hitting at an angle of 20 degrees to the line of the fence at a speed of 112 km/h, so that the car remains close to the side of the fence. Upon impact the bolts fracture and the fence is displaced laterally as the energy is dissipated through the knocking over of the posts and the flattening of the beam. A clearance of about 1.2 m is desirable behind the single beam fence to allow for safe deflection; a double beam barrier (one with a beam on each side of a standard post) reduces the safe deflection space to about 0.6 m, while further stiffening can be achieved by shortening the post spacing.

Special attention needs to be paid to the end treatment of a beam fence, as an end impact is particularly dangerous. This is best overcome by ensuring that the end of the fence is continued beyond the sphere of influence of traffic.

The growth in the numbers of heavy commercial vehicles, plus the increase in 1983 of the authorised maximum gross weight of these vehicles from 32.5 t to 38 t, led to the development of a stronger steel fence (consisting of four parallel open box beams set in pairs on either side of the posts at heights of 610 mm and 1020 mm), and of a concrete crash barrier (consisting of 3 m long precast-concrete units fixed by six dowel pins per unit onto a concrete foundation flush with the carriageway), for use at locales subject to large flows of heavy commercial vehicles. These barriers[9] can contain the heaviest vehicles in general use on British roads under realistic impact conditions while being as forgiving as possible to light vehicles. A cross-section of this concrete fence is shown in Fig. 19.12.

Most *concrete barriers* in use throughout the world, including that in Fig. 19.12, are variants of the 'safety-shape' crash barrier used on New Jersey freeways since about 1955, and their performance characteristics are different from those of steel beam fences. The metal fence is designed to deflect upon impact and the major redirection and deceleration force is provided by the friction generated between the body of the vehicle and the steel beam(s). With properly contoured concrete barriers, however, redirection is accomplished by the vehicle wheels and not by the body, while energy absorption at shallow impact angles results from compression of the suspension system and not from deformation of the vehicle body. Additional advantages claimed for concrete barriers include that they: (a) provide a base for lighting and sign supports; (b) provide some protection against headlamp glare; (c) partially shield adjacent land areas from road noise (especially noise from the tyre–carriageway interaction); (d) are highly visible, damage-proof, and require little maintenance; and (e) are particularly suitable for use on urban roads where central reservation widths are restricted.

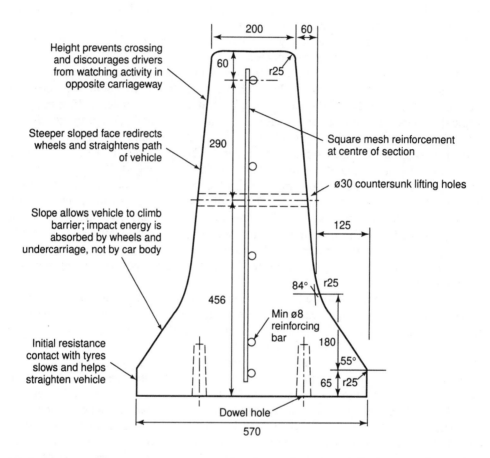

Fig. 19.12 Performance characteristics and cross-section dimensions of the precast-concrete barrier developed for heavy-duty localised use in Britain (dimensions in mm)

The main advantage of a *cable fence* is that, because of its flexibility, it can 'slowly' decelerate an erring vehicle and redirect it easily along a path parallel to the fence. Its satisfactory operation is dependent upon the correct height of impact being maintained, so that the wheels of vehicles do not climb over the cable instead of being deflected by it. It cannot be used at locations with narrow clearances because of the cable's potential for deflection under impact (typically 1.8–3.5 m). While not in general favour in modern road design, flexible cable fencing is probably most suitable at sites where minimal air resistance is needed, e.g. to prevent snow accumulations on mountain roads.

19.5.9 Crash-attenuation devices

Many accidents involve collisions between out-of-control vehicles and individual fixed objects adjacent to the carriageway. The basic approaches to resolving this problem are to either (a) remove or relocate the objects, (b) use frangible objects (if they are manufactured), or (c) 'protect' the objects with a section of barrier fence or a crash attenuator.

The simplest solution to this roadside hazard problem is to ensure that no fixed objects are within, say, 10 m of the edge of the carriageway. In practice, however, this is not possible, even in the case of new roads. In the case of essential light and electricity poles and signposts, the number of potential accidents can be reduced by the usage of increased spacing, encouraging joint use by different utility authorities, and/or selective resiting. The severity of accidents can be reduced by the use of frangible light poles and signposts that break near the bottom under impact. Breakaway electricity poles have also been developed which reduce accident severity without causing the downfall of electric wires.

The removal/relocation of fixed objects such as bridge abutments, piers or parapets, ends of retaining walls, heavy overhead sign supports, etc., is impracticable, and impact-attenuation devices (known as *crash cushions*) have been developed in the USA for installation prior to such fixed obstructions, to reduce the severity of collisions with these massive objects. Some of the types of crash cushions are:

- an array of empty steel drums that absorb impact energy by progressive crushing
- nylon cells containing water which dissipates the impact energy by being forced out at controlled rates through orifices
- lightweight vermiculite concrete 'helicells' that absorb the impact energy by controlled crushing
- a V-shaped array of frangible plastic barrels (with each row containing increasing amounts of sand) that are designed to shatter on impact
- tyre-sand inertia barriers composed of sand-filled scrap tyres that are scattered as the momentum of the vehicle is transferred to the cushion.

In Britain only low priority has been given to research into the application of crash-attenuation devices to reduce accident severity.[10]

19.5.10 Noise and noise barriers

In 1972 a national survey found that 9 per cent of the adult population of England were seriously bothered by traffic noise at home, and 16 per cent were bothered when out in the community. In urban areas noise produced by road traffic is now the major and most pervasive source of community noise. Consequently, it is not surprising that natural and/or manufactured noise barriers are now a common feature of the cross-sections of new major roadways in sensitive locations.

Britain's *Noise Insulation Regulations* require a road authority to provide noise insulation works if the construction of a new or improved road results in: (a) the combined expected maximum traffic noise level (i.e. the 'relevant' noise level) from the new or altered road together with other traffic in the vicinity being not less than 68 dB(A)L_{10}(18h); (b) the relevant noise level being at least 1.0 dB(A) more than the total traffic noise level existing before the roadworks were begun; and (c) the contribution to the increase in the relevant noise from the new/improved road being at least 1.0 dB(A). Many major roadworks therefore require before-and-after noise analyses which involve assessing current traffic-noise levels and predicting future levels before, and often after, the interposition of natural or artificial noise attenuators.

Various reliable techniques for assessing traffic noise levels are used in many countries. The British procedure[11] for estimating current or future noise at a reception point (e.g. a dwelling) is summarised in Fig. 19.13. This process relies mainly upon a proven series of formulae linking noise levels with various factors. In exceptional circumstances, e.g. if the calculations involve extrapolation beyond the quoted validity ranges of the formulae or if overall noise levels (including extraneous noise from non-traffic sources) are required, recourse is made to measurement (see Chapter 13). In the case of a single carriageway road the source of the traffic noise is taken as a line 0.5 m above the carriageway and 3.5 m in from the nearside carriageway edge, and the reception point is normally taken as 1 m from the most exposed window or door in the building facade under consideration.

Figure 19.13 shows that the procedure is divided into five stages. The first stage involves dividing the road scheme into a small number of segments so that the noise variation within any one segment is less than 2 dB(A).

Stage 2 involves each segment being treated as a separate noise source, and estimating the basic noise level at a standard reference distance of 10 m from the nearside carriageway edge, for a single carriageway. In the case of a divided road the noise level produced by each carriageway is treated separately, and the source line and effective carriageway edge used in the distance correction for the far carriageway are assumed to be 3.5 m and 7 m in from the far kerb, respectively.

The fundamental relationship used to estimate the basic $L_{10}(18h)$ noise level, in dB(A), at the standard reference distance is

$$L_{10}(18h) = 29.1 + 10\log_{10} Q \qquad (19.34)$$

where Q = flow of vehicles between 6 a.m. and 12 midnight. This formula assumes that traffic is travelling at an average speed of 75 km/h on a level road with a particular type of surface, and that the proportion of heavy vehicles of unladen weight >1.525 t is zero; the result must therefore be adjusted to take account of variations from these conditions. The correction, in dB(A), for a different mean traffic speed (V km/h) and percentage of heavy vehicles ($p\%$) on level roads is given by

$$\text{Correction}(\pm) = 33\log_{10}\left(V + 40 + \frac{500}{V}\right) + 10\log_{10}\left(1 + \frac{5p}{V}\right) - 68.8 \qquad (19.35)$$

where V is normally taken from Table 19.5. If the road is on a gradient of G per cent the presence of heavy commercial vehicles will lower the mean speed below that on the flat; the speed-value then used in the above calculation is that taken from Table 19.5 minus ΔV, where

$$\Delta V = \left[0.73 + \left(2.3 - \frac{1.15p}{100}\right)\frac{P}{100}\right] \times G \qquad (19.36)$$

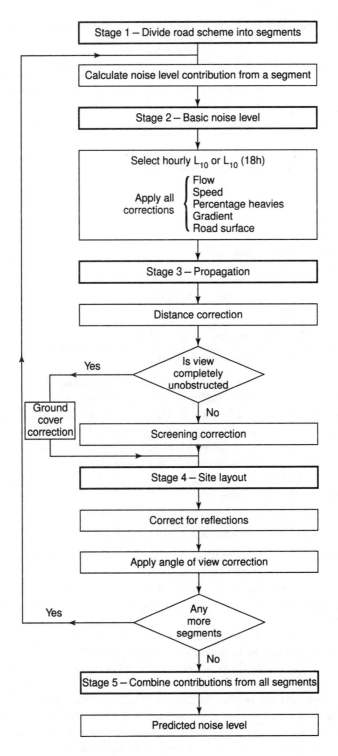

Fig. 19.13 Flow chart for estimating noise from road schemes[11]

The correction in dB(A) for the extra traffic noise on a gradient (G per cent) is

$$\text{Correction} = +0.3G \qquad (19.37)$$

Note that this gradient correction is separate from the correction for traffic speed above. In the case of one-way traffic schemes or carriageways that are treated separately the correction only applies to the uphill flow.

Table 19.5 Predicted traffic speeds normally used in noise level calculations

Road type	Speed (V km/h)
Roads not subject to a speed limit of <96.5 km/h:	
● special roads (rural)*	108
● special roads (urban)*	97
● all-purpose dual carriageways*	97
● single carriageways > 9 m wide	88
● single carriageways ≤ 9 m wide	81
Roads subject to a speed limit of 80 km/h:	
● dual carriageways	80
● single carriageways	70
Roads subject to a speed limit of <80 km/h but >48 km/h:	
● dual carriageways	60
● single carriageways	50
Roads subject to a speed limit of ≤48 km/h: all carriageways	50

* Excluding slip roads which are estimated individually

In the case of road surfaces the noise level needs to be adjusted for a number of factors, e.g. whether the texture is randomly distributed chippings (as in bituminous surfacings) or transversely aligned (as for concrete surfacings) and, for bituminous surfacings, whether they are essentially impervious to water or have an open structure with rapid drainage properties. For surfacings that are impervious to water and when the mean traffic speed used in the calculation of the speed and percentage-heavy-vehicles correction calculation in Equation 19.35 is ≥ 75 km/h, the following corrections, in dB(A), are applied to the basic noise level:

for concrete surfaces: correction = $10 \log_{10}(90TD + 30) - 20$ (19.38)
for bituminous surfaces: correction = $10 \log_{10}(20TD + 60) - 20$ (19.39)

where TD is the texture depth (as measured by the Sand-Patch Test). If the mean speed is <75 km/h, 1 dB(A) is subtracted from the basic noise level.

If the carriageways are surfaced with pervious macadams then 3.5 dB(A) is subtracted from the basic noise level for all traffic speeds, as these surfacings have different acoustic properties.

The final noise levels determined at Stage 2 for the various road segments, after the application of the above corrections, are then taken as the basic noise levels resulting from the current traffic or from the predicted traffic in the design year (usually 15 years hence), as appropriate.

Additional corrections are applied in Stage 3 which take into account the effects of distance from the source line to the reception point, the nature of the intervening ground surface, and the screening provided by any obstacles between the road and the reception point. (Space limitations do not make it possible to detail these corrections here; see reference 11 regarding how they are calculated and applied.)

In general, the further the distance to the reception point the greater the noise reduction. If the ground surface is all or partially grassland, cultivated fields or plantations, there will be an additional reduction due to ground absorption. Dense vegetation that is near a road can be very effective in attenuating traffic noise; for instance, a 10 m deep vegetation belt close to a road can achieve an attenuation 5 dB(A)L_{10} more than the same depth of grassland.

Calculation of the screening correction may involve estimating the impact of existing or proposed intervening obstructions such as buildings, walls, purpose-built acoustic barriers, etc.

Stage 4 of the assessment process described in Fig. 19.13 requires correcting the noise level from each segment at the reception point to take account of site layout features, e.g. reflections from building facades and other hard surfaces, propagation down side-roads, and size of segment. Stage 5 involves ensuring that the contributions from all the source segments comprising the road scheme are brought together.

Purpose-built barriers

The main ways of alleviating the noise nuisance from roads are by increasing the insulation of buildings and/or by constructing intervening barriers. In the case of a building, for example, a closed double-glazed window (100 mm cavity) can result in a perceived noise reduction of the order of 25–38 dB(A).

Purpose-built noise barriers are most usually dense timber fences, concrete walls, or landscaped earth mounds. These barriers are at their most effective if they are (a) sited close to the reception point or to the noise source (best), (b) made continuous and without gaps, and (c) long enough to block the 'line of sight' between the roadway and the area being protected. Barriers should be as high as practicable, but are normally 1–3 m high; those less than 1 m are limited in value, but barriers higher than about 3 m often appear as major intrusions upon the landscape. The minimum mass to be used in a purpose-built barrier in order to achieve a specified noise attenuation can be obtained from

$$M = 3 \text{ antilog}\left[\frac{(A-10)}{14}\right] \tag{19.40}$$

where M = barrier mass per unit area (kg/m^2) and A = potential noise attenuation, dB(A) (taken as positive).

Grassed or planted earth mounds are often used as aesthetically pleasing noise barriers in rural-type environments. Dense plantings of broad-leaf evergreens and deciduous shrubs, together with some conifers, have also been proposed[12] as a barrier that is both attractive in appearance and effective in reducing traffic noise.

Whatever the type of barrier, a freely flowing alignment with the top edge following the general ground slope is preferred. Also, effective barriers begin and end at existing features, e.g. at bridges, walls or fences.

19.6 Safety audits

Understanding why, where and when road accidents occur are matters of fundamental interest to the traffic engineer. The knowledge gained about accidents over the past 40 years in particular has been translated into numerous sustained and successful road safety and traffic management programmes to reduce the number and severity of accidents on existing roads. It has also influenced the design of new or improved roads so as to make them safer to use.

During the 1980s there was an increased recognition of the importance of specifically considering safety issues during the development of road schemes, and of having new road designs systematically, formally, and independently audited to ensure that unsafe features are not introduced into new schemes. In other words, a greater emphasis began to be placed on the need to prevent accidents before they occur rather than on fixing the problem later.

The main objective of a safety audit is to ensure that a road scheme operates as safely as possible when it comes into use. Depending upon the size of the project, the audit requires a review team, which contains at least one member who is an expert in road safety and accident investigation and is separate from and independent of the road design team, to carry out a careful methodical examination of the detailed operational characteristics of the proposed geometric design. The successful application of safety principles in the audit process hinges on adherence to two main requirements:[13] a systematic organisation of auditing in planning, design and construction, and a well-defined auditing procedure that refers to standard checklists (as an aide memoire).

From an organisational aspect it is critical that there be good communication between the safety auditor(s) and the designer(s). It is not the auditor's role to redesign the scheme nor to implement changes but rather to pay particular heed to the safety needs of all road users (pedestrians and cyclists as well as motorists) and *formally* to bring any remedial design requirements to the attention of the designer(s). If any advice is rejected by the design team, an explanation should be provided.

From a procedural aspect it is important that safety audit procedures be incorporated into the various stages of the critical path network for a road scheme; for example: as an input to the feasibility/initial design of a scheme; on completion of draft plans or preliminary design; during, or on completion of, detailed design and before the preparation of contract documents; and prior to the opening of a scheme.

The Department of Transport now requires safety audits to be carried out on all new trunk road and motorway schemes (see references 14 and 15). Although safety audits are not mandatory for new schemes that are not under the Department's control, local authorities are increasingly requiring such audits to be carried out on projects under their control. At the time of writing, formal road safety audits are practised only in Great Britain among the countries of the EU.

19.7 References

1. *Highway link design*. Departmental Advice Note TA43/84. London: Department of Transport, Dec. 1984.
2. Glennon, J.C., Effect of alignment on highway safety. *State of the art report 6*. Washington, DC: The Transportation Research Board, 1987, 48–63.
3. Hosking, J.R., and Woodford, G.C., *Measurement of skidding resistance: Part 1, Guide to the use of SCRIM*. TRRL Report LR737. Crowthorne, Berks: Transport Research Laboratory, 1976.
4. *Highway link design*. Departmental Standard TD9/93. London: Department of Transport, 1993.

5. Brock, G., *Road width requirements of commercial vehicles when cornering.* Report LR608. Crowthorne, Berks: Transport Research Laboratory, 1973.
6. Zegeer, C.V., and Deacon, J.A., Effect of lane width, shoulder width, and shoulder type on highway safety. *State of the art report 6.* Washington, DC: The Transportation Research Board, 1987, 1–21.
7. Marquis, E.L., and Weaver, G.D., Roadside slope design for safety. *Transportation Engineering Journal of the ASCE*, 1976, **102**(TE1), 61–73.
8. Walker, A.E., and Chapman, R.G., *Assessment of anti-dazzle screen on M6.* Report LR955. Crowthorne, Berks: Transport Research Laboratory, 1980.
9. *Safety fences and bridge parapets: TRRL papers for the 1986 TRB annual meeting.* Research Report 75. Crowthorne, Berks: Transport Research Laboratory, 1986.
10. Lawson, S.D., Some approaches to roadside crash protection in the U.S. and their potential for application in the U.K. *Traffic Engineering and Control*, 1988, **29**(4), 202–209.
11. Department of Transport, *Calculation of road traffic noise.* London: HMSO, 1988.
12. Huddart, L., *The use of vegetation for traffic noise screening.* Research Report 238. Crowthorne, Berks: Transport Research Laboratory, 1990.
13. *IHT guidelines for the safety audit of highways.* London: The Institution of Highways and Transportation, 1990.
14. *Road safety audits.* Departmental Advice Note HA42/94. London: Department of Transport, 1994.
15. *Road safety audits.* Department Standard HD19/94. London: Department of Transport, 1994.

CHAPTER 20

Intersection design and capacity

C.A. O'Flaherty

Intersections, where two or more roads meet, are points of potential vehicle conflict. The dangers to pedestrians and riders of two-wheeled vehicles are also great at intersections. On heavily travelled rural and urban roads, the concentration of vehicles at an intersection can be of such import that it controls a road's/road network's capacity. Consequently, maximising road user safety and ensuring that the available capacity adequately meets operational traffic-flow needs are the two key considerations when designing an intersection.

20.1 Types of intersection

Intersections can be divided into the basic forms shown in Fig. 20.1. From a design aspect these intersections can also be divided according to whether they are uncontrolled, priority controlled (Stop, Give Way), space-sharing (i.e. roundabouts), time-sharing (i.e. traffic-signal controlled), or grade-separated (including interchanges).

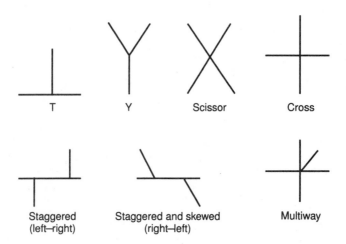

Fig. 20.1 Basic intersection forms

20.2 Overview of the design process

⌈The at-grade intersection design process involves[1] data collection of both traffic and site conditions, the preparation of preliminary designs from which a layout is selected, and the development of the final design using appropriate design standards.

Traffic data gathered for design purposes normally include peak period traffic volumes, turning movements and composition for the design year, vehicle operating speeds on the intersecting roads (these are needed for sight-distance/speed-change lane design) pedestrian and bicycle movements (these affect the layout/traffic control design), public transport needs (e.g. bus priority measures and bus stop locations affect the layout/traffic control design), special needs of oversize vehicles (the selected design may have to cope with the occasional heavily loaded commercial vehicle with a wide turning path), accident experience (if an existing intersection is being upgraded), and parking practices (especially in built-up areas).

Site data collected typically include topography, land usage, and drainage and related physical features (natural and manufactured), public and private utility services (above and below ground), items of special interest (e.g. environmental, cultural and historical features), horizontal and vertical alignments of intersecting roads (existing and future), sight distances (and physical features which limit them), and adjacent (necessary) accesses.

The *preliminary design* phase is essentially an iterative one. It involves preparing a number of possible intersection layouts and generally examining each in terms of its operating characteristics (especially safety and capacity), ease of construction and likely capital cost, and environmental and local impacts that might affect the design selection. The most promising of the rough layouts are then selected for further development and analysis (including road user and vehicle operating costs, if appropriate), refined and examined in greater detail until that considered most suitable for the intersection is selected for detailed design and the preparation of final construction plans and specifications.⌉

20.2.1 Design principles

Underlying most design standards relating to the preparation of safe layouts of at-grade intersections are a number of basic design principles.

1. Minimise the carriageway area where conflict can occur

Large uncontrolled carriageway areas within intersections (especially T- and Y-intersections) provide greater opportunities for collisions resulting from unanticipated vehicle manoeuvres (e.g. see Fig. 20.2(a)). However, traffic-island channelisation (using elongated islands in this instance) can be used to constrain the turning vehicles to defined paths and reduce the size of the potential conflict area.

2. Separate/reduce points of conflict

This can be achieved by prohibiting certain traffic movements at an intersection (e.g. see Figs. 24.1, 24.2 and 24.3), by using auxiliary diverging lanes to separate multiple manoeuvres within an intersection (Fig. 20.2(e)), or by using two separated (staggered)

intersections instead of a single more complicated one (Fig. 20.3(b)). In Britain a right/
left stagger (whereby crossing traffic turns right out of the minor road, proceeds along
the far side lane of the major road and then diverges left without stopping) is safer than
a left/right stagger if two T-junctions replace a dangerous four-way intersection (Fig.
20.3(a)). The reasons for this are (a) drivers of entering minor road vehicles can wait at the
T-junction and judge when it is safe to cut across the on-coming traffic in the nearside
lane(s), and (b) they do not have to wait on the main road prior to diverging left into the
farside minor road. A left/right stagger will operate effectively provided that an auxiliary
right-turn lane is provided at the centre line of the major road to protect right-turning vehicles that have to be stored prior to cutting across the opposing traffic stream(s).

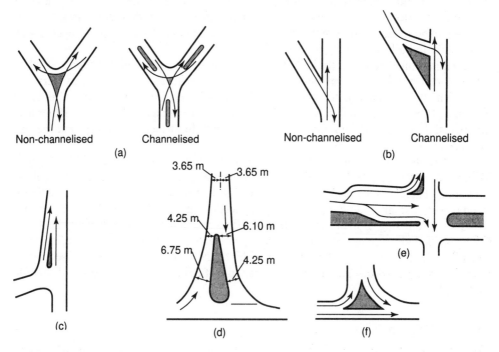

Fig. 20.2 Channelisation techniques illustrating some basic intersection design principles

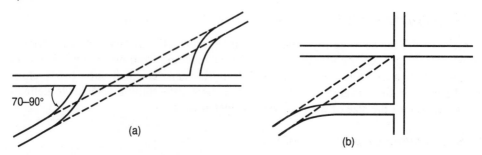

Fig. 20.3 Staggered intersections: (a) converting a 4-way scissors intersection into
two T-junctions with a right/left stagger, and (b) converting a multi-way intersection
into a 4-way junction and a T-junction (not to scale)

3. Traffic streams should merge/diverge at flat angles and cross at right angles

Weaving, merging and diverging manoeuvres should always be carried out at low relative speeds so that, if collisions occur, they are less severe. (Note: Relative speed is the vectorial speed of convergence of two vehicles in a conflict manoeuvre, e.g. two vehicles that are each travelling at 80 km/h in directly opposite directions have a relative speed of 160 km/h, but their relative speeds are reduced to 42 km/h and 14 km/h if they travel in the same direction and merge at 30 degrees and 10 degrees, respectively.) Figure 20.2(c) suggests how the provision of a long speed-change lane with a flat intersecting angle enables vehicles to accelerate and merge easily with the through traffic stream when a safe entry gap appears. Merging angles of 10 to 15 degrees are optimum; longer gaps are required at greater angles and some form of traffic control over the entering traffic is then desirable.

Figure 20.2(b) illustrates the dangerous situation where two opposing traffic streams intersect at about 30 degrees. However, swinging the minor road about so that it intersects the major road at close to 90 degrees ensures that the driver on the minor road is in a better position to estimate traffic speeds in either direction on the major road and enter it more safely. Also, the time and distance required by minor road vehicles to cut across the conflicting traffic streams are reduced.

4. Encourage low vehicle speeds on the approaches to right-angle intersections

Minor road vehicles intending to cut across major road traffic should approach the intersection slowly so that they can easily stop and give way to through traffic. Figure 20.2(b) also suggests how low approach speeds are encouraged by causing traffic to bend substantially prior to the intersection, while Fig. 20.2(d) illustrates how the same effect can be achieved by the use of a traffic island to funnel vehicles into a gradually narrowing opening. Funnelling also has the advantage that it prevents overtaking from taking place at or close to the intersection.

5. Decelerating or stopping vehicles should be removed from the through traffic stream.

This principle is illustrated in Fig. 20.2(e). Separating the traffic streams reduces the number and severity of rear-end crashes. It also ensures that the through traffic is not held up by turning vehicles.

6. Favour high-priority traffic movements

The operating characteristics and layout of an intersection should deliberately favour the intended high-priority movements. This principle, which is also reflected in Fig. 20.2(e), generally improves intersection capacity as well as safety.

7. Discourage undesirable traffic movements

Traffic islands and corner radii can be used to discourage motorists from taking undesirable travel paths, and encourage them to take defined ones. For example, Fig. 20.2(f) illustrates how a deliberately placed and shaped traffic island can be used to force vehicles entering or leaving a one-way traffic stream to turn left only and follow the correct direction of travel.

8. Provide refuges for vulnerable road users

Pedestrians (including handicapped persons) often need to cross a road in two separate manoeuvres. Properly sited traffic islands have the added advantage that they can be used as refuges by these vulnerable road users, especially at intersections on wide roads.

9. Provide reference markers for road users

Drivers should be provided with appropriate references at intersections, e.g. Stop/Give Way lines which indicate where, say, the lead vehicle in a minor road traffic stream should stop until a suitable entry gap appears in the main road stream.

10. Control access in the vicinity of an intersection

No driveways should be permitted within the immediate area of influence of a newly designed intersection. If such access points already exist and closure is not possible for practical reasons, then channelisation techniques should be used to prevent entering vehicles from crossing the traffic flow, i.e. the vehicles entering the intersection from the driveway should always merge with the nearside traffic stream.

11. Provide good safe locations for the installation of traffic control devices

The possible use of traffic control devices should always be considered; for instance, the design of an intersection to be eventually controlled by signals may differ from one requiring channelisation and signs.

12. Provide advance warning of change

Drivers should never be suddenly faced with the unexpected. Advance signing that warns of intersections ahead should be provided on minor roads leading to controlled intersections, on all roads where visibility is restricted prior to an intersection, and on high-speed roads where it is desirable to cause vehicles to slow.

13. Illuminate intersections where possible

Priority for lighting at night should be given to intersections with heavy pedestrian flows and/or with heavy vehicular flows, at roundabouts and where raised channelisation islands intrude on what might be considered the 'natural' vehicle pathways, and where an intersecting road already has lighting.

20.2.2 Other design considerations

Design vehicle

The turning capabilities of the design vehicle influence the shape of kerblines and traffic islands, as well as the width of the carriageway, at at-grade intersections, because the off-tracking of the rear wheels of large vehicles require larger corner radii and extra lane widths to enable these vehicles to negotiate intersections easily and safely without having to stop and carry out complicated manoeuvres. British practice is to provide for the swept turning path of long vehicles where they can reasonably be expected to use an intersection.[2] The swept turning paths are normally generated by a 15.5 m long articulated vehicle with a single axis at the rear of the trailer; this gives a turning width that is greater than that for all other vehicles permitted by the *Vehicle Construction and Use Regulations*. Where no provision is made for long vehicles, minimum corner radii of 6 m and 10 m are used in urban and rural areas, respectively.

Parking

An important consideration in urban locales is whether parking is to be permitted on the intersecting streets. If parking is allowed, it may be appropriate to use a smaller corner radius as the wheel tracks on the approach arm will usually be further from the kerb. If moving vehicles use the parking lane, however, a wider swing will be made by heavy vehicles into the intersecting street unless a large corner radius is provided.

Approach alignment

The alignment prior to and at an intersection should enable the intersection to be seen and appreciated by the approaching driver. The optimum locations for intersections occur in gentle sag curves or on sections of straight level road, where there is good forward visibility. The available sight distance should never be less than the desirable minimum safe stopping sight distance for the approach speed. Intersections on the inside of sharp curves or on the crests of hills should be avoided because of restricted sight distances.

Types of channelisation

As noted previously channelisation involves the use of directional traffic islands to divert vehicles into definite travel paths so that the safe movement of traffic is facilitated, vehicle conflict points are reduced, and traffic friction points are minimised. The traffic islands may be raised above or flush with the carriageway. Raised islands are most commonly used where there is a need for them to be very clearly seen, e.g. where they protect pedestrians, prohibit certain vehicle movements, and/or provide for traffic control devices. Flush or 'ghost' islands are usually defined by carriageway markings; they tend to be used in constrained locations where there is inadequate space for raised islands (a raised island should have a width of at least 1.2 m and an area of at least 4.5 m²), on high-speed single carriageway roads to delineate turning/storage lanes, and at locations where snow removal is a major consideration.

Auxiliary lanes

Auxiliary turning lanes are often provided at at-grade intersections to improve their safety and capacity. They are normally not necessary at intersections where traffic volumes and speeds are low, e.g. in urban residential areas or on minor rural roads.

Auxiliary left-turn lanes are most usually provided (if space is available) at the approaches to intersections that have heavy left-turning traffic volumes, or that are under traffic signal control, or that are on high-speed roads where it is desirable to remove the left-turning vehicles from the through traffic stream before they slow down. If part of the left-turn lane is separated from the through lane(s) by a corner traffic island (see Fig. 20.2(e)) it is often termed a 'slip' lane.

The length of an auxiliary left-turn lane is mainly dependent upon its prime function. On high-speed rural roads, for example, the length is mainly controlled by comfortable deceleration considerations; however, on urban roads, where speeds are lower, the dominant need may be for vehicle storage (particularly at signal-controlled intersections) and the full speed-change length may not be provided. On urban roads care must also be taken to ensure that the queue of through vehicles in the adjacent lane during peak periods is not longer than the left-turn auxiliary lane, thereby preventing access to it. All left-turn auxiliary lanes should be provided with entry tapers to facilitate their usage.

Vehicles using left-turn slip lanes may be required to give way to traffic on the intersecting road. Alternatively, when it is desirable to have free-flowing traffic conditions, the slip lane may lead into an acceleration lane which allows the left-turning vehicles to accelerate to the speed of traffic on the intersecting road and merge safely with it. As much as possible of the acceleration lane should be flush with and parallel to the main carriageway; its length is dependent upon the speed at which the merging vehicles negotiate the slip-lane turn, and the gradient of the acceleration lane, and the volume and speed of the traffic on the main road. The acceleration lane should be ended with a merge taper; also, for safety reasons, a merging vehicle that cannot find a gap in the main traffic stream before running out of lane length should be able to overrun onto a shoulder.

Auxiliary right-turn lanes are most usually provided (if space is available) at the approaches to at-grade intersections that are on dual carriageways with central reservations, or that are traffic signal-controlled, or that are on high-speed and/or high-volume single carriageway roads where right-turning vehicles are a safety hazard and/or cause traffic congestion.

In built-up areas the length of a right-turn lane is most usually determined by vehicle storage needs during peak traffic periods; however, this length should be compared with the queue length in the adjacent through lane to ensure that entry to the right-turn lane is not blocked. On major high-speed urban roads, the storage component of the right-turn lane may be preceded by an appropriate deceleration length. Entry into the right-turn lane should be via a tapered section.

Ideally the length of the auxiliary right-turn lane on a rural road should be based on the deceleration length required to slow a vehicle to a stop from the design speed of the road. On many roads, especially low-volume single carriageways with good visibility, this is neither practical nor necessary, however, and the deceleration distance is based on the change from the design speed down to an assumed turning speed, e.g. 20 km/h.

Sight distance

Adequate visibility in both the horizontal and vertical planes is fundamental to safe intersection design. Thus, drivers of vehicles approaching an intersection along a minor road should have unobstructed views to the left and right along the major road for a distance (at least equal to the safe stopping sight distance) which is dependent upon the speed of traffic on the main road, so that they may judge when it is safe to merge with or cross the traffic in the nearside lane(s) of the major road.

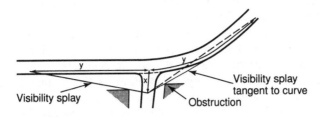

Fig. 20.4 Visibility splays at a major/minor intersection

Similarly, drivers on the major road should be able to see any vehicles emerging from the minor road in time to take evasive action should it be necessary.

British practice in respect of visibility splays is reflected in Fig. 20.4. This shows that visibility should be available on a line (between two points 1.05 m above the carriageways) from a point x metres along the centre line of the minor road (measured from the continuation of the line of the nearer edge of the main road's running carriageway) to a point on the nearer edge of the main road's running carriageway at a running distance of y metres from the centre line of the minor road. If the visibility line lies partially within the major road carriageway (see the dashed right-hand splay in Fig. 20.4) it is made tangential to the nearer edge of the major road's running carriageway. If the major road has dual carriageways and the central reservation gap is wide enough to store minor road vehicles waiting to turn right, then the standard visibility splay to the left is not required; however, the central reservation to the left must be clear for a distance of y metres. The distance y varies according to the main road's design speed and ranges from 295 m to 70 m for speeds of 120 km/h and 50 km/h, respectively. The distance x is normally 9 m; however, it may be reduced to 4.5 m for lightly trafficked simple intersections or to 2 m in exceptionally difficult circumstances.

Spacing

The question of what are appropriate intersection spacings is complex and confusing and there is little international agreement on appropriate values. In principle, short road lengths between intersections are desirable in residential areas to provide ready access to homes and to ensure that speed and traffic volumes remain low, i.e. the intersections act as 'chokes' on traffic. The converse applies to high-quality roads whose primary function it is to provide for safe rapid movement over longer distances between and within urban areas.

A major factor controlling intersection spacings on high-speed high-volume roads is the minimum distance required to allow safe weaving to occur. This problem arises between adjacent intersections, between successive merging and diverging lanes at intersections, and on links within free-flow interchanges. Carriageway lengths between roadside service areas and intersections on major roads also constitute weaving areas. British practice in respect of the spacing of grade-separated intersections and interchanges on high-speed high-volume rural highways is that they should normally be at least 2 km apart (although in extreme cases, where traffic forecasts are at the lower end of the range for the carriageway width, spacings down to 1 km may be acceptable).

Generally, considerable savings in accidents are achieved by reducing the number of lightly trafficked accesses and minor road connections onto major roads.[2] Research has established that the addition of one T-intersection (or one commercial access) per kilometre also reduces the average speed of traffic by about 0.7 km/h and 0.5 km/h on normal and wide single carriageway rural roads, respectively.

The vehicle used for design purposes determines the minimum spacing between intersections which, of necessity, are close together. Thus, for example, the minimum spacings used for simple right/left staggers and dual carriageway right/left staggers in a staggered T-intersection are 50 m and 60 m, respectively; these relate to the critical situation where an 18 m long drawbar trailer combination (i.e. the longest vehicle normally encountered on British roads) enters the intersection by one minor road and leaves by the other.

20.3 Priority intersections

The majority of existing intersections in Britain are priority controlled. At these intersections the minor road traffic gives way to that on the main road and only enters the main road traffic stream during 'spare time' gaps. These intersections are normally controlled by Stop or Give Way signs and markings on the minor road; at less important intersections markings may only be provided. The main advantage of all major/minor intersections is that through vehicles on the main road are not delayed.

Simple priority T-intersections (including staggered) intersections do not have any ghost or physical islands in the major road or traffic islands in the minor road approach. Their usage is appropriate for most new accesses and new minor intersections on single carriageway rural roads when the two-way AADT on the minor road is less than about 300 vehicles, and for existing rural and urban accesses and junctions when the two-way AADT on the minor road is less than 500.

Ghost island T-intersections have a painted hatched-island in the middle of the single carriageway main road to provide a diverging lane and waiting space for vehicles turning right (in Britain) into the minor road; they also assist vehicles turning right from the minor road to complete this manoeuvre in two stages. These intersections tend to be used where a right-turning accident problem is apparent and/or the two-way AADT on the minor road exceeds about 500 vehicles.

T-intersections with single-lane dualling have a physical island in the middle of the single carriageway main road. This provides an offside diverging lane and shelter area for vehicles turning right from the main road, and a safe central waiting area for vehicles turning right from the minor road. With single-lane dualling only one through lane is provided in each direction on the main road, thereby preventing overtaking and excessive speeds through the conflict areas. Single-lane dualling is best used on rural roads with good overtaking opportunities between intersections, and when the volume of traffic is greater than that which can be adequately handled by a ghost island intersection.

Priority T-intersections are also used on all-purpose dual two-lane rural roads (but never on dual three-lane roads) when the two-way AADT on the minor road does not exceed about 3000 vehicles. The major/minor intersections are formed by widening the central reservation on the major road and providing a diverging lane and waiting space for right-turning vehicles entering (and leaving) the minor road.

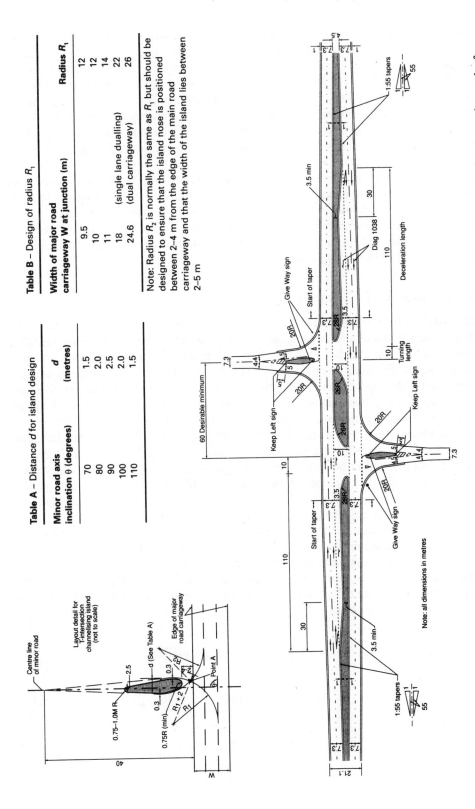

Fig. 20.5 A right/left staggered intersection with a 10 m-wide physical island on a dual 2-lane rural road with 1 m edge strips[2]

Table A – Distance *d* for island design

Minor road axis inclination θ (degrees)	*d* (metres)
70	1.5
80	2.0
90	2.5
100	2.0
110	1.5

Table B – Design of radius R_1

Width of major road carriageway W at junction (m)	Radius R_1
9.5	12
10	12
11	14
18 (single lane dualling)	22
24.6 (dual carriageway)	26

Note: Radius R_2 is normally the same as R_1, but should be designed to ensure that the island nose is positioned between 2–4 m from the edge of the main road carriageway and that the width of the island lies between 2–5 m

Ideally, channelisation islands should be provided in the minor road approaches to major/minor intersections on all but the most lightly trafficked minor roads. As well as providing guidance to long vehicles engaged in turning manoeuvres they also forewarn motorists on the minor road that there is an intersection ahead and (in urban areas) they assist pedestrians, especially the elderly, when they are crossing wide and/or busy minor roads. Central islands, preferably, should be developed symmetrically about the centre line. However, in many instances this is not possible for practical reasons, e.g. where there is need to avoid land intrusion, and the islands are then located asymmetrically to suit the needs of the site.

For safety reasons *priority crossroads* are appropriate only at very low minor road traffic flows, and are no longer used in Britain for new intersections on single carriage-way rural roads or on dual carriageways. Staggered intersections are now used instead of crossroads when a minor road crosses a main road. Figure 20.5 shows the recommended layout for a typical right/left staggered intersection on a rural dual carriageway road with a 120A km/h design speed.

20.3.1 Capacity of a T-intersection

The capacity of a priority T-intersection is primarily dependent upon the ratio of the flows on the major and minor roads, the critical (minimum) gap in the main road traffic stream acceptable to entering traffic, and the maximum delay acceptable to minor road vehicles. As traffic builds up on the main road, the headways between vehicles decline, fewer acceptable gaps become available, and delays to vehicles on the minor road increase accordingly, theoretically to infinity. Field measurements on single carriageway roads indicate that the critical time gaps accepted by minor road vehicles at the head of a queue average about 3 seconds for left-turn merging with, and 4–5 seconds for right-turn cutting of, the traffic stream in the nearside lane of the main road. Empirical research[3] has resulted in predictive capacity equations for T-intersections which were derived from traffic flow measurements and from certain broad features of junction lay-out. This empirical approach has been adopted by the Department of Transport for making decisions on major/minor intersection selection and layout design.[2,4]

A T-intersection has six separate traffic streams (see Fig. 20.6(a)), of which the through streams on the major road (C–A and A–C) and the left-turn stream off the major road (A–B) are generally assumed to be priority streams and to suffer no delays from other traffic, while the two minor road streams (B–A and B–C) and the major road right-turn stream (C–B) incur delays due to their need to give way to higher priority streams. Predictive capacity equations for the three non-priority streams are as follows:

$$q^s_{B-A} = D\{627 + 14W_{CR} - Y[0.364q_{A-C} + 0.114q_{A-B} + 0.229q_{C-A} + 0.520q_{C-B}]\}$$

$$\text{(20.1)}$$

$$q^s_{B-C} = E\{745 - Y[0.364q_{A-C} + 0.144q_{A-B}]\}$$ (20.2)

$$q^s_{C-B} = F\{745 - 0.364Y[q_{A-C} + q_{A-B}]\}$$ (20.2)

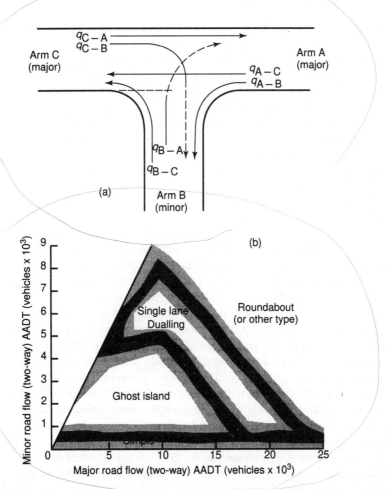

Fig. 20.6 Major and minor road flows applicable to various types of T-intersection on single carriageway roads:[4] (a) flows and notations to be used in conjunction with Equations 20.1, 20.2 and 20.3, and (b) levels of junction used with different flow combinations

where $Y = [1 - 0.0345W]$
$D = [1 + 0.094(w_{B-A} - 3.65)][1 + 0.0009(V_{rB-A} - 120][1 + 0.006(V_{lB-A} - 150)]$
$E = [1 + 0.094(w_{B-C} - 3.65)][1 + 0.0009(V_{rB-C} - 120]$
and $F = [1 + 0.094(w_{C-B} - 3.65)][1 + 0.0009(V_{rC-B} - 120]$

The superscript s (e.g. q^s_{B-A}) denotes the flow from a saturated stream, i.e. one in which there is stable queuing. The geometric parameters w_{B-A} and w_{B-C} denote the average widths of each of the minor road approach lanes for waiting vehicles in streams B–A and B–C, respectively, measured over a distance of 20 m upstream from the Give Way line; w_{C-B} denotes the average width of the right-turn (central) lane on the major road, or 2.1 m if there is no explicit provision for right-turners in stream C–B. The parameters V_{rB-A} and V_{lB-C} denote right and left visibility distances, respectively, available from the minor road while V_{rC-B} is the visibility available to right-turning vehicles waiting to turn right from the major road. W is the average major road carriageway width at the intersection; in the case of dual carriageways and single carriageways with ghost or raised islands, W excludes the width of the central (turning) lane. W_{CR} is the average width of the central reserve lane, at the intersection, on a dual carriageway road.

All capacities and flows are in passenger car units per hour (pcu/h) and distances are in metres. One heavy goods vehicle is considered equivalent to two pcu for calculation purposes. Capacities are always positive or zero; if the right-hand side of any equation is negative, the capacity is taken as zero. The ranges within which the geometric data are considered valid are: $w = 2.05 - 4.70$ m, $V_r = 17 - 250$ m, $V_l = 22 - 250$ m, $W_{CR} = 1.2 - 9$ m (dual carriageway sites only), and $W = 6.4 - 20$ m.

Design Reference Flow

The measure most commonly used in Britain to assess the adequacy of the capacity available to a non-priority traffic stream is the ratio of the hourly demand flow (called the Design Reference Flow or DRF) to the capacity; this ratio is called the Reference Flow to Capacity (RFC) ratio. For the satisfactory operation of any given approach lane it is generally considered that the RFC-ratio should not exceed 0.85.

When selecting an hourly demand flow to represent the DRF-value the function of the road should always be taken into account. British practice is to use the 200th highest hourly flow in the design year to examine intersection designs on recreational roads, the 50th highest hourly flow on inter-urban roads, and the 30th highest hourly flow on urban roads (which have relatively little seasonal variation). This means, of course, that in the course of the design year it can be expected that the design reference flow used at any given intersection will be exceeded and some congestion experienced; however, this is accepted on the grounds that it would be economically and/or environmentally undesirable to attempt to design for the highest hours in the design year.

In the case of an existing intersection the DRF-values are often determined from manual counts (including classifications and turning movements) of the existing flows which are grossed up to the design year, using factors given in the *National road traffic forecasts* (Table 17.13), to produce a range of high and low future year flows. In the case of a new intersection on a new road, or where the traffic flows through an existing intersection are expected to change significantly in the future (due to other network changes or the relief of congestion at the existing intersection), the future year design reference flows are derived from the traffic forecasting model used in the design of the new road.

Figure 20.6(b) shows for single carriageway roads the different types of T-intersection that are normally considered for use with different combinations of major and minor road traffic flows. The ranges suggested take into account geometric and traffic delays, entry and turning stream capacities, and accident costs.

T-intersections are normally unlikely to be cost effective on continuous all-purpose two-lane dual carriageways when the two-way AADT on the minor road exceeds about 3000 vehicles.

20.3.2 Delay

An estimate of the total 24-hour delay due to congestion, D_{24x}, at an existing T-intersection can be estimated from the empirically derived equation[5]

$$D_{24x} = \frac{D_{-3}}{8P_3^{\ 2}}$$

(20.4)

where D_3 = total intersection delay (h) during the peak three hours, and P_3 = ratio of flow in the peak three hours to the 24-hour flow. (Note: This formula assumes that delays are inflicted only on minor road vehicles, which have to yield priority to the major road traffic streams.)

20.4 Roundabout intersections

A roundabout is a form of channelised intersection in which vehicles are guided onto a one-way circulatory road about a central island. Entry to the intersection is controlled by Give Way markings and priority is now given to vehicles circulating (clockwise in Britain) in the roundabout.

The main objective of roundabout design is to secure the safe interchange of traffic between crossing traffic streams with the minimum delay.[6] The operating efficiency of a roundabout depends upon entering drivers accepting headway gaps in the circulating traffic stream. Since traffic streams merge and diverge at small angles and low relative speeds, accidents between vehicles in roundabouts rarely have fatal consequences.

20.4.1 Usage of roundabouts: general

Roundabouts are most effectively used at at-grade intersections in urban or rural areas that have all or a number of the following characteristics:

- high proportions and/or high volumes of right-turning traffic
- it is neither practical nor necessary to give priority to traffic from any particular road
- a disproportionate number of accidents involve crossing or turning movements
- the use of Stop or Give Way signs results in unacceptable delays to minor road traffic
- they are the cause of less overall delay to vehicles than traffic signals
- there is a significant change in road standard, e.g. from a dual to a single carriageway.

Roundabout intersections are normally not appropriate at sites:

- where inadequate space or unfavourable topography does not permit a satisfactory geometric design to be developed
- where traffic flows are unbalanced, e.g. at major/minor T-intersections, and it is not desirable or necessary to cause delays or deflections to all traffic
- that follow steep downhill approaches
- that have heavy volumes of vehicular traffic but are also heavily used by cyclists and/or pedestrians (especially if there are significant proportions of old, young or infirm pedestrians) and there are not satisfactory surface crossings, or subway crossings, to meet the needs of these vulnerable road users
- where it is anticipated that reversible lanes may be employed during peak traffic periods
- between traffic-signal controlled intersections which could cause queuing back into the roundabout exits or that are subject to linked control.

Although roundabouts are most commonly used in urban locales and on non-high-speed rural roads, they can also be used on high-speed rural roads provided that (a) they are not on dual three-lane all-purpose roads, (b) the visibility normally available on each approach road is not less than the desirable minimum stopping sight distance, and (c) the necessary speed reduction at/through the intersection is expected and accepted by drivers.

Research has shown that drivers on high-speed roads can be encouraged to slow down by placing a series of 0.6 m wide thermoplastic 'bars' (90 in total) across the approach lane(s) for about 400 m before the roundabout, with each spaced at exponentially reduced intervals, e.g. 7.69 m between the first and second bars and reducing to 2.73 m between the 89th and 90th bars. This marking arrangement gives drivers the illusion that their speeds are increasing as they approach the roundabout which then encourages them to slow to a more realistic speed more quickly, thereby reducing the risk (as well as the severity) of an accident at the intersection.

20.4.2 Types of roundabout

The three main types of roundabout intersections used in Britain are the normal, mini-, and double roundabouts.

The *normal roundabout* has a kerbed central island 4 m or more in diameter and, usually, three or four access arms that have flared approaches to allow multiple entry lanes. This type of roundabout performs well at four-way intersections, and particularly well at T-intersections provided that the traffic demand is well balanced between the arms.

A *mini-roundabout* has a one-way circulatory carriageway about a flush or slightly raised (domed) circular island that is less than 4 m in diameter and is all white and reflectorised; the intersection arms may or may not have flared approaches. Mini-roundabouts are never used on high-speed roads; they are used most effectively to improve existing intersections in urban areas, on roads subject to a 48 km/h speed limit which are experiencing capacity and safety problems, and where expansion space is very restricted. The circular domes are easily overrun by the wheels of large vehicles and thus they cannot be used to contain street furniture. Because of the quick decision-making by drivers at mini-roundabouts, the slower-moving pedal cyclists can be at particular risk; hence this type of intersection is not normally used (i.e. traffic signals are preferred) where substantial numbers of cyclists are expected and a subway crossing cannot be provided for the cyclists.

A *double roundabout* has two normal (with flared approaches) or mini-roundabouts that are either contiguous or connected by a central link road or kerbed island (Fig. 20.7). Cases where double roundabouts are especially useful include:[6]

- at existing crossroads, where the double roundabout is used to separate opposing right-turn movements allowing them to pass nearside to nearside (Fig. 20.7(a))
- at awkward sites, e.g. scissors intersections (Fig. 20.7(b)) where a normal roundabout would require extensive realignment of the approaches or excessive land take
- at an existing staggered intersection where their usage avoids the need to realign one of the approach roads

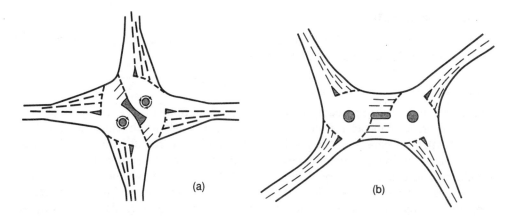

Fig. 20.7 Illustrations of double roundabouts: (a) contiguous, and (b) with a short central link road

- to join two parallel routes separated by physical features, e.g. by a river or railway line
- at overloaded single roundabout where, by reducing the circulating flow past critical entries, the overall capacity is increased
- at intersections with more than four entries, to achieve better capacity with acceptable safety characteristics and a more efficient use of space and/or less land take.

Roundabout variants

Variants of the above three main types include the ring, signalised, and grade-separated roundabouts.

A *ring intersection* has the usual one-way circulation about a very large central island replaced by two-way circulation about the central island and three-arm one-way mini roundabouts and/or traffic signals are installed at the junction of each approach arm with the circulatory carriageway. The drivers on the two-way circulatory system are required to give way to other vehicles in the roundabout. The conversion to a ring junction is an effective solution for a large roundabout with congestion problems at its entries. Many large conventional 'weaving' roundabouts which were constructed prior to the introduction of the 'give way to the right' rule in 1963 have now been converted to ring roundabouts.

A *signalised roundabout* has traffic signals installed on one or more approach roads of a normal roundabout. The signals may be operated full-time or, more commonly, part-time, e.g. during morning and/or evening peak periods. The signals are installed when there is an overloading or an unbalanced flow at one or more entries, or there are high circulatory volumes, which threaten the self-regulating nature of the intersection so that vehicles have to be provided with forced opportunities to enter the roundabout.

A *grade-separated roundabout* has one or more of its arms linked with a road at a different level via an interconnecting slip road(s).

Also, it can be argued that a *gyratory system* is a form of roundabout composed of a series of one-way roads about a 'central island' containing normal land use activities.

20.4.2 Design features

High capacity and safety within a roundabout intersection are encouraged by layout features which maximise the use of the space available.

Entry width

The entry width of the approach arm at the roundabout (dimension e in Fig. 20.8(a)) is the largest single factor, apart from approach carriageway half-width (dimension V in Fig. 20.8(a)), affecting capacity. It is good practice therefore to add at least one extra lane width to the lanes on each entry approach; generally, however, not more than two lanes should be added and no entry should be more than four lanes wide.[6] Flares on the approaches to roundabouts should be such that the maximum entry widths are not greater than 10.5 m and 15 m for single and dual carriageway approach roads, respectively. A typical flare length l' (see Fig. 20.8(b)) on a rural road is 25 m; it may be as low as 5 m in urban areas. As well as increasing capacity, a flared entrance makes it easier for a vehicle to be passed in the event of breakdown; also the turning of long vehicles is eased.

Lanes at the Give Way line should be as wide as practicable but never less than 3 m.

Entry angle

The best entry angle (i.e. the optimum conflict angle between entering and circulating streams) is about 30 degrees (see Figs 20.8(e) and (f)). Very acute merging angles encourage high-speed merging and should be avoided.

Exits

Similarly to entrances, exits should be easy for vehicles to negotiate. As a general rule, each exit should allow for an extra traffic lane over that of the link downstream. On new roundabouts the exit arms are often staggered from their 'opposite' entry arms, so as to ensure good entry deflection; this also tends to result in a reduced size of roundabout.

Entry path curvature

For safety reasons it is essential that vehicles slow down on the approaches to a roundabout and traverse the intersection at a relatively low speed. Speed reduction through the intersection is enhanced by use of an entry path curvature which ensures that through vehicles are deflected to the left as they enter the junction. This is most easily achieved by the use of a large central island which forces the driver to deflect and slow down. However, if the central island is too small to provide sufficient deflection to the left, additional deflection should be generated by the use of enlarged traffic islands or subsidiary traffic-deflection ghost islands in the entry.

No part of a through vehicle's design path (assumed 2 m wide) should be closer than 1 m to the central island, or on a radius greater than 100 m, i.e. the curvature corresponding to approximately 50 km/h with a sideways force of 0.2 g.

Circulatory carriageway

Ideally the circulatory carriageway should be circular in plan and have a constant width that is normally 1.0–1.2 times the maximum entry width (but not more than 15 m). For safety reasons its full width should be visible to the right by drivers entering the inter-section (measured from 1.05 m high points at the Give Way line and from the centre of the farside lane at a distance of 15 m back from the Give Way line on each approach road) and ahead by drivers on the circulating carriageway (measured along its centre line) for a distance appropriate to the size of the roundabout, as shown in Table 20.1.

Table 20.1 Desired visibility distances at roundabouts in Britain[6]

Inscribed circle dia. (m)	Visibility distance (m)	Inscribed circle dia. (m)	Visibility distance (m)
<40	Whole junction	60–00	50
40–60	40	>100	70

Design vehicle

The vehicle normally used to develop the layout for a roundabout is a 15.5 m long artic-ulated heavy commercial vehicle with a single axle at the rear of the trailer. If the largest circle that can be inscribed within the intersection outline (see Fig. 20.8(a) and (c)) is called the inscribed circle (IC), then the smallest IC-diameter that will accommodate this design vehicle at a normal roundabout is 28 m; a mini-roundabout must be used if the IC-diameter is smaller. However, it is usually hard to meet the entry deflection requirement at a normal roundabout if the IC-diameter is less than about 40 m; the usual practice in this instance is to install a hard-surfaced low profile central island that pro-vides adequate deflection for standard vehicles but can be overrun by the rear wheels of articulated vehicles and trailers.

Left-turn lanes

Segregated left-turn lanes are often used at roundabouts between an adjacent entry and exit if space is available and more than 50 per cent of the entry flow, or more than 300 veh/h in the peak hours, intends to leave the roundabout at the first exit after entry. Segregation is achieved by either an elongated traffic island or road markings. Pedestrians should not be expected, or allowed, to cross segregated left-turn lanes defined by mark-ings.

Crossfall

Superelevation is unnecessary on circulatory carriageways. However, a crossfall (typi-cally 2 per cent) is required to drain surface water, and the correct siting and spacing of gullies is critical to efficient drainage. The application of appropriate superelevation at entrances and exits can help drivers to negotiate associated curves as they enter and leave roundabouts.

Lighting

A roundabout should be illuminated at night, for safety reasons. Its layout should be simple, clear and conspicuous so that its important design features are easily seen and understood in the night as well as during the day.

20.4.3 Capacity of a roundabout

The capacity of a roundabout as a whole is a function of the capacities of the individual entry arms. The capacity of each arm is defined as the maximum inflow when the traffic flow at the entry is sufficient to cause continuous queuing in its approach road. The main factors influencing entry capacity are the approach half-width, and the width and flare of the entry, while the entry angle and radius also have small but significant effects. The predictive equation used with all types of single at-grade (mini- or normal) roundabouts is

$$Q_e = k(F - f_c Q_c) \quad \text{when } f_c Q_c \leq F \tag{20.5}$$
$$= 0 \quad \text{when } f_c Q_c > F \tag{20.6}$$

where Q_e = saturation or capacity entry flow (pcu/h); Q_c = circulating flow across the entry (pcu/h); $k = 1 - 0.00347(\phi - 30) - 0.978[(1/r) - 0.05]$; $F = 303x_2$ where $x_2 = v + (e - v)/(1 + 2S)$ and $S = 1.6(e - v)/l'$; and $f_c = 0.210t_D(1 + 0.2x_2)$ where $t_D = 1 + 0.5/(1 + M)$ and $M = \exp[(D - 60)/10]$

The symbols e, V, l', S, D, ϕ and r are described in Table 20.2 and Fig. 20.8. Q_e and Q_c are in pcu/h, and one heavy goods vehicle is assumed equivalent to 2 pcu for computation purposes.

Table 20.2 The limits of the parameters used in the roundabout capacity equation

Geometric parameter	Symbol	Unit	Practical limits
Entry width	e	m	4 – 15
Approach half-width	V	m	2 – 7.3
Average effective flare length	l'	m	1 –100
Sharpness of flare	S	–	0 – 2.9
Inscribed circle diameter	D	m	15 –100
Entry angle	ϕ	deg	10 – 60
Entry radius	r	m	6 –100

Figure 20.8(a) indicates the entry and circulating flows, Q_e and Q_c, respectively. Note that Equation 20.5, which is empirically derived, clearly shows that (because of the off-side priority rule) entry capacity decreases as the circulation flow Q_c increases. In the case of a four-arm roundabout Q_c is the sum of the entry flows from the two previous arms which have not already exited the roundabout before passing the arm whose entry capacity is being determined.

The entry width e in the equation is measured from the point A along the normal to the nearside kerb. The approach half-width v is measured along a normal from a point in the approach upstream from any entry flare, from the median line (or from the offside edge of the carriageway on dual carriageways) to the nearside kerb. The diameter D in Fig. 20.8(a) refers to the size of the biggest circle that can be inscribed within the intersection outline; Fig. 20.8(c) also illustrates the determination of D in the extreme case of a double roundabout at a scissors crossroads.

The average effective flare length l' is measured along the curved line CF' in Fig. 20.8(b); CF' is parallel to BG and at a distance $(e - v)/2$ from it. The sharpness of flare S is a measure of the rate at which extra width is developed in the entry flare, i.e. small S-values correspond to long gradual flares and big ones to short severe flares.

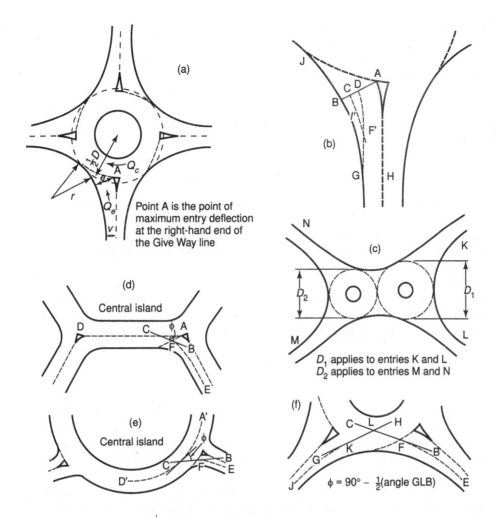

Fig. 20.8 Parameters used in the predictive equation for roundabout entry capacity[4]

The entry angle ϕ acts as a proxy for the conflict angle between the entering and circulating traffic streams. Figures 20.8(d) and (e) show ϕ for conventional roundabouts with straight and curved circulatory carriageways, respectively; for all other roundabouts ϕ is defined by the equation shown in Fig. 20.8(f), where the angle GLB is measured on the side facing away from the central island. (Details of the construction involved in determining ϕ are available in reference 4.)

The entry radius r is measured as the minimum radius of curvature of the nearest kerbline at entry (Fig. 20.8(a)).

Design Reference Flow

When designing a roundabout intersection the entry capacity for each arm of a trial layout is compared with the expected hourly traffic flow for the design year, i.e. with the Design Reference Flow (DRF). As with T-intersections the Reference Flow/Capacity

Ratio (RFC-ratio) is an indicator of the likely performance of an intersection under the future year traffic loading. However, the capacity formula was developed from multiple regression analyses of observations at a large number of sites, and gives capacities that are average values of very broad distributions. Due to site-to-site variation there is a standard error of prediction of entry capacity by the formula of plus or minus 15 per cent. Thus if an RFC-ratio of about 85 per cent occurs it can be expected that queuing will theoretically be avoided in the chosen design year peak hour in five out of six cases (schemes).

20.4.4 Delay at roundabout-modified T-intersections

Delays at a conventional T-intersection are normally inflicted on minor road vehicles, which have to yield priority to the major road traffic streams. At an offside-priority roundabout intersection, however, the previously unimpeded straight-ahead major road traffic has both to reduce speed and to be prepared to give way to minor road vehicles on its right.

The delay to previously unimpeded main road traffic due to the physical layout (including markings) at an intersection with a roundabout is known as the fixed or geometric delay, and is present throughout the whole day. For a mini-roundabout in rural and semi-rural locales the fixed delay can be simply estimated from the formula[5]

$$D_{24g} = \frac{Q_{24}(1 - 2R)D_g}{6000} \tag{20.7}$$

where D_{24g} = total anticipated 24-hour fixed delay (h), Q_{24} = anticipated total daily intersection flow (veh/24-h), R = ratio of minor to total flow (daily average), and D_g = average geometric delay per vehicle (hundredths of a minute) obtained from the empirical equation $D_g = 0.432V - 15.5$ in which V is the average of the mean of the approach and exit speeds (km/h).

Most congestion delay occurs during peak traffic periods. The following relationship provides an estimate of the total 24-hour congestion delay at a mini-roundabout:

$$D_{24c} = \frac{2 \times 10^{-4} Q_{24}^2 P_3}{Q_p} \tag{20.8}$$

where D_{24c} = total anticipated 24-hour congestion delay (h), Q_{24} = anticipated total daily intersection flow (veh/24 h), P_3 = ratio of flow in the three highest hours to the 24-hour flow, and $Q_p = 0.8K(\Sigma W + A^{1/2})$. Here ΣW = sum of the basic road widths used by traffic in both directions on all approaches (m), A = area within the intersection outline (including islands) which lies outside the area of the basic crossroads (m²), and K = site factor depended on the roundabout type and arms, e.g. 60 for a single-island three-arm roundabout.

If the combined delay $D_{24g} + D_{24c}$ anticipated at a proposed mini-roundabout intersection is significantly less than the calculated delay to minor road traffic at a priority-controlled intersection, it suggests that a changeover to a roundabout should be seriously considered.

20.5 Traffic signal-controlled intersections

Intersections under traffic signal control operate on the basis that separate time periods are allotted to conflicting traffic movements so that each can make safe and efficient use of the carriageway space available. Traffic signals are usually installed only at at-grade intersections in built-up areas.

20.5.1 Advantages/disadvantages of traffic signal control

Properly installed and operated signals have the following advantages:

- they are accepted by the public as providing for the orderly movement of vehicles and pedestrians at conflict locations
- they reduce the frequency of vehicle–pedestrian and right-angle vehicle–vehicle accidents
- they can increase the traffic-handling capacities of congested non-signalised intersections
- they can be programmed to increase the people-handling capacities of intersections (by giving priority to pedestrians and public transport vehicles)
- under conditions of favourable intersection spacing they can be coordinated to provide for nearly continuous vehicle progression through intersections in linked or area-wide urban traffic control schemes
- they can be programmed to give priority to movements through particular arms of an intersection
- their capital costs and land take needs are usually less than for roundabouts of similar capacity.

Disadvantages of traffic signals include:

- they usually result in an increase in rear-end vehicle collisions (more so with fixed-time signals than vehicle-actuated ones)
- they can increase total delay (and operating costs) to vehicles, particularly in uncongested non-peak periods
- signal installations need to be continuously maintained and their operations monitored
- signal failure, although infrequent, can lead to serious and widespread traffic difficulties especially during peak traffic periods.

20.5.2 Usage

The decision as to whether a traffic signal should be used at a particular intersection depends very much upon the conditions prevailing at the site, such as:

- the type of intersection (Note: three- or five-way intersections are better treated by roundabouts than by traffic signals, especially when the flows are balanced and the junction is Y-shaped)
- how close the intersection is to the next intersection(s), and whether it is within the linking orbit of an urban traffic control scheme. (Note: About two out of every five traffic signals in Britain are eventually expected to be part of area-wide schemes)

- the number and types of accident previously experienced (if it is an existing inter-section)
- the speed, volume and composition of vehicular traffic. (Note: On well trafficked dual carriageways, for similar flows on both roads a signalised intersection will generally have more accidents than a roundabout[6])
- the (turning) destinations of the traffic. (Note: When the proportion of right-turning vehicles (in Britain) is greater than, say, 30–40 per cent a roundabout may be preferred)
- the volumes of pedestrians and cyclists. (Note: Traffic signals represent much less of a hazard for cyclists than roundabouts)
- the site topography. (Note: Signalised intersections located on the crests of hills may be difficult to perceive for drivers approaching uphill while those following steep downhill approaches will require carriageway surfacings with high skid resistance)
- the land take available.

20.5.3 Some design considerations

Through vehicles operating at intersections controlled by traffic signals should be provided with short paths. *Turning vehicles* should be provided (where space permits) with auxiliary lanes where they can slow down and be stored clear of the through traffic. If storage lanes are not provided or, if provided, they are of inadequate length, right-turning vehicles in particular are likely to delay through vehicles.

For *safety* reasons, drivers on any given approach should always be able to see at least two signal heads, usually a primary and a secondary (or reinforcing) signal. In Britain the primary signal is located 1–2 m beyond the stop-line as seen in the direction of the traffic flow, and normally on the nearside of the carriageway. On a high-speed approach and/or on a wide approach where it is difficult to see the primary signal, a duplicate primary signal is also placed at the end of, and beyond, the stop-line on a central island that is proud of the carriageway (in the case of a single carriageway) or in the central reservation/median (in the case of a dual carriageway). On wide one-way streets a duplicate primary signal is also normally provided on the offside of the carriageway.

In the case of a dual carriageway road controlled by traffic signals the opposing right-turn lanes at intersections should be opposite each other. This arrangement facilitates non-hooking turning movements during the major road traffic signal phases.

A primary signal is normally mounted on the footpath or verge, or on a wall (if there is no verge and footpath space is limited), or on a traffic island which is located to protect a slip-road operating under priority control (which allows left-turning vehicles to bypass the signal safely). A secondary signal, which may transmit information additional to, but not in conflict with, that of the primary signal, e.g. a right-turn arrow, is usually on the far offside of the intersection.

Advice on the layouts and operation of various types of traffic signal-controlled intersections in Britain is available in the literature.[7, 8] Principles underlying intersection design are discussed in Chapter 26, in relation to Figs 26.1 and 26.2.

20.5.4 Capacity of a signal-controlled intersection

The main factors affecting the capacity of a single isolated intersection controlled by traffic signals are the characteristics of the intersection layout, the traffic composition and its straight-through and turning needs, and the setting of the signal control.

Nowadays virtually all traffic signals that are not included in area-wide urban traffic control (UTC) systems are equipped with vehicle detectors linked to a switching device (either an electro-mechanical switch or, more recently, a programmable microchip) located in a control box at the side of the road. In Britain the detectors are most usually inductive loops buried in the carriageway surface, with the loop which activates the green period for approaching vehicles positioned 40 m from the stop-line. Once a green period is activated its length can be extended by additional loops located 12 m and 25 m from the stop-line; the extension provided by each loop is normally 1.5 s, based on a minimum vehicle approach speed of 32 km/h. In peak morning and evening traffic periods vehicle-actuated signals normally operate as fixed-time signals whereby the lengths of green time given to all traffic streams during signal cycles have fixed maximum values, so as to ensure that waiting vehicles are not stopped indefinitely.

In Britain the usual sequence of signal aspects is red, red/amber (standardised at 2 s), green and amber (standardised at 3 s). Ideally, capacity design involves determining the optimum cycle length (see Chapter 26), and the durations of the red and green periods in this cycle, that will maximise the total flow through the intersection with the minimum average delay to all vehicles using the intersection. The ultimate intersection capacity is then the sum of the maximum flows that can pass through the intersection approaches during the green plus (part of the) amber periods given to each approach during this cycle, expressed in vehicles per hour (veh/h) or in its equivalent passenger car units per hour (pcu/h). However, if the optimum cycle length (C_o) is too long, waiting drivers can become very impatient at not receiving a green aspect for the approach at which they are in a queue; thus the optimum cycle length is normally limited to a maximum of 120 s, as there is relatively little gain in efficiency through using a longer C_o-value.

In intersection design the practical capacity of a signal-controlled intersection is normally taken as 90 per cent of the ultimate capacity.

When a green aspect is given to a traffic stream waiting in, say, a lane of an approach at a signalised intersection, the vehicles that are early in the queue take some time to accelerate to the constant running speed at which most of the queue discharges into the intersection at a more or less constant rate. At the end of the green time given to this traffic stream, some of the vehicles in the queue make use of the amber period to cross the stop-line into the intersection as the discharge rate falls away to zero at the start of the red period. The term *saturation flow* is used to describe the maximum rate of discharge of the traffic stream across the stop-line that can be sustained from a continuous queue of vehicles during the green plus amber time. That part of the green plus amber time which, when multiplied by the saturation-flow rate, gives the maximum number of vehicles that can enter the intersection is known as the effective green time, and the difference between the actual green plus amber time and the effective green time is called the lost time (see Chapter 26 for an example of the determination and usage of these times).

Observations at intersections throughout Britain[9] have concluded that the straight-through saturation flow for a non-nearside lane of 'average' width (3.25 m) at an approach is about 2080 pcu/h, and for a nearside lane the figure is 1940 pcu/h. The equivalent pcu-values used in these determinations were: 1.00 for cars and light vans, 0.2 for pedal cycles, 0.4 for motorcycles, 1.5 for medium commercial vehicles, 2.3 for heavy commercial vehicles, and 2.0 for buses or coaches.

These studies also concluded that the contributions from individual lanes to the saturation flow at multi-lane stop-lines where all lanes were saturated could be treated independently, and that the total saturation flow for an approach at a signalised intersection could be predicted as the sum of the individual lane predictions.[9]

For individual lanes containing only straight-ahead traffic, it was also found that saturation flows (a) decrease with uphill gradient by 2 per cent per 1 per cent of gradient, but are unaffected by downhill gradient, (b) increase with lane width, by 100 pcu/h per metre of width, and (c) are 6 per cent lower in wet road conditions than in the dry.

For individual lanes containing unopposed turning traffic, saturation flows decrease for higher proportions of turning traffic and lower radii of turn. Thus, for *unopposed streams in individual lanes* in dry weather conditions, the saturation flow S_1 (pcu/h) is given by

$$S_1 = \frac{(S_0 - 140d_n)}{\left(1 + \frac{1.5f}{r}\right)} \tag{20.9}$$

where d_n = nearside lane dummy variable (d_n = 0 for non-nearside lanes, and = 1 for nearside lane or single lane entries); f = proportion of turning vehicles in a lane; r = radius of curvature of vehicle paths (m); and $S_0 = [2080 - 42d_G G + 100(w_l - 3.25)]$ where d_G = gradient dummy variable (d_G = 0 for downhill entries, and = 1 for nearside lane or single lane entries), G = gradient (%), and w_l = lane width at entry (m).

For individual lanes containing mixtures of straight-ahead traffic and opposed right-turning traffic the saturation flow is affected by the traffic intensity on the opposing arm; the number of storage spaces available within the intersection which right-turners can use without blocking straight-ahead traffic; and the number of signal cycles per hour. Thus, for streams containing *opposed right-turning traffic in individual lanes* in dry weather conditions, the saturation flow S_2 (pcu/h) is given by

$$S_2 = S_g + S_c \tag{20.10}$$

where S_g = saturation flow in lanes of opposed mixed turning traffic during the effective green period (pcu/h) and S_c = saturation flow in lanes of opposed mixed turning traffic after the effective green period (pcu/h).

The value S_g in Equation 20.10 is obtained from

$$S_g = \frac{(S_0 - 230)}{[1 + f(T - 1)]} \tag{20.11}$$

where T = through car unit factor of a turning vehicle in a lane of mixed turning traffic (i.e. each turning vehicle is equivalent to T straight-ahead vehicles), obtained from

$$T = 1 + \frac{1.5}{r} + \left[\left[\frac{\frac{12x_o^2}{\{1 + 0.6N_s(1 - f)\}}}{1 - (fx_o)^2}\right]\right] \tag{20.12}$$

where X_o = the ratio of the flow on the opposing arm to the saturation flow on that arm (the degree of saturation), obtained from the equation

$$X_o = \frac{q_o}{\lambda n_l s_o} = \frac{q_o c}{g S_o} \tag{20.13}$$

where q_o = flow on opposite arm (veh/h of green time – excludes non-hooking right-turners), λ = proportion of the cycle which is effectively green for the signal phase under consideration (i.e. $\lambda = g/c$), n_l = number of lanes on opposing entry, s_o = saturation flow per lane for opposite entry (pcu/h), c = cycle time (s), and g = effective green time(s); N_s = number of storage spaces available inside the intersection which right-turners can use without blocking following straight-ahead vehicles; and S_o, f and r are as defined before.

The value S_c in Equation 20.10 is obtained from

$$S_c = \frac{3600 P (1 + N_s)(fx_o)^{0.2}}{\lambda c} \tag{20.14}$$

where $P = 1 + \Sigma(\alpha_i - 1)p_i$ such that α_i = pcu-value of vehicle type i and p_i = proportion of vehicles of type i in stream; and N_s, f, X_o, λ and c are as defined previously.

For multi-lane approaches the saturation flow S_{app} (pcu/h) in dry weather conditions is given by

$$S_{app} = \Sigma_{\text{all lanes, } i} S_i \tag{20.15}$$

where S_i is the saturation flow across the stop-line for the ith lane predicted by Equations 20.9 and 20.10, as appropriate.

20.6 Intersections with grade-separations

These intersections are composed of a system of connector roads which, in conjunction with a grade-separation or grade-separations, provide for the interchange of vehicular traffic between two or more roads on different levels. They can be divided into two major groups:[10–12] grade-separated intersections and interchanges.

A *grade-separated intersection* uses slip roads to connect a major dual carriageway road with a lesser category of road, and has an at-grade roundabout, signalised or priority-controlled junction at the non-mainline end of each slip road. The slip roads are normally one-way and have low design speeds, typically 70 and 60 km/h in rural and urban areas, respectively; however, a slip road longer than 0.75 km is designed as an interchange link.

An *interchange* does not have any at-grade intersections; it provides turning traffic with the capability of uninterrupted movement between intersecting mainline carriageways via interchange links (these links have design speeds of, typically, 85 and 70 km/h for rural and urban roads, respectively) and a succession of diverging and merging manoeuvres on the mainline carriageways. Good interchange design minimises the number of conflict points, and uses signs and markings to ensure that the paths between them are easily understood by drivers.

A slip road or interchange link that turns through 270 degrees is also called a loop.

20.6.1 Usage

The capital costs of grade-separated intersections and interchanges can be great, and some situations where their usage is justified follow.

1. *At intersections involving a motorway.* A motorway, with its complete control of access, is the highest category of road and its intersection with another motorway automatically justifies the construction of an interchange to ensure the free movement of turning high-speed traffic. A grade-separated intersection is justified where a motorway crosses a lesser quality road with which it is desired to connect.
2. *To eliminate traffic bottlenecks*, e.g. where mainline traffic flows through an important at-grade intersection are heavy and an appropriately sized roundabout or signal-controlled intersection is not feasible or economic.
3. *For safety reasons*, e.g. in some locations land is relatively cheap and the construction of a grade-separated intersection as part of a high-speed dual carriageway road can be expected to reduce the accident rate significantly as compared with a conventional at-grade intersection.
4. *For economic reasons*, i.e. where the long-term operating benefits from freer and safer traffic movement clearly outweigh the higher capital and maintenance costs.
5. *For consistency reasons*, e.g. where all other intersections on the road in question or on similar category roads in the area are grade-separated.
6. *For topography reasons*, i.e. at a site where land is available and the topography levels are such that the construction of an at-grade intersection would be as expensive.

In situations other than the above the decision as to which type of at-grade or grade-separated intersection to use is most commonly based on the relative economic and environmental advantages of the various intersection types under consideration. In the case of new roads the choice of intersection is especially difficult because of the sensitivity of the economic analysis to delay costs; relatively small errors in the estimates of future traffic flows and intersection capacity can have a significant impact on delay estimates. (Formulae for estimating geometric delays at various types of intersection are available in the literature.[13])

Grade-separated intersections in Britain are most usually associated with dual carriageway roads with individual carriageways carrying in excess of 30 000 AADT.

Proposals to grade-separate two two-way single carriageway roads are generally treated with caution as experience has shown that they give the impression that the carriageway carrying the main traffic flow is part of a high-speed dual carriageway and this engenders a level of confidence in that carriageway that cannot be justified. For similar reasons a grade-separated intersection should not be provided on dual carriageways within about 1 km of the changeover from single carriageway standard.

The earthworks, concrete structures and relatively great land take usually associated with grade-separated intersections and interchanges normally have detrimental environmental and visual impacts. Consequently, and apart from the normal economic justification, their usage in built-up areas in particular usually requires considerable additional justification.

20.6.2 Example grade-separated intersections and interchanges

There are a huge number and variety of intersections with grade-separation in use throughout the world today. The following examples relate to some recommended for use in Britain.[10, 12]

Grade-separated intersections

The simplest type of four-arm two-level intersection with grade-separation is the full *diamond* (Fig. 20.9(a)). It requires only one bridge and four one-way slip roads. The slip roads have direct high-speed entrance (merging) and exit (diverging) connections with the carriageways carrying the main traffic flows through the intersection, i.e. the main-lines, but the at-grade intersections on the minor road are staggered and priority-controlled or signalised. The dumb bell type of roundabout (Fig. 20.10(a)) can also be adapted to fit a diamond (and a half cloverleaf) intersection. The diamond layout is most appropriately used where the turning movements from the major road onto the minor road are relatively low, because the capacities of the two intersections on the minor road which involve the exit slip roads are often limited. One or more of the slip roads may be omitted from the full diamond if the demand for particular turning move-ments is low. Ideally, the mainlines pass beneath the minor road; vehicles leaving the mainlines can then use gravity on the uphill slip roads to lose speed before coming to the minor road intersections, and vice versa for vehicles entering the mainlines. The diamond's configuration is very easily understood by drivers. Because of its narrow width its usage is applicable to urban areas. However, care needs to be taken with the design of the two exit slip road minor road intersections to ensure that 'wrong-way' movements are prevented.

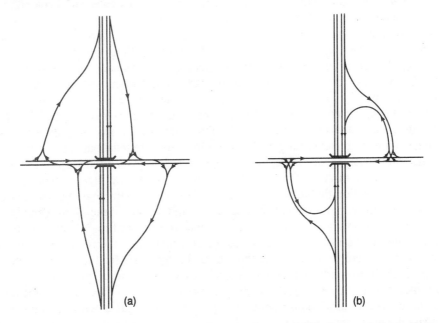

(a)   (b)

Fig. 20.9 Examples of grade-separated intersections: (a) full diamond, (b) half clover-leaf

A *half cloverleaf* grade-separated intersection is used in Britain at similar flow levels as the diamond. This type of intersection is especially applicable when the site conditions are such that the use of all four quadrants of the intersection is not possible. The configuration shown in Fig. 20.9(b), which has its 270 degree (entrance to the mainline) loops in advance of the intersection, requires only one bridge; also vehicles can leave the high-speed mainline via an uphill slip road, while the downhill loops can be designed as exits from the lower-speed minor road. For loops leading onto and off all-purpose dual carriageways the minimum radii used are 30 m and 50 m, respectively. The at-grade intersections on the minor road normally utilise ghost islands, and they may be signal-controlled in urban areas; alternatively, dumbbell roundabouts may be used.

At one time full cloverleaf interchanges (with loops in all four quadrants) were commonly constructed in the United States to enable vehicles to transfer between two high-speed high-volume roads. However, these interchanges are now rarely constructed as experience has shown that, unless provided with long spacings between mainline entrances and exits (with consequent large land takes) they have fairly poor safety and operational histories. This is because vehicles leaving a particular loop to enter a mainline have to accelerate and weave their way through other high-speed vehicles attempting to decelerate and leave the same carriageway via the adjacent existing loop in the following quadrant.

The *two-level dumbbell roundabout* (Fig. 20.10(a)) is an intermediate grade-separation layout between the diamond and the two-bridge roundabout: it has greater traffic capacity and a lesser land take than the diamond and a lesser cost (only one bridge) and a lesser land take compared with the two-bridge roundabout.

The *two-bridge two-level roundabout* (Fig. 20.10(b)) is possibly the most common type of grade-separation currently in use in Britain. Vehicles enter and leave the main road via diagonal slip roads in a manner similar to that for a diamond intersection with grade-separation; however, this intersection has a greater traffic capacity and is therefore commonly used when the turning movements from the mainlines are greater than can be easily handled by a conventional diamond layout. This roundabout intersection also makes U-turns easy to carry out; this is particularly important in rural areas, where intersections are usually spaced far apart. Another advantage of the grade-separated roundabout is that it can often be adapted from an existing at-grade roundabout that is overloaded; the grade-separation frees the movement of the mainline through traffic while the minor road and turning traffic is confined to the roundabout above. This layout is particularly advantageous when the intersection has more than four approach arms.

The *five-bridge three-level roundabout* (Fig. 20.10(c)) is a higher level of grade-separated intersection that is typically used to provide for vehicle movement between two non-motorway main roads. With this configuration one of the main roads is sited below the roundabout and the other above it; thus through traffic on both mainlines is uninterrupted while turning vehicles use the slip roads and the roundabout to move from one main road to the other. Although the bridge costs for this type of intersection are high, it uses less land take and carriageway area, and usually costs less, than a full interchange. However, if the turning movements exceed those used in the design over the course of time, then operational problems associated with queuing on the roundabout entries may emerge.

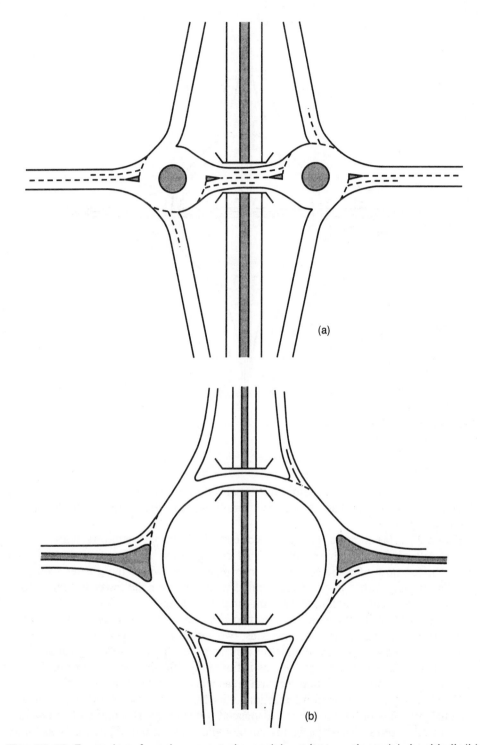

Fig. 20.10 Examples of grade-separated roundabout intersections: (a) dumbbell, (b) conventional two-level, and (c) three-level

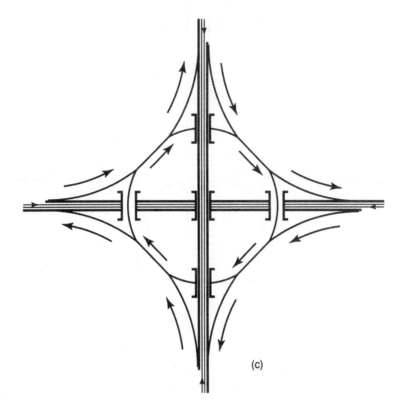

(c)

Fig. 20.10 Continued

Interchanges

Interchanges are described as semi-directional if they contain loops, and all-directional if without loops. Loops leading onto or off motorways in Britain normally have a minimum radius of 75 m.

Figure 20.11(a) shows a three-arm two-level single-bridge semi-directional T-interchange (also known as the *trumpet*). This configuration is used to provide for the uninterrupted flow of vehicles between two main roads where the greater volume of mainline turning traffic is given preferential treatment over the lower turning volume from the lesser road, which uses the semi-directional loop.

Figure 20.11(b) illustrates a four-arm two-level *'cyclic' interchange* which has four semi-directional reverse curve 'loop' roads, two successive diverges off and one merge onto each mainline, and a low number of conflict points. Its land take is great and there is a high structural content; however, as it fits easily into the topography this form of interchange is a suitable solution for rural areas where land is not at a premium.[10]

When two grade-separated intersections with high traffic flows are close together, potential weaving problems can be associated with the short lengths of intervening mainline carriageway. These problems can be avoided by the use of link roads, as shown in Fig. 20.11(c) between an all-directional interchange and a diamond grade-separated intersection. In this context the link road, also known as a *collector-distributor road*, is

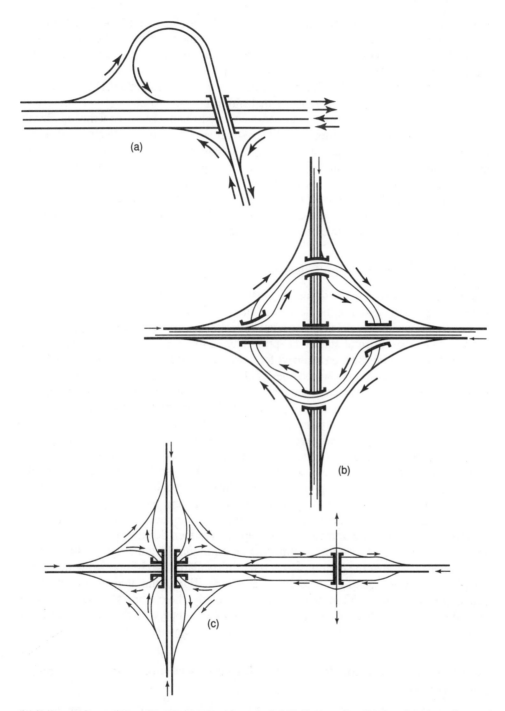

Fig. 20.11 Examples of interchange layouts: (a) T or trumpet, (b) two-level cyclic, and (c) four-level all-directional connected by link roads with a diamond grade-separated intersection. (Note: for reasons of space only two of the four bridges are illustrated in (c))

a one-way connector road adjacent to but separated from the mainline carriageway travelling in the same direction, which is used to connect the mainline to the local highway network where successive direct connections cannot be provided to an adequate standard because the intersection spacing is too close.

The four-arm four-level *all-directional interchange* shown in Fig. 20.11(c) is an example of the highest form of interchange that might be used to connect two motorways where land is expensive. This layout has been used relatively infrequently in Britain, however, even though it requires a relatively small land take, has no loops, its structural content is relatively low and concentrated at a central location, and vehicle operating costs through the intersection are minimised. Its major disadvantages are that it is visually highly intrusive, and has a great number (16) of merging and diverging points.

All of the interchanges shown in Fig. 20.11 exhibit the desirable characteristic that vehicles entering the mainline carriageway do not have to weave their way through diverging vehicles seeking to exit the same carriageway.

20.6.3 Overview of design process

The process[11] involved in the design of most interchanges and intersections with grade-separation is summarised in the flow-chart in Fig. 20.12.

The first steps involve determining a network strategy, fixing a design year, and deciding whether urban or rural standards apply. This is followed by decisions regarding an initial network and intersection strategy; for instance, should all connections be made with the crossing roads, and if not which ones? Following the derivation of the low- and high-growth design year traffic flows, the decision is taken (or confirmed) as to whether the intersection should be designed to all-purpose road or motorway standards.

The next step (and possibly the first of many iterative ones) is to look at the lane requirements. If the proposed layout cannot be adjusted to cope with future traffic demands (including those associated with maintenance work needs) then an alternative network and intersection strategy will need to be devised, such as reducing the number of intersection accesses or using link roads to reduce the frequency of direct access points along the mainline, thereby easing weaving problems and promoting free flow. Link roads may also be used where it is unsafe or impossible to make direct connections.

The proposed merge and diverge arrangements are then determined and the weaving sections are examined to ensure that they meet minimum length criteria. The next step involves ensuring that the appropriate geometric standards can be applied with the proposed intersection spacings, lane gains and lane drops, and that an effective and economic signing system can be implemented. Finally, the continuity of lane provision and lane balance is checked to ensure that it is able to cope with the anticipated traffic demand and that the objectives of the intersection strategy are met.

20.6.4 Some design considerations relating to weaving sections, merges and diverges

The most efficient form of grade separation is that which presents drivers with the minimum number of clear unambiguous decision points as they traverse any at-grade component of the intersection and when merging and diverging.[10] Consistency of design at successive intersections is also important, and the closer the intersections the more

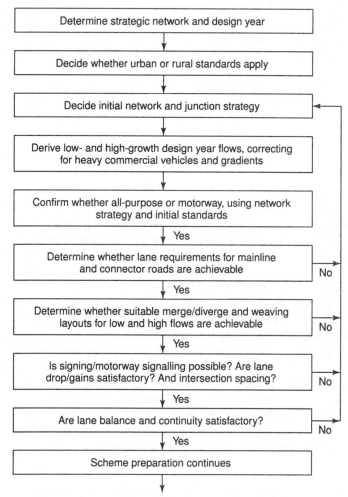

Fig. 20.12 Flow-chart of the grade-separated intersection/interchange design process[11]

critical it is that the same design speed, and road marking and signing practices, be adopted.

The following discussion relates to design practice in Britain for weaving sections, and for merges and diverges.

Weaving sections

Weaving takes place on roads where traffic streams merge, diverge and cross when travelling in the same direction between intersections that are close together. For safety reasons, therefore, intersections with grade-separation on *rural* motorways and all-purpose dual carriageways should be more than 3 km and 2 km apart, respectively, as successive merges and diverges normally do not interact, and weaving does not occur, above these distances. Below these spacings, however, weaving of vehicles is assumed to occur between traffic streams travelling in the same direction, and the design objective then is to ensure that vehicular movement from one lane to another takes place safely, gradually and at small intersecting angles. Ideally, weaving lengths on motorways and all-purpose roads should be at least 2 km and 1 km, respectively; however, in exceptional cases, and when the traffic forecasts are at the lower end of the range for the carriageway in question, absolute minimum weaving distances as low as 1 km and 450 m, respectively, are used.

Within the above length limits the carriageway width required to ensure safe weaving on non-urban high speed (120/100A km/h) roads can be obtained from

$$N = \frac{1}{D}\left[Q_{nw} + Q_{w1} + Q_{w2}\left(\frac{2L_{min}}{L_{act}} + 1\right)\right] \qquad (20.16)$$

where N = required number of traffic lanes; D = maximum mainline flow per lane (veh/h) upstream of the weaving section (= 1800 veh/h and 1600 veh/h for motorways and all-purpose dual carriageways, respectively); Q_{nw}, Q_{w1} and Q_{w2} = the hourly design flows, i.e. total non-weaving, major weaving and minor weaving flows (veh/h), respectively, as shown in Fig. 20.13; L_{min} = desirable minimum weaving length (m) (normally = 2 km and 1 km for motorways and all-purpose dual carriageways, respectively); and L_{act} = actual weaving length available (m) where $L_{act} \geq L_{min}$.

Each hourly design flow used in non-urban grade-separation design relating to weaving sections (and to merges and diverges) is based on the 50th or 200th highest hourly flow on inter-urban or recreational routes, respectively, in Britain. The flows are normally estimated by (a) multiplying each 24-hour AADT flow by the appropriate factor in Table 20.3 and dividing by 24, (b) correcting the value obtained for heavy goods vehicle content and gradient, using Table 20.4, and (c) projecting the adjusted figure forward to the 15th year after the opening of the grade-separation (the design year). In Table 20.4 the average mainline gradient is established over a 1 km section composed of 0.5 km on either side of the merge or diverge nose tip; the merge connector road gradient is based on the average of the 0.5 km section before the nose tip.

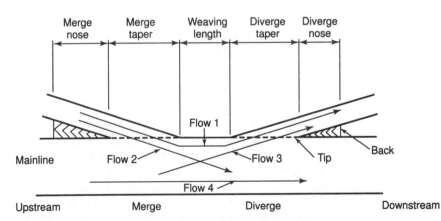

Flow 1 + Flow 4 = Non-weaving flow Q_{nw}
Greater of Flow 2 or Flow 3 = Major weaving flow Q_{w1}
Lesser of Flow 2 or Flow 3 = Minor weaving flow Q_{w2}

Fig. 20.13 Terms used in weaving calculations and design (not to scale)

Table 20.3 Peak hour flow factors[14]

Highest hour	Main urban*	Inter-urban*	Recreational inter-urban*
10th	2.837	3.231	4.400
30th	2.703	3.017	3.974
50th	2.649	2.891	3.742
100th	2.549	2.711	3.381
200th	2.424	2.501	3.024

*These factors need to be divided by 24 to give an hourly flow

Table 20.4 Percentage corrections applied to the mainline and connector design hour volumes, for varying gradients and heavy goods vehicle contents[14]

Heavy goods vehicles (%)	Mainline gradient (%)		Merge connector gradient (%)		
	<2	>2	<2	2–4	>4
5	–	+10	–	+15	+30
10	–	+15	–	+20	+35
15	–	+20	+5	+25	+40
20	+5	+25	+10	+30	+45

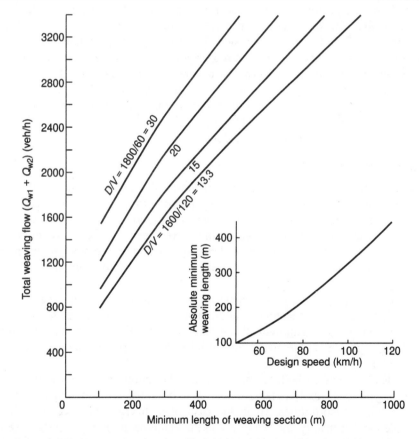

Fig. 20.14 Minimum weaving length relationships used in the design of urban motorways and all-purpose dual carriageways with speed limits less than 97 km/h[11]

Note that Equation 20.16 shows that when $L_{act} = L_{min}$ the minor weaving flow has an impact on the traffic demand equivalent to three times its numerical value, but that this is reduced if the actual weaving length is greater than the desirable minimum. Use of the formula invariably results in the addition of one or two auxiliary lanes to the mainline.

For urban motorways or all-purpose dual carriageways with speed limits less than 97 km/h, the predicted hourly design flows used in Equation 20.16 (after correcting for commercial vehicle contents and gradients) are based on the 30th highest hour flows. The maximum mainline flows are again taken as 1800 and 1600 veh/h per lane for urban motorways and all-purpose dual carriageways, respectively. In this instance, however, the total weaving design flows ($Q_{w1} + Q_{w2}$) are inserted in the large graph in Fig. 20.14 to obtain a minimum weaving length, L_{min}, for the chosen mainline lane capacity to design speed ratio (D/V) as applied upstream of the weaving section. This L_{min} is then compared with the absolute minimum length required for the design speed, determined from the smaller graph in Fig. 20.14, and the greater of the two L-values is taken as the minimum weaving length to use in the design. The width of the weaving section is then determined from Equation 20.16.

When the number of lanes has been calculated for either urban or rural conditions, a fractional part will inevitably require an engineering-based decision to round up or down. Sometimes, the position of an intersection can be varied to increase or decrease the weaving length so that the fraction converges toward a whole number of lanes. If the fractional part is large and there is a considerable weaving flow, rounding up may be favoured; however, rounding down may be more appropriate if the fraction is small and the weaving flow is low. Rounding down may also be appropriate on an urban road that is used every day by the same (commuting) drivers (and where extra land take is expensive and hard to get), as compared with a recreational route where rounding up might be preferred. The degree of reliability placed on the predicted flows is another factor which can influence the decision to round up and down.

Merges and diverges

The merge and diverge areas on a mainline carriageway should be designed so that drivers approaching a particular section of a grade-separation from upstream (i.e. either on the mainline or from a connector) are able to leave that section with the same degree of comfort as when they entered it. This comfort (and safety) criterion normally imposes a lane-balance requirement that the number of lanes upstream of the section and the number provided downstream, either on the mainline or on a connector road, not differ by more than one.

If, in the case of a merger, the joining flows are greater than single-lane capacity (and therefore justify a two-lane connector), an additional lane is normally added to the mainline as a lane gain. Also, the individual merging areas of the joining lanes are then separated by a ghost island at the merge and sufficient length provided between them to allow the mainline vehicles to adjust.

When the design flows on both the mainline and the joining connector are at near-capacity for long periods, auxiliary lanes may need to be added to the mainline to ensure continuous merging opportunities. If, in the case of a diverge, the diverging flows exceed a single lane's capacity and the flows on the mainline are high, then the diverging vehicles will need to be provided with auxiliary lanes so that they have prolonged opportunities to leave the mainline carriageway. Where queuing on the connector is in

Table 20.5 Connector road cross-sections for use with various connector design flows[11]

	Mainline road class		Connector road cross-section		
			For slip road:		For interchange link/loop merge/diverge
	All-purpose	Motorway	Merge	Diverge	
Peak corrected design flow on connector road, veh/h	0–800	0–900	Single lane 3.70 m (with 2.30 m hard shoulder)		Single lane 3.70 m with hard shoulder
	800–1200	900–1350	Single lane 3.70 m (with 3.30 m h/s)	Two 3.00 m lanes (with 1 m hardstrip	
	1200–2400	1350–2700	Two full lanes in 7.30 m standard carriageway		Two full lanes in 7.30 m standard carriageway
	2400–3200	2700–3600		Two full lanes: 7.30m	

danger of extending back onto the mainline, extended auxiliary lanes may have to be provided to ensure that the queues occur off the mainline.

British practice regarding the connector road cross-sections to use with various connector design flows is given in Table 20.5. Further details regarding the hardstrip and hard shoulder standards to be used in conjunction with Table 20.5 are available in the literature.[11]

In British practice, the *merge layout* used in the design of most grade-separated junctions (but excluding major interchanges – see reference 12) is normally obtained from Fig. 20.15(a) after inserting the projected corrected hourly design flows into the merging diagram in Fig. 20.15(b). Note that where, for reasons of continuity, the mainline capacity is in excess of the design flows, and a merging design flow of more than one lane capacity is anticipated, then layout C in Fig. 20.15(a) may be substituted for layout E; normally, however, with such a large flow expected to merge, a lane would be added to the mainline.

The procedure used to obtain an appropriate *diverge layout* is the same as that for merges except that the corrected diverge and downstream design flows are inserted into Fig. 20.16(b) to select a recommended diverge layout from Fig. 20.16(a).

Geometric design parameters used with the merging and diverging layouts in Figs 20.15 and 20.16, respectively, are shown in Table 20.6. Note that the numbers in brackets (1 to 6) that head the columns in Table 20.6(a) and (b) refer to the same dimension numbers in brackets in Figs 20.15(a) and 20.16(a), respectively.

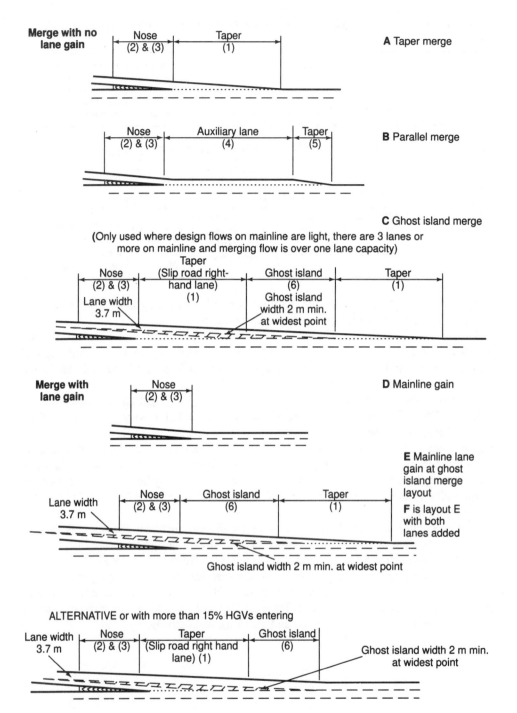

Fig. 20.15(a) Determining an appropriate merge layout at a grade-separation:[11] recommended merge lane layouts. (Note that the figures in brackets refer to columns in Table 20.6(a))

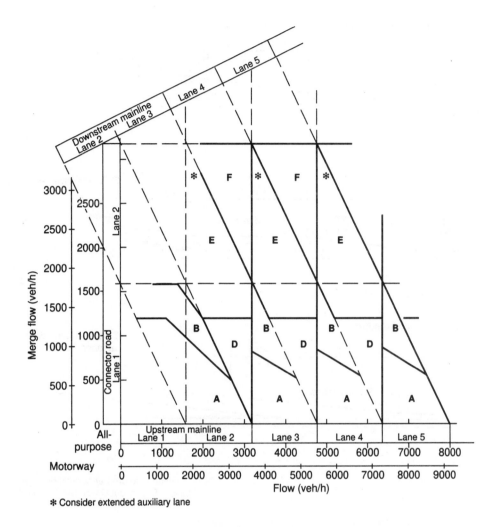

* Consider extended auxiliary lane

Fig. 20.15(b) Determining an appropriate merge layout at a grade-separation:[11] merging diagram

Diverge with no lane drop

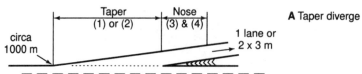

A Taper diverge

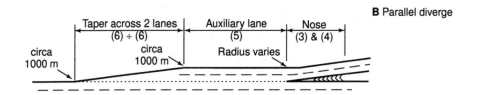

B Parallel diverge

Diverge with lane drop

C Mainline lane drop at taper diverge

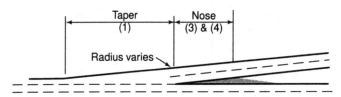

D Mainline lane drop at parallel diverge
E Layout E is layout D with two lanes off

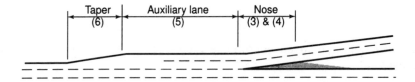

Fig. 20.16(a) Determining an appropriate diverge layout at a grade-separation:[11] recommended diverge lane layouts. (Note that the figures in brackets refer to columns in Table 20.6(b))

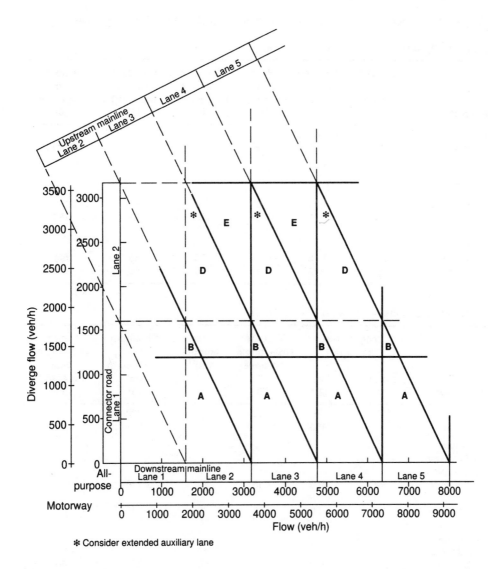

* Consider extended auxiliary lane

Fig. 20.16(b) Determining an appropriate diverge layout at a grade-separation:[11] diverging diagram

Table 20.6 Geometric design parameters for merging and diverging lanes[11]

(a) Merging lanes – see also Fig. 20.15(a)

Road class	Length of entry taper (m) (1)	Taper for min. angle at nose (2)	Nose length (m) (3)	Minimum auxiliary length (m) (4)	Length of auxiliary lane taper (m) (5)	Ghost island length (m) (6)
Rural motorway	205	1:40	115	230	75	180
Rural dual carriageway design speed:						
120 km/h	150	1:30	85	190	55	150
≤100A km/h	130	1:25	75	160	55	150
Urban road speed limit:						
96 km/h	95	1:15	50	125	40	n/a
≤80 km/h	75	1:12	40	100	40	n/a

(b) Diverging lanes – see also Fig. 20.16(a)

Road class	Length of exit taper (m)		Taper for min. angle at nose (3)	Nose length (m) (4)	Minimum auxiliary lane length (m) (5)	Length of auxiliary lane taper (m) (6)
	1 lane (1)	2 lane (2)*				
Rural motorway	170	185 (150)	1:15	80	200	75
Rural dual carriageway design speed						
120 km/h	150	150 (120)	1:15	70	170	55
≤ 100A km/h	130	130 (110)	1:15	70	150	55
Urban road speed limit:						
96 km/h	95	110 (90)	1:15	50	125	40
80 km/h or less	75	90 (75)	1:12	40	100	40

*Figures in brackets in this column refer to 2 × 3.00 m lanes

20.7 References

1. *Guide to traffic engineering practice: Part 5 – Intersections at grade*. Sydney: National Association of Australia State Road Authorities, 1988.
2. *Junctions and accesses: The layout of major/minor junctions*. Advice Note TA20/84. London: Department of Transport, 1984.
3. Kimber, R.M., and Coombe, R.D., *The traffic capacity of major/minor priority junctions*. Report SR582. Crowthorne, Berks: Transport Research Laboratory, 1980.
4. *Junctions and accesses: Determination of size of roundabouts and major/minor junctions*. Advice Note TA23/81. London: Department of Transport, 1981.
5. Marlow, M., *Conversion of rural and semi-rural major/minor T-junction to offside priority*. Report LR883. Crowthorne, Berks: Transport Research Laboratory, 1979.
6. *The geometric design of roundabouts*. TD16/93. London: Department of Transport, 1993.
7. *Junction layout for control by traffic signals*. Advice Note TA/18/81. London: Department of Transport, 1981. (Reprinted February 1989.)
8. *Pedestrian facilities at traffic signal installations*. Advice Note TA15/81. London: Department of Transport, 1981. (Reprinted March 1990.)
9. Kimber, R.M., McDonald, M., and Hounsell, N.B., *The prediction of saturation flows for road junctions controlled by traffic signals*. Research Report 67. Crowthorne, Berks: Transport Research Laboratory, 1986.
10. *Layout of grade separated junctions*. TA48/92. London: HMSO, 1992.
11. *Layout of grade separated junctions*. TD22/92. London; HMSO, 1994.
12. *The design of major interchanges*. TD39/94. London: HMSO, 1994.
13. McDonald, M., Hounsell, N.B., and Kimber, R.M., *Geometric delay at non-signalised intersections*. Report SR 810. Crowthorne, Berks: Transport Research Laboratory, 1984.
14. *Traffic flows and carriageway width assessment*. TD20/85. London: Department of Transport, 1985.

CHAPTER 21

Introduction to computer-aided design of junctions and highways

G.R. Leake

21.1 Role of computer-aided design

The rapid development of computer packages in recent years has transformed the highway and junction design process. No longer do large numbers of repetitive calculations have to be laboriously carried out by hand. Instead increasingly powerful and compact computers, together with increasingly sophisticated software packages, which now include computer graphics, make it possible to investigate a greater number of design options, and to undertake the necessary design calculations more reliably and in a much shorter period of time.

In recent years it has become necessary to include a comprehensive environmental assessment of any major highway alignment or junction layout scheme as part of the design process. An important element in such an assessment is the visual impact which will be created. The development of computer graphics has been an important input into such assessments, not only because it has replaced the 'artist's impression', which tended to be somewhat idealistic, but also because it enables the visual impact of a scheme to be studied from a large number of different positions ranging from the viewpoint of a driver to that of a resident living adjacent to the proposed scheme. Furthermore, the development of computer graphics programs can be of considerable help to decision-makers since a picture or diagram can often replace many words and numerical tables, and get over a message or impact in a much more satisfactory and clear-cut way.

Computer-aided design (CAD) packages have a number of important roles to play.

1. Because they enable repetitive calculations to be carried out quickly and accurately, the designer can investigate a wide range of possible solutions and determine their consequences.
2. It follows from the above that any modifications to a particular scheme involving recalculations can also be easily and quickly carried out.
3. Computer packages are capable of storing considerable volumes of design data. These can be readily recalled if necessary. Furthermore, if any design modifications are carried out, the new information goes into the memory and replaces the original information. Three advantages follow from this:

(i) the information in the memory store is always up to date
(ii) information on various aspects of a design such as horizontal and vertical align-
     ment calculations, drainage alignment details including inlet and manhole
     locations, lighting column positions, details of public utility services runs and
     depths are stored, and so any potential design conflicts can be identified by retriev-
     ing and plotting from appropriate data sets (e.g. a lighting column coincident with
     a drainage pipe, or inadequate clearance under a bridge after adjusting the road
     vertical alignment)
(iii) all the information can be stored centrally, but accessed from remote terminals,
     including any on the construction site.
4. The visual appearance of a scheme can be determined from a wide range of viewing
   positions. Normally the proposed scheme is set against the developmental background in order that an assessment can be made of the visual impact (if any) on the
   local area. It is also possible to assess the extent to which landscaping and planting
   will reduce the visual intrusion.

In the UK a number of analytical programs have been produced for junctions including those by the Transport Research Laboratory (TRL) such as PICADY[1] – priority
junctions; ARCADY[2] – roundabouts; and OSCADY[3] – traffic signals. These programs
enable the designer to determine alternative junction layout dimensions and methods of
control capable of dealing with the anticipated vehicle and pedestrian flows. For each
alternative, the program produces estimates of total vehicle delay and the anticipated
accident rates. From this information, and other data such as the estimated cost of implementing each alternative, the designer can choose the 'best' solution. However, it should
be noted that the above packages are restricted to analysis only and do not include
detailed design and graphics facilities.

Many commercial design packages exist for determining the best alignment for a
length of road. MOSS[4] is one of the best known of these packages in the UK, and also
has the capability to deal with junction design. An early version of such packages was
NOAH, developed in the 1970s by the TRL.[5, 6]

21.2 What is CAD?

When using any design package it is important to realise its limitations. For example,
although the computer can carry out calculations rapidly and accurately, it can only do
these in the way specified within the program. It cannot be creative nor can it usually
take decisions except in very simple situations. What it can do, however, is to provide
information as an input into the decision-making process. This information will often
include not only numerical comparisons between alternative schemes, such as capital
cost, but also subjective comparisons, such as the extent of visual intrusion.

In all computer applications it is important to know precisely the analytical basis of
the design program, and its limitations. It should be remembered that the onus of inter-
preting any outputs always lies with the designer using the program.

When using any CAD packages it is important to ensure that all the input information is soundly based and up to date. This will range from information on site conditions
such as locations and types of property, existing accesses, topographic information and
existing road and junction layout details (since these impose possible constraints on

what alternative proposals might be developed) to the design year traffic flow predictions. If the input data is inaccurate or unreliable, then the outputs will also be poor, and this can lead to sub-optimal layouts being produced and adopted. The old adage still holds good in respect of any CAD analysis: 'rubbish in, rubbish out'.

A CAD system will normally consist of the four main components listed below.[7]

1. *Input devices.* These are the means by which the operator inputs instructions and/or data into the computer, and include keyboards and data disks.
2. *Processor.* This is the 'black box' containing the computer logic which uses the software program for undertaking the design calculations.
3. *Data storage.* This may be contained on a hard disk within the computer, or on floppy disks which can be stored separately. The stored data will include direct inputs, such as existing road levels, as well as design calculations.
4. *Output devices.* These include VDU screens, printers and plotters and enable the outputs to be scanned (VDU screen) or reproduced as a permanent copy (printer, plotter).

A single computer can either operate in isolation or be linked to other computers via a modem or the normal telecommunications network. These linkages can be within the same office, or with outside locations. This enables information and drawings to be transmitted quickly without the necessity of physically moving large amounts of paper.

21.3 Data input requirements

Highway design CAD programs and packages can be considered under two broad headings – junction design and road alignment design. Some CAD packages deal only with one or other of the above design areas, and for clarity and convenience the data input requirements of these two groups will be covered separately. However, it should be noted that more comprehensive packages, such as MOSS[4] (Section 21.4.2), embrace both.

21.3.1 Junction design programs

Although the details of the data input required will vary between programs, and these should be consulted,[1, 2, 3] the following basic information will always be necessary:

- number of entry arms at the junction
- start and finish times of each modelled run
- design year traffic flow data, including turning movements and the proportion of heavy vehicles for each modelled time period
- pedestrian flows across the approach arms for each modelled time period
- geometric details for each alternative layout proposal to be investigated. The precise requirements will depend on the type of junction (priority, roundabout, traffic signals) and can be found in references 1, 2, and 3. To be included, however, will be the number of traffic lanes on each approach arm, lane widths, turning radii, gradient details and visibility distances.

21.3.2 Road alignment design packages

Although the precise data requirements will vary between commercial packages, the broad data input requirements are essentially the same. Hence the data input requirements of the NOAH program, the details of which are available in published documents,[5, 6] will be used to indicate what these broad data requirements are. These requirements are summarised below.

1. Digital ground model – this gives the ground heights at the nodes of a 50 m square grid covering the geographic area of interest.
2. Details of the road cross-section.
3. Full range of engineering unit costs covering items such as earthworks, road pavement, drainage and culvert materials, and land costs.
4. Traffic flow data along the various sections of the new road together with turning traffic flows and through traffic information at all the junctions during the first year of operation, broken down by vehicle type.
5. A set of design speed and geometric design standards appropriate to the type of road being considered. This should also include information on minimum acceptable values (e.g. minimum gradients to ensure satisfactory drainage, maximum gradients to ensure satisfactory operating speeds by heavy goods vehicles).
6. An initial road alignment in a form acceptable to the program, against which the first generated alignment can be compared.

Most of the remaining data required is input in the form of strings, where a string is a sequence of points, often defining a polygon, that approximates to the feature described.

7. Prohibited area strings – these specify areas within which the road may not encroach.
8. Cost area strings – these define areas where the cost is proportional to the land area taken.
9. Cost enclosure strings – where a fixed charge is made whenever part of the area is taken (an example would be an enclosure around an electricity pylon where the cost would be that of relocating the pylon).
10. Level control strings – these specify upper and lower bound levels for the proposed new road where it crosses the string.
11. Culvert strings – these indicate the flow area, flood level and bed level of the stream to be crossed. A suitable culvert is then automatically designed.
12. Bridge strings – these specify the feature to be crossed (road, river, railway) and whether the crossing is to be an overbridge or underbridge. A control on the bridge level is automatically generated to ensure adequate clearance.
13. Junction strings – these specify where junctions will be located along the new road.

If an optimal alignment is to be found, then it is necessary to input into the program an appropriate objective function which is to be minimised. In NOAH a simple 'monetary' function was used of the following form:

$$\text{Objective function} = \text{traffic cost} + \text{construction cost} \qquad (21.1)$$

where traffic cost = sum of the drivers' costs over 30 years discounted to the year of construction.

21.4 Outputs from CAD programs

21.4.1 Junction design programs

The three UK programs which have been highlighted as typical junction design analysis programs (PICADY, ARCADY and OSCADY) have somewhat different outputs. For clarity and brevity, the output from the roundabout program, ARCADY,[2] will be described as being typical of the types of output to be obtained.

The most important output relates to the calculated capacity of each entry arm of the junction over the modelled time period. This is illustrated in the printout in Fig. 21.1 and shows the typical ARCADY output for one time period.

	Queue and delay information for each 15-min time segment						
Time	Demand (veh/min)	Capacity (veh/min)	Demand/ capacity (RFC)	Pedestrian flow (peds/min)	Start queue (vehs)	End queue (vehs)	Delay (veh min/ time segment)
07.00–07.15							
Arm A	9.00	21.35	0.422		0.0	0.7	10.5
Arm B	9.60	27.63	0.347	3.0	0.0	0.5	7.8
Arm C	8.20	14.92	0.550	1.7	0.0	1.2	16.9
Arm D	8.10	34.08	0.238		0.0	0.3	4.6

	Effect on capacity (pcu/min) of marginal changes in:			
Marginal change	Entry width (.1 m)	Flare length (m)	Inscribed circ diam (m)	Entry angle (10°)
Arm A	0.227	0.105	0.021	−0.870
Arm B	0.287	0.053	0.027	−1.142
Arm C	0.294	0.034	0.022	−0.623
Arm D	0.343	0.027	0.028	−1.341

Fig. 21.1 ARCADY output for a typical 15-minute time period[2]

The ratio of the entry flow to the entry capacity (RFC) for each entry arm is of prime importance. Practically this should be ≤ 0.85 if long queues are not to form on an approach.

Information is also available on the anticipated queue length on each arm at the beginning and end of each time period. Fig. 21.1 shows, as is to be expected since it has the highest RFC value, that Arm C has the longest anticipated vehicle queue.

A third important output shown in Fig. 21.1 is the total vehicle delay (veh.min) occurring on each arm during each time period. This is vital information when comparing alternative layouts, since one of the major objectives when seeking the optimal design layout is to minimise the total delay at the junction.

A fourth important output from ARCADY, and one which is useful to the designer when modifying the layout when searching for the optimal solution, is the effect on the entry capacity (which the designer is attempting to maximise within the physical constraints of the site) of marginal changes in particular layout parameters. For the situation illustrated, the Fig. 21.1 output shows the changes in entry capacity (pcu/min) for each of the entry arms resulting from marginal changes in entry width, entry flare length, inscribed circle diameter, and entry angle. With the exception of the entry angle, increases in the parameter increase the entry capacity, with the entry width being the most important variable, since significant increases in the entry capacity can be obtained from relatively small increases in entry width.

Finally, an important part of any economic assessment of alternative junction layouts is the future safety record. ARCADY predicts future accident levels, as shown in the printout in Fig. 21.2. This shows not only the anticipated vehicle and accident frequencies on each entry arm and for the junction as a whole, but also likely changes in these accident frequencies resulting from marginal changes of the more important layout design parameters. For fuller details of the outputs from ARCADY, as well as those from PICADY and OSCADY, the reader is referred to the original references.[1, 2, 3] Visual outputs, which do not form part of the above analytical programs, form part of the MOSS system covered in the next section.

Accidents output

	Accident frequencies (/yr)	
Arm	Vehicle	Pedestrian
A	0.41	0.06
B	0.70	0.05
C	0.88	0.09
D	0.47	0.03
Total	2.47	0.22

	Effect on vehicle accident frequencies (/yr) of marginal changes in:				
Marginal change	Entry flow (000/24 h)	Circulating flow (000/24 h)	Centre island diameter (m)	Entry path curvature (.01/m)	Approach curvature (.001/m)
A	0.047	0.021	0.000	−0.066	−0.002
B	0.086	0.024	0.000	−0.029	−0.008
C	0.097	0.038	0.000	−0.060	−0.007
D	0.060	0.013	0.000	−0.001	−0.009

	Effect on pedestrian accident frequencies (/yr) of marginal changes in:		
Marginal change	Pedestrian flow (000/24 h)	Entry flow (000/24 h)	Circulating flow (000/24 h)
A	0.029	0.002	0.002
B	0.122	0.002	0.002
C	0.088	0.002	0.002
D	0.058	0.001	0.001

Fig. 21.2 Vehicle and accident frequencies predicted by ARCADY[2]

21.4.2 Road alignment design packages – MOSS

MOSS[4] is an interactive graphics system, and is probably the most widely used highway alignment and junction design program in the UK. The system includes facilities such as data capture, analysis, design, contract preparation, drafting and visualisation, and can be used for the preliminary and detailed design of a wide range of design projects, such as highway alignments, at-grade junctions and multi-level interchanges, road widenings, estate road layouts, and environmental screening and landscaping proposals.

Designers working on road alignment and junction design projects are always faced with a number of constraints, such as the need to avoid existing properties; adhering to specified design standards; ensuring that important aesthetic and environmental requirements are maintained; minimising construction costs; and maintaining stated construction time schedules. Interactive design programs, such as MOSS, enable designers to develop alternative proposals that satisfy specified site constraints, to evaluate alternatives rapidly, and to present conclusions in a visual form which is easily understood by non-engineering decision-takers, such as local authority councillors. Furthermore, at any stage of the design process plans, sections, profiles and contours can be automatically generated. Similarly, when the design has been finalised, areas, volumes, setting out data and contract drawings can be easily produced. Standard alignment drawings and junction plans, produced by MOSS, are shown in Figs 21.3 and 21.4.

Visualisation is now an important feature of junction and highway alignment design. Interactive graphic design, an important feature of systems like MOSS, enables continual visual checking of designs, as they are developed, to be carried out. With the increasing

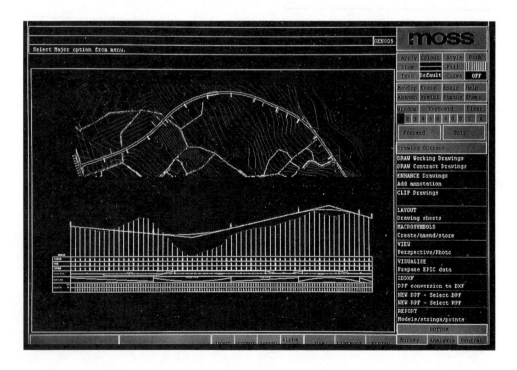

Fig. 21.3 Typical horizontal and vertical alignment details of a road section using MOSS[4]

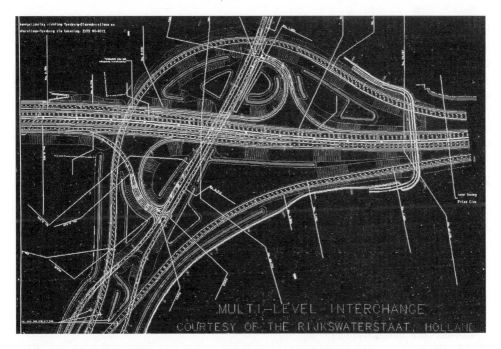

Fig. 21.4 Plan layout of a multi-level junction using MOSS[4]

Fig. 21.5 Perspective view of a local estate road using MOSS[4]

interest in the environmental consequences of road schemes, computer-generated perspective views are an invaluable aid to decision-makers. Perspective views can be produced in colour, with shading, surface texturing and realistically modelled objects such as bridges, lighting columns, traffic signs and planting. Such treatment introduces a level of realism to the generated perspective previously unachievable. In addition animation techniques can now be used to simulate vehicle and pedestrian movements and thereby add a further degree of realism to the process. A typical perspective view produced by MOSS is shown in Fig. 21.5.

21.5 References

1. *PICADY 3 user guide*. Crowthorne, Berks: Transport Research Laboratory, 1992.
2. *ARCADY 3 user guide*. Crowthorne, Berks: Transport Research Laboratory, 1993.
3. *OSCADY 3 user guide*. Crowthorne, Berks: Transport Research Laboratory, 1993.
4. *MOSS – highway design*. Croydon: MOSS Systems Ltd.
5. Broughton, J., *Design of a by-pass using horizontal alignment optimisation program NOAH*. LR 925. Crowthorne, Berks: Transport Research Laboratory, 1980.
6. Davies, H.E.H., and Broughton, J., *Horizontal alignment optimisation: Test of program NOAH on a motorway scheme*. LR 894. Crowthorne, Berks: Transport Research Laboratory, 1979.
7. MacPherson, G., *Highway and transportation engineering and planning*. Harlow, England: Longman, 1993.

CHAPTER 22

Design of off-street parking facilities

C.A. O'Flaherty

Parking space is required to store vehicles when they are not in use. All motor vehicle trips terminate eventually at some form of parking facility so that drivers can go about their business. Business development is promoted by the availability of convenient parking and discouraged by a lack of parking.

Prior to the 1950s parking in Britain was generally provided for by utilising spare roadway space and, where available, the backyards and alleyways associated with buildings. Up until then it was only in the central areas of the largest cities that it was normally considered necessary to associate the provision of off-street parking with the initiation of new developments. The construction of a formal parking space at a residence was usually done at the behest of the home owner and not of the local authority. Nowadays, however, the demand for parking space is so great and the detrimental effects of on-street parking in terms of road congestion and accidents are so significant, especially in built-up areas, that all local authorities have strong policies and local ordinances relating to the provision of parking spaces.

The purpose of this chapter is to discuss factors relating to the use of parking standards, and to the location and design (including capacity) of off-street car parking facilities.

22.1 Car parking standards

Local authorities now normally specify the number of parking spaces to be provided with a new development when granting planning permission. In some instances this will be the *maximum* number of parking spaces that is permitted, in other cases it will be the *minimum* number that must be provided. Central London, for example, is provided with high-quality rail and bus systems and its roads are normally so congested that most people use public transport to access it (e.g. see reference 1); thus Central London is a classic example of a central-area locale where a limit would be placed on the number of parking spaces permitted, because any additional parking provision would simply encourage more car usage and make the roads more congested. The cores of the central areas of large cities are other examples of locations where maximum parking limits might also be imposed. By contrast, the outer reaches of large towns and smaller towns with good access roads and poor public transport services would typically have strict requirements regarding the minimum number of spaces that must be provided with a new development. The data in

Table 22.1, which are extracted from parking standards specified by Lancashire County Council,[2] are some examples of minimum requirements for stand-alone developments; they illustrate the differences in the method of specification, number and type of spaces required by different activities. Different local authorities often have different parking standards; further, their application in a given area may be affected by local mitigating circumstances.

In Table 22.1 the term *operational parking* is used to refer to the essential parking spaces needed for vehicles that are regularly and necessarily involved in the operation of the new development; these spaces include commercial vehicles servicing the buildings. *Non-operational parking* refers to the spaces provided for vehicles that do not necessarily have to park or wait precisely at the premises in question, e.g. spaces for commuting employees as well as for shoppers, business callers and tourists. (Note: Only

Table 22.1 Car parking standards for various land uses (extracted from reference 2)

Land use	Operational parking	Non-operational parking
1a Family housing	None	Residents – 1 garage space per house Visitors – 1 space per house
1b Single person housing	None	Residents – 3 spaces per 4 dwellings Visitors – 2 spaces per 4 dwellings
1c Homes for the elderly	None	Resident staff – 1 garage space Day staff – 1 space per 3 staff Visitors – 1 space per 6 residents
2 Shops	Minimum space for loading/unloading (m²): GFA* Space ≤ 500 50 ≤ 1000 100 ≤ 2000 150 For each extra 500 m² GFA, add 50 m² space	Staff – 1 space per 100 m² or 1 space per managerial staff + 1 space per 4 other staff Customers – 'normal' shops, 1 space per 25 m²; carpet etc. warehouses, 1 space per 20 m²; DIY superstores and garden centres, 1 space per 15 m² (including external sales area); superstores > 2000 m², 1 space per 10 m²
3 Banks	25 m² (minimum)	Staff – 1 space per 3 staff Customers – 1 space per 10 m² of net public floor space of banking hall
4 Offices	Minimum space for loading/unloading (m²): GFA Space ≤ 500 50 ≤ 1000 100 ≤ 2000 150 ≤ 3000 200 For each extra 1000 m² GFA, add 25 m² space	Staff – 1 space per 25 m² GFA or 1 space per 3 staff Visitors – 10% of staff need
5 Industrial premises/warehouses	Minimum space for loading/unloading (m²): GFA Space ≤ 100 70 ≤ 235 140 ≤ 500 170 ≤ 1000 200 ≤ 2000 300 for each extra 1000 m² GFA, add 50 m² space	Staff – 1 space per 50 m² GFA (industrial premises)/1 space per 200 m² GFA (warehouses) visitors – 10% of staff need (industrial premises only)
6 Petrol stations	5 spaces if station has an automatic car wash	Staff – 1 space per 2 employees at busiest time

Table 22.1 (continued)

Land use	Operational parking	Non-operational parking
7 Hotels, motels and public houses	Minimum space for loading/unloading (m²): *GFA* — *Space* ≤ 500 — 140 ≤ 1000 — 170 ≤ 2000 — 200 For each extra 1000 m² GFA, add 25 m² space	Resident staff – as for (1a) Non-resident staff – 1 space per 3 staff at the busiest time Resident guests – 1 space per bedroom Bar customers – 1 space per 4 m² of net public space in bars plus 1 space per 8 m² in beer garden
8 Restaurants and cafés	Minimum space for loading/unloading (m²): †*DFA* — *Space* ≤ 100 — 50 ≤ 250 — 75 ≤ 500 — 100	Resident staff – as for (1a) Non-resident staff – 1 space per 3 staff at the busiest time Diners – 1 space per 2 seats or 4 m² in the dining area
9 Dance halls and discotheques	50 m² (minimum)	Staff – 1 space per 3 staff at the busiest time Performers – 3 spaces Patrons – 1 space per 10 m² of net public floor space
10 Cinemas	50 m² (minimum) + 2 spaces at main entrance to set down/pick up patrons	Staff – 1 space per 3 staff at the busiest time Patrons – 1 space per 4 seats
11 Bingo halls	50 m² (minimum) + 2 spaces at main entrance to set down/pick up patrons	Staff – 1 space per 3 staff at the busiest time Patrons – 1 space per 10 seats
12 Swimming baths	50 m² (minimum)	Staff – 1 space per 2 staff normally present Patrons – 1 space per 10 m² of pool area
13 Places of worship	50 m² (minimum) + 2 spaces at main entrance to set down/pick up worshippers	Worshippers – 1 space per 10 seats
14 Museums/public art galleries	50 m² (minimum)	Staff – 1 space per 2 staff normally present Visitors – 1 space per 30 m² of public display space
15 Hospitals	Minimum space for loading/unloading (m²): *GFA* — *Space* ≤ 1000 — 200 ≤ 2000 — 300 ≤ 4000 — 400 ≤ 6000 — 500 For each extra 1000 m² GFA, add 50 m² space	Staff – 1 space per doctor/surgeon + 1 space per 3 other staff Outpatients/visitors – 1 space per 3 beds
16 Doctor's surgeries/clinics/health centres	1 space per doctor (minimum)	Other staff – 1 space per 2 other staff at the busiest time Patients – 2 spaces per consulting room
17 Schools	30 m² (minimum) – primary/nursery schools 50 m² (minimum) – secondary schools/sixth form colleges	Staff – 3 spaces per 4 staff normally present Visitors – primary/nursery schools, 2 spaces; secondary schools (a) ≤ 1000 pupils, 4 spaces, (b) > 1000 pupils, 8 spaces; sixth form and tertiary colleges, 1 space per 10 students

*GFA = gross floor area †DFA = dining floor area

operational parking would normally be specified for new buildings in the cores of most central areas in large towns in Britain today.)

Table 22.1 also illustrates that the land use characteristics are the determinant of the unit used to specify the parking provision. For example, the dwelling unit is normally the basis for residential requirements (except in the case of multi-family apartment buildings when the number of bedrooms per dwelling unit might be the basis). Office building and retail shopping requirements are usually specified on a gross floor area basis even though, in the case of shopping, there is no relationship between shopper concentration and employment that relates to all types of shopping. Seating availability is normally the basis for public assembly areas such as cinemas, theatres and restaurants.

Parking standards vary considerably from land use to land use. As noted previously they can also vary from locale to locale within a country, depending upon such factors as local planning policy, development density, and the quality of the public transport system. Similar land uses in similar locales in different countries can also use different parking standards, depending upon the economic conditions and degree of motorisation experienced as well as the degree of sophistication exhibited at governmental level in respect of transport planning and processes.

Parking generation rates are not static and parking standards should therefore not be regarded as fixed indefinitely. For example, over the past 20 years there has been a dramatic decline in the average length of patient stay at hospitals and a shift to outpatient treatment; this has resulted in an increased parking demand at hospitals. The advent of the computer and better working conditions have resulted in a significant increase in the floorspace per employee provided in office buildings and this affects parking standards expressed on a floor area basis. These and many other changes in the work force, in transport, in leisure activities, etc., emphasise the importance of reviewing parking standards on a regular basis.

22.2 Locating off-street car parking facilities

Ideally, off-street parking for a new development should be directly associated with the buildings which it is intended to serve. In some instances the desired/required on-site parking may not be possible because of lack of space (i.e. the overall space need, including access and manoeuvring space, per parked car in a surface car park is typically 20–25m²), and it is then necessary to locate the parking facility elsewhere. In such instances, the site selected should be within an acceptable walking distance of the development(s). Current American experience[3] is that this walking distance should not exceed 90 m to 180 m (see also Table 8.1 for walking distances acceptable to elderly people and those with disabilities).

Sometimes circumstances may make it possible for a parking facility to be shared between two developments so that each has its designated number of spaces and the costs are shared. In other instances, a smaller common site may be shared by two developments which have non-concurrent parking demands; for instance, banks, professional offices and shops tend to have their peak parking demands during the day whereas theatres, bars and discotheques have theirs at night.

The location of the principal parking generators (i.e. the main destinations of the travellers after parking) is probably the major criterion influencing the siting of a public car park. If it is to thrive, then the car park must be sited conveniently for the parkers it is intended to serve. As a general rule, the larger the town the further people are willing to walk from their cars to their destinations, and the acceptable walking distance generally correlates with the trip purpose and length of time parked (see for example Table 7.1); however, parkers will not walk long distances to or from car parks in the wet or if their walk is through distressed areas where they may feel unsafe. In some instances the preliminary investigations may indicate that a large proportion of the users of the car park are likely to come from a particular part of the urban area; in this case the car park should be located on the side of the central area that is toward their origins. Not only will this location be more attractive to the potential patrons of the facility but it will also have the added benefit of causing less traffic congestion within the town centre.

In large cities a peripheral car park involved in a park-and-ride scheme (see also Chapter 7) should be sited so as to be attractive to its potential users, e.g. upstream of major road congestion and adjacent to a major public transport route on the side of the urban area relating to the origins of the motorists being targeted. A park-and-ride car park should not be too close to large daytime generators of parking demand as it might then be utilised by local all-day parkers, thus obviating its special purpose. Smaller park-and-ride car parks are often located on existing paved land which is well serviced by existing access roads, e.g. at stadium, shopping centre or church car parks whose 'normal' peak activities occur during evenings or weekends.

The location and size of a car park should always be related to the capability of the surrounding street system to handle safely and efficiently the entering and leaving vehicles. The road access problem is one of the main reasons why it is generally held to be more desirable to have several car parks with lower numbers of spaces strategically located throughout a central area rather than a few of very high static capacity, unless they are constructed in conjunction with special access roads. Ideally, off-street car parks should be sited so that their entrances and exits are on lightly trafficked (preferably one-way) streets leading to high-capacity roads. They should be sited at mid-block locations rather than corner ones, and no exit or entrance should be within 50 m of a street junction so as to avoid interference with or from the intersection control.

A sloping hill site may be appropriate for a parking garage as it may allow direct access/egress at different levels without the need for ramps. However, such car parks require extra walking effort from their customers, and thus may be shunned by older or unfit people and shoppers who have to carry packages.

The locations of car parks should be compatible with adjacent land uses; they should not obviously and adversely impact upon nearby environments. Sites should be fairly flat and well drained so as to minimise construction costs.

22.3 The design-car concept

The first step in the functional design of an off-street parking facility is to determine the vehicle to be used for design purposes.

Car sizes and capabilities vary from country to country, and from manufacturer to manufacturer within a country. To cope with the size variations, it is normal practice to design for at least one hypothetical standard car which reflects the space requirements for the majority of cars in a given country. Particular parking facilities may also have some spaces with critical dimensions capable of providing for the largest cars. Odd-shaped sites may have 'left-over' spaces marked for small cars.

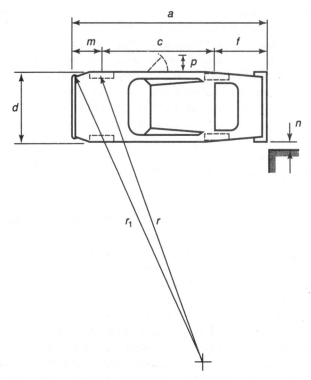

Fig. 22.1 Critical design-car dimensions used in off-street car park design in Britain (to be read with Table 22.2)

Table 22.2 Critical design-car dimensions used in off-street car park design in Britain (to be read with Fig. 22.1)

Dimension	Symbol (as Fig. 22.1)	Dimension value (m)		
		Small car	Standard car	Large car
Length	a	4.100	4.750	6.100
Width	d	1.600	1.800	2.000
Height	–	1.500	1.700	2.200*
Door opening clearance				
– normal	p	0.500	0.500	†
– minimum	p	0.400	0.400	†
Wheelbase (worst cases)	m	†	0.900	†
	c	†	2.900	3.700
	f	†	1.100	†
Nearside clearance				
– normal	n	0.200	0.200	†
– minimum (piers etc.)	n	0.150	0.150	†
Turning radius				
– kerb	r	5.500	6.500	7.500
– wall	r_1	†	7.000	†
Ground clearance		0.130	0.100	0.050*

* Specialised vehicles; † Generalised dimensions not available

Table 22.2 and Fig. 22.1 show some critical dimensions used in Britain that were originally published in 1971.[4] Since then car dimensions have changed; but investigations from time to time have indicated that the changes are not sufficient to justify altering these design standards.[5] Table 22.2 includes turning capabilities, which influence both aisle width and corner radii; to minimise these radii, it is usual to assume that cars travel at slow speeds within the car park. Ground clearance and end clearance are important in relation to ramp design and entrances and exits, if grade changes which cause a vehicle's ends or centre to ground are to be avoided.

It is likely that about 95 per cent of the cars in Britain have dimensions within those for the standard car, and essentially all within those for the large car.

22.4 Surface car parks

Most parking facilities are surface car parks, mainly because of the ease with which they can be quickly and economically designed and constructed when suitable sites become available. Surface car parks have no ramps, stairs, lifts, structural columns or roofs; consequently, they are more space efficient, driver sightlines are normally clearer, users feel more secure when walking to/from their cars, there are no ventilation or tall vehicle problems, and more generous layout dimensions can often be used, as compared with multi-storey structures.

The safe and effective operation of an off-street car park is primarily dependent upon its layout. Generally, the functional design seeks to maximise vehicle capacity and ease of manoeuvrability and circulation as well as pedestrian safety and convenience. The most effective operational layout is normally determined by site-specific conditions although, as a general rule, rectangular sites are most easily adapted to meet these criteria.

The type of operation, i.e. whether self-parking or attendant-parking, is also important. Self-parking, which is most common, means that each car is driven only by the parker and the car can be locked when left, delays in retrieving parked vehicles are minimised, and insurance costs are lower for the car park operator. The parking area required per vehicle is usually more with a self-parking layout than with an attendant-parking one, especially when the car park is heavily used by short-stay shoppers (who normally require wider spacing between parked cars). Attendants may also double-park cars in the circulation aisles to increase the static capacity of the car park; this would not be permitted in a self-parking facility.

Whether a fee is to be charged and, if so, the collection method to be employed also affects the functional design of a car park.

It is useful to describe the factors affecting the functional design of a car park in a step-by-step operational context. However, since a surface car park is obviously similar to one floor of a multi-storey facility, many of the considerations discussed here also apply to a parking garage.

22.4.1 Entrances

Traffic volumes on the access street(s) and the closeness of the car park to busy intersections may limit the locations available for use as entrances. Entrances should be far back from a busy intersection so that queuing vehicles do not interfere with the intersection operation or control. Entrances on one-way streets should be upstream of exits so that entering queues of cars do not block the exits. If the access street is two-way, it may be necessary to provide a central right-turning lane in the street where entering vehicles can be stored until gaps appear in the opposing traffic stream.

A right-hand turn (in Britain) from a one-way street places drivers on the inside of the turn, thus providing them with better visibility and better judgement as to car placement as they enter the car park.

Cars entering off-street car parks should not have to cross footpaths that are heavily used by passing pedestrians.

The capacity of an entrance at a car park is determined by the angle of entry, the type of control used, and the freedom of internal circulation. If the users are 'regulars' and know the car park operation well, the entrance capacity is usually increased. In general, the greater the entrance width (measured at the throat limit where kerbs are constructed) and radius, the easier it is for motorists to access the car park. Entrances with sharp turns have lower capacities than straight-in approaches.

Whether access to a car park is controlled and if so the form of control used has a considerable effect on entrance capacity. Control can be exercised by simply locking the entrances (and exits) with padlocked chains that are removed at certain times, or by using barrier systems which can be operated by passcards or an attendant. If the car park is fee-paying various methods of operation are used. (The following are in addition to those used to control on-street parking, e.g. meters/vouchers/cards (see Chapter 24) which can also be used to control off-street parking usage.)

Pay-on-entry/free-exit operation usually requires the payment of a flat fee to a cashier or into a coin-operated machine, and is best used at facilities which lend themselves to this form of charging, e.g. at special events, all-day parking, or car parks which may be

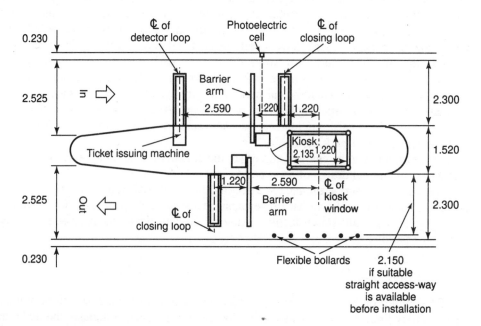

Fig. 22.2 Typical layout of a combined entry/exit system at a car park.[5] (Note: All dimensions are in metres)

unattended when the driver returns. This form of operation normally requires a reservoir area prior to an entrance to handle surges in car arrivals and avoid queues on the servicing street.

Free-entry/pay-on-exit operation is the most usual way of collecting fees at fixed- and variable-charge surface car parks (and multi-storey or underground parking garages). With a variable-charge operation the driver collects a time ticket through the car window from an automatic or push-button ticket issuing machine at the entrance; the ticket is presented to a car park supervisor through the car window when exiting, and the requested fee is then paid for the time parked. Exit cashiering reduces (but does not eliminate) the reservoir requirement at the entrance; however, it increases the reservoir space requirement within the car park prior to the exit.

Free-entry/pay-before-exit operation also requires drivers to collect a time-ticket from a dispenser on entering the car park. However, before returning to their cars at the end of their business they divert to a convenient central cashier point in the car park and pay there for the time parked; the drivers then enter their cars and leave the facility after (momentarily) presenting some form of receipt for manual or, more usually, machine inspection at the exit. As no cash transaction takes place at either the entrance or the exit, this method of fee collection results in a most efficient entrance/exit operations and minimises potential reservoir requirements.

Table 22.3 gives maximum flows measured at entrances to and exits from controlled car parks by the Transport Research Laboratory.[6]

Table 22.3 Suggested maximum input and output flows at car park entrances and exits for use in design[6]

Operation condition	Maximum flow (cars/h)
Entrance lane	
– Tight left-hand turn and take ticket	350–450+
– Straight approach and take ticket	650–670
– Tight left-hand turn only (no ticket taken)	575–970
Exit lane	
– Pay variable fee	180–200
– Pay fixed fee	270

Another common method of control used at fee-paying car parks is *pay-and-display*. This involves the driver purchasing a parking voucher for the desired length of parking time from a ticket dispensing machine immediately after parking; the voucher is then displayed on the windscreen where it can be inspected by a patrolling warden. This method of operation is very efficient in that no barrier arms or fee payments are required at the entrances or exits. However, to be successful it requires stringent monitoring and this can be staff-expensive (see also Chapter 24).

Entrances and exits at small car parks should be located together if the facility is fee-paying, as one central supervisor can then monitor both entering and exiting vehicles (see Fig. 22.2).

If a car park is operated only with attendant parking, the 'entrance capacity' is directly related to the number of attendants. It is reported[7] that American studies determined rates of acceptance of 8–16 cars/h per attendant (and delivery rates of 13–17) at central area attendant-operated car parks.

22.4.2 Traffic circulation

Traffic circulation within a car park should favour incoming vehicles over exiting cars or service vehicles. Circulation can be either one-way or two-way, depending upon the size and shape of the facility and the parking angle. Figure 22.3 shows some alternative circulation movements. Generally, one-way circulation with vehicles driving directly into parking bays (or 'stalls') will ensure the most efficient traffic flow, with the minimum number of conflict points, when the parking angle is less than 90 degrees. Two-way traffic circulation is usually best with 90-degree parking.

The most desirable internal circulation arrangement enables incoming motorists to drive in a continuous movement past all potentially available/vacant parking bays. It should also enable them to recirculate easily as necessary; thus dead-end aisles should be avoided. For safety reasons, the layout should discourage speeding by having shorter rather than longer aisles; thus, aisles should never exceed about 100 m in length in large car parks.

Very large surface car parks, e.g. at regional shopping centres, may have circulation roadways (on which no parking is permitted) within their confines. These roads, which may be one- or two-way, are typically located at the site perimeter and are connected with the circulation aisles.

Pedestrian circulation and passenger pick-up/loading should be given priority over vehicle circulation on roads that are immediately adjacent to building frontages, for safety reasons.

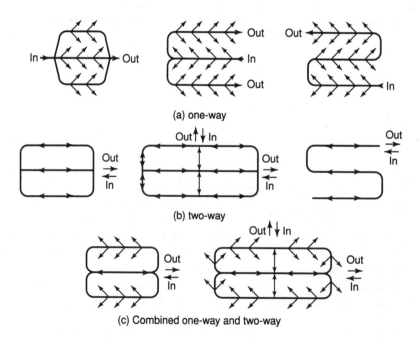

Fig. 22.3 Alternative circulation patterns in a car park: (a) one-way, (b) two-way, and (c) combined one- and two-way

22.4.3 Parking stalls and aisles

When developing the layout of a car park it is usual for the designer to work with 'stalls' (or 'bays'), aisles which provide access to the stalls, and stall and aisle combinations called modules (or 'bins'). A parking module is the clear width provided for the parking of vehicles; thus, it may be composed of either one-way or two-way service aisles with parking stalls on either or both sides. The module width (dimensions *m* in Fig. 22.4) is influenced by the boundary conditions, i.e. whether there are walls on each side, stalls on each side, or a wall on one side and a parking stall on the other. Aisles with stalls on both sides are termed 'double-loaded' aisles; those with stalls on one side only are 'single-loaded' aisles. Single-loaded aisles are inefficient in respect of space usage per vehicle and are avoided in car park layouts where possible.

An aisle should be sufficiently wide to allow a car to park/unpark in one manoeuvre. The design width varies mainly with the angle of parking and with the stall width. (The effective aisle width is obviously also related to stall length.) Generally, 90-degree parking stalls with two-way aisles parallel to the long dimensions of the site, and 60- to 80-degree interlocking stalls with alternating one-way aisles, are the most efficient in terms of car park space usage.

Other advantages of 90-degree parking are: there is greater freedom of traffic circulation and (in large car parks) it is not necessary to have a circulation road adjacent to a building, thus minimising the potential for pedestrian–vehicle conflicts; the aisle width

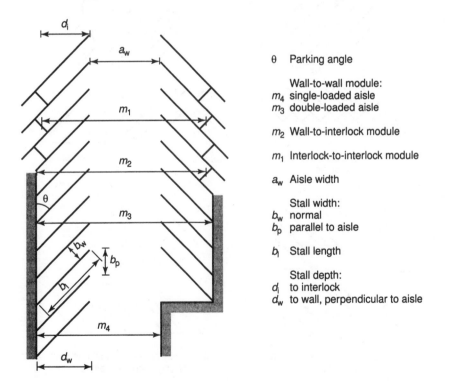

Fig. 22.4 Basic dimensional elements of a functional parking design (adapted from reference 3)

facilitates pedestrian movement and the passing of stopped cars by other vehicles; travel distances are minimised for cars seeking vacant stalls and for cars unparking and exiting the car park; and one-way direction signs are not necessary.

Disadvantages of right-angle parking are that a 90 degree turn is required to park/unpark (which is difficult for some people) and there is the potential for head-on and sideswipe collisions in the aisles.

Advantages of angle parking are: parking/unparking is easier for most drivers, and the delays to other vehicles are minimised; the one-way operation ensures that head-on and sideswipe accidents are avoided, while the number of potential conflict points at intersections are minimised; rear doors can be opened without interference from adjacent cars (particularly important in shopping centres); the flatter the parking angle the less the aisle width required; and angled stalls may be appropriate where the site dimensions do not allow an integral number of 90-degree parking spaces.

Disadvantages of angle parking tend to be the converse of the advantages cited previously for 90-degree parking. Also, a very flat angle layout is less efficient in rectangular car parks as it usually results in space wastage.

Figure 22.5 shows some possible layouts associated with interlocking parking stalls. Figures 22.5(a) and (c) are conventional arrangements; wheel stops are normally provided at the 'butt' ends of the spaces to ensure that vehicles do not encroach into

(a) one-way aisles, butt stalls, 60° angle	(b) one-way aisles, drive-through stalls, 60° angle
(c) two-way aisles, butt stalls, 60° angle	(d) one-way aisles, herringbone stall arrangement, 45° angle

Fig. 22.5 Some interlocking parking arrangements

opposing parking stalls or collide with each other when parking. The drive-through arrangement at Fig. 22.5(b), whereby parkers back into a stall, or drive into opposing spaces when parking and then exit into the aisle next to the one used for entering, is normally only used in special event or large industrial commuter car parks; this pull-through arrangement expedites the exit flow of cars during peak periods. The herringbone (or 'nest') arrangement at Fig. 22.5(d), which is used only with 45-degree parking, is very economical in space utilisation; it is now rarely used in practice, however (particularly in shopper parking facilities) as the head-to-side parking has the potential for significant collisions between adjacent vehicles even though wheel-stoppers are used.

Table 22.4 Some aisle and module widths used with 4.75 m long parking bays

Parking angle (deg)	Aisle width (m)	Module width (m) for stall widths of	
		2.4 m	2.5 m
45 (1-way)	3.600	13.712	13.853
60	4.200	14.827	14.927
90	6.000	15.500	15.500
90 (2-way)	6.950	16.450	16.450

Table 22.4 shows some aisle and module widths recommended[5] for use in Britain. The stall width (measured at right-angles to the stall line-markings) is affected by the clearance needed for motorists to get in and out of their cars; this clearance can be dependent upon the agility of the vehicle occupants as well as on the trip purpose. Extra-wide stalls are desirable for shoppers with packages at convenience grocery stores or in locations adjacent to walls or pillars, whereas narrow ones may be acceptable for commuters to industrial or office developments located on high-cost land or in attendant-parking facilities. Widths actually used vary, with 2.3 m, 2.4 m and 2.5 m wide stalls often being used in long-stay, general, and short-stay car parks, respectively.

Wider stalls are used for physically handicapped motorists; 3.6 m is the preferred minimum width for wheelchair users. Such stalls are normally located on level ground close to building entrances and exits (where ramps and lifts are also normally located), and each is identified with special signing and marking. Barrier-free pathways should be available between these spaces and the building entrances/exits, and all kerbs should be provided with ramps mountable by wheelchairs.

The number of wide stalls provided for the handicapped is most usually between 1 and 3 per cent of the total number of car park spaces, except at car parks frequented by people with disabilities, e.g. at hospitals, when a greater proportion (say 3–4 per cent) may be required. The initial number provided may need to be adjusted subsequently, based on experience.

A parking stall should be able to accommodate nearly all cars expected to use the car park. A bay length of 4.75 m is commonly used to cater for the standard design car (although 5.0 m has also been used).

22.4.3 Exits

Vehicle pathways to the nearest exits of large car parks should be signposted so that vehicles can leave quickly and efficiently. In large car parks also, exits should preferably be separated from entrances so that the potential for vehicle–vehicle and vehicle–pedestrian conflict, both inside and outside the parking facility, is minimised. In some car parks subject to tidal-flow traffic movements during the morning and evening peaks and which use a combined exit plus entrance, it may be possible to use a three-lane entrance-exit design with a reversible centre lane, depending upon the control/fee-payment method employed and the capacity of the service street.

As with entrances, exit design should take into account width, turning radii, angle of approach to the egress street (90 degrees is most common), direction(s) of traffic flow, and distance to the next intersection on the street. Street traffic is always given priority over exiting car park vehicles; in large car parks Stop signs may be required to ensure this and, in extreme instances, traffic lights. Traffic islands and signs at exits can be used to ensure that wrong turns are not taken into one-way streets. Adequate sight distances should always be provided at exits from car parks onto streets.

Measured capacities of exits operating under a lifting barrier/pay-from-car type of fee control indicates that this type of exit operation can handle about 200 cars per hour per lane when variable fees are paid, and up to 270 cars per hour per lane when fixed fees are paid (Table 22.3). If the exit is uncontrolled and at 90 degrees to the street its optimum capacity can be determined as for a conventional T-intersection (see Chapter 20); however, the capacity so determined may have to be reduced if there is interference from pedestrians on a footpath past the exit.

Although disliked by some designers, step-down kerbs at the footpath–exit interface have the merit of warning pedestrians of the conflict area and may prevent a serious pedestrian–vehicle accident.

If the traffic on the service street is very heavy it may be necessary to provide storage space after the exit barrier and prior to the street, where cars can wait until acceptable gaps appear in the traffic stream; if a reservoir space is not provided and it is needed, then the exit may not be able to achieve its maximum output. A detailed procedure for estimating the reservoir sizes required at car park exits is available in the literature.[8] In practice, however, a reservoir of 2–3 car lengths is usually adequate at exits from most car parks.

22.4.4 Pedestrian considerations

Ideally, pedestrian entrances, exits and walkways should be separated from travelways used by vehicles, and sited so as to reflect pedestrian desirelines through the car parks. The routes used by pedestrians after leaving their cars are to the entrances of the facilities they wish to use, e.g. shops and offices. If the parking facilities are large and the pedestrian flows are great the establishment of formal walkways may be appropriate; if possible these paths should be raised above the carriageways as this allows pedestrians to scan the car park, get their bearings more easily, and see and be seen by drivers of moving vehicles. If the walkways cannot be raised they should be clearly defined by a change of surface material and/or line markings. Widths vary but pathways about 2.4 m wide allow two people to walk abreast, while permitting car bumper overhang.

In practice, people walking to and from their cars commonly use access aisles. In general, safety is improved if the parking layout encourages pedestrians to walk along the aisles rather than across them.

Car parks should be well lit, well signed, well drained, kept clean, and attractively landscaped.

Good lighting – which implies that pedestrians and vehicles are visible at all times, and that there are no stalls in dark corners – is essential if night-time usage is desired; it assures patrons regarding their personal safety and reduces vandalism and pilfering, as well as enabling drivers to manoeuvre their cars safely into and out of parking stalls. British practice is to have illumination levels of 5–20 lux in surface car parks. High-mounted floodlighting can be effective in large surface car parks; however, care must then be taken to ensure that the spillage of light into adjacent properties is limited to that which is appropriate for the environment. Lighting columns should not be sited where they are vulnerable to being hit by parking vehicles; preferably, they should be located about the outside perimeter of the car park, and in kerbed traffic islands at the ends of rows of parking. Lighting columns should never be sited on lines between adjacent (parallel) stalls or where they might interfere with the ease of pedestrian movement.

Public car parks are usually paved with bitumen or concrete. Paved surfaces are preferred by patrons, especially when on foot; they also ensure a more efficient car park operation, e.g. through the use of aisle and stall markings. Further, there is no dust in warm weather, cleaning (and, in cold climates, snow removal) is easier, and good drainage is facilitated.

Clear signposting and, in large car parks, the coding of aisles and the identification of sub-areas by numbers, letters or pictorial signs help pedestrians to find their cars on returning from shopping, work, etc.

Nowadays large car parks are often divided into smaller compartmental areas with the aid of landscaping. This breaks up the large paved area and makes the parking facility more attractive and, usually, a better fit with the surrounding physical environment. Landscape shrubs, hedges, and mounds should generally be of low height (*ca* 1.25 m maximum) and able to withstand concentrated fumes from car exhausts. Trees chosen for their shading capabilities in sunlight should not be of the varieties that drop sap that will damage car surfaces or great quantities of blossom or pod that will clog drains. Sight distances should be protected against shrubbery blockage, particularly at aisle intersections and at entrances and exits. (Reference 9 provides useful advice regarding plantings suitable for use in outdoor car parks in Britain.)

22.5   Off-street commercial vehicle parking

As with off-street car parks, the layout of a commercial vehicle parking facility is dependent upon the vehicles using it. However, unlike car parks, it is not practicable to design layouts for heavy commercial vehicles on the basis of a standard vehicle. Rather, the design vehicle(s) selected for any given facility will vary according to local knowledge of the vehicles normally expected to use the site.

Generally, however, rigid vehicles tend to be used for deliveries (single and multi-drop) in urban areas and therefore tend to predominate at shopping centres and the like. Articulated vehicles tend to be used for the longer inter-city trips and are often seen at motorway rest areas as well as at trunk lorry parks and the depots of long-distance hauliers.

Heavy goods vehicle lengths vary according to whether the vehicle is rigid or articulated, as well as according to manufacturer and intended function. Typically, overall lengths range from about 8 to 11 m for rigid vehicles, and from 12 to 15.5 m for articulated vehicles. Thus the minimum stall length, as would be used in a 90-degree parking layout, should be the vehicle length plus 0.5 m.

Many heavy goods vehicles have widths in excess of 2 m. As the maximum vehicle width normally permitted in Britain is 2.5 m (for refrigerated vehicles 2.58 m), it is appropriate to design layouts for a 2.5 m wide vehicle. On this basis a minimum stall width of 3.3 m is sufficient to allow for wing mirrors and the opening of the cab door, although in practice a stall width of 3.5 m is often used.

The aisle width varies according to the angle of parking and vehicles' turning capabilities. Experience suggests that single row, 45-degree parking with one-way aisles gives an effective layout for most parking facilities. Single-row parking allows a vehicle to make a left-hand turn from the one-way aisle into the parking stall (which is at 45 degrees to the aisle); when unparking the vehicle then exits the stall by simply driving forward into a parallel one-way same-direction aisle. (With this arrangement, of course, all traffic circulation takes place on a perimeter road.)

Table 22.5 shows dimensions appropriate for use with a 45-degree layout. Stall and aisle dimensions for different lengths of vehicle (excluding trailers drawn by rigid or articulated vehicles) and parking angles are available in the literature;[10] these data enable a layout to be chosen which best suits the size and shape of the site available and the vehicles expected to use it.

Table 22.5 Recommended dimensions for use in a one-way 45-degree lorry park[10]

Vehicle length (m)	*Stall length (m)	Stall width (m)	Aisle width (m)
8	11	3.3	6.8
9	12	3.3	7.1
10	13	3.3	7.7
11	14	3.3	7.9
12	15	3.3	7.9
13	16	3.3	8.6
14	17	3.3	9.5
15	18	3.3	10.8

* Stall length = vehicle length + vehicle width (for 45-degree parking only) + 0.5 m

A particular parking challenge arises at large activity centres which generate significant numbers of commercial vehicle visits (e.g. deliveries, garbage collection) as part of their normal operations. For example, at the central Milton Keynes Shopping Building (see Table 22.6) most commercial vehicle arrivals (> 80 per cent) occur during the normal

Table 22.6 Commercial vehicle trips generated and median parking times at central Milton Keynes Shopping Building in June[11]

	SIC group description*									
	A	B	C	D	E	F	G	H	I	J
Generation (veh/h/100 m²)	1.222	0.734	0.264	0.980	0.439	3.104	0.407	0.673	1.295	2.201
Length of stay (minutes)	22	10	16	25	15	21	13	11	17	111

*A = food; B = confectioners, etc.; C = clothing, footwear and leather goods; D = household goods; E = other; F = market traders; G = mixed retail; H = service; I = catering; J = licensed trade

working day and, during the peak morning periods, these vehicles often arrive at 20-second intervals or less.[11]

The parking/docking arrangements for commercial vehicles at office blocks, hospitals, hotels, regional shopping centres, etc., should be screened from those for the parking of cars. The access points and internal circulation routes used by commercial vehicles should also be separate, and they should not interfere with car and pedestrian movements. Loading areas 3.5 m wide by 16.0 m and 12.0 m long are normally the minimum required for articulated and rigid vehicles, respectively. If deliveries/pick-ups take place under cover, provision will need to be made for clearance of the maximum height of vehicle expected (normally 4.2 m in Britain).

As rear unloading/loading is more efficient than side loading, the commercial vehicle circulation and loading systems should be designed for right-side driver viewing (in Britain) when backing a vehicle into a docking area, prior to unloading/loading.

22.6 Types of multi-storey car park

Most multi-storey car parks are located where land costs are so high, e.g. in or about the central areas of large towns, that it is more economical to build vertical facilities rather than purchase additional land for surface parking. Multi-storey facilities can be single-purpose, i.e. used only for parking, or multi-purpose, i.e. containing other uses (e.g. offices or shopping) within the same structure; however, the more expensive the land the greater the likelihood of a parking facility being designed for multi-usage purposes.

At some locations the geography of the land, e.g. at a hillside site, may make it practical to build a multi-storey car park and allow direct entry to a number of parking levels. In some cases there is no need to connect the various levels and each can be designed as a surface car park; in others, the need for internal vehicle circulation may make ramp connections necessary.

There are so many types of multi-storey car park that it is difficult to classify them. Basically, however, any parking garage can be described by a combination of three groupings: by vertical location, i.e. whether the parking facility is under or above ground; by the interfloor travel system, i.e. whether vehicles travel between floors via ramps, sloping floors or mechanical lifts; and by the person who parks the vehicle, i.e. whether vehicles are privately parked or attendant-parked.

Structurally, multi-storey facilities can also be divided according to whether they are of long- or short-span construction. Operationally, long-span construction is preferable because (a) it is more adaptable to future changes in car design, (b) it allows more parking stalls per floor area, (c) there are fewer columns, so that parking/unparking takes place more quickly with less damage to cars, (d) visibility is improved for both drivers and pedestrians, and (e) drivers prefer column-free parking. The main disadvantage of a long-span facility is its higher cost.

22.6.1 Underground car parks

The main benefits of an underground car park are the conservation of expensive surface land, the preservation of the visual amenity (especially important in historic locales), and the capability of sharing construction costs with non-parking (often above-ground) usages. Because of the last criterion many underground parking facilities are part of multi-purpose facilities. The main disadvantage of underground parking is its high construction cost, i.e. typically up to twice the cost per car space as in an equivalent above-ground multi-storey facility.

Major contributors to the extra expense associated with underground car parks include the high costs of:

- extensive excavation (magnified if rock is encountered)
- the relocation of public utilities
- concrete retaining walls on all sides
- the installation of ongoing dewatering systems (very costly in ground subject to a high water table)
- the need for a major load-bearing and waterproof 'roof' slab
- artificial lighting (to near daylight quality)
- artificial ventilation (to remove dangerous exhaust fumes and add air for circulation)
- special fire-fighting provisions.

These and other factors generally constrain the maximum sizes of underground car parks to three or four levels.

With an underground car park there is a special operational design need for the provision of directional signs and markings. Drivers and pedestrians get disoriented when underground and need clear directions to guide them to the appropriate exit when leaving the car park. Other than this the development of the layout for an underground facility is similar to that for an above-ground one.

22.6.2 Ramped garages

Ramps are used to allow vehicles to travel between adjacent floors in multi-storey (above- or below-ground) car parks. The ramps may be straight and/or curved. They may be separated from the access aisles or fully or partially integrated with them; clearway ramp systems provide interfloor travel paths that are entirely separated from conflicting parking/unparking manoeuvres, but adjacent parking ramp systems require drivers to share aisles with parking/unparking cars. In practice, large multi-storey car parks often have a combination of ramp systems, with adjacent parking ramp systems being used by entering vehicles and clearway ones by exiting vehicles.

Clearway ramp systems

These systems provide drivers with the safest interfloor movement with the least delay, and lend themselves to speedy entrance to/egress from large parking facilities heavily used by, for example, commuters. However, they are rarely used in multi-storey car parks on small sites as the ramps would require too great a proportion of the available space.

Figure 22.6 shows (in schematic form) some ramp systems used to enable vehicles to circulate between the floors of multi-storey car parks. Figure 22.6(a) illustrates a clearway-type opposed straight ramp system. The ends of the one-way ramps are on the ground floor, facing in opposite directions. This system lends itself to a fairly narrow site where vehicles can enter from and exit onto different, parallel, streets.

Figures 22.6(b) and (c) illustrate two-way concentric clearway ramps internal and external to the parking structure, respectively. These ramps are usually helically curved (spiral) to save space and provide gentler grades. Concentric spiral ramps are most commonly used, however, to provide for one-way express egress from the various parking floors. Vehicles can travel either clockwise or anti-clockwise on these one-way systems; a clockwise movement is generally preferred in Britain (and in other countries where drivers sit on the right of the vehicle) as it places drivers on the inside of turns, thereby enabling better and safer vehicle control. When two-way traffic is handled on a single spiral, the outer lane is normally used for up movements as it has a larger radius and a more gentle grade; the up and down movements are usually clockwise and anti-clockwise, respectively.

Adjacent parking ramp systems

These types of system are more susceptible to accidents and delays than clearway ramp systems. Nonetheless they are often preferred by self-parking patrons, particularly when they are entering the multi-storey car park. Adjacent parking ramp systems also lend themselves to a more efficient use of space on smaller sites.

Figures 22.6(d) and (e) show adjacent parking straight ramp systems with the up and down movements separated and non-separated, respectively. These systems usually have the ramp wells (on one side only) along the structure's longer side as this results in gentler ramp grades. Vehicles using these systems follow an elliptical path, most of which is two-way on a level (full) floor surface. The basic straight ramp system in Fig. 22.6(e) has an adjacent entrance and exit on the same street.

Alternatively, in small rectangular high-cost sites the adjacent parking ramps may lead to split-level floors. With this design the multi-storey car park is essentially constructed in two sections, with the floors in one section staggered vertically by a half-storey from those in the other section, and short straight ramps are used to connect the half-floors. The split-level car park can also be designed so that the floors overlap, i.e. so that the bonnets of the cars below are tucked in under the rear ends of the cars above; this design increases the space efficiency (provided vans are not expected to use the facility) and may render useable otherwise unusable narrow sites. The ramp operation may be one-way with two-way level aisle movements (Fig. 22.6(f)), or two-way with two-way level aisle movements (Fig. 22.6(g)) or one-way aisle movements (Fig. 22.6(h)).

Figure 22.6(i) is an interesting example of an adjacent parking ramped facility; it is a basic sloping-floor system under two-way circulation, with level end cross-aisles. There are no separate narrow ramps with this design; rather the floors on which the cars are parked are themselves sloped so that each long aisle forms a gradual ramp leading to the next floor. If two sloping-floor units are placed end to end, it is possible for the multi-storey garage to have one-way traffic flow operation with 90-degree or angled parking along the long aisles at every level (Fig. 22.6(j)); with this design the traffic flow can change from up to down, and vice versa, in the level centre section where the two units meet.

Because of the lack of interconnecting narrow ramps, a sloping-floor facility can provide more parking stalls than a comparable split-level design. However, as the vehicles are parked across the slope of the floor, the floor gradient must be restricted (usually to 3–5 per cent) to allow for easy parking and pedestrian walking; hence, these facilities tend to suit long sites. The natural slope of the land normally does not take up all the waste space created at ground level by the long sloping floor; it is usually desirable, therefore, to have a partially sunk basement section of parking so as to get the full use of the site.

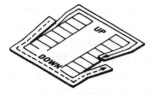

(a) Clearway system: one-way, straight ramps

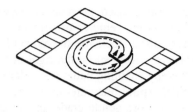

(b) Clearway system: two-way, curved, internal ramps

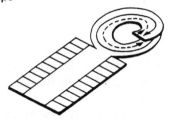

(c) Clearway system: two-way, curved, external ramps

(d) Adjacent parking system: one-way ramps

(e) Adjacent parking system: two-way ramps

(f) Adjacent parking system: split-level floors, two-way aisles, one-way ramps

(g) Adjacent parking system: split-level floors, two-way aisles, two-way ramps

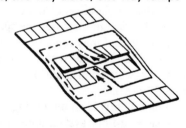

(h) Adjacent parking system: split-level floors, one-way aisles, two-way ramps

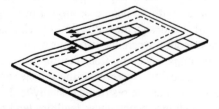

(i) Adjacent parking system: sloping floors, two-way level cross aisles

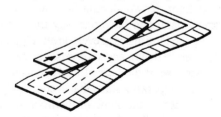

(j) Adjacent parking system: sloping floors, one-way level cross-aisles

Fig. 22.6 Diagrammatic examples of ramped multi-storey car parks

Self-parkers generally like sloping-floor facilities because they do not have to negotiate narrow ramps. However, they become confused and displeased if they have to turn through more than six continuous circuits.[3] Parking is facilitated by the use of adjacent parking; however, congestion can occur at the lower levels of large facilities during peak egress periods if clearway spiral ramps are not provided.

The long floors of a multi-storey sloping-floor car park give an undesirable effect in external elevation. To overcome this aesthetic effect, a construction has been developed which uses warped (hyperbolic paraboloid) floors to give a horizontal external line to the building, while containing the internal circulation features of a split-level facility in combination with the efficiency of a sloping-floor parking system.

22.6.3 Attendant-parking ramped facilities

With attendant-parking multi-storey (and surface) car parks the motorist drives the car to a reservoir adjacent to the entrance, leaves the key in the ignition, and an attendant then drives the car to a parking bay. The reverse operation takes place when the motorist returns to retrieve the car.

The big advantage of an attendant-parking operation is that lesser layout standards can be used by the professional parkers; this results in a reduction in the space required per car as compared with that of a facility in which the parking stalls, aisle widths, ramp grades, and turning radii are designed to cater for the varied capabilities of self-parking patrons. The headroom provided can also be lower for attendant-parking facilities. Thus, the initial capital cost per parking bay for an attendant-parking facility is significantly less than that for a self-parking one, although the cost of labour to operate the facility is higher. The decision as to whether to construct an attendant-parking structure is then dependent on whether the extra labour costs are offset by the additional income from the extra number of parking stalls.

Good attendant-parking operation encourages the parking activity to be spread throughout the multi-storey facility so as to reduce the possibility of collisions. Long-stay cars are often placed in rear rows of double-parking or in stalls adjacent to ramps, so as to leave the most accessible spaces for short-term higher-turnover parking. However, the key operating area in an attendant-parking facility is the entrance reservoir; its function is to absorb excess traffic during peak inbound flows, when cars arrive at a rate greater than the attendants can store them. If it is assumed that

$$A = 60N/t \qquad (22.1)$$

where A = average storage rate (cars/h), N = number of attendants on duty, and t = time required by an attendant to make a round trip (minutes), it can be seen that an inadequate reservoir will require the employment of extra attendants if congestion from traffic backup and delays are to be avoided.

22.6.4 Attendant-parking mechanical garages

With these multi-storey car parks the vehicles do not use their own power to enter the parking stalls, lifts take the place of ramps, the headroom required is for cars rather than people (which enables more cars to be parked per unit height of building), and automation is employed either wholly or partly to park the vehicles. Because of their space efficiency mechanical garages have tended to be constructed on high-cost land with limited space availability.

Modern mechanical garages have electronic controls and fast reliable lifts so that a vehicle can be stored or delivered in less than one minute. The lift systems can move laterally, as well as vertically, so that each tier of parking bays can be reached by more than one lift, thus minimising the problems of individual lift breakdown. Some garages also have counter-balanced lifts which automatically rise, without power, when empty and descend, under automatic braking control, when loaded; this enables parking bays to be discharged in an electrical power emergency (previously a cause of much customer frustration and bad public relations).

Mechanically operated car parks are expensive and have high maintenance costs. They also have difficulty in coping with surges of inbound (unless large reservoirs are provided) and outbound traffic. The cost factors, the teething problems and bad publicity associated with the installation of the mechanical and electrical equipment, congestion problems caused by cars queuing in service streets in locales subject to peak parking demands, and general aesthetic considerations, have all combined to ensure that very few new mechanical garages are now constructed.

22.7 Self-parking multi-storey car parks: some design considerations

Many of the design factors previously discussed in relation to off-street surface car parks also apply to multi-storey facilities, and so they will not be repeated here. The following additional material is of relevance to the design of multi-storey (above- and below-ground) car parks.

Notwithstanding the many types of ramped, self-parking multi-storey car park there is no one type that can be said to be superior to others under all circumstances. The choice at any given location is normally governed by site shape and dimensions, access and egress requirements, parker characteristics, and cost considerations.

Maximum ramp grades are mainly limited by safety considerations, and by the psychological effect on drivers, with the braking/hill-climbing abilities of cars being of secondary importance. Excessively steep ramps hinder traffic movement and, where open to the weather, can be dangerous when wet. In Britain the recommended[5] maximum gradient for a straight ramp with a rise of up to 1.5 m is 1 in 7; for a longer ramp with a rise greater than 1.5 m it is reduced to 10 per cent. For helical ramps the gradient along the centre line should not exceed 10 per cent if the rise is not greater than 3 m, or 1 in 12 if the rise is more than 3 m.

A change in surface planes occurs where a ramp joins a parking floor. If the change of 'breakover' angle is too sudden the ends or centres of cars may scrape the floor. The rule of thumb used to avoid this is to apply a transition grade of half the ramp gradient over a minimum blending distance of at least 3 m at both the top and bottom of the ramp. The intersection lines should also be rounded.

The preferred minimum widths between confining kerbs on one-way straight and curved ramps are 3.0 m and 3.65 m, respectively; the kerbs on either side should be at least 0.3 m wide by 0.1 m high. A median kerb (usually 0.5 m wide) is essential on two-way curved ramps to confine cars to their proper lane. End sections of a straight ramp, where a car starts or completes a turn, are normally flared (by about 0.6 m over the length of the vertical transition gradient) to allow the rear wheels to turn to a straight line. The preferred minimum radius of the outside kerb of a curved ramp is 9 m, although the design car can cope with a 7.5 m radius.

Superelevation of up to 10% may be incorporated into the surfacings of curved ramps to make turning easier.

The minimum vertical clearance or headroom at entrances and exits, at ramps, and under beams and projecting fixtures such as signs, lighting, sprinklers, etc., is normally 2.05 m. If it is desired to accommodate taller vehicles, e.g. caravans, a greater clearance (say 2.3 m) is provided on the ground floor and at the entrances and exits.

If columns are used, there should be at least three parking stalls between adjacent columns. Columns should not be closer than 0.8–1.0 m to an aisle; if set closer the practical effect is to reduce the stall width for entering vehicles.

A clearance of about 0.230 m should be provided if a parking stall is adjacent to a wall.

Cul-de-sac aisles should be avoided. If they are necessary the capacity of a cul-de-sac should normally not exceed six stalls.

Ramp structures should be as open as possible to provide good sight distances and reduce closed-in impressions. However, perimeter walls flanking ramps and parking floors should be designed to limit a driver's view of the surroundings outside the car park, as this can detract from concentration on driving; this is especially important when travelling on parking levels higher than nearby buildings.

Rain, ice and snow can create hazardous driving conditions on exposed ramps (and circulation routes) in multi-storey car parks in colder climatic zones. Where justified this problem can be overcome by using embedded electric cables to warm the surface of the travelway. The heating is automatically activated by control probes set in the carriageway which energise the cables when an appropriate combination of surface temperature and moisture is sensed.

Because they are generally open to the elements, as well as being subject to water carried in by cars, the floors of car parks need to be waterproofed and sloped to provide falls to gully outlets. The most effective waterproofing system is one that gets the water into drains quickly. A gradient of about 2 per cent is normally considered appropriate, but up to 5 per cent is used.

The adequacy of the reservoir storage areas provided at entrance and exit points is key to the successful design of a self-parking multi-storey car park, as otherwise cars will back up into the surrounding streets, or within the facility, during peak periods. A simple rule is that a reservoir should hold at least two cars and preferably more. Standard queuing and probability theory formulas are available[3, 12] to provide more exact estimates of reservoir requirements, dependent upon arrival and handling rates at different types of entrances.

Multi-storey car parks should be well lit to enable the safe movement of pedestrians and cars and to facilitate the locating of parking stalls. Desirable standards of lighting are given in Table 22.7. Note that higher lighting levels are required at entrances and exits to compensate the driver for the lighting change between the outside and inside of the building.

Good ventilation is essential in a multi-storey car park to minimise any damage to health that might arise from excessive concentrations of exhaust gases, and to avoid the risk of fire and explosion arising from inflammable or explosive vapours. As many multi-storey parking facilities are open-sided this is not often a great problem; generally, it is assumed that a ventilation problem does not exist if an open space equivalent to 2.5 per cent of the total floor area is provided on each long elevation of the car park. The problem is perhaps most critical in underground car parks where, generally, it is desirable

Table 22.7 Lighting levels suggested for self-parking multi-storey car parks

Area	Illumination level (lux)
Parking bays	30 (at floor level)
Access aisles	50
Ramps	50 (at vertical walls)
Entrances and exit	150 (at floor level)

to install alarm equipment (especially at locations where cars have to queue) which will automatically start up the ventilation plant when, for example, the carbon monoxide level reaches a predetermined amount.

Lifts should be provided in all multi-storey car parks that have three or more floors above ground level. They are preferably installed in banks of two or more, as close as possible to the principal pedestrian generators serviced by the parking facility. In large car parks the lifts should also be located within about 100 m of all parking bays.

Lifts can be raised/lowered either hydraulically or electrically. Hydraulic lifts are generally limited to heights of about 15 m; above five floors (faster) electric lifts are always used. As a rule-of-thumb two lifts are normally provided in a multi-storey car park with up to 600 stalls, plus one extra lift for every additional 600 spaces or substantial parts thereof.[3]

22.8 Fee-collection control and audit

Early items to consider when designing a public off-street car parking facility are the fee-collection method to use, and the audit system to employ to ensure that the fees are collected and passed on to the car park operator. Decisions taken at this stage affect the location and number of entrances and exits, as well as the financial viability of the car park.

Parking thefts are a not uncommon occurrence where proper control procedures are not implemented. To help overcome illegal activities the basic steps given below have been recommended[13] as essential to the successful financial operation of a car park.

- There should be an accurate count made of all vehicles entering and leaving the car park. No patron, executive or floor staff member should be able to enter or leave the parking facility by vehicle without being counted.
- No vehicle should be allowed to enter the car park without the driver taking a ticket that can be surrendered when leaving. A record of the entry should be available in the cashier booth, or a positive vehicle registration count made with a recording device.
- All time clocks used in the control system should be operated accurately and synchronised with each other. They should be of a type designed to ensure that an unauthorised person cannot reset them.
- Some kind of validating device, cash register or fee calculator should be used to record all exit transactions. Tickets should be stamped with appropriate information so that each cashier-attendant can be held accountable for his or her transactions.
- Ticket-issuing machines should be operated so that it is impossible for any car to be given more than one ticket. Additional supplies of tickets should be kept in a secure place.

- Parking barrier-arms or similar preventative devices should be used to stop cars from entering through an exit or exiting through an entrance.
- A display of the fee charged should be visible to all car parkers.

Vehicle detectors, e.g. inductive loop detectors (see Chapter 13), are commonly used in car parks to register the presence of a vehicle in order to count it, to send an impulse to a ticket dispenser, and/or to activate an entrance or exit gate. In large car parks differential counters located at entrances and exits normally keep a running score of the number of vehicles inside at any given time; thus when all the parking bays are nearly full, the car park can be closed until more spaces become available. It is better to advise potential parkers that the facility is closed than allow them to enter, circulate, and cause internal congestion.

Tickets issued by ticket dispensing machines are, most commonly, made of paper and stamped with the time and date of entry and, in the case of a large car park, the entrance of issue. With computer-operated control systems these data are usually encoded on magnetic-sensitive tickets. Whatever the type of ticket used, the subsequent action is the same: before the car exits the car park the ticket is inserted into a time clock which marks the exit time and date, after which the length of time parked is calculated, the fee owed is displayed, and the fee is paid by the parker into a machine or to a cashier-attendant.

The parking charge is usually shown in an illuminated display at the exit booth that is connected to a recording device such as a cash register. Whatever data are punched into the register, or read by the computer, are then automatically visible to the driver as a confirmation of the charge being levied.

A parking 'gate' (see Fig. 22.2 and reference 14) consists of a metal cabinet containing electrical equipment, a belt or drive system, and a 2–3 m long barrier arm which prevents cars from making unauthorised movements into/out of the car park. The removal of a ticket from the ticket issuing machine causes the arm to rise so that a vehicle is able to enter the facility; as the car passes a detector beyond the arm another signal causes the gate to close. Of course, in the case of a fixed-charge automated car park, the entrance (or exit) is controlled by a barrier arm which is automatically raised when the fee (or short-life token) is inserted into the appropriate machine.

Off-street car parks are often designed to share usage between short-stay parkers who pay cash and long-stay parkers who pay on a monthly, quarterly or annual basis. Ideally, the long-stay contract parkers are provided with separate access and egress gates; alternatively the normal entry and exit points are designed to cater for their needs. In either case care must be taken to ensure that a patron is not able to switch tickets or that a cashier cannot classify a fee-paying vehicle as a non-fee-paying one.

A simple method of ensuring proper control in this type of car park operation is to issue each contract vehicle with an identifying non-transferable, non-removable sticker for the windscreen. Alternatively (or in addition), the contract parker is issued with a plastic card which, when inserted into a reader at the entrance, causes a special ticket to be issued; when exiting, the driver presents this ticket to the cashier-attendant together with the plastic card (for identification). Instead of a ticket being issued when the plastic card is inserted into the reader at the entrance, the card itself may instead be encoded to indicate that it is in use; the card must then be neutralised by the exit reader before the driver can use it at the entrance again.

A basic requirement of all fee-charging car parks is that the cashier-attendants be held accountable for their own operations. When, say, an exit cashier goes on duty, it should be necessary for his or her (coded) personal key to be inserted into the booth's equipment before it can be operated; the booth's audit-recorder then notes the switch-on time and date, cashier number, and other relevant auditing information. At the end of any period of time, e.g. the cashier-attendant's shift, it should be possible to obtain a print-out of the number of vehicles that have gone through each exit, the number of transactions recorded on the cash register (or other recording device), and an accumulative total of fees charged. As a basic check the number of vehicles that have gone through the exit should always correspond with the number of cash transactions recorded for the time period considered.

22.9 References

1. *Social Trends 25*. London: HMSO, 1995.
2. Lancashire County Planning Department, *Car parking standards*. Lancashire County Council, 1986.
3. Weant, R.A., and Levinson, H.S., *Parking*. Westport, Conn: The Eno Foundation for Transportation, 1991.
4. Department of the Environment, *Cars and housing/2*. London: HMSO, 1971.
5. Joint Committee of the Institution of Structural Engineers and the Institution of Highways and Transportation, *Design recommendations for multi-storey and underground car parks*. London: Institution of Structural Engineers, 1984 (2nd edition).
6. Ellison, P.B., *Parking: Dynamic capacities of car parks*. Report LR221. Crowthorne, Berks: Transport Research Laboratory, 1969.
7. Pline, J.L. (ed.), *Traffic engineering handbook*. Englewood Cliffs, NJ: Prentice Hall, 1992.
8. Layfield, R.E., *Parking: Reservoir sizes at car park exits*. Report 39UC. Crowthorne, Berks: Transport Research Laboratory, 1974.
9. McCluskey, J., *Parking: A handbook of environmental design*. London: Spon, 1987.
10. Brannam, M., and Longmore, J.D., *Layout of lorry parks: Dimensions of stalls and aisles*. Report SR83UC. Crowthorne, Berks: Transport Research Laboratory, 1974.
11. Rowlands, J.B., and Wardley, J.G., Shopping centres: Deliveries and servicing. *Highways and Transportation*, 1988, **35** (4), 16–21.
12. Salter, R.J., *Highway traffic analysis and design*. London: Macmillan, 1989 (2nd edition).
13. *Parking revenue control*. Transportation Research Circular No. 184. Washington, DC: Transportation Research Board, June 1977.
14. *BS 4469: Car parking control equipment*. London: British Standards Institution, 1973.

CHAPTER 23

Road lighting

C.A. O'Flaherty

The lighting of traffic routes and of roads in residential and shopping areas has long been an important part of the road transport scene, especially in urban areas. It is reported[1] that in 1405 citizens of London were required to hang burning 'lanthorns' outside their homes. From about 1700 regularly spaced oil lamps in the streets of large towns were paid for out of local rates for lighting and cleaning. Better standards were achieved in the nineteenth century with the use of gas lighting. The introduction of electric lighting in 1913, combined with the growth in numbers and usage of motor cars, saw the initiation of a new era in respect of road lighting. Today, lighting is a major consideration in the design and operation of roads in built-up areas.

The purpose of this chapter is to provide a general background to road lighting. The actual design process is complex and lengthy, and space does not permit its detailed coverage here. The reader is referred to the *British Standard for Road Lighting*[2] for details of design practice in Britain.

23.1 Objectives

From the traffic engineer's viewpoint the primary objective of road lighting is to improve the safety of roads at night by providing good visibility conditions for all road users, whether they be drivers, motorcycle or bicycle riders, or pedestrians. When installing lighting at any given road location, however, care must be taken to ensure that it does not become a cause of potential danger to any other transport mode, e.g. when adjacent to airports.

A second reason for installing road lighting is to promote better traffic flow at night, by providing improved delineation of road geometrics, safer overtaking opportunities, and easier observance of traffic management measures. A third reason for having road lighting is to enhance the amenities of the general area being illuminated; while lighting is provided for the primary reason of improving road safety, it should not detract from buildings and monuments in the vicinity.

Numerous before-and-after accident studies have shown that the use of road lighting on traffic routes results in a reduction in night-time accidents. For example, one study[3] which examined the effect of varying the quality of lighting on dry two-lane roads in built-up areas concluded that an increase in average carriageway luminance of one candela per square metre (within the range of average luminance values of 0.5–2.0 cd/m²) resulted in a 35 per cent decrease in the night-to-day accident ratio. British practice when carrying out economic appraisals[4] to determine whether or not to install road lighting on trunk roads or motorways is to assume, on average, that after-dark accidents will be reduced by 30 per cent following installation.

In urban areas the greatest beneficiaries of accident reductions attributed to road lighting installations are often pedestrians and cyclists.

Analysis of various studies suggests that road deaths and injuries at night can be significantly reduced if good lighting is installed at the following critical areas of driver decision-making:

- driveway entrances and exits
- intersections and interchanges
- bridges, overpasses and viaducts
- tunnels and underpasses
- guide sign locations
- hills and curves with poor geometrics and/or sight distances
- busy roads in urban areas
- rest areas and connecting accesses
- at-grade railway crossings
- elevated and depressed road sections.

23.2 Lighting terminology

Before examining the basic considerations underlying the provision of road lighting, it is helpful to define some of the terms in common usage. These may be divided into two main groups: *photometric terms*, which relate to light and its measurement units, and *lighting installation terms*, which relate to the physical equipment and its layout on the roadway.

23.2.1 Photometric terms

The following definitions are given as simply as possible, rather than seeking to make them absolutely precise. The units and abbreviations are those normally used in Britain.

Light. This term is used to describe the radiant energy that is capable of being perceived by the eye.

Luminous flux. This is the light given out by a light source or received by a surface irrespective of the directions in which it is distributed. The unit of luminous flux is the lumen (lm), which is the flux emitted through a unit solid angle (a steradian) from a point source having a uniform luminous intensity of 1 candela. The total flux from this point source is 12.566 lm (i.e. 4π lm).

Luminous intensity. This describes the light-giving power (candlepower) of a light source in any given direction. The unit of luminous intensity is the candela (cd), where 1 cd is 1 lumen/steradian. The candela is an international standard related to the luminous intensity of a black body (which absorbs all radiation incident upon it) at the temperature of freezing of platinum.

Lower hemispherical flux. By this is meant the luminous flux emitted by a light source in all directions below the horizontal. The *mean hemispherical intensity* is the downward luminous flux divided by 2π, i.e. the average intensity in the lower hemisphere.

The ratio of the maximum luminous intensity to the mean hemispherical intensity is called the *peak intensity ratio*.

Illuminance. This is the density of luminous flux incident on a surface. The unit of illuminance is the lux (lx) where 1 lux = 1 lumen per square metre (lm/m^2).

Luminance. This is the luminous intensity reflected from a unit projected area of an illuminated surface in a given direction. Whereas illuminance is a measure of the amount of luminous flux falling on a surface, luminance is a measure of the amount of flux which the area reflects toward the eye of the observer. The unit is the candela per square metre (cd/m^2). Thus if a very small portion of a surface has a luminous intensity of 1 cd in a particular direction, and if the projection of that portion on a plane perpendicular to the given direction has an area of D square metres, then the luminance in that direction is $1/D$ cd/m^2. Average luminances recommended for traffic routes in Britain are shown in Table 23.1. Also shown in this table are the recommended overall and longitudinal uniformity ratios of luminance. The *overall ratio*, U_O, is the ratio of the minimum to the average luminance over a defined area of the road surface viewed from a specified observer position. The *longitudinal ratio*, U_L, is the ratio of the minimum to the maximum *luminance* along a longitudinal line through the observed position on the carriageway.

Table 23.1 Lighting requirements for traffic routes[2]

Example roads	Maintained avg. luminance, L (cd/m^2)	Uniformity ratios	
		Overall, U_O	Longitudinal, U_L
High speed roads. Dual carriageway roads	1.5	0.4	0.7
Important rural and urban traffic routes. Radial roads. District distributor roads	1.0	0.4	0.5
Connecting less important roads. Local distributor roads. Residential area major access roads	0.5	0.4	0.5

Luminosity. Sometimes loosely termed *brightness*, this is the visual sensation indicating that an area appears to emit more/less light. While correlating approximately with the term *luminance*, it is not measurable.

Beam. The beam is the portion of the luminous flux emitted by a lantern which is contained by the solid angle subtended at the effective light centre of the lantern containing the maximum intensity but no intensity which is less than the maximum intensity.

23.2.2 Installation terms

Lighting installation. This term refers to the entire equipment provided for the lighting of a road section. It comprises the lamp(s), lantern(s), means of support, and the electrical and other auxiliaries.

Lamp. The light source or bulb is called the lamp.

Luminaire. Also known as a *lantern,* this is a complete lighting unit consisting of the lamp together with its housing and such features as wiring terminals, a socket to support and position the luminaire, specular (mirror-type) reflectors within the luminaire body, and prismatic refractor elements (known as refractors) which form the glass or plastic enclosure.

Lighting column. Also known as a *lighting standard,* this is the pole that is used to support a luminaire. A *bracket,* also known as a *mast arm,* is an attachment to a column or other structure (e.g. a wall) that is used to support a luminaire. The horizontal distance from the point of bracket entry to the luminaire and the centre line of the column is called the *bracket projection.*

Mounting height. This is the nominal vertical distance between the carriageway surface and the photometric centre of the luminaire, i.e. the centre of the lamp (see Fig. 23.1). Mounting heights in Britain are normally 8 m, 10 m or 12 m, with 10 m being that which is most commonly used on urban traffic routes. 12 m is used on wide or heavily trafficked routes where advantage can be taken of a longer spacing between luminaires, while 8 m

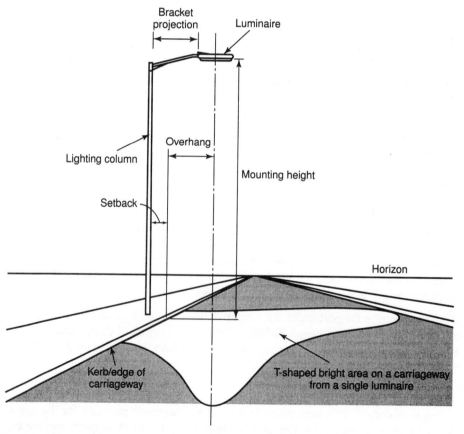

Fig. 23.1 Some terms used in road lighting installations

tends to be used on narrow roads (e.g. local distributor and access roads to residential areas), on traffic routes where greater heights would be out of scale with adjacent buildings, and on urban routes with many intersections (i.e. where no reduction in the number of luminaires would result from the use of a greater height).

Overhang. This is the horizontal distance between the photometric centre of the luminaire and the adjacent edge of the carriageway. It is taken as positive if the luminaire is in front of the edge, and negative if it is behind the edge.

Spacing. This is the distance between successive luminaires in an installation, measured parallel to the centre line of the carriageway (see Fig. 23.2). The successive luminaires may or may not be arranged on the same side of the carriageway.

Lighting arrangement. The pattern according to which luminaires are sited in plan view is termed the *arrangement* (see Fig. 23.2). In a *staggered* arrangement the lanterns are spaced alternately on either side of the carriageway. When the lanterns are placed on either side but opposite each other the pattern is termed an *opposite* arrangement. With a *central* arrangement the lanterns are sited on an axial line overhanging the centre of

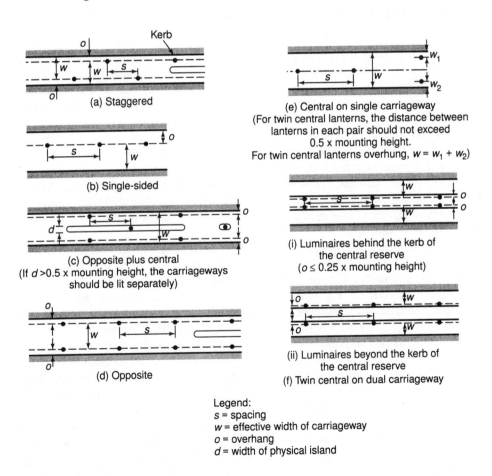

Fig. 23.2 Road lighting arrangements

the carriageway (in the case of a single carriageway) or close to the middle of the central reservation (in the case of a dual carriageway road). A *single-side* arrangement is one in which the lanterns are placed on one side only of a carriageway.

Width of carriageway. This is the distance between kerbs or carriageway edges as measured at right-angles to the centre line of the carriageway. The *effective width of carriageway* is that which it is intended to make bright. The relationships between carriageway width, effective width, and overhang are also shown in Fig. 23.2.

Setback. This is the minimum distance between the face of a lighting column and the kerbline/edge of the carriageway. The recommended *absolute minimum* setbacks for roads with 80, 100 and 120 km/h design speeds are 1, 1.25 and 1.5 m, respectively; the recommended setbacks for 50 and 80 km/h roads are 0.8 and 1.5 m, respectively.

Geometry. The geometry of a lighting system is the general term used to refer to the interrelated linear dimensions and characteristics of the system, i.e. the spacing, mounting height, effective width, overhang, and arrangement.

23.3 Basic means of discernment

The basic aim when designing road lighting is different from that of interior lighting. With interior lighting the aim is to reproduce daylight as closely as possible so that the forms and textures of objects are clearly seen. With road lighting detail is relatively unimportant; what is critical is that the road user be able to discern clearly the presence and movement of any object on or adjacent to the carriageway that may be a potential hazard.

An object is visible to a road user if there is sufficient contrast of luminosity between it and its background, or between different parts of the object. While object colour and size are also important, night-time visibility for a driver in a road/traffic situation mainly depends upon contrast. Thus the normal design of a road lighting installation is usually based upon arranging luminaires so as to have a uniformly bright carriageway surface (and the immediate surrounds) against which a darker object, e.g. a pedestrian, may be seen in *silhouette*.

Road lighting does not make dark objects bright (as do vehicle headlights); instead the usual effect is to make the carriageway surface more luminous so that the object is seen in *dark silhouette*. It is only occasionally that the luminosity of an object is greater than that of the surface, e.g. if a pedestrian dresses in white clothing and stands just behind a luminaire and is highly illuminated on the side facing an oncoming driver; then the object is said to be seen in *reverse silhouette*, i.e. by direct vision.

The performance requirements for average luminance shown in Table 23.1 will ensure that the road is generally bright enough to reveal objects adequately in dark silhouette. The overall uniformity ratio ensures that no part is so dark that it is ineffective as a background for revealing objects, while the longitudinal uniformity ratio ensures that pronounced visual patchiness of the lighted carriageway surface is avoided.

When discernment of an object is by means of variations in luminosity over its surface rather than by contrast with its background, then it is said to be visible by *surface detail*. Such discernment, which underlies the design of illuminated roadside traffic signs, requires a high degree of illumination on the side of the object facing the observer.

23.3.1 Discernment by dark silhouette

Each luminaire in a lighting installation on a straight section of empty road contributes a characteristic T-shaped *bright patch* on the carriageway surface (see Fig. 23.1). Note that the head of the patch stretches across the road but reaches only a relatively short distance behind the luminaire, while the tail extends toward the observer. In an ideal design the luminaires are sited to ensure that the patches link up to cover the entire road surface so that it appears to the driver that the carriageway has high luminosity, and any intrusive object is seen in silhouette.

The luminance of a point on the carriageway surface is primarily influenced by the illuminance at the point, the angles of incidence and of observation, and the reflecting characteristics of the carriageway surface.

The bright patch has a number of interesting, albeit complex, characteristics.

- The tail of the patch always extends toward the observer, and the farther he or she is from the luminaire the more clearly defined is the shape of the patch.
- If the carriageway adjacent to a luminaire is looked at from two different observation points two different bright patches are seen.
- The point of maximum luminous intensity occurs neither in the head of the patch nor at the end of the tail, but somewhere in between.
- Under dry carriageway conditions the shape of the patch depends very much upon the reflection characteristics of the carriageway surface. At one extreme, a rough-textured harsh surface with white stones produces a broad bright head to the 'T' and a very small tail; the average luminance is high with a reasonable overall uniformity, but longitudinal uniformity will be poor if the patches do not join together along the carriageway. At the other extreme, a dark-coloured polished stone in the carriageway surface reduces the luminance of the head and produces a long bright tail, so that the overall pattern consists of longitudinal streaks of light; however, the average luminance is not low as the loss in brightness in the head is generally compensated for by the bright streak of the tail.
- On wet carriageways, the lightness of the surface is of less importance and the macro-texture is more important. If the surface is smooth and wet from rain a patch with hardly any head is produced and the tail becomes very long and tries to become as luminous and as thin as the luminaire; hence the intense light streaks commonly noted on a very wet carriageway. The more coarse-textured the carriageway surface the greater is its ability to maintain a uniform luminance under rain conditions (also, less spray is thrown up by tyres and the skid resistance is higher).
- The lower the mounting height the shorter the tail of the patch, the greater the luminance of the central part of the patch, and the lower the luminance of the outer regions of the carriageway.
- Since the tail of the patch is formed by light rays which leave a lantern near to the horizontal, cutting off these flux shortens the length of the tail.

Figure 23.3 shows the general shapes of bright patches obtained for a number of the conditions experienced on roadways.

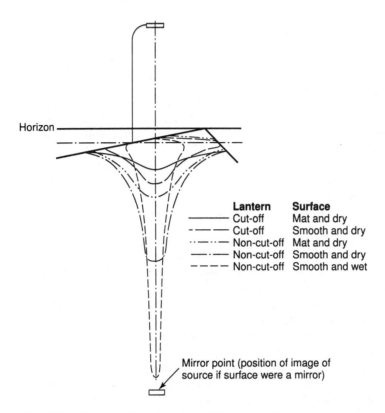

Lantern   Surface
————   Cut-off   Mat and dry
– — – —   Cut-off   Smooth and dry
–·—··—   Non-cut-off Mat and dry
—·—·—   Non-cut-off Smooth and dry
– – – –   Non-cut-off Smooth and wet

Mirror point (position of image of
source if surface were a mirror)

Fig. 23.3 Perspective diagrams of bright patches shown in contours of equal luminance formed on the carriageway by a single luminaire, according to the type of its light distribution curve and the nature of the carriageway surface (after reference 7)

23.3.2 Discernment by reverse silhouette

On many road sections seeing by silhouette against a luminous carriageway background can be difficult, e.g. at crests of hills, at sharp bends, and at intersections. The driving task is also more complex at these sites and drivers are faced with additional visual tasks as they approach and negotiate these accident-prone locales. A higher level of direct lighting is normally provided at these locations, therefore, so that objects in the danger areas and their immediate surrounds are seen by direct vision or reverse silhouette.

At intersections, for example, the level of illumination and its uniformity should be such that the driver is able to discern clearly the existence of the intersection, the presence of any pedestrians or obstructions such as parked cars, the positions and forms of any traffic islands, kerbs and road markings, the directions of the roads, and the movement of any vehicles in the vicinity. Although it is not a primary function of the lighting to indicate the routeing of traffic, good lighting design on the approach roads will often provide such guidance through an intersection. At no time should the carriageway luminance through an intersection be less than that provided on the carriageways of the main approach roads.

At intersections (including grade-separations and interchanges) where merging and diverging take place at high speeds particular care must be taken to light directly all areas of traffic and geometric complexity, e.g. all acceleration and deceleration lanes, slip and loop roads, and curves and hills. Eye adaption is important at these locations also, and this can often be achieved by the use of lamps of varying wattage rather than by luminaire spacing.

23.4 Glare

If the total light output from a lamp is allowed to radiate with uniform luminous intensity in all directions, much of the flux will be wasted, particularly that which goes upward. To overcome this road luminaires are provided with a redirectional apparatus which limits the directions in which the flux is distributed. A bright patch having a particular carriageway luminosity is then obtained by varying the luminous intensity distribution so that higher intensities are generally emitted at greater angles from the vertical. However, the vertical angle at which the maximum luminous intensity is emitted may affect the safety of road users because of the glare effects to which drivers can be subjected. Glare can be described as light which reduces the ability to see or produces a sensation of visual discomfort.

Glare which produces a loss of visual efficiency is termed *disability glare*; this has been likened to the production of a uniform luminance veil over the whole visual field which reduces the contrast between the object and the background. When an object is at the threshold of visibility (i.e. when it is just visible and there is no disability glare), it will blend into the background in the presence of disability glare. The percentage by which the background luminance has to be increased to make the object just visible again is termed the *threshold increment* (TI). The TI-value depends on the light distribution from the luminaire between 70–90 degrees in elevation in the vertical plane at which the luminance is observed.

A driver cannot normally see more than about 20 degrees above the horizontal because of the cut-off effect of the roof of his or her vehicle. Thus, when a vehicle is approaching a luminaire the luminous intensities emitted at angles less than about 70 degrees from the vertical cannot reach the driver's eyes and cause glare and, therefore, the intensities need not be limited below this angle. However, glare will be caused by light leaving the luminaire within about 20 degrees of the horizontal, and to minimise this the luminaire can be designed to emit relatively low luminous intensities within this 20-degree glare zone.

A TI-value of ≤ 15 per cent is recommended for British roads on which it is desirable to minimise glare, e.g. on high-speed roads in rural areas where the absence of buildings bordering the road tends to worsen the effect of glare (i.e. light-coloured buildings increase the background luminance and reduce disability glare). In most other lighting installations it is recommended that the TI-value not exceed 30 per cent; this less stringent glare control generally allows more scope in relation to the siting of lighting columns and greater spacing between luminaires, thereby permitting lower capital and running costs.[2] Special requirements are often necessary near airports (where the luminous intensity above the horizontal normally has to be limited), and in environmentally sensitive areas where spill light is considered to be visually intrusive.

Discomfort glare (also known as *psychological glare*) describes the sensations of distraction and annoyance experienced when glare sources are in the field of view. While discomfort glare in itself has little or no effect on visual performance in the short term,[5] the application of the above TI-controls for disability glare will normally ensure that this other type of glare is also adequately controlled.

23.5 Lamps

The choice of the light source involves such considerations as lamp life and efficacy, capital and operating costs, reliability and routine maintenance requirements and, in historical/tourist locales and environmentally sensitive areas, colour appearance and rendering capability.

The most widely used lighting sources on traffic routes are the low-pressure sodium vapour lamps (types SOX and SLI). These lamps, with their familiar monochromatic yellow colour, have very high luminous efficacies for the energy consumed. However, their colour rendering properties are very poor. Typical wattages are 90–100 W for traffic route lighting, and 35–50 W on other roads.

Another widely used light source is the high-pressure sodium lamp (type SON). This lamp provides a warm sunlight appearance and its usage is particularly applicable on routes where the rendering of colours is important, e.g. on prestige routes, in shopping centres, and in historical towns and tourist areas. The high-pressure sodium lamp is less efficient (in terms of lumen output per watt of energy) than the low-pressure sodium lamps but it has a longer life. This lamp is also smaller and good optical control is easier to achieve.

There are significant installations of high-pressure (MBF) mercury vapour lamps in existence also. While these white-light lamps give good colour rendering, they are less efficient than either of the sodium types and their use in Britain now tends to be limited to traffic signs, subways, underpasses and tunnels.

Typical *average lives* of high-pressure sodium vapour, low-pressure sodium vapour, and high-pressure mercury vapour lamps are 21 000–27 000 h, 15 000 h, and 14 000 h, respectively.[5] In practice, however, bulk lamp changes are carried out as a maintenance activity well before burn-out time, e.g. every 12 to 24 months in Britain.

It is appropriate to note here that the *hours of operation* of road lighting in Britain are from about 30 minutes after sunset to about 30 minutes before sunrise.

A road with lighting should be lit during all the hours of darkness, independently of traffic flow. The practice of extinguishing certain luminaires when the traffic flow is low should not be encouraged as it does not fulfil the lighting needs of drivers and increases the likelihood of a collision with a darkened column. The lighting also serves the needs of pedestrians, cyclists, emergency services and public security, and these interests would be detrimentally affected if lights were extinguished during darkness.

23.6 Luminaires

Modern luminaires tend to have linear light sources that are mounted horizontally, and to have a non-symmetrical light distribution with main beams that are axially opposed and directed along the linear length of the carriageway. The efficiency with which a luminaire (with its reflectors and refractors) can control and redirect light is largely affected by the size of the lamp in relation to the overall dimensions of the lantern, and the smaller the lamp the easier it is to control optically.

In British practice, the selection of luminaires is resolved by dividing them into two classes, based on their glare-control capabilities: low threshold increment (LTI) and moderate threshold increment (MTI). The LTI and MTI luminaires produce threshold increments that are not greater than 15 per cent and 30 per cent, respectively, for at least 90 per cent of the geometrics quoted in each of the design tables for that luminaire (see reference 2).

23.7 Mounting height

As noted previously, luminaires are normally mounted 8 m, 10 m or 12 m above the carriageway surfaces of traffic routes. In villages and housing developments, however, they may be as low as 5.7 m (the minimum permissible clearance). It is advantageous visually if the heights of the lighting columns and luminaires do not exceed those of nearby buildings.

The mounting heights at intersections are normally not less than those on any approach road, but may be increased if this results in a reduction in the number of lighting columns. High mast (<18 m) lighting is commonly used at grade-separated intersections; the dual objectives in this instance are to light a number of roads at different levels and to reduce the numbers of lighting columns (thereby improving both road safety and the visual appearance of the intersections during daylight).

23.8 Luminaire arrangements

The design of a road lighting installation basically involves locating the luminaires so that the bright patches overlap to give the surface of the carriageway an average luminance which, in conjunction with acceptable overall and longitudinal uniformities, is suitable for the road usage (see Table 23.1). Of the lighting arrangements used for this purpose (Fig. 23.2) those which are most commonly used on single carriageway roads are single-sided, staggered, and central plus opposite, with the choice being primarily influenced by the effective carriageway width. Central arrangements are now most commonly associated with dual carriageways.

23.8.1 Single carriageway road arrangements

The use of a *suspended central arrangement* on a single carriageway road appears at first to be very attractive, in that the light patches overlap very well to produce a uniformly lighted road centre with luminosity falling off to the side. However, this arrangement may also give rise to potential accident situations. For example, because the sides of the road are in comparative darkness, pedestrians, cyclists or vehicles entering a main road may be well into the carriageway before being noticed by the main road drivers; the luminous centre tends to invite a driver to increase speed and to 'hog' the crown of the road; and on busy roads maintenance crews will be exposed to danger, as well as obstructing traffic, when servicing the luminaires. Consequently, suspended central arrangements are only used over single carriageway roads when special conditions justify them; for instance, they might be used over narrow carriageways that are flanked with trees on either side.

Single-side arrangements are normally discouraged on wide single carriageways as they cause one side of the road to be well lit while the other side is left in comparative darkness. This results in an unsafe tendency for drivers to veer toward the lighted side of the roads; it also creates the conditions for vehicles, pedestrians and cyclists entering the dark side of a main road carriageway not to be noticed until too late to prevent an accident. Consequently, single-side arrangements are mostly used on narrow streets.

An *opposite arrangement,* which is essentially two independent single-side systems, tends to be used with very wide single carriageways. This arrangement produces a very uniform carriageway luminosity, albeit at the cost of increased expenditure on extra luminaires, columns, etc.

Overall, a *staggered arrangement* is now generally deemed to give the 'best-value' light distribution on most normal single carriageway roads in Britain.

23.8.2　Dual carriageway road arrangements

A *twin central arrangement* is normally the preferred system in Britain for use with a dual carriageway road where the distance between the outer kerbs is not excessive. Locating the lighting columns in the central reservation reduces the number of supports by one half from that required for, say, a single-side arrangement on each carriageway. Also, since there is only one row of columns there is only need for one run of electrical conductor, and the consequent savings in material and construction costs are significant.

Situations where the use of a single central arrangement on a dual carriageway is not appropriate include: (a) where the central reservation is to be used by public transport vehicles, (b) where there is no central crash barrier within which the luminaire supports can be encased to ensure that the collision dangers associated with the presence of fixed objects are not increased, (c) where the central reservation is too wide or the two carriageways are on separate alignments, and (d) where there is significant pedestrian activity on footpaths adjacent to the road. In these situations each carriageway should be lit as if it were a separate road.

A special situation may also arise in relation to dual carriageways in urban areas where the roadway is depressed, and the use of wall mountings may be more appropriate.

23.9　Overhang, bracket projection and setback

Experience would suggest that it is not normally necessary for luminaires to overhang carriageways that are less than about 9 m wide. On wider roads with side lighting, over-hanging may be necessary to light the centre of the carriageway, which might otherwise appear unduly dark. Generally, however, the *overhang* should be as small as possible, so as to ease luminaire maintenance problems. Experience also suggests that the overhang should not exceed about 1.8 m on heavily trafficked urban streets as footpaths and kerbs may be left in shadow.

The *bracket projection* should be as small as possible, and should never exceed 25 per cent of the mounting height, to limit vibration.[2] A long bracket, which might be necessitated by a large setback and use of the maximum permissible overhang, presents a poor appearance and should only be used for safety reasons or when surrounding objects effectively hide its full length. Large arc or quadrant brackets are usually more conspicuous than straight ones, because they contrast with the straight lines of adjacent roofs. However, a straight horizontal bracket gives the illusion of sagging, so that a straight rising bracket is generally preferred.

Lighting columns should be set back from the edge of the carriageway so as to minimise the opportunity for vehicles to collide with them at high speed. Recommended *setbacks* on roads with various design speeds are given in Section 23.2.2. On roads flanked with footpaths, however, care needs to be taken to ensure that the setback does not interfere with the free passage of visually disabled persons, perambulators or wheelchairs; this may mean that on footways up to 3 m wide, the columns may have to be sited on the back of the path to provide the maximum useable width, as well as to minimise the opportunity for vehicle collisions.

23.10 Spacing and siting

Column *spacing* is normally calculated from photometric data tables relating to the luminaires under consideration (e.g. see reference 2, Part 2). Typically, however, the spacing is three to five times the mounting height for most traffic routes, and six to eight times the mounting height on more minor routes.[6] On sharp bends and at intersections the spacings have to be closer (and the illumination greater) than on straight sections of road.

There are many constraints on the *siting* of lighting columns, ranging from overground/underground obstructions to their effects on the occupation and maintenance of properties. For example, columns are better placed at the junctions of two buildings rather than in front of either building, especially if they are buildings or monuments of architectural or historical interest. Nor should they interfere with scenic views. If a column is sited close to an overbridge, care has to be taken to ensure that the main beam from the luminaire is not obstructed by the bridge.

As a general rule, lighting columns are sited first at difficult locations such as intersections and bends, and the pattern necessary for uninterrupted sections of road is then added. The layout should always be examined in perspective to check that the array of luminaires forms a pattern that gives guidance to traffic, particularly at complex intersections and winding roads. The importance of this guidance role in areas subject to fog should not be underestimated.

23.11 References

1. Road lighting: State-of-the-art report. *Municipal Engineer*, 1984, **2**, 171–83.
2. *BS 5489: Road lighting*. London: British Standards Institution, 1990.

   Part 1: Guide to the general principles
   Part 2: Code of practice for lighting for traffic routes
   Part 3: Code of practice for lighting for subsidiary roads and associated pedestrian areas
   Part 4: Code of practice for lighting for single-level road junctions including roundabouts
   Part 5: Code of practice for lighting for grade-separated junctions
   Part 6: Code of practice for lighting for bridges and elevated roads
   Part 7: Lighting for underpasses and bridged roads (Group E)

Part 8:   Code of practice for lighting for roads near aerodromes,
railways, docks and navigable waterways
Part 9:   Lighting for town and city centres and areas of civic
importance
Part 10: Code of practice for lighting for motorways

3. Scott, P.P., *The relationship between road lighting quality and accident frequency.*
Report LR929. Crowthorne, Berks: Transport Research Laboratory, 1980.
4. *Appraisal of new and replacement lighting on trunk roads and trunk road motorways.* TA49/86. London: Department of Transport, 1986.
5. *Guide to traffic engineering practice, Part 12: Roadway lighting.* Sydney: National Association of Australian State Road Authorities, 1988.
6. Ruston, B., Street lighting. *Highways and Transportation*, 1987, **34** (11), 29–32.
7. Waldran, J.M., International recommendations for public thoroughfares. *Traffic Engineering and Control*, 1966, **7** (12), 753 and 755.

PART IV

Traffic management

CHAPTER 24

Regulatory measures for traffic management

C.A. O'Flaherty

Regulatory traffic management has its basis in law, and uses mandatory and prohibitory traffic signs and markings (see Chapter 28) to inform drivers regarding what they must and must not do in relation to speed, movement and waiting. Regulatory measures of particular interest include speed limits, restrictions on turning movements, the closure or one-way operation of streets, tidal-flow operation of major roads, priority for high-occupancy vehicles, and on-street waiting restrictions and parking control.

24.1 Speed limits

The evidence is overwhelming that the number and severity of road accidents increase with vehicle speed. Properly implemented maximum speed limits modify speeding behaviour and the downward change generally reduces the number and severity of accidents by (a) shortening the distances required to stop vehicles safely (see Chapter 19), (b) ensuring that more moving vehicles maintain at least the minimum desirable clearance distances between them to avoid rear-end collisions, e.g. 24.4 m and 73.2 m for cars travelling at 48 km/h and 90 km/h, respectively, (c) maximising the available skid resistance (the skidding resistance offered by a wet carriageway surface decreases as a vehicle's speed increases), and (d) making it easier for pedestrians to judge when it is safe to cross a road. Other benefits of maximum speed limits include a reduction in fuel consumption, and a smoother traffic flow resulting from a greater proportion of vehicles travelling at similar speeds.

Two types of speed limit are in use throughout the world: absolute and prima facie limits. An *absolute speed limit* is that above which it is illegal to drive, irrespective of the traffic, roadway, weather, or other conditions prevailing. While this type of limit is preferred by enforcement police (as it is difficult for the motorist to challenge a violation in court), motorists can feel disgruntled at apparently unreasonable speed restrictions being placed on obviously higher-speed roads. A *prima facie speed limit* is that above which motorists are assumed to break the law; however, they may argue in court that their speed was safe for the conditions prevailing at the time. This is a flexible limit in that police can adjust their enforcement according to their view as to whether the conditions prevailing are safe; however, it relies on the police to use judgement as to what is safe, and this is more easily challenged in the courts.

The worldwide trend is toward using absolute speed limits. In Britain, for example, the mandatory speed limits imposed are 96 km/h on rural single carriageway roads, and 112 km/h on rural motorways and dual carriageway roads. The national speed limits for urban roads are 48 km/h (30 mile/h), 64 km/h (40 mile/h), 80 km/h (50 mile/h), and 96 km/h (60 mile/h) for roads wih design speedss of 60B, 70A, 85A and 100A km/h, respectively. In mainland Europe the speed limits for rural roads vary considerably, e.g. from 80 to 130 km/h for motorways and all-purpose dual carriageways and from 70 to 110 km/h for single carriageway roads. The spread of speed limits for urban roads in Europe is more consistent, from 48 to 60 km/h.

Other worldwide trends that are discernible in respect of speed limits are:

- limits are applied throughout the year
- different upper limits are used on different types of road and in different environmental locations
- different upper limits are applied to different types of vehicles on the same high-speed roads, e.g. speed limits for heavy commercial vehicles on rural highways are typically 8–16 km less than for passenger cars
- lower limits (proposed to reduce the number of overtaking/rear-end accidents) are rarely used other than on motorways
- different upper limits may apply at a given location at different times of the day, e.g. adjacent to schools or on road sections subject to fog.

The greatest hindrance to motorist observance of a speed limit is the application of a general restriction where it is obviously inappropriate. Factors which should be taken into account when establishing a speed limit include (a) the design speed of the road, (b) prevailing vehicle speeds, (c) road cross-section, curvature, gradient and (quality of) surfacing, (d) frequency and spacing of intersections, (e) traffic volume and composition, (f) accident history, (g) presence/absence of pedestrians and/or of parking/unparking vehicles, and (h) traffic control devices that affect/are affected by vehicle speeds. As a guide, the imposed speed limit should not be much less than the 85th percentile speed distribution measured at (b) above, and it should never be greater than the design speed of the road section.

While it is generally accepted by the public that drivers should not be left to decide appropriate speeds for themselves, and that speed limits are therefore necessary, speed infringements will occur regularly if the limits are not enforced. Observance is encouraged by police enforcement using hand-operated radar speed-meters (most usual), radar-operated speed-cameras (more economical), or patrol cars (most expensive, but most effective when long distances have to be policed, on congested roads, and/or in foggy weather). Self-policing is also encouraged by the use of traffic calming methods in built-up areas, and by using rumble strips and bar markings on high-speed roads in non-urban areas (see Chapter 25).

24.2 Restriction of turning movements

Congestion and accidents caused by right-turning vehicles (for left-hand rule-of-the-road) at signal-controlled intersections are usually coped with by inserting an extra phase or early cut-off and late start arrangements in the signal cycle. In some instances it may be preferable to ban right-turning vehicles at a critical intersection during all or part of the day, rather than attempt to provide directly for this movement.

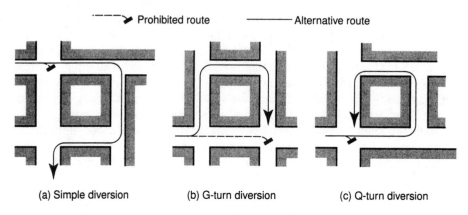

(a) Simple diversion   (b) G-turn diversion   (c) Q-turn diversion

Fig. 24.1 Rerouting right-turning traffic

Figure 24.1(a) shows the diversion of the right-turning movement to an intersection further along the road where there is more capacity. This routing is applicable to a difficult right turn from a minor to a major road as this right-turn movement then takes place at a minor–minor intersection.

Figure 24.1(b) shows a diversion to the left before the critical intersection. This 'G-turn' is applicable to a right turn off a major road as it changes it to a left turn off the major road and a straight-over movement at the critical intersection. As it also involves two right turns at minor intersections, care must be taken to ensure that these do not create extra problems. Careful signing is essential to ensure that motorists do not overshoot the initial left turn.

Figure 24.1(c) shows a diversion to the left after the critical intersection that requires three left turns. This 'Q-turn' is the least obstructive diversion, but requires vehicles to travel twice through the critical intersection, thereby increasing the total volume of traffic handled there.

For the above procedures to work alternative routes must be available that are suitable in terms of both engineering and the environment. Buses, because of their need to serve particular routes, are extra-sensitive to these turning restrictions; thus careful consideration should always be given to exempting buses from right-turn bans.

In some instances it may be desirable to close off access between a minor road and a major road entirely, particularly if the minor intersection has a high accident rate and/or is located between two major junctions that are close to each other, e.g. within 300 m. This action can be expected to improve journey times and reduce accidents on the main road; it also facilitates the use of linked traffic signals.

24.3 One-way streets

One-way traffic operation is a simple regulatory tool available for the relief of traffic congestion which does not require expensive policing. Its most effective usage is on streets in and about the central areas of towns.

Advantages of one-way operation include:

- road capacity is increased, with the amount depending upon the conditions prevailing locally, e.g. the distribution of the previous two-way flow, the street width, and the turning movements at intersections

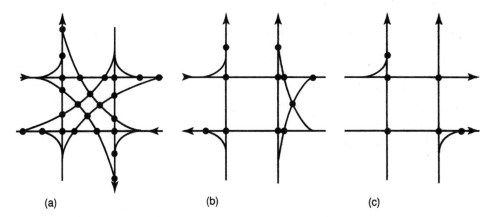

Fig. 24.2 Potential points of conflict at an intersection: (a) two two-way streets, 24 conflicts, (b) one one-way and one two-way street, 11 conflicts, and (c) two one-way streets, 6 conflicts

- odd lanes which were not usable previously can be fully utilised
- slow-moving/stopped vehicles are more safely overtaken
- turning movements at intersections are not delayed by opposing traffic
- journey times and delays are reduced through more efficient traffic-signal timing and higher vehicle speeds
- the linking of traffic signals is facilitated
- public transport operation is more reliable
- overall there is generally a reduction in vehicle–vehicle (see Fig. 24.2) and pedestrian–vehicle (see Fig. 24.3) accidents at intersections, and mid-block head-on accidents are eliminated
- parking is facilitated (in terms of increased numbers of places and ease of parking).

Disadvantages of one-way operation include:

- vehicles generally travel further to reach their destinations
- public transport stops for the opposing direction of travel have to be relocated, and walking distances to/from stops increased
- the severity of non-head-on accidents is increased because of higher speeds
- mid-block weaving accidents may increase
- non-local motorists may become confused and additional signs, markings, channelisation and signal indications may be required to deal with unanticipated vehicle movements
- displaced traffic may have to be routed to a complementary street through residential areas, thereby causing increased speeds, loss of amenity and more accidents in these locales.

One-way street operation is most easily introduced into an urban area with a gridiron street pattern; with radial and linear street systems suitable complementary streets may not be available to take displaced traffic. Complementary streets need to be close, preferably not more than about 125 m apart; neither should they be at such different levels that heavily trafficked connecting roads have steep gradients. Ideally each complementary pair of streets should converge to form a Y-intersection. Care should be taken to ensure that both roads do not begin/end on a cross-street with insufficient capacity.

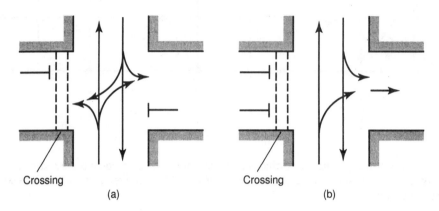

Crossing   Crossing
        (a)   (b)

Fig. 24.3 Pedestrian crossings: (a) on two-way streets the crossing may need a separate signal phase to enable safe crossing of the road by pedestrians, and (b) on the one-way street the crossing is fully protected when the side traffic is moving

A one-way scheme needs to be thoroughly signed at all points where the motorist makes a decision. 'No-entry' signs are essential at all terminals of one-way streets, and 'One-way' or 'Two-way' signs should be placed at the entrances and exits of appropriate intersections within the scheme. Supplementary 'No left turn' or 'No right turn' signs should also be displayed.

The planning and data collection necessary for the introduction of one-way traffic operation will vary according to the size and complexity of the proposed scheme. An excellent overview of the required studies is available in the literature.[1]

24.4 Tidal-flow operation

Tidal-flow operation is a traffic management tool whereby the total carriageway width is shared between two directions of travel in near proportion to the flow in each direction. The number of lanes assigned to each direction of travel varies with the time of day so that extra capacity is provided to the heavier traffic flow during, typically, peak commuter periods.

The great advantage of tidal-flow operation is that extra capacity is provided on the same road at the time required and, unlike one-way streets, traffic in the minor direction does not have to move to a complementary street. Its usage is particularly applicable to heavily trafficked bridges, tunnels and radial roads in urban areas.

Disadvantages of tidal-flow schemes include:

- their implementation can be expensive
- poor implementation can result in increasing numbers of head-on accidents
- central pedestrian refuges have to be removed
- no-parking restrictions are normally imposed

- right turns from minor roads may have to be banned
- bus stops and laybys may have to be removed if single lanes are used to carry the minor flow
- this type of operation normally cannot be applied to roads divided by central reservations.

All but three of the tidal-flow systems in use in Britain in 1987 used two or three lanes to carry traffic in the peak direction and only one lane to carry the minor flow, during peak periods;[2] all work well but are vulnerable to stoppages on the single lane if clearway restrictions are not rigorously enforced. Two-way tidal-flow operation is most readily applied to five-lane carriageways, with three lanes applied to the heavier flow during peak periods; during off-peak periods the median lane is converted into a temporary central reservation. On a six-lane carriageway, four lanes are given to the heavier flow during the peak periods; during off-peak hours traffic is allowed on three lanes in each direction.

The method most used in Britain to advise drivers and pedestrians regarding the lanes operating in each direction at stated times is to locate overhead signs at strategic points along the route. These signs range from the conventional informatory signs (including fibre-optics) to rotating prism signs which display appropriate messages at different times of the day.

24.4.1 Controlling tidal-flow operation

Traffic lights (usually in conjunction with an overhead sign) suspended over each reversible lane at the beginning and end of the tidal-flow section, and as necessary at intersections along the route, are the most flexible means of controlling lane usage under tidal-flow operation. The signals may be of the normal pattern or, more desirably, of a pattern comprising, say, a green arrow or a red cross spelled out in energised retro-reflectors so that drivers facing a downward-pointing green arrow can enter the lane beneath the arrow but cannot do so when facing a red cross. Draining of a lane prior to its usage by contraflow traffic occurs in conjunction with progressively changing lights; detectors set into the carriageway normally monitor the draining of vehicles from the variable lanes.

Movable barriers of various type are also used to separate lanes of opposing traffic under tidal-flow operation. The simplest dividers are traffic cones that are hand-laid along lane lines, or flexible posts inset into the carriageway along the lane lines. These dividers, supplemented with signs, are very effective; however, they are labour intensive and costly on an ongoing basis and are therefore best used for temporary applications, e.g. at roadworks.

Moveable 'rigid' barriers may be economically justified on very heavily trafficked routes, when the capital and maintenance costs are favourably balanced against savings in accidents, property acquisition and road construction. A 2.8 km long movable concrete lane barrier has been operating in Paris since 1986. The recently opened Jersey-type barrier across the Auckland Harbour Bridge[3] is composed of 2200 1 m long concrete block units that are hinged together and shifted using a self-propelled transfer vehicle. The transfer vehicle is fitted with a line of rollers which runs under the top of the T-profile of each unit, picks them up, conveys them in an S-curve, and puts them down exactly one lane across from their original position.

Whatever the type of control used, an essential criterion for the successful use of a tidal-flow operation is that the terminal conditions are able to cope with the traffic from the additional lane(s).

24.5 Priority for high-occupancy vehicles

The use of high-occupancy vehicle (HOV) facilities, which focus on increasing the *person-movement* efficiency of a road or travel corridor, is aimed at addressing traffic congestion and environmental concerns as well as reducing the delays to buses and increasing their reliability. A variety of HOV treatments are currently in operation worldwide; they include busways on separate rights of way, exclusive lanes, and priority for HOVs at intersections. HOV facilities that are open to private car pools as well as public buses are prevalent in North America, whereas elsewhere in the world (including Britain) the emphasis is on bus usage.[4] However, in Britain access to HOV operation is often given to taxis (and emergency vehicles) and sometimes to cyclists and disabled drivers.

24.5.1 Busways/bus roads

Guided bus systems are in use in a small number of cities (e.g. Essen, Germany) whereby specially designed buses operate in regular service on local streets and then access a special busway by engaging lateral guidewheels into the kerbing on a concrete guideway. However, most busways involve conventional buses operating on segregated bus roads in high-volume urban corridors where their usage results in time savings for the buses. When a new town, or an urban motorway, is being built the opportunity is more easily taken to reserve the continuous strip of land required by a busway; for instance, Britain's Redditch and Runcorn, which were designated new towns in 1964, have exclusive bus roads of 13 km and 21 km long, respectively, as a consequence of their master plans.

If 2.5 m is taken as the bus width and 0.75 m clearance is needed on either side, the minimum width of a single-lane busway is 4 m. A one-way carriageway width of 6 m is needed to pass a disabled bus while a width of 6.75 m allows comfortable two-way operation. Ideally bus roads should be grade-separated from general purpose roads; if this is not feasible traffic signals should be used to give priority to the buses at at-grade intersections.

A *bus-only street* is a form of bus road that is created by banning all vehicles except buses and (usually) taxis and necessary access vehicles from an existing street during all or part of the day. Particular care has to be taken with the signing of bus-only streets to ensure their safe usage by pedestrians. Bus-only streets are most usually applicable to busy destination streets (e.g. London's Oxford Street) or between adjacent distributor roads, or between cul-de-sacs in residential areas to allow buses to cross quickly between housing precincts.

24.5.2 Priority lanes for buses

Bus priority lanes are of two types: with-flow and contra-flow. With *with-flow* operation the buses operate on a reserved lane in the direction of the normal traffic flow. The priority lane is normally the kerb lane; however, on heavily travelled arterial roads in cities or on an inter-city motorway, the lane next to the central reservation may be used when there are no bus stops, as it results in less interference to buses on long runs. In practice, most with-flow lanes are fairly short (often <250 m in Britain) and are taken up to and through intersections or are stopped short of the intersection.

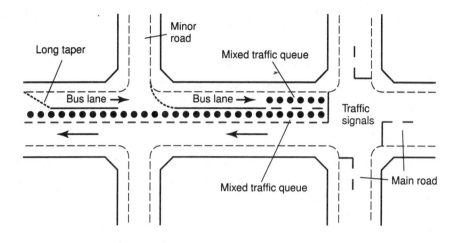

Fig. 24.4 General layout of a pre-intersection with-flow bus lane (not to scale)

Figure 24.4 shows a bus lane that is terminated back from the stop-line at an intersection to allow mixed traffic into the lane. Ideally this setback (which is usually about 50–80 m long) should be just great enough to allow the full capacity of the approach road to be used throughout the green period in a cycle during which the queue on the road does not clear. This arrangement ensures that buses can 'jump the queue' with minimal disruption to traffic flow at the intersection. The start of the bus lane ideally should have a long gradual taper to allow non-priority vehicles to merge into adjacent lanes safely.

While kerbside-reserved lanes increase the journey speeds and reduce the travel times of buses (which usually attracts new riders), they also (a) require rigorous policing to prevent their usage by non-priority vehicles, (b) cut across commercial vehicle access to kerbside properties, (c) may increase congestion (and queue lengths) for other traffic and encourage its diversion into adjacent non-suitable streets, and (d) may initially increase accidents.

Contra-flow priority lanes are usually installed in the kerbside lane of one-way streets and used only by buses. In general they shorten bus routes, as compared with one-way operation; save bus running time, increase service reliability, and lower bus operating costs; reduce bus-passenger walking distances; and increase bus ridership. Disadvantages of contra-flow operation include: the cost of additional signalling/signing/channelisation to obviate problems resulting from the reintroduction of conflicts at intersections; the capacity of the traffic system may be reduced; the design of linked-signal progression (which is very efficient in a one-way system) may have to be compromised to give buses reasonable progression in the opposite direction; and problems of loading and unloading commercial vehicles are hard to solve.

In contrast with with-flow priority lanes contra-flow lanes are essentially self-policing. They are also more easily justified than with-flow lanes when traffic congestion is not great. While with-flow lanes may operate for 24 hours or for the working day (7 a.m. to 7 p.m.) or during peak periods only, contra-flow lanes almost invariably operate for the full 24 hours of every day.

Clear signing and lane marking is essential for the safe and efficient operation of both with-flow and contra-flow priority lanes.

24.5.3 Priority for buses at intersections

A simple way of giving priority to buses is to exempt them from turning prohibitions at intersections. Often the number of buses involved is so small that the resultant interference is negligible, e.g. the buses may be able to use longer intergreen periods to cross in the face of opposing traffic.

Priority for buses at isolated intersections can also be ensured by favouring traffic streams known to contain buses over other streams when calculating normal delay-minimising signal timings. One objective way of doing this is to include the average vehicle occupancy in the calculations so that person-delay is minimised. In the case of linked signals the relative timings of adjacent signals on bus routes may be adjusted to give a favourable progression to buses.

If buses are equipped with transponder units which can be detected by a signal controller, they can be given priority when it is needed. For example, priority-by-extension can be caused to occur when a signal is already green and it needs to be extended until the bus has negotiated the intersection. Priority-by-recall can be provided when a signal is showing red and it needs to be recalled to green as soon as is safely possible. Priority-by-hurry-call refers to the situation where a signal is red but the duration of the various stages during which it remains so is reduced to the minimum allowed when a bus seeks priority.

24.6 Waiting restrictions and parking control

The main functions of a road are to provide for the safe and efficient movement of people and goods in vehicles, and to provide access for adjacent properties. Thus, ideally, all stopping and parking of vehicles should take place off the carriageway. In practice, of course, this is not possible and vehicles are permitted to stop and to park at the kerbside unless it is specifically prohibited.

Although convenient for motorists, on-street parking contributes to accidents. For example, a major study of road accidents in 32 U.S. cities concluded[5] that 18.3 per cent involved parking either directly or indirectly, while 20.7 per cent of the pedestrians involved in road accidents in Britain in 1983[6] (two thirds of whom were children or of retirement age) were masked by stationary vehicles.

Kerb parking also reduces the traffic capacity of a street as part of the carriageway width is lost for movement purposes and frictions are caused by parking manoeuvres in the lane next to the parking lane.

Road locations where stopping/waiting/parking should not be permitted for safety and/or capacity reasons include:

- on approaches to or departures from major intersections and at-grade rail crossings
- at or immediately adjacent to pedestrian crossings, fire hydrants, and public transport lanes and stops

- in front of driveways or alleyways
- where double 'no-overtaking' lines are provided at the centre of the carriageway
- at locations which might interfere with the movement of emergency vehicles, e.g. at hospitals, ambulance and fire stations
- in tunnels, on bridges, or on narrow streets.

24.6.1 Clearways

Waiting and parking restrictions are often also applied to long lengths of busy arterial road during peak traffic hours, in the direction of the predominant flow, in order to increase capacity. These clearways are controlled by edge-of-the-carriageway markings and signs which indicate the length of the no-stopping zone and the times of operation. In large cities the operating times for clearways need to take into account that the times of peak periods move progressively with the traffic toward/away from the central area. In large urban areas also, and on some rural roads where it is difficult to define when the restrictions should not apply, some clearways may operate for the full 24 hours.

24.6.2 Bus stops

Bus stops are provided at the kerb to enable passengers to be picked up and discharged at locations which meet their grouped desires. Very often this latter requirement results in bus stops being sited just prior to or after intersections, with consequent interference to the efficient operation of the junctions. Ideally these bus stops should be in tapered laybys that are clear of the carriageway, so that boarding/alighting can occur in safety and the stopped buses cause the minimum interference to moving vehicles. In practice, however, buses using such laybys on busy roads have difficulty in re-entering the traffic stream (which works against bus priority policies). At a busy bus stop close to an intersection this problem may be obviated by extending the full-width bus bay through the intersection, so that it becomes an extra priority lane. However, this is a very expensive solution and more commonly partial bus bays are provided which do not cater for the full widths of stopped buses but enable other vehicles to pass them; the buses are then able to re-enter the traffic stream more easily.

Bus stops in urban areas are generally located 300–500 m apart, and in residential areas are ideally not more than a five-minute walk from any individual home. The sites should be selected so that the need for passengers to cross a road to change buses is avoided where possible, and the walking distances between bus stops at interchange points are minimised. For safety reasons a bus stop that is close to a pedestrian crossing should normally be sited *after* the crossing.

Bus stop zones should be clearly marked on the carriageway and signed at the kerb. However, in Britain these carry no mandatory significance for drivers of other vehicles unless reinforced by an appropriate waiting restriction or by a Bus Stop Clearway Order which makes it illegal for other vehicles to stop at a bus stop between 7 a.m. and 7 p.m.

24.6.3 Commercial vehicle loading/unloading

The loading and unloading of goods for businesses should preferably happen off-street at loading docks, service yards and alleyways. In central areas and suburban centres, however, there is normally a need for kerb space where commercial vehicles can service business premises during the working day. If this space is not provided, double-parking will often occur with consequent major interference to traffic flow.

Loading and unloading zones should be marked at the kerb and signed with details of the times when they are to be used only by commercial vehicles. These zones are often located as extensions to existing no-waiting zones at intersections, alleyways, driveways, fire hydrants, etc., so that they can be more easily accessed by goods vehicles. The length of kerb reserved for loading and unloading will vary according to need. Usage at any given location should not be restricted to abutting premises as it is a public waiting space. The hours during which goods vehicles may use the reserved space will also vary with need; however, loading/unloading normally should not be allowed on busy streets during peak traffic periods. (Note: Most deliveries tend to take place in the morning.)

24.6.4 Time-limit car parking at the kerb

In central and suburban areas it is common practice to ration the available spaces so that parking preference is given to persons whose activities primarily contribute to the development of these areas, e.g. shoppers and visiting business people. This usually means the introduction of parking schemes to control the lengths of time that vehicles stay. Successful schemes usually contain the following four features:

1. *They utilise parking control mechanisms that are easily understood by motorists and result in reductions in parking durations, increases in dynamic capacity, and improved parking availability.*

The optimum utilisation of parking spaces is generally considered to be about 85 per cent, with about 15 per cent of the spaces being available within reasonable walking distances of the principal traffic generators during the peak parking-demand period. The degree of utilisation in any instance is regulated by the allowable parking durations and the charges imposed. Time-limits commonly imposed are 15 minutes for parking zones near errand-type establishments, two hours in most town centres (with 40 minutes in the 'cores' of most large central areas), and four hours on the edges of town centres and in residential areas adjacent to large generators of all-day parking. Before-and-after parking studies (see Section 13.7) are used to ensure that the parking time-limits imposed, and the parking fees (if any), have the desired effects upon parking durations, capacity, and availability (e.g. see reference 7).

Parking control used to be carried out by displaying the authorised time-limit on signs (still current practice) and then using a police officer or traffic warden – complete with watch and chalk – to mark a tyre on each parked vehicle during a regular patrol. The officer then returned to the scene after the passage of the authorised time and checked for, and issued a legal summons to, marked vehicles which had overstayed the time-limit. The inefficiency of this method of control, and the fact that parking was always free, led to the eventual development of user-pay control systems that are in common usage today, e.g. parking meters, pay-and-display parking vouchers, and pre-paid parking cards.

Parking meters are timing mechanisms activated by insertion of a coin. They control individual parking spaces and can be set to allow parking for a single fixed period of time or, dependent upon the coinage inserted, for varying lengths of time up to the authorised time-limit. As soon as the motorist has used up the allotted time, a violation signal is displayed on the meter which is easily seen by a patrolling traffic warden, who is authorised to issue a fixed-penalty ticket.

With *pay-and-display control* a parking voucher for the desired length of time, up to the time-limit, is obtained from a conveniently located kerbside ticket dispensing machine, following insertion of the appropriate fee (usually in coinage but also, more recently, using pre-paid decrementing magnetic cards[8]). The voucher, which has printed upon it the time it expires, is then displayed by the motorist, often on the car's dashboard or, preferably, by affixing it to the inside of the car windscreen where it is most easily monitored by a patrolling warden.

With *pre-paid parking cards* (see Fig. 24.5) the cards are usually purchased, either singly or in books, at designated governmental or retail outlets, such as local authority

Fig. 24.5 Pre-paid parking card issued by West Sussex County Council

offices, petrol stations and newsagents. Each card has a prescribed value and is valid for a fixed parking duration, e.g. 20 or 40 minutes, or one or two hours. Upon arrival at the kerbside the motorist pierces (or scratches) five locations on the card which describe the month, date, day, hour and minute (to the nearest 5 minutes) of the start of parking. Each marked card, or combination of cards, is then displayed for inspection by trapping its top in the window nearest to the kerb.

The *parking disc* is a form of pre-paid card which is obtained free from the designated outlets. A time-marked circular disc within a pocket attached to the card is rotated by the motorist until the time of arrival at the kerbside is displayed in an aperture at the front of the card. The disc, which is re-usuable, is then affixed to the inside of the windscreen. With the parking disc, and the regular pre-paid parking card, the patrolling traffic warden observes the time at which parking was started and then, by checking a watch, determines whether a car is parked in excess of the permitted length of time – the sign-displayed time-limit in the case of parking discs, and the lesser of the sign-displayed time-limit or purchased time on the card(s) in the case of the regular pre-paid cards.

Table 24.1 summarises the main advantages and disadvantages of the various parking control systems that are currently in wide usage.

Table 24.1 Comparison of various parking control systems

System	Advantages	Disadvantages
Parking meters	Simple to use Well proven Generate revenue Easiest system to supervise	Expensive to install/maintain and to adjust to new charges Aesthetically unsightly Obstruct the footpath Can attract vandals
Pay-and-display parking vouchers	Simple to use Small number of machines are less environmentally intrusive Lower capital and maintenance costs than for rows of meters, and fee structure is easily changed Generate revenue	Impose short walking distances Supervision is more difficult than for meters Additional signing required Can attract vandals
Parking cards	No intrusive equipment No capital maintenance or vandalism costs Revenue collection (from outlets) is easy Fee is easily changed	Usage is complex for some Creates problems for visitors Greatest risk of cheating and requires more careful supervision Fees are shared with sales outlets

2. The parking arrangements make efficient use of the carriageway and cause the minimum interference to moving traffic.

In most British towns this usually means using parallel parking. Of the two arrange-ments shown in Fig. 24.6 that composed of pairs of, say, 4.9 m bays with a 1.2 m manoeuvring space between each pair is to be preferred as it (a) reduces the kerb length required for two cars by 10 per cent while still allowing 6.1 m manoeuvring space for each car, (b) permits adequate parking space for long cars, and (c) reduces the number of unsightly pedestals at the kerb if meters are used.

The use of angle parking instead of parallel parking can be a cost-effective way of providing additional spaces within a given length of kerb. However, the greater the angle to the kerb the wider the carriageway width necessary for the extra manoeuvring required. Converting from parallel to angle parking is generally associated with increased numbers of accidents resulting from (a) additional vehicles being involved in parking manoeuvres, (b) visibility problems for drivers engaged in the back-out manoeuvre, and (c) additional rear-end accidents in the traffic stream as motorists stop suddenly upon noticing exiting vehicles backing into traffic lanes.

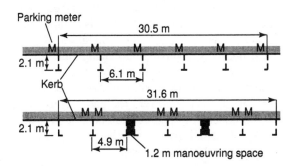

Fig. 24.6 Alternative parallel parking arrangements at the kerb

3. *The controlled zones are clearly defined, and signing and parking instructions leave no doubt as to the conditions under which parking may occur.*

The general principle is that signs should at least be displayed at both ends of a controlled length of kerb, which inform the moving motorist regarding the method of control, the times and days of operation, and the maximum parking time allowed. Detailed information regarding the payment of fees is usually provided at the control equipment or on supplementary signs.

4. *The parking regulations are rigorously enforced.*

Adequate resources for enforcement are essential to the success of any parking scheme.[9] Most enforcement is now in the hands of patrolling wardens empowered to 'ticket' a vehicle in breach of the regulations. The enforcement process is enhanced if unpaid fixed-penalty tickets are automatically registered as fines with a surcharge (commonly 50 per cent), which are enforceable through the courts.

In recent years the use of highly visible wheel-clamping (in conjunction with a heavy supplementary fine) has increased as a means of stopping illegal parking at sensitive locations. If the illegal vehicles cause an obstruction, e.g. on a clearway, they are often towed away and impounded until the fine and cost of towing is paid. The use of wheel-clamping is now being extended to target motorists who persistently ignore parking tickets (a problem that is very common in large cities); new technology in the form of hand-held ticket issuing machines is now available which also allows a database of persistent evaders to be stored and accessed on the street by the parking-control officer.

24.6.5 Residents' parking

A parking problem that has received greater recognition in recent years is the lack of off-street parking spaces at many inner-city and suburban homes. In Britain, for example, nearly half the housing (and in Greater London nearly two thirds) was built before 1945, when car ownership levels were low and parking space was not a priority for most people. It has been estimated that the number of cars kept on-street or on verges in 1989 was about 4.8 million (28 per cent of the total), and that this number may double by 2025.[10]

Householders with cars expect to park outside their homes. While that expectation is adequately met in very many of the older areas, problems most usually arise in established high-density residential areas immediately surrounding such locales as town centres, centres of recreation and entertainment, large hospitals and industrial complexes, educational institutions for older students, established shopping centres, and suburban railway stations. In these locales there is active competition for parking space between the residents and the users of the traffic generators.

Tackling the problem normally involves introducing time-limit parking control but granting the residents exception from its operation.[10] The maximum time-limit selected will vary according to the nature of the generator but should be short enough to discourage kerb-parking by non-residents; it can vary from zero (no parking) to, say, four hours (to deter commuters). Residents are issued with permits (to attach to windscreens) which exempt their vehicles from the time-limit and any parking charges imposed. The residents' parking permits may be free or a fee may be levied to assist in covering the cost of the scheme.

Detailed advice regarding the introduction of residents' parking schemes is available in the literature.[11]

24.7 References

1. Pline, J.L. (ed.), *Traffic engineering handbook.* 4th edition. Englewood Cliffs, NJ: Prentice-Hall, 1992.
2. McKenna, M.N., and King, T.A., Review of tidal flow systems. *Traffic Engineering and Control,* 1987, **28** (10), 544–7.
3. Leak, M.J., Hawkins, N.V., Sansom, E.P., and Dunn, R.C.M., The movable lane barrier on Auckland's Harbour Bridge: Problems, solutions and features. *Proceedings of the Australian Road Research Board,* 1992, **16** (4), 173–87.
4. Turnbull, K.F., International high-occupancy vehicle facilities. *Transportation Research Record 1360,* 1992, 126–137.
5. Seburn, T.J., Relationship between curb uses and traffic accidents. *Traffic Engineering,* 1967, **37** (8), 42–47.
6. *Road accidents in Great Britain 1983.* London: HMSO, 1984.
7. Poxon, M.J., On-street card parking scheme. *Highways and Transportation,* 1989, **36** (8), 13–15, 17.
8. Pickett, M.W., and Gray, S.M., *Magnetic pre-payment cards in Rushmoor.* Project Report 29. Crowthorne, Berks: Transport Research Laboratory, 1993.
9. Cullinane, K., and Polak, J., Illegal parking and the enforcement of parking regulations: Causes, effects and interactions. *Transport Reviews,* 1992, **12** (1), 49–75.
10. Balcombe, R.J., and York, J.O., *The future of residential parking.* Project Report 22. Crowthorne, Berks: Transport Research Laboratory, 1993.
11. Wadsworth, W.S.A., Residents' parking. *Municipal Engineer,* 1988, **5**, 149–160.

CHAPTER 25

Physical methods of traffic control

C.A. O'Flaherty

By physical methods of control are meant those procedures which are essentially self-policing in terms of their influence on vehicular behaviour. They include the application of what is now known as traffic calming to existing roads, as well as the use of pedestrian crossings and precincts, and cycle routes.

25.1 Traffic calming

Traffic calming is a generic term used in Britain to describe changes to the horizontal and/or vertical alignments of existing roads in built-up areas, e.g. in residential and shopping areas, in order to reduce the speeds of motor vehicles. In a wider sense it has also been defined[1] as a transport policy concept which includes a strong promotion of pedestrian, public and bicycle transport, as well as a reduction in average speeds in built-up areas. Traffic calming can also be used to control vehicular speeds in outdoor recreational areas.[2]

25.1.1 Development of traffic calming

It is sometimes said that traffic calming had its genesis in *Traffic in Towns*,[3] which was published in 1963. In this seminal report considerable emphasis was placed on determining the amount of traffic that, say, a residential street could carry before the environmental conditions became intolerable. It was then proposed that traffic management techniques be used to ensure that unwanted vehicles were diverted from these streets so that their environmental capacities (which were much less than their traffic capacities) were not exceeded. Traffic management techniques subsequently used to divert traffic from residential streets onto major traffic routes included street closures, turning bans, creation of pedestrian precincts, and curbing the use of traffic 'rat-runs' using signals, stop signs, etc. What was never considered in that era, however, was the physical redesign of existing streets deliberately to slow down traffic; this was virtually impossible to contemplate at that time.[4]

It is only since the mid-1980s that the concept of changing the physical alignment(s) of the street, and thereby its appearance, so as to discourage its usage by heavy volumes of high-speed traffic and to improve the amenity of its environs, began to be accepted in

Britain. This acceptance resulted from a growing concern amongst the populace about the impacts of traffic upon the environment. Transport planners and traffic engineers also noted the success that their counterparts in mainland Europe had in controlling the use of the motor vehicle in residential areas following physical changes to the streetscape,[5] especially in Woonerven[6] (literally 'living yards') residential areas in the Netherlands where the emphasis was placed on turning streets into spaces to be shared by the pedestrian, the cyclist, and the motor vehicle (which was initially constrained to a speed below 15 km/h in these areas).

At the present time applications of traffic calming in Britain tend to be identified with residential areas, both in suburbia and in inner cities, where traffic capacity is not an issue. Traffic calming schemes are also used in villages without bypasses (to return the villages to their occupants), in town centres (to encourage their greater use by pedestrians), and in bypassed towns and villages (to redefine surplus road space for use by other groups of road users). In the future, however, it is possible that their use will encompass corridor-level traffic restraint where capacity is an issue, i.e. traffic calming at this level could be used to discourage traffic growth along these corridors while at the same time avoiding the transfer of traffic onto adjacent local streets.[7]

25.1.2 Objectives

The main objectives underlying traffic calming are to:

- reduce the higher speeds of vehicles in the traffic stream(s)
- create road conditions which encourage motorists to drive carefully and calmly
- remove extraneous car and commercial vehicle traffic from the road being calmed
- improve amenity and enhance the environment
- reduce accident numbers and severity.

The key objective is that of reducing high vehicle speeds; if that is achieved the other objectives will generally be attained also. Achievement of these objectives normally results in a general improvement in the quality of life in the environs associated with the calmed road and often encourages their positive redevelopment.

Slower speeds reduce the likelihood of accidents since all road users, whether they be drivers, cyclists or pedestrians, have more time to judge and react to each other's actions. The higher a vehicle's speed the greater the distance it needs to stop. The higher the traffic speed (especially if it is above about 48 km/h) the more likely it is that pedestrians, especially the elderly, seeking to cross a road will misjudge the gaps available in the traffic stream.

When a pedestrian–vehicle (or cyclist–vehicle) collision occurs, the lower the speed of the vehicle the more likely it is that the pedestrian (or cyclist) will survive the accident. For example, if an accident occurs at 70 km/h, the likelihood that the pedestrian will be killed is estimated at 83 per cent; at 50 km/h the likelihood of a fatal injury is still 37 per cent, whereas at 30 km/h it reduces to 5 per cent.[1]

Vehicle speeds, traffic volumes, the proportions of heavy vehicles in the traffic streams, and the design of the road all influence the level of noise pollution experienced at the footpath. The lower the traffic speed (down to about 30 km/h) the lower the noise levels, assuming that the traffic volumes and the commercial vehicle proportion remain constant.[8] In practice, of course, roads with slow traffic speeds also make it more likely that vehicles that are tempted into using them as through routes will divert to higher-speed routes.

The level of air pollution is reduced if motor vehicles are driven carefully and calmly. In general, the higher the driven speed in built-up areas, the greater the amounts of acceleration, slowing down, and braking that are required. All of these increase air pollution.

If vehicles travel at lower speeds the road is less intimidating to other users. Further, at 30 km/h or below, vehicles do not need as much road space as when driven at high speeds. This 'spare' space can often be reclaimed and planted or, perhaps, given over for use by public transport, cyclists or pedestrians. If vehicles approach intersections at low speeds, the layouts of the junctions can often be changed to allow for easier and safer usage by pedestrians and cyclists.

The traffic calming concept is not necessarily assured of success if only traffic engineers are involved in the development of ideas for its implementation in a given area. *Its likelihood of success is considerably increased if it incorporates the preferences and requirements of the people living in the locale undergoing treatment,* many of whom will be car users. Thus, community participation through public meetings, questionnaire surveys, the organisation of exhibitions and the like, is central to the successful planning and design of a traffic calming scheme.

25.1.3 Engineering elements

Engineering elements commonly used in traffic calming projects in Britain[9] are listed in Table 25.1. Some of these physically restrain road users, others psychologically encourage them to behave in a certain way. Some elements can be designed to function differently for different road users; for instance, a flat-topped road hump may be used to reduce vehicle speeds and, if well located relative to pedestrian desire lines, it will also serve as an easy-to-negotiate pedestrian crossing between kerbs. A mix of elements is often used in traffic calming schemes; for instance, a combination of the more severe measures may be used on roads that are unsuitable for through vehicles, and less severe ones on other roads.

Table 25.1 Engineering elements commonly used in traffic calming projects to meet objectives (a) to (e) specified at the start of Section 25.1.2

Engineering element	Objective					Engineering element	Objective				
	a	b	c	d	e		a	b	c	d	e
Road humps	x	x	x		x	Rumble strips	x	x			x
Speed tables	x	x	x		x	Different surface					
Cushions	x	x	x		x	treatments		x		x	
Road narrowing/						Gateway treatments	x	x		x	x
throttles	x	x	x	x	x	Road markings	x	x			x
Chicanes	x	x	x	x	x	Landscaping		x		x	
Footway build-out				x	x	Electronic information/					
Central islands	x	x			x	enforcement	x				x
Traffic management											
measures	x		x		x						

Road humps

The most effective measures used to lower vehicle speeds are vertical deflectors, commonly known as *road humps*. Variants of the standard road humps are known as *speed tables* and *cushions*.

Various road hump designs are used in different countries. Basically they can be divided into those that are narrow enough in the direction of travel to be straddled by the wheels of all normal vehicles, and wide humps that cannot be straddled except by a minority of large vehicles. Narrow humps administer a sharp jolt to the vehicle suspension, unless the vehicle is travelling at a low speed, i.e. when the crossing time is long enough for the vehicle body to deflect upward as each axle passes over the hump. While narrow humps can be crossed at quite high speeds without undue discomfort, the driver is usually afraid to do so for fear of damaging the vehicle or losing control. Wide humps result in a less severe ramp effect and have a longer crossing time; a greater height may also be used without fear of grounding low-slung vehicles.

Road humps are most normally used in Britain on roads that have a 48 km/h (30 mile/h) speed limit although, in recent years, they have also been used to traffic-calm areas subject to 32 km/h (20 mile/h) limits. The *standard road hump* has a circular cross-section (see Fig. 25.1(a)) with a chord length (in the direction of travel) of 3.7 m and a height of 50 mm (minimum) to 100 mm (maximum). This hump may be extended from kerb to kerb across the full road width (in which case alternative, more expensive drainage arrangements must be made), or with a drainage gap of 200 mm adjacent to each kerb. Typical 85th percentile and mean vehicle crossing speeds for 100 mm humps average 29 km/h and 22.5 km/h, respectively.[10] Speeds midway between 100 mm high humps are given by

$$V_1 = 16.73 + 0.087S \quad (r = 0.80 \text{ and } S.E. = 1.59) \tag{25.1}$$

$$V_2 = 12.10 + 0.092S \quad (r = 0.87 \text{ and } S.E. = 1.28) \tag{25.2}$$

where V_1 = 85th percentile speed (mile/h), V_2 = mean speed (mile/h), S = hump separation (m), r = correlation coefficient, and $S.E.$ = standard error (mile/h).

Standard 100 mm high humps give better speed reductions than lower humps, for a given spacing. British practice is for them not to be more than 150 m or less than 20 m apart, and the first in the series must be 40 m or less from a low-speed feature such as an intersection or bend. Inter-hump spacings vary according to the speeds desired at the midway point; however, spacings of about 40 m at a typical site will achieve an average 85th percentile midway speed of 32 km/h.

The heights of humps are often varied within a traffic calming scheme. Thus hump heights of 50 to 75 mm may be used on bus routes or on roads likely to be used by emergency services.

Within the road hierarchy, standard road humps are commonly used on residential access roads when vehicle speeds are excessive and severe reduction measures are required, and on distributor roads where lower speeds are required. There need be no loss of parking spaces on roads with humps as vehicles are able to park on them.

Disadvantages of standard circular road humps include:

- they are often opposed by bus companies and emergency services, depending on the types and frequency of vehicles that they use
- cyclists often try to use the tapers to negotiate humps which taper down short of the kerb

Fig. 25.1
Examples of road humps: (a) a standard hump with a circular profile, and (b) a flat-top raised intersection

(a)

(b)

- the humps can be so successful that they lead to increased traffic on other adjacent roads, which may not be desirable
- certain long-wheelbase, low-clearance vehicles may ground when crossing
- very many humps may be required on a long road.

Trapezoidal flat-top road humps (also known as *raised tables* and *speed tables*) may be used as an alternative to circular profile humps. They are often regarded as being more environmentally acceptable to pedestrians, as they provide flat crossing places between kerbs. As with circular humps these are typically 50 mm (minimum) to 100 mm (maximum) high and may extend across the full road width or have a 200 mm drainage gap adjacent to each kerb. The profile used in Britain is composed of a 2500 mm (minimum) long flat top with two ramps, each at least 600 mm long; the length at the carriageway level is thus 3.7 m (minimum). Typical vehicle crossing speeds for 100 mm high tables with ramps of 1:10 to 1:12 are 27.4 km/h (85th percentile speed) and 21 km/h (mean speed). Steep ramps of 1:6 will give greater speed reductions at the humps but are less acceptable to drivers and residents as they are severe and may cause grounding of low vehicles. The speed-separation relationships for flat-top road humps that are 100 mm high and have average slopes of 1:10 are:

$$V_1 = 12.95 + 0.107S \quad (r = 0.93 \text{ and } S.E. = 1.70) \tag{25.3}$$

and
$$V_2 = 11.06 + 0.090S \quad (r = 0.91 \text{ and } S.E. = 1.61) \tag{25.4}$$

where V_1, V_2, S, r and $S.E.$ are as defined for standard circular road humps.

Flat-top humps are used on the same types of road as standard circular humps. Particular locations where they are deemed to be especially useful are outside schools and shops, at zebra and pelican crossings (especially across roads where high proportions of the pedestrians crossing are people with disabilities or with prams), and at entrances to 32 km/h (20 mile/h) zones (preferably with the addition of narrowing and coloured surfaces in order to highlight the 'gateway' effect).

Raised intersections (see Fig. 25.1(b)) are a development of the flat-top hump. They are used at intersections which are known to be dangerous, but where major reconstruction is not judged to be justified or viable. The raised area is often extended at least 6 m into side roads so that the leading vehicles can be level on the immediate approaches to the intersection. This also allows Give Way markings to be placed in their conventional position.

Speed cushions, although relatively new in Britain, have been used in mainland Europe for many years where they are regarded as being user-friendly to buses, cyclists and emergency vehicles. Cushions are a derivative of the flat-topped hump and are often used with chicanes.

Instead of extending from kerb to kerb, a speed cushion is typically placed in a lane so as to end about 1 m from a kerb and designed so that bus wheels (on the front axle at least) are able to straddle it. Cars, because of their narrower wheelbase, must put at least one wheel on the cushion. Research has shown that cushions can lead to grounding problems for some limousines and low sports cars if the cushion height is more than 75 mm or if the approach ramps are steeper than 1:8. Suggested[10] overall lengths (in the direction of travel) range from 2 m to 3.7 m. Overall widths in excess of 2 m cause problems for buses, while platform widths of less than 1.4 m are likely to have a reduced effect. Side ramps steeper than 1:4 are considered dangerous for motorcycles. Bicycles are normally able to pass to the left of each cushion.

Build-outs

As the name implies, a build-out describes a feature extending into the carriageway, on one side of the road only, which narrows the road. A build-out can be at footpath level or substantially higher and containing, for example, planted material. It may be connected with the footpath/verge or, more usually, separated from it by a drainage space.

From a safety aspect the length of time that a pedestrian spends on the carriageway when crossing a road should be as short as possible. This can be assured by increasing the effective width of the footpath by constructing a build-out so as to narrow the road at the crossing point(s). A very effective result can be obtained by combining a *footway build-out* and a flat-topped road hump.

If two build-outs are constructed opposite each other, the construction is often called a *pinch-point* or a *throttle*. This is a very effective self-enforcing speed control mechanism when combined with a road hump.[11]

Fig. 25.2 Chicane in a residential street

A *chicane* is a narrowing formed by locating build-outs alternately on each side of the carriageway, with each chicane consisting of two or more build-outs. If the build-outs are closely spaced, say 10–15 m apart, they can be very effective at reducing speeds; however, buses and other large vehicles can find it difficult to manoeuvre through chicanes of this spacing. Wider-spaced chicanes depend more on opposing traffic to provide effective speed control. A chicane (see Fig. 25.2) is normally not a suitable place for people to cross a road, as a driver's attention is usually concentrated on negotiating the element rather than in looking out for wayward pedestrians.

Build-outs are usually connected to a verge or footway. Thus a series of build-outs located on one side of a road can be used to provide protected roadside parking spaces for cars.

Rumble devices

Rumble strips, *rumble areas* and *jiggle bars* are attention-getting raised areas that are extended across the carriageway. They are usually less than 15 mm high, and are of a colour contrasting with the road surface (but not white, to avoid confusion with road markings). Strips and bars are usually installed in groups (see Fig. 25.3) with a clear length of road between each group. The spacings between strips/bars within a group are usually decreased as the distance to the locale being warned about decreases.

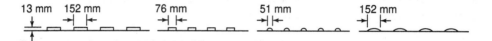

Fig. 25.3 Jiggle bars (diagrammatic)[10]

The function of a rumble device is to alert drivers to road signage warning of the need to slow down for particular hazards. It does so by sending severe vibrations through a crossing vehicle, making the ride feel very uncomfortable for the driver and passengers.

Rumble devices are claimed to be very effective in reducing speeds and accidents, especially when used at locales where drivers are accustomed to high speeds and need to be alerted to a major change in conditions, e.g. at the end of a high-speed dual carriageway that terminates on a roundabout and/or at the entrance to a village.

The main disadvantage of rumble devices is the noise that they produce: one rumble area (composed of roadstones held in an epoxy resin binder) has been measured to produce an increase in noise level of 10 dB(A) at most speeds in cars and vans. The noise effect limits their usage to locations which are not sensitive to a sound detriment of this nature; typically, these locales are more than 200 m from residential properties, hospitals and similar establishments. Rumble devices are also not popular with cyclists and motorcyclists and a gap of 0.75–1 m is often left between the end of a device and a kerb to allow these road users to have clear passage.

Gateways

These are provided at locales where it is desired to inform the driver of a change in character of the roadway, e.g. at the start of a traffic calming scheme or at the entry to a village. They usually consist of structures carrying signs at the side of the road and even above it. While not necessarily effective speed-reducing devices in their own right, they will act as such when combined with other measures, e.g. pinch-points or rumble strips.

Other elements

Many other engineering elements are used to reduce vehicle speeds in traffic calming schemes. For example, *false roundabouts* have been used on long straight roads at locations where there are no side road connectors. *Overrun areas* of a contrasting colour, but usually constructed to form a sloping (< 15 degrees) raised surface, are often located adjacent to a kerb either at the near side of a road or at a traffic island; they give the appearance of a narrowing of the carriageway while allowing large vehicles easily to negotiate the overrun element. *Islands* are often used in conjunction with gateways at the start of traffic calming schemes. Offset islands have been used to help create chicanes while protecting a cycle lane.

In many instances *landscaping* can greatly add to the aesthetic value and acceptability of a traffic calming scheme to the public. However, not all schemes lend themselves to landscaping; for instance, plantings of trees or high shrubs should not be used at locations where pedestrians cross a road or where children might play on the carriageway.

25.2 Pedestrian priority

The overall aim of traffic calming is to tame the motor vehicle so that its usage at particular locales is compatible with cyclists and pedestrians. In the case of pedestrian priority the aim is to subordinate the motor vehicle to the pedestrian at particular locales.

25.2.1 Pedestrian considerations

Pedestrians are the largest single group of road users and walking is the most used transport mode (because it is involved in all modes of travel). Pedestrians encompass people of both sexes and of all ages and socio-economic groupings. They include people of various degrees of physical fitness, including the disabled. There is no test that has to be passed in order to become a pedestrian on a public way. Pedestrians are very vulnerable to serious injury when involved in a collision with a motor vehicle.

The plan view of the average adult male person occupies an area of about 0.14 m². However, a module size of 0.21 m² (= 460 mm by 610 mm body ellipse) is commonly used to determine practical *standing capacity*.[12] This allows for body sway, the fact that many pedestrians carry personal articles, and natural psychological preferences to avoid bodily contact with other persons.

Walking speeds vary over a wide range, generally determined by age of pedestrian, purpose of trip, crowd density and other traffic impediments. Figure 25.4 shows basic walking relationships determined by various researchers. These data suggest that the maximum *walking capacity* of a footway is about 25 ped/min/ft-width (82 ped/min/m), and this occurs when the average module size is 5–9 ft²/ped (0.46–0.84 m²/ped). As a comparison, Table 25.2 summarises one set of recommendations regarding pedestrian flows to be used for design purposes in various (non-footpath) types of public spaces.

In practice most walkways and pedestrian-operated traffic devices are designed for the 'normal' pedestrian. This can result in the young, the elderly, and the disabled being disadvantaged when walking, and special care should therefore be taken to review any design to ensure that their needs are taken into account. For example, pedestrians with poor eyesight require a strong contrast between the road and footways; kerbs or strips of textured paving can provide this physical differentiation. People with hearing problems rely on seeing vehicles in order to cross a road safely; thus an unobstructed view from the side of the road is particularly important for these pedestrians. Elderly people walk more slowly, and so pedestrian phases in signal operations need to be designed to take this into account at individual intersections. Wide roads may require centrally located

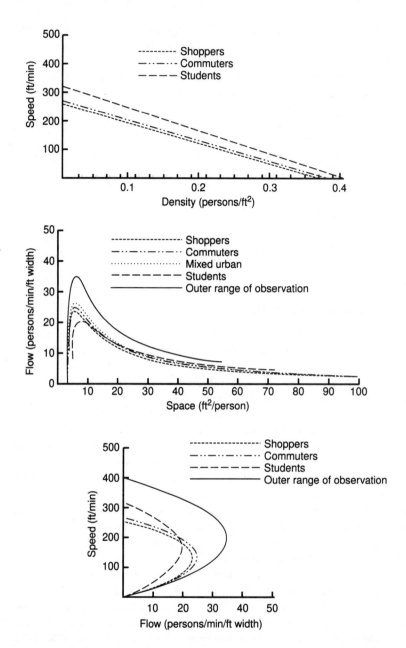

Fig. 25.4 Relationships between pedestrian speed and density, flow and space, and speed and flow for various groups of pedestrians[24]

raised refuges (normally 1.8 m wide) on the carriageway at pedestrian crossing points, to assist persons who must walk slowly because of natural infirmities or when escorting children. Footpaths and carriageway surfaces at crossings should be stable, not slippery or subject to undulations; if composed of flags or block paving, these should not be broken or have raised edges.[13]

Table 25.2 Suggested level of service for horizontal pedestrian movement[12]

Level of service*	Module size (m²/ped)	Flow rate (ped/min/m)	Sample applications
A	>3.3	23	Public buildings or plazas without severe peaking
B	2.3–3.3	23–33	Transport terminals or buildings with recurrent but not severe peaks
C	1.4–2.3	33–49	Design level for heavily used transport terminals, public buildings or open space where severe peaking and space restrictions limit design feasibility
D	0.9–1.4	49–66	Found in crowded public spaces where continual alteration of walking speed and direction is required to maintain reasonable forward progress
E	0.5–0.9	66–82	Used only where peaks are very short, e.g. sports stadia or a railway platform as passengers disembark. A need exists for holding areas where pedestrians may seek refuge from the flow
F	0.5	variable up to 82	The flow becomes a moving queue, and this is not suitable for design purposes

*For an explanation of the level of service concept (in relation to vehicles) see Chapter 17

25.2.2 Footways

Ideally, footways should be planned as a secondary network of pedestrian ways that are completely separate from vehicular traffic. In practice, however, the segregated footway approach is most readily implemented in new 'green-field' residential developments, whereas existing built-up areas have to make maximum use of footpaths adjacent to the carriageway.

Historical British practice[14] with respect to footpaths is summarised in Table 25.3. Footways flanking shopping frontages in town centres should be sufficiently wide to allow for both free movement and window-shopping without people being jostled. One guide that has been used to check the adequacy of such footpaths is that their capacity should be 33–49 ped/min/m after deducting approximately 1 m 'dead width' in shopping streets and 0.5 m elsewhere.

At dangerous and/or congested locations, e.g. alongside district and primary distributor roads, at busy intersections on local distributor and access roads, at bus stops, opposite school and recreation area exits, and adjacent to pedestrian crossings, *guardrails* are often used to prevent pedestrians from intruding onto the carriageway. These are normally inset about 0.5 m from the kerb in order to give adequate clearance for passing vehicles and provide a place of refuge for persons on the wrong side of the railings. Widening of the footpath may be necessary before guardrails are installed. High-visibility guardrails, i.e. through which children can see, should be considered at locales heavily used by young people. To be effective at busy intersections and pedestrian crossings, guardrails may have to be extended for, say, 10 m on either side of the crossing. They are often extended for the length of the zig-zag markings on either side of a busy zebra crossing.

Table 25.3 General British practice for footways adjacent to roads

Road type	Minimum footway widths
Primary distributor	No footways on urban motorways; 3 m on all-purpose roads*
District distributor	3 m in principal business and industrial districts*; 2.5 m in residential districts*
Local distributor	3 m in principal business and industrial districts*; 2 m in residential districts*
Access roads	*Principal means of access*: 3 m in principal business districts*; 2 m in industrial and (normally) in residential districts*; 3.5–4.5 m adjoining shopping frontages *Secondary means of access*: 1 m verge instead of footway in principal business and industrial districts; 0.6 m verge instead of footway in residential areas

*If no footway is required, verges at least 1 m wide may be provided

25.2.3 At-grade pedestrian crossings

Pedestrians wishing to cross a road must either wait for gaps in the traffic stream or walk to the nearest formal crossing location. At-grade crossings, which are the most common type of formal crossing, are composed of (a) uncontrolled crossings, e.g. zebra crossings, (b) light-controlled crossings, e.g. conventional traffic signals with/without pedestrian phases, and pelican and puffin crossings, and (c) person-controlled crossings, e.g. police-controlled and school crossings.

Uncontrolled crossings

A *zebra crossing* is simply a 'mid-block' uncontrolled portion of the carriageway where the pedestrian has legal priority over the motor vehicle. The crossing strip is outlined by parallel lines of studs and marked with alternate black and white thermoplastic stripes parallel to the centre line of the road. The beginning and end of each crossing are marked by flashing yellow beacons. A Give Way line is located about 1 m from the crossing on either side. Longitudinal zig-zag centre and edge lines are placed on the carriageway for nearly 19 m on either side of the crossing strip in Britain.

A motor vehicle must give way to a pedestrian who steps onto a zebra crossing, and this precedence continues while the pedestrian is on the carriageway. (In practice, however, drivers often move off after the pedestrian has passed the appropriate part of the crossing.)

The underlying purposes of the British zig-zag lines (introduced in 1971) are to: (a) discourage pedestrians from crossing the road adjacent to the crossing; (b) indicate to drivers where they are not permitted to park or overtake other vehicles; and (c) improve driver and pedestrian visibility. (Note: In 1970, it was determined that about 25 per cent of pedestrian accidents at crossings were associated with overtakings.)

Although relatively cheap to install, and very economic to operate, zebra crossings are not suitable at the following locations: where vehicle flow is high and fast-moving, as pedestrians have difficulty in establishing precedence; where heavy continuous pedestrian flows may cause excessive delay to vehicular flow, e.g. opposite railway stations or on busy shopping streets; or where an urban traffic control (UTC) system is in operation.

In 1981 there were about 13 000 zebra crossings in Britain. There are now about 8000 in use.[15]

Light-controlled crossings

In most motorised countries pedestrians are encouraged to cross with the green lights when opposing vehicular flow is brought to a standstill at *conventional traffic signal-controlled intersections*. If the pedestrian movements are heavy, additional pedestrian-only stages may be incorporated into the signal phasing; these are normally activated by push-buttons located on signal posts. Usually this involves a pedestrian-precedence period on one or more arms while non-conflicting traffic movements are allowed to continue; at some junctions, e.g. in central shopping areas, all vehicular traffic is stopped and pedestrians are given precedence on all arms at the one time.

The pedestrian stages provided at conventional signals should be long enough to ensure that all pedestrians can cross in safety; however, this may require a substantial lengthening of the signal cycle if vehicles are also to be handled properly. This can mean that the signal cycle becomes so long that impatient pedestrians will not wait for the period allotted to them, and will try to cross during traffic phases. Alternatively, if the cycle length is reduced to satisfy pedestrian requirements, the free movement of vehicular traffic may be excessively impaired. At signal-controlled intersections in Britain the provision for pedestrians is severely limited by the difficulty in ensuring that there is adequate vehicle capacity, and this is reflected in the relatively large number of intersections still without pedestrian stages. It is reported[15] that there are about 8000 signal-controlled intersections in Britain and that fewer than 30 per cent of these are believed to incorporate pedestrian stages on one or more arms.

Pelican (pedestrian light-controlled) crossings were introduced into Britain by regulation in 1969, and were intended to overcome some of the disadvantages of the zebra crossing. Pedestrian risk reasons aside, these crossings (of which there are now about 9000), are preferably sited away from intersections to minimise driver confusion as to whether the intersection is signal-controlled.

Pelican signals remain at green to drivers and red to pedestrians until a pedestrian activates a push-button to secure a crossing phase; vehicles must then stop for a red signal, even if there are no pedestrians on the crossing. For drivers, the main difference between the pelican and conventional intersection signals is a flashing amber period before the green; this means 'give way to pedestrians – proceed only if the crossing is clear'. For pedestrians on the footpath, the difference is a flashing 'green figure' period before the start of the red period; this means 'do not start to cross – the lights are about to change'. The flashing green figure also retains priority for pedestrians already on the carriageway while allowing vehicles to proceed as soon as the crossing is clear.

Initially, all pelican signals operated on a fixed-time basis, and then vehicle-actuation was initiated on roads with 85th percentile speeds >56 km/h (in 1974). In the late 1970s the use of vehicle-actuated pelican signals was extended to all roads subject to a 48 km/h speed limit. The principle of vehicle-actuation means that, following the expiry of a pre-set minimum green to vehicles, the change to pedestrian priority is initiated when a suitable gap in the vehicular stream is detected or when a vehicle maximum running time expires. (The maximum running time, as originally specified, started from when the pedestrian pushed the button and was intended to prevent long delays to pedestrians.)

A pelican signal has an upper limit on the time that a crossing may be occupied by pedestrians; this reduces the delays to vehicles in locales with high pedestrian flows. Vehicle-actuated pelicans are used at locations with high vehicle approach speeds and/or where there are significant numbers of elderly pedestrians. Textured pavements, ramps, build-outs, and various audible signals have been introduced to assist disabled persons at pelican crossings. Pelican signals operating on a fixed-time basis can be included in urban traffic control (UTC) systems so that there is, usually, one pedestrian precedence period in each cycle.

The *puffin (pedestrian user-friendly intelligent) crossing* is an improved form of signalised crossing that is now on trial in Britain (since 1990/91). As with pelican crossings the puffin signals are push-button activated; however, pressure-sensitive mats or active infra-red detectors at the kerbside where the people wait allow unwanted pedestrian stages to be cancelled. Also, passive infra-red detectors can monitor people on the carriageway and enable extra green time to be allotted to crossing pedestrians if this is necessary.

Pedestrian risk

The great majority of pedestrians crossing roads do so near intersections, and three out of five of these use formal pedestrian crossings where they are available. Women use crossings more than men, and the elderly use them more than other adults, i.e. nearly three quarters of pedestrians in the over-60 age group use crossings, and two thirds of women and 51 per cent of men in the 15–59 age group use them.[16]

Formal crossings are the safest places at which to cross roads, but the carriageway areas beside them can be very hazardous. The highest risk is found on carriageway sections within 50 m of light-controlled crossings that are close to intersections,[16] i.e. where the complexity of decision-making is greatest for both the pedestrian and the driver. In this context a light-controlled crossing refers to any type of signal-controlled facility, such as pelicans, signals with a separate or protected pedestrian phase activated by a push-button, and signals that make no special provision for pedestrians.

25.2.4 Segregated crossings

Ideally all pedestrian crossings should be grade-separated from the vehicular carriageway as these types of crossing eliminate the possibility of pedestrian–vehicle conflict, and cause no delay to vehicles. Segregated crossings are most appropriately used where high pedestrian flows have to cross high-speed roads carrying high volumes of vehicular traffic, at roundabouts where the concentration of vehicles is so high that there are insufficient gaps for pedestrians to cross, and at busy intersections where pedestrian volumes are so heavy that the use of surface crossings only would interfere excessively with vehicular flow.

Whatever the advantages of a segregated crossing, it is of little value if it is not used by pedestrians in preference to crossing the road at-grade. Factors which influence their use are (in order of importance to pedestrians) directness of the route, ease of negotiation, interest of specific features, general environmental appeal and, lastly, safety. Assuming that it is clean, well designed and well illuminated, pedestrians will still only use a segregated crossing provided that the route via the crossing is shorter and quicker than the surface route. Subways which are judged safe (i.e. secure) by pedestrians generally require only a small time-saving in order to ensure 100 per cent usage. It is normally not possible to realise almost complete usage of a footbridge until the crossing time is about three quarters that of the surface route. Guardrails are therefore very often provided adjacent to the vehicular carriageway prior to the accesses to segregated crossings so as to lengthen the pedestrian path via the surface route and encourage usage of the segregated way.

Vandalism can be a major concern affecting the use of subways in built-up areas. Wide approaches with no spaces for concealment, alignments with good through visibility, and good vandal-proof lighting, all within the view of passing pedestrians and passing traffic, are design features which help assuage pedestrians' fears for their personal security when using subways. In some instances, the installation of physical barriers on subway approach ramps may be necessary to prevent some motorised vehicles from using subways. (In Britain pedestrian subways that are less than 23 m long are at least 3 m wide by 2.3 m high, while those longer than 23 m are at least 3.3 m wide by 2.6 m high.) Access ramps to subways are usually built with grades of less than 5 per cent if they are likely to be used by significant numbers of disabled pedestrians, heavily laden shoppers, elderly people, and/or people with prams, pushers or strollers.

A footbridge is the least suitable form of pedestrian crossing for people who have difficulty negotiating steps or long and/or steep ramps. The change of level experienced by pedestrians using footbridges is usually greater than that via subways, due to the greater headroom requirement for vehicles. Thus, it is British practice to provide a footbridge only where an at-grade crossing or a subway is deemed unsuitable. The access to the footbridge, whether by ramp or stair, is normally on the line of the main pedestrian flow. If the access is immediately adjacent to the carriageway it is sited so that descending pedestrians face the oncoming vehicular traffic. British practice is for the clear width of bridge or access ramp/stair to be 1.8 m or greater, depending upon the peak pedestrian flows as follows: (a) 300 mm of width per 20 ped/minute on the level or on gradients less than 5 per cent, or (b) 300 mm of width per 14 ped/minute on steps or ramps on gradients greater than 5 per cent.

Details of good design features of subway and footbridge crossings are readily available in the literature.[17, 18]

25.2.5 Pedestrianisation in town centres

Most existing shopping and commercial centres grew up along main traffic routes and at their intersections. They thrived, mainly because of the then ease of their accessibility. Now, as traffic congestion increases, the attractiveness of many of these older centres is diminishing. The desirability of pedestrianising parts of these areas, or of particular lengths of road within them, has gained much momentum over the past two decades as a means of retaining the attractiveness of these centres and of reasserting the role of the pedestrian within a safer, more user-friendly environment. In this context, pedestrianisation can be described as the extreme face of traffic calming.

Pedestrianisation schemes inevitably involve controversy, and good public consultation is essential to the likely acceptability of any proposed scheme to the people who are most affected by it, e.g. building owners and tenants, chambers of commerce, emergency and public utility services, public transport and delivery operators. Successful schemes generally seek to satisfy the following criteria:[19]

- to bring about an improvement of the shopping, commercial and leisure environment of the area
- to lead to an improvement in the economy and environment of an established town centre, enabling it to compete with edge-of-town and out-of-town centres
- to have full regard to the traffic engineering and road safety implications both in respect of movement within the immediate vicinity of the pedestrianised area and

over the highway network as a whole

- to contribute to an improvement in the safety, accessibility and comfort of pedestrians.

The *full pedestrianisation* of a street or group of streets means that displaced traffic, especially through vehicles, must be handled by other links in the road network. This requires consideration of the capability of these links (especially intersections) to cope with the extra loads. Where additional traffic capacity is required in the network, it may be possible to achieve this by prohibiting parking, and/or introducing one-way operation, and/or improving intersection operation in conjunction with upgraded traffic control.

The impact of the displacement of public transport vehicles can often be minimised by ensuring that there is provision for convenient services at the perimeter of the pedestrianised area. The delivery and service needs of businesses have to be seriously considered; for instance, if some businesses do not have rear accesses it may be possible to meet their needs by the provision of strategically sited loading areas at the perimeter in conjunction with the use of trolleys. Persons with disabilities also need to be considered to ensure that special parking spaces and bus stops are within reasonable access distance, e.g. within 50 m on a route open to the weather.

With full pedestrianisation all motor vehicles are excluded from the treated area, with exemptions only for emergency and statutory undertaker maintenance vehicles. This makes it possible for carriageways and kerbs to be removed and the whole area to be paved to the same level, with consistent surface textures, from wall to wall. Significant landscaping and amenity improvements are usually also carried out. The designated streets are normally left uncovered, although sheltered areas may be extended.

In many instances it may be impractical to carry out full pedestrianisation, and consideration may be given to carrying out *partial pedestrianisation*, e.g. allowing access to vehicles carrying disabled persons, or public transport vehicles, or delivery vehicles during certain hours of the day. In some instances it may be still possible to remove the carriageways and kerbs; in other cases the appropriate answer may be to narrow the existing roads and widen the pedestrian walking areas, so that only essential vehicles seek to use the access ways. Slow speeds by vehicles are critical to the success of partial pedestrianisation schemes. When the vehicles are slowed to pedestrian speed, and it is clear to drivers that they are in the pedestrians' domain, there is ample evidence to show that vehicles do not constitute a significant accident hazard.

25.3 Cyclist priority

It has been reported[20] that there are 13–17 million pedal bicycles in Great Britain, and that 11 million and 3.6 million people claim to use a bicycle at least once per year and once per week, respectively. Cycling is nonetheless a minority mode of transport in Britain; throughout most of the 1980s and early 1990s cycling accounted for only 1 per cent of total passenger-kilometres and about 2.5 per cent of all journeys. In 1993 cyclists accounted for about 8 per cent of all road casualties and 5 per cent of those killed in road accidents.

Cyclists range in age from the very young to the very elderly, and they do not need licences to ride their bicycles. Bicycles cannot legally be ridden on motorways, footpaths, zebra crossings or conventional pelican crossings. They can be ridden on bridleways and, normally, on most roadways – but not against the traffic flow on one-way streets (unless special contra-flow lanes are provided).

The patterns of cycle use are complex. It is possible to model cycle trips in particular areas, but experience suggests that the fit is not so good when the model is applied elsewhere. Generally, cycle trips may be grouped as: (a) neighbourhood trips that are short, often made by children for pleasure, or by children or adults on errands; (b) recreation trips, very often outside a neighbourhood; and (c) commuting trips, e.g. to work or to educational institutions.

The hilliness of an area is a major factor affecting cycle usage, particularly for commuting and recreational purposes. Volume of vehicular traffic, surface quality and scenery are important when choosing cycling routes, and safe, quick, direct routes on smooth surfaces are important for commuting. Safety is more important to women cyclists than to men, on average.

Notwithstanding the obvious desirability of separating cyclists from vehicular traffic one major study[20] found that there was no evidence to support the hypothesis that the 80 per cent of cycle owners who do not use their bicycles regularly are deterred from cycling by the absence of safe routes.

25.3.1 Cycle routes

Generally, cycle routes can be divided according to their function. Thus, *strategy routes* are the long-distance routes that link large areas (e.g. urban regions) together; as such they are the cycling equivalent of a primary distributor road system and therefore tend to abut or complement such roads. *Area routes* serve as links between local areas and surrounding shared open spaces such as river parks and sports fields. *Recreational routes* provide for pleasure cycling within a recreational area; they also link parks and other recreational areas. *Neighbourhood routes* are short-distance links that, typically, radiate from schools, the local shopping centre, or community facilities.

Three criteria have been described[21] as essential when planning non-recreational routes for cyclists in an urban area: all routes should be as direct as possible, and certainly as direct as alternative footpaths or roads; the network should link all the major points of attraction, e.g. schools, shopping and employment; and cycle routes should be separate from major roads and, where possible, from the local road system. Applying these criteria, a basic network of neighbourhood routes can be considered as a series of spokes radiating from a centre, with the lengths of the spokes being determined by typical journey distances for purposes associated with the centre; then when spokes from adjoining centres coincide and meet, a continuous area or strategy route is formed.

In practice, the designation of cycle routes usually involves combining low-volume back-streets that are conducive to safe cycling with marked cycle lanes on existing wide roads and segregated cycle tracks. Where pedestrians and cyclists share the same right of way their carriageways should preferably be separated either by a minor level difference or a physical barrier.

Safe crossings at main roads are essential features of designated cycle routes. Preferably cyclist crossings should be carried out via segregated crossings (often in conjunction with pedestrians – see references 17 and 18). If this is not possible for economic reasons, there is ample evidence to show that the sharing of signal-controlled at-grade crossings by cyclists and pedestrians can be carried out satisfactorily, with no problems of conflicts or use of the equipment. A new cycle/pedestrian crossing based on the light signal-controlled crossing used at signal-controlled intersections has been developed and tested by the Transport Research Laboratory.[22] Known as the *TOUCAN crossing* (because the two can cross at the same time in the same space), it has additional push-buttons, a modified push-button plate, and a green cycle signal next to the green figure signal. The TOUCAN crossing can also be installed with inductance loop detectors capable of picking up pedal cycles.

Traffic calming mechanisms in existing residential areas, and the pedestrianisation of shopping and business centres (sometimes in conjunction with traffic calming), are activities conducive to the establishment of cycle routes. Cyclists need to be provided with safe and convenient access to strategically located cycle racks adjacent to shops and other destination facilities in pedestrianised areas. Pedestrianisation measures should never result in cyclists being forced to use busy distributor roads; instead a narrow defined cycle track should be provided through the pedestrians' area as it is general impractical to expect cyclists to dismount and walk through.

Table 25.4 Summary of Redway cycle track design standards used in Milton Keynes[25]

Item	Standard
Design speed	25 km/h
Width	3 m standard; 5 m for lengths of high activity
Horizontal alignment	Minimum radius = 6 m
Gradients	≤1:12 for 15 m, ≤1:15 for 30 m; and ≤1:20 for 100 m
Visibility at road junctions	Dependent on category of road. Range varies from 4.5 m × 70 m at a main road to 2.5 m × 25 m in a cul-de-sac with traffic calming humps
Pavement construction	20 mm red macadam wearing course, 40 mm open-texture macadam basecourse, 125 mm lean-mix or 70 mm macadam base, and 150 mm granular subbase (type 1)
Lighting	Group B road lighting. Light source = 50 watt high pressure sodium. Column spacing = 25 m to 30 m
Road/Redway junction design	Twin yellow bollards on Redway (1.8 m clear opening) visible to Redway and road users. 'Give Way' markings and tactile paving on Redway

An inventory of the existing urban infrastructure may result in the identification of footways, bridlepaths, towpaths, tracks in parks, disused rights of way, and sometimes even bridges and tunnels not on the road network that can be integrated with new cycle tracks. The development of segregated tracks is a much easier process in new towns; for instance, about 200 km of segregated cycle tracks, called *Redways*, currently criss-cross the fabric of the new town of Milton Keynes. Table 25.4 summarises the design standards used in the Redway cycle system.

Reference 23 is an excellent source book relative to the design of cycle lanes and cycle tracks.

25.4 References

1. Hass-Klau, C., *The theory and practice of traffic calming: Can Britain learn from the German experience?* Rees Jeffreys' Discussion Paper 10. Oxford: Transport Studies Unit, Oxford University, 1990.

2. Windle, R., and Hodge, A.R., *Public attitude survey – The New Forest traffic calming programme*. PR 14. Crowthorne, Berks: Transport Research Laboratory, 1993.
3. Buchanan, C., *et al.*, *Traffic in towns*. London: HMSO, 1963.
4. Buchanan, C., *Traffic calming and 'Traffic in towns' – the relationship*, in Foreword to reference 1.
5. Kjemtrup, K., and Herrstedt, L., Speed management and traffic calming in urban areas in Europe: A historical view. *Accident Analysis and Prevention*. 1992, **24** (1), 57–65.
6. *Woonerf: A new qpproach to environmental management in residential areas and the related traffic legislation*. The Hague: Royal Dutch Touring Club, 1980.
7. Katz, R.J., and Smith, N.S., Traffic calming, pedestrians and bicyclists, *Proceedings of the 7th National Conference of the Australian Institute of Traffic Planning and Management*, held at Sydney on 26–27 May 1994, 1–13.
8. *Sharing the main street*. CR 132. Sydney: Roads and Traffic Authority of New South Wales and Federal Office of Road Safety, Nov. 1993.
9. Bicknell, D., Traffic calming. *Proceedings of the Institution of Civil Engineers – Municipal Engineers*, 1993, **98**, 13–19.
10. Webster, D.C., *Road humps for controlling vehicle speeds*. PR 18. Crowthorne, Berks: Transport Research Laboratory, 1993.
11. *Traffic calming regulations*. Traffic Advisory Leaflet 7/93. London: Department of Transport, August 1993.
12. *Guide to traffic engineering practice: Part 13 – Pedestrians*. Sydney: Austroads, 1995.
13. Leake, G.R., May, A.D., and Pearson, D.I., Pedestrians' preferences for footway maintenance and design, *Highways and Transportation*, 1991, **38** (7), 5–10.
14. *Roads in urban areas*. London: HMSO, 1966 (as corrected in Technical Memorandum H12/73, Oct. 1973).
15. Hunt, J., The operation and safety of pedestrian crossings in the United Kingdom. *Proceedings of the 17th Australian Road Research Board Conference*, 1994, Part 5 – Road Safety, 51–64.
16. Grayson, G.B., Pedestrian risk in crossing roads: West London revisited. *Traffic Engineering and Control*, 1987, **28** (1), 27–30.
17. *Subways for pedestrians and pedal cyclists: Layouts and dimensions*. TD 36/93. London: HMSO, 1993.
18. *Design criteria for footbridges*. BD 29/87. London: Department of Transport, 1987.
19. *Pedestrianisation guidelines*. London: Institution of Highways and Transportation, 1989.
20. Harland, G., and Gercans, R., *Cycle routes*. PR 42. Crowthorne, Berks: Transport Research Laboratory, 1993.
21. Robinson, K., Cycle routes in Peterborough, in *Cycling as a mode of transport*. SR 540. Crowthorne, Berks: Transport Research Laboratory, 1980, 97–106.
22. Morgan, J.M., *TOUCAN crossings for cyclists and pedestrians*. PR 47. Crowthorne, Berks: Transport Research Laboratory, 1993.
23. *Guide to traffic engineering practice: Part 14 – Bicycles*. Sydney: Austroads, 1995.
24. *Highway capacity manual* (3rd edition). Special Report 209. Washington, DC: Transportation Research Board, 1994.
25. Ketteridge, P., and Perkins, D., The Milton Keynes redways. *Highways and Transportation*, 1993, **40** (10), 28–31.

CHAPTER 26

Signal control at intersections

M.G.H. Bell

26.1 Hardware

The control of traffic at intersections by lights dates back to 1913 in Cleveland, Ohio, although the current format of three lamps showing red, yellow and green is believed to date back to 1918 in Detroit and New York.[1,2] In the UK, the first manually operated signals were installed in 1925 in London and the first automatic system was installed in 1926 in Wolverhampton. Since then, traffic signals have become all pervasive, successfully regulating traffic in all major cities round the world.

At the intersection, the hardware consists of a set of *signal heads* and *vehicle detectors* connected by cable to a *signal controller*. The signal controller may be connected to a *traffic computer* or possibly to other signal controllers in the vicinity.

The signal heads are located on columns. Each signal head has at least three lamps, one red, one yellow and one green. In the UK and other European countries, the three lamps of the signal head are arranged vertically with red at the top (for visibility over the greatest distance), green at the bottom and yellow in the middle. Directly behind the lamps there is a black board to provide a background against which the lamps will be clearly visible. Where the signal head is beside (rather than above) the road, the bottom of the signal head is about 2 m from the ground. In Japan, by contrast, the lamps are arranged horizontally and located above the road. There may also be green or red arrows for separately signalled turning movements.

In some cases, buses and trams may be signalled separately using a set of black and white indications. Pedestrians often have their own signals, generally showing a red stationary pedestrian or a green walking pedestrian (in Ireland there is also a yellow pedestrian). In some cases, cyclists may also have their own signals, showing a red, yellow or green cycle. While there are variations in practice around the world, these are generally not likely to confuse the international traveller.

The modern signal controller is a microprocessor-based device which is programmed to control the signal indications. The most important function of the signal controller is to ensure safety by preventing combinations and sequences of signal indications that could be dangerous to the road users. Constraints on combination and sequence are stored in a non-erasable form (specifically in ROM, or Read Only Memory) in the signal controller.

In an increasing number of cases, the control of the intersection is responsive to the demands of the road users as indicated by vehicle detectors of various kinds. In Europe and North America, the most common form of vehicle detector is the inductive loop, consisting of a loop of wire located just below the road surface which is connected by cable to the detector circuitry housed in the signal controller cabinet. By monitoring changes in inductance, the detector (loop plus circuitry) acts as a metal detector. The loop is polled regularly, somewhere between every 1/4 to 1/100 of a second, as to whether or not it is occupied (i.e. whether or not there is a vehicle over the loop). From this binary information, the following basic data may be derived: a *vehicle arrival*; a *vehicle departure*; the *headway* (the time elapsed since the last vehicle departed); and the *time occupied* (the time elapsed since the current vehicle arrived, before it has departed).

From this basic data, the following statistics may be derived: *speed* (calculated from the time difference between the arrival of a vehicle at two closely spaced loops); *occupancy* (the proportion of a unit of time that a loop is occupied); and *flow* (the number of vehicles to arrive at the loop during a unit of time). The following two examples illustrate how the binary detector data may be converted into usable traffic data.

Example 26.1

Suppose an inductive loop is polled every 0.25 s and returns the following data, where 1 signifies that the loop is occupied and 0 that the loop is unoccupied:

0 0 0 0 1 1 1 1 1 1 0 0 0 1 1 1 1 1 1 1

After 1 s, a vehicle arrives at the loop and departs from the loop 1.5 s later. A period of 0.75 s then elapses before the next vehicle arrives at the loop. Speed can only be estimated from the data from one loop if one knows the length of the detection zone in the direction of travel, which will be slightly greater than that for the loop itself, and the length of the vehicle. If the vehicle is 5 m long and the detection zone is 1 m long, the vehicle covers a distance of 6 m in 1.5 s, suggesting a speed of 4 m/s (or 14.4 km/h). Over the 5 s interval, the occupancy is 65 per cent and the flow is two vehicles.

Example 26.2

Suppose there is a second loop in the same lane yielding the following data:

0 0 0 0 0 1 1 1 1 1 1 0 0 0 0 1 1 1 1 1

If the leading edges of the two loops are 2 m apart, with the second loop downstream of the first, the speed of the first vehicle was in the region of 8 m/s instead of 4 m/s. The vehicle must therefore have been closer to 10 m long than to 5 m long. The second vehicle appears to be travelling slower than the first, covering the 2 m in 0.5 s, suggesting a speed in the region of 4 m/s. Note that, because of the binary nature of occupancy data, precise estimates of vehicle speeds and lengths are not possible.

Other technologies are available for vehicle detection, such as microwave, passive infra-red, active infra-red, magnetism and ultrasound. Ultrasonic detectors, suspended above the carriageway, are commonly found in Japan where the prevalence of metal structures beneath the road limits the usefulness of inductive loop detection. Microwave detectors are frequently used at temporary installations.

There are also pressure sensitive devices that may be used to register the passage of axles, such as pneumatic tubes, pressure switches, triboelectric cable and piezoelectric cable. Piezoelectric cable may also be used to weigh the axle load approximately, a function that is useful for certain regulatory purposes (like checking for overweight vehicles). When used in combination with vehicle detectors, axle detectors may be used for vehicle classification. A combination of a double loop (one loop for each side of the lane) and two triboelectric axle detectors situated inside the loops are used to collect traffic census data for the UK Department of Transport. The inductive loops respond to the presence or absence of a vehicle while the axle detectors can measure its speed, the number of axles and the distances between the axles. The detector data may be combined to estimate factors such as the size of the overhang at the end of the vehicle, and the height of the floor of the vehicle may also be measured. As a result, 20 classes of vehicle may be identified.

Where certain kinds of vehicle (like public transport and emergency service vehicles) are to be given priority over other kinds, *selective detection* may be used. A number of technologies are available for this. With the exception of contact switches for trams, these technologies require the priority vehicle to carry a *transponder* of some kind. A widely encountered form of selective detection uses a buried loop and inductive coupling (a proprietary system known as VETAG is an example of such a system). Infra-red and microwave are alternative media for short-range communication, requiring the installation of roadside beacons.

Another possibility, which is becoming increasingly popular, is the use by buses and trams of radio transmission to make requests for green on the approach to an intersection. This can occur either by a data channel to the public transport control centre, which then relays the request for priority to the traffic signal control system, or by low power transmission directly to the relevant signal controller, which, in this case, must be equipped with a radio receiver.

26.2 Intersection design

The signal-controlled intersection is a location in the road network where road users of different types are forced to share a common road surface. Through the operation of the traffic signals, the usage of the intersection is shared between the road users. The job of the traffic engineer is to design both the intersection and the signal control so that the intersection operates in a way that is objectively safe, efficient and perceived by the road users to be fair.

The mix of road users is an important aspect of intersection design. This varies not only from location to location within a city, but also from country to country. For example, in many Chinese cities, the problem is to control a large number of pedal driven vehicles and pedestrians alongside buses, trolley buses, trucks, taxis and a limited but growing number of private cars. By contrast, in the cities of northern Europe, the problem is typically to control a large number of private cars, trucks and pedestrians alongside a limited number of trams and buses requiring priority at the intersection. The road user groups considered here are: *cars and trucks*; *buses and trams*; *cyclists*; and *pedestrians*.

Any intersection consists of three or more *approaches*, each of which consists of one or more *lanes*. Where there is more than one lane, these are separated by a broken white line painted on the surface of the road, each lane being 3 to 5 m wide. At the intersection, each lane ends at a stop-line, a thicker unbroken white line across the end of the lane indicating the point beyond which the car at the front of the queue should not proceed when the signal is red. Where a lane is reserved for one or more movements, this is indicated by a one- or two-headed arrow painted in white in the centre of the lane on the approach to the stop-line. These arrows should be located sufficiently far from the stop-line to avoid being obscured by the queue (this should be possible, as every lane should be long enough to contain the queue most of the time).

The lanes channel the traffic on each approach into *streams*, which may consist of one or more lanes. A stream is a set of one or more movements that always receive green together. Each stream is controlled by its own *signal group*, which is a set of *signal heads*. Even though it may consist of more than one lane, each stream may be regarded from the control point of view as a single queuing process. The intersection must be designed so that there is adequate space for queues and so that clearly visible locations can be found for the signal heads. The suggested space requirement for a queue is 1.2 multiplied by the mean arrival rate over the cycle (namely, the mean arrivals between the start times of two consecutive green periods for the stream in question) multiplied by 6 m per vehicle. Concerning visibility, a distance of 70 m is recommended if the maximum permitted speed is 50 km/h or a distance of 125 m if the maximum permitted speed is 70 km/h.[3]

Practice regarding the location and design of signal heads varies between countries but the principles are the same, namely that the signal indications should be clearly visible to the stream to which they pertain, particularly to the vehicles at the front of any queue at the stop line, and that they should not be misleading to drivers, cyclists or pedestrians in other streams. Sometimes this requires the use of hoods or vertical slats over the lamps to reduce the horizontal *angle of visibility*. The principles of intersection design are illustrated by the following two examples.

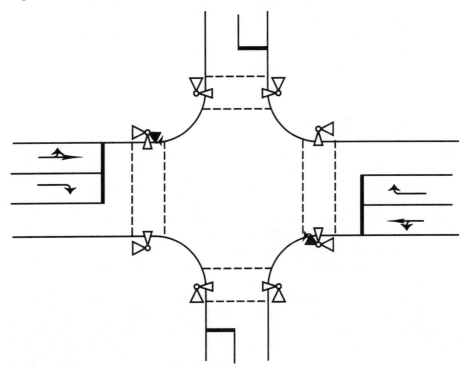

Fig. 26.1 Simple four-arm intersection

Example 26.3

Consider the four-arm intersection as shown in Fig. 26.1. The lane should be between 3 m and 5 m wide. The position where the first vehicle in the queue is to stop is indicated by the stop-line. The primary signal head is generally located at the roadside close to the stop-line. In Germany, the primary signal head is between 2.5 m and 3.5 m downstream of the stop-line, and so is clearly visible to the vehicles at the front of the queue. In the UK and France, the primary signal head may be closer to the stop-line and may not be visible to the drivers at the front of the queue. In the UK, a secondary signal head is generally located on the far side of the intersection, either on the same side of the road as the primary signal head or diagonally opposite it, depending on which location is more visible to the relevant drivers and least likely to mislead other drivers. For a straight-ahead movement there are generally two signal heads, while for a separately signalled turning movement there is usually only one signal head. In France, the secondary signal head, in a smaller format, is located on the same column as the primary signal head but lower down so that the driver at the front of the queue can see it. In Japan, the signal head is generally suspended above the stop-line with the lamps arranged horizontally. The secondary signal head, when provided, not only assists the drivers at the front of the queue but also offers a duplication of the indication, which is particularly useful when a bulb burns out.

The pedestrian crossing is located about 1 m downstream of the stop-line and is between 3 m and 12 m wide. The pedestrian signal is located on the far side of the crossing, and the indications consist of a walking green pedestrian or a standing red pedestrian. Where the crossing is close to a corner, the corner should not be too rounded, otherwise vehicles turning into the crossing will be less likely to stop, they will be travelling faster and the pedestrians will have further to go, necessitating a longer clearance time for them (pedestrians are generally assumed to walk at between 1.2 m/s and 1.5 m/s for the purposes of calculating clearance times).

Example 26.4

Consider next a complex intersection with *traffic islands* as shown in Fig. 26.2. The traffic islands separate the opposing streams, provide refuge to pedestrians and offer a location for a second signal column close to the stop-line. If the lanes are carrying different streams then the two signal heads at the stop-line would belong to different signal groups. In the UK, a suitable location would be found on the far side of the intersection for a secondary signal head, serving those drivers at the front of the queue who wish to go straight on. Where vehicles making an *opposed turning movement* are stored in the centre of the intersection, the secondary signal head may carry an additional lamp showing (when lit) a green arrow. The green arrow indication would be given when safe to do so, namely when all opposing flows (including pedestrians) had been stopped and had cleared the intersection.

The treatment of the pedestrians on the traffic island differs between countries. In the UK, offset (or staggered) crossings are common. The pedestrians cross to the island when encouraged to do so by the green pedestrian indication. They then move down the traffic island and wait for the green pedestrian indication on the far side of the road in order to complete their crossing. The storage on the island should be adequate. On average, 0.5 m² are required per pedestrian.[3] Waiting pedestrians should be protected from

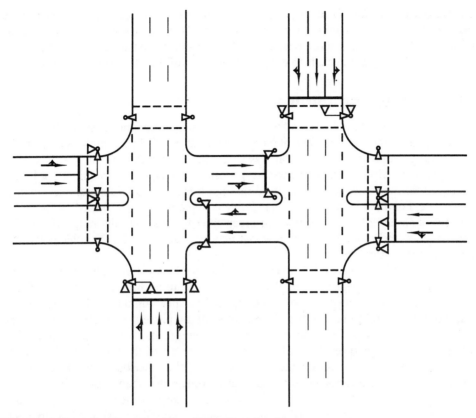

Fig. 26.2 Complex intersection with traffic islands

accidental incursion onto the road by guardrails. The physical offset of the two crossings reduces the possibility of the pedestrians confusing the indication on the far side of
the road with the indication on the traffic island. In Germany, the practice is to prefer a
crossing in one stage without the need for pedestrians to wait on the traffic island and
without a physical offset. This removes the possibility of pedestrians on the traffic island
misinterpreting a red pedestrian indication on the far side of the road as a recommenda-
tion to clear the crossing when it really means stop. However, when the road is too wide
for a one-stage crossing, the traffic island should be at least 4 m wide to separate the
two stages clearly.

Where priority for public transport is required, lanes may be reserved for this use. In
the UK, bus lanes are frequently encountered on major arterial routes. These may terminate at, or shortly before, the stop-line. Trams may have a reserved lane, either beside
the road or in the central reservation. In addition to reserved lanes, public transport may
be given priority treatment within the signal control. It should be noted, however, that
priority treatment within the signal control is very much less effective without a reserved
lane on the approach to the intersection.

Cyclists are sometimes treated separately through the provision of cycle lanes, occasionally with their own signal indications. The stop-line for cyclists is generally located
ahead of that for other vehicles to improve the visibility of cyclists to drivers and to give

the cyclists a head start. Usually cycle lanes are located by the roadside, but where turning vehicles may pose a hazard to cyclists, the cycle lane may be brought into the centre of the road. Sometimes the advanced cycle stop-line is extended across the whole approach, allowing the cyclists to congregate in front of the other vehicles on red.

26.3 Safety and fairness

Safety and fairness are linked as safety depends on user compliance to signal indications and compliance rests to some extent on perceived fairness. While compliance by drivers is generally required by law, unfair laws fall into disrepute and their enforcement becomes more difficult. Compliance by pedestrians may not be required by law. A degree of fairness is generally achieved through a *maximum red time* and a *minimum green time*. For drivers, the maximum red time is in the region of 120 s, while for pedestrians it is around 60 s. The minimum green time is usually in the range 10 to 15 s, depending on the expected traffic flow, but may be reduced to 5 s for vehicle-actuated traffic signal control.

The end of a right of way for a stream is anticipated by a yellow indication. The duration of the yellow depends principally on the maximum allowable speed of approach, for example 3 s for a speed limit of 50 km/h, 4 s for a speed limit of 60 km/h and 5 s for a speed limit of 70 km/h.[3] For the comfort of standing passengers on trams and buses, rather longer yellow periods may be required, ranging from 3 s for a speed limit of 30 km/h up to 8 s for a speed limit of 70 km/h. As a result of the lower speeds, cyclists require only a 2 s yellow period and pedestrians no yellow period at all (in Ireland, the yellow pedestrian indication denotes the clearing time rather than a forewarning).

In many countries the start of a right of way for a stream is denoted by a short combined yellow and red indication of around 1 s (and no more than 2 s) to allow drivers to get into gear. Some drivers may also have switched off their engines during red, although there is a debate at present about whether this practice is environmentally beneficial. For trams, buses or cyclists, when separately signalled, there is no combined yellow and red indication. Pedestrians generally have no yellow lamp. In Italy, the combined yellow and red indication is used at the end, rather than the start, of the right of way.

The calculation of the *clearance times* is critical to the safe operation of the intersection. This is the time that needs to elapse between the end of right of way for one stream (start of red) and the start of right of way of another *incompatible* stream (start of green). The paths of two incompatible streams intersect at a point referred to as the *conflict point* (see Fig. 26.3). Streams that are permitted to cross each other, for example some opposed turning movements, are referred to as *semi-compatible* and have no clearance time. The clearance time (t_z) can be decomposed into the *changeover time* $(t_{\ddot{u}})$, the *clearing time* (t_r) and the *entry time* (t_e), whereby:

$$t_z = t_{\ddot{u}} + t_r - t_e \qquad (26.1)$$

and is rounded to the nearest second. The clearance time is calculated separately for each road user group included in the stream and then the largest value is retained in the clearance time matrix (the matrix shown in Table 26.1 relates to the signal plan shown in Fig. 26.4).

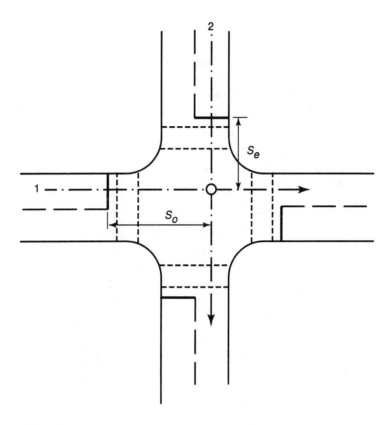

Fig. 26.3 Conflict points

The *clearing distance* (s_o) is the distance from the stop-line to the conflict point for the stream losing right of way plus one vehicle length (typical vehicle lengths would be 0 m for cycles and pedestrians, 6 m for vehicles and 15 m for trams). The *entry distance* (s_e) is the distance from the stop-line to the conflict point for the stream gaining right of way. Given clearing and entry distances, there are a number of ways in which the clearing and entry times are calculated, depending on the situation and road user groups involved. The changeover time is an allowance made for those vehicles that are unable to stop during the yellow period, and lies somewhere between 0 s and 3 s (typical values would be 0 s for a tram or bus always stopping just before the stop-line and for pedestrians, 1 s for cyclists, 2 s for turning vehicles when not separately signalled, and 3 s otherwise). Drivers forced to make a technical violation of a red signal are caught in the *dilemma zone*; should they try to stop or try to pass the intersection as swiftly as possible? The following examples illustrate the calculation of clearance times.

Example 26.5

Consider two streams, where the stream losing right of way has a clearance distance of 13 m for vehicles travelling straight on and the stream gaining right of way has an entry distance of 11 m. The clearance time is calculated as follows. Reasonable clearing and entry speeds would be 10 m/s and 11 m/s respectively. The changeover time would be 3 s

Table 26.1 A clearance time matrix

Ending signal group \ Clearance time (s)	Starting signal group														
	K1	K2	K3	K4	K5	K6	K7	R1	F1	F2	F3	F4	F5	F6	FR7
K1		4			4	4		4						7	
K2			5	8	5	4		5	2		8	8			
K3	6	4			4			1		4	4				6
K4		2				2							4		
K5		3	6			5		6						3	
K6	6	6		12	7			4			7	7			6
K7															
R1		2	6			2			8					2	
F1	10	7			6			5							
F2			6												
F3		4	6			3									
F4		4				3									
F5				5											
F6	8				8			10							
FR7			3			6	4								

and an average vehicle length would be 6 m. Thus $t_{\ddot{u}} = 3$ s, $t_r = (6 + 13)/10 = 1.9$ s and $t_e = 11/11 = 1$ s. Hence $t_z = 3 + 1.9 - 1 = 3.9$ s, or 4 s when rounded to the nearest second.

Example 26.6

Consider a pedestrian crossing that is 20 m long which is losing right of way to a stream of vehicles at an adjacent stop-line. The clearance time is calculated as follows. As pedestrians should not start to cross after the red pedestrian indication is shown, $t_{\ddot{u}} = 0$ s. The clearing distance is 20 m as no allowance is made for the size of pedestrians. Pedestrians move at between 1.2 m/s and 1.5 m/s, depending on their average age (the young and the old move more slowly). Taking a value of 1.3 m/s gives $t_r = 20/1.3 = 15.4$ s. The entry distance of the vehicles can be assumed to be 0 m because of the adjacency to the stop-line of the pedestrian crossing, so $t_e = 0$ s. Hence $t_z = 0 + 15.4 - 0 = 15.4$ s, which rounds to 16 s.

Due to the large number of rules of thumb employed in their calculation, the clearance times should where possible be checked in practice. The validated values are then stored in the signal controller in permanent Read Only Memory (ROM) where they cannot be accidentally erased by a power failure. The signal controller ensures that the clearance times are never violated.

Vulnerable road users (pedestrians and cyclists) need to get to the conflict point 1–2 s ahead of the conflicting semi-compatible stream of vehicles. This leads to an offset between the starts of the right of way for the two semi-compatible streams.

26.4 Control variables

The sequence of signal indications usually encountered by a stream of vehicles (green → yellow → red → red and yellow → green) has been referred to previously. As both the yellow and the combined yellow and red periods are fixed according to criteria previously discussed, they do not constitute control variables. The green and red periods, however, can be varied, subject to the clearance times, maxima and minima, also discussed previously.

To give each stream green in sequence would be wasteful of intersection capacity, because some streams are compatible or semi-compatible with others. The streams are therefore grouped into *stages* (referred to in some countries as *phases*) which can receive green simultaneously. Where there are alternative ways to group the streams, streams with similar green time requirements should be grouped together. It is generally advantageous to keep the number of stages to a minimum, as the transition from one stage to the next incurs a loss of capacity, due to the clearance times that must elapse as well as the start and stop losses. For a four-arm intersection, two stages are preferred, unless this causes problems for opposed turning flows. If some turning flows require protection against opposing flows in order to clear, three stages may suffice, otherwise four stages will be required (an alternative, of course, would be to ban the problematic turns). A requirement for priority for public transport may also lead to the definition of extra stages.

Where there are more than two stages, the *stage sequence* will be determined by factors such as the need for vehicles in the intersection to clear, calls for public transport priority, the need for *green waves* for pedestrians or cyclists within the intersection, and the need for particular streams with heavy demands to receive green in more than one stage. In vehicle-actuated signal control, stages for which there is no demand may be dropped until they are demanded.

The *intergreen period* is the time between the end of one stage and the start of the next, namely the duration of the *stage transition*. This should, in general, be as short as possible and will be determined by the maximum clearance time that must elapse and any required staggering of the green starts to protect vulnerable road users. Within the intergreen period, there will generally be some freedom to choose the timing of the starts and ends of right of way for those streams not affected by the maximum clearance time. The precise times, relative to the start of the intergreen period, at which streams lose or gain right of way are referred to as the *stage transition structure*.

There are two forms of control. The prevalent form is *stage-based control*, where the *stage durations* (the periods during which signal indications remain unchanged), and possibly also the stage sequence, are determined subject to the maximum red and minimum green constraints. The stage transition structures, which incorporate the clearance time and other constraints, are not varied.

The alternative is *group-based control*, where the allocation of green time is to the streams directly. In this case, the stage transition structures may be varied. Group-based control constitutes a finer level of control than stage-based control, but implies rather more control variables. For a simple four-arm intersection with no protected turning movements there would be two stages and therefore two variables under stage-based control, but four streams and therefore four variables under group-based control. While modern controllers are capable of group-based control, most software offers only stage-based control.

In both stage- and group-based control, the determination of signal timings is generally done in two steps. First a *cycle time* is determined that offers adequate capacity. Then, within the cycle time, the green time is allocated to the streams or stages. The result is a *signal plan*. Fig. 26.4 shows a signal plan relating to the clearance time matrix given in Table 26.1. The time of each change of signal indication relative to the start of the cycle is shown. Changes of signal indication occur only during the stage transitions.

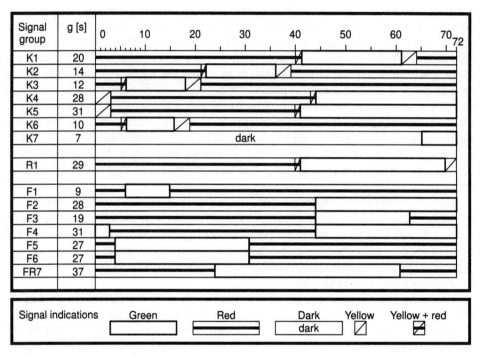

Fig. 26.4 Example signal plan

26.5 Capacity

Control essentially refers to the allocation of intersection *capacity* to streams in a way that is safe, fair and efficient. Of importance is the determination of the capacity of a stream of vehicles when permanently given green. This is conventionally equated with the flow across the stop-line while a queue is discharging, referred to as the *saturation flow*, as in practical terms this will be the capacity of the stream. It is worth noting in passing that significantly higher flows than the saturation flow have been observed on straight sections of motorway.

Numerous measurements of the saturation flow have been made, and it has been found to be influenced by many factors, in particular the *lane width*, the *curvature* of the path of the stream, the prevailing *weather and visibility* conditions, and the *mix* of vehicle types in the stream. However, one remarkable finding has been the relative constancy of the saturation flow over time. For a given set of circumstances defined by the factors just mentioned, the flow of vehicles across the stop-line after the start of green appears to remain relatively constant so long as a queue is discharging. For a single lane of average width and a straight (as opposed to a curved) path, the saturation flow has been found to lie in the range 1800 to 2000 veh/h.

Saturation flow may be measured directly by repeatedly counting the number of vehicles passing the stop-line in short intervals (of say 6 s) after the start of green. When this is plotted as a frequency histogram, it can be observed that after the first interval a plateau is reached corresponding to the saturation flow. The following example illustrates the calculation of saturation flow.

Example 26.7

Suppose vehicles are counted crossing the stop-line for the first six intervals of 6 s from the start of green over 24 cycles. The total counts are given in Table 26.2.

Table 26.2 Hypothetical traffic counts at a stop-line

Interval	0–6 s	7–12 s	13–18 s	19–24 s	25–30 s	31–36 s
Count (veh)	51	75	79	74	77	39
Rate (veh/h)	1275	1875	1975	1850	1925	975

Leaving out the first interval, where the flow is building up, and the last interval, where the flow has passed saturation, the estimated saturation flow is 1906 veh/h.

Example 26.7 illustrates that the flow at the start of the green period does not reach saturation instantaneously. It is also the case that the flow does not end instantaneously at the end of the green period. For the calculation of delay and stops, it is helpful to define a period of *effective green* during which it is assumed that vehicles are serviced (pass the stop-line) at the saturation flow rate and which gives the same effective capacity as the actual green period. The area under the curve in Fig. 26.5 constitutes the actual capacity. The effective green time is found by constructing a rectangle enclosing the same area as the curve and having a height equal to the expected saturation flow rate. The effective green time is found in practice to be about 2 s less than the actual green time plus the yellow time. This difference is referred to as the *lost time* per green plus yellow period.

Example 26.8

Suppose that the data presented in Table 26.2 relate to a sequence of 24 cycles where the queue never completely discharged during the green period and that there was no flow across the stop-line after 36 s from the start of green as the red time would have commenced. The estimated saturation flow was 1906 veh/h or 0.53 veh/s. On average, 16.46 vehicles are serviced per cycle, giving an effective green time of 31.06 s. If the green plus yellow periods each cycle sum to 34 s, the lost time on average was 2.96 s.

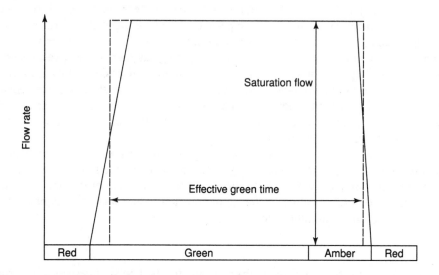

Fig. 26.5 The effective green time

The time that is not effective green is referred to as *effective red time*. Control is therefore a binary variable with two states, effective green or effective red. For a given stream, if g is the effective green time, r the effective red time and c the cycle time, then $c = r + g$. If q is the expected inflow of vehicles, then the expected vehicle arrivals over the cycle is cq. If s is the saturation flow, then the available capacity over the cycle is sg. The *degree of saturation*, ρ, is defined as follows:

$$\rho = \frac{cq}{sg} \tag{26.2}$$

If $\rho > 1$ the stream is overloaded and g/c should be increased until $\rho < 1$. As the inflow will to some extent be random, ρ should not exceed a threshold generally taken to be 0.9.

Every time the right of way switches between two incompatible streams, time is lost through the clearance and lost times. Suppose an intersection has three mutually incompatible streams which have to be given green in sequence. Let L be the total lost time per cycle, composed of the three clearance times and the three lost times. Then the cycle time must be chosen such that:

$$c \geq L + \frac{cq_1}{s_1} + \frac{cq_2}{s_2} + \frac{cq_3}{s_3} \tag{26.3}$$

The minimum cycle time that keeps the degree of saturation of each stream to 0.9 is

$$c = \frac{0.9L}{0.9 - Y} \tag{26.4}$$

where $Y = \sum_i q_i / s_i$.

Where streams are combined in stages, only the critical stream for each stage (that is the stream with the highest q/s ratio) should be included in Y. The longer the cycle time the greater the capacity of the intersection, but also the longer the queues. Where the geometry of the road network imposes maximum queue size constraints on an intersection, this may place an upper limit on the permissible cycle time. A useful rule of thumb is:

$$c \le \frac{h_i}{6q_i} \tag{26.5}$$

where h_i is the queue capacity of stream i (in m), 6 is the average length of a vehicle (in m/veh) and q_i is the inflow on stream i (in veh/s). A well-known expression for a good cycle time obtained by Webster[4] is:

$$c = \frac{1.5L + 5}{1 - Y} \tag{26.6}$$

where as before $Y = \sum_i q_i/s_i$ with only the critical streams included in the summation. Normally the cycle time will lie in the range 30–90 s, although in certain circumstances it may be as large as 120 s. Maximum red times will also place an upper limit on the maximum cycle time. The following example illustrates the calculation of the cycle time.

Example 26.9

Consider a four-arm intersection with four streams and two stages (that is, with no protected opposed turns). Each approach has two lanes and therefore a saturation flow of 3600 veh/h. The flows, saturation flows and their ratios are given in Table 26.3.

Table 26.3 Hypothetical flows and saturation flows

Stream	1	2	3	4
q (veh/h)	1100	1000	650	1500
s (veh/h)	3600	3600	3600	3600
q/s	0.31	0.28	0.18	0.42

Streams 1 and 2 belong to stage 1 while streams 3 and 4 belong to stage 2. Streams 1 and 4 are the critical streams. $Y = 0.31 + 0.42 = 0.73$. The clearance time matrix is given in Table 26.4.

Table 26.4 Clearance time matrix

From/to	1	2	3	4
1			4 s	2 s
2			3 s	4 s
3	2 s	3 s		
4	3 s	2 s		

The intergreen period from stage 1 to stage 2 is 4 s and from stage 2 to stage 1 is 3 s. Thus the total intergreen period per cycle (equal to the total duration of the stage transitions per cycle) is 7 s. To this must be added the lost time for the stages, say 4 s per cycle, giving $L = 11$ s. The minimum cycle time is 58.2 s. A better cycle time as given by Webster's expression would be 79.6 s.

26.6 Performance

The optimisation of control, whether on-line or off-line, requires some measure of performance. There are a number of measures, each reflecting different aspects of system performance. The weight given to each measure will reflect political decisions about public concerns. The *number of stops, delay, travel time* and *queue length* are the more commonly encountered measures. Reducing the number of stops improves the comfort of standing passengers on buses or trams, reduces the noise and exhaust emissions, reduces the probability of collisions (head to tail), improves the efficiency (particularly for heavy goods vehicles), and reduces fuel consumption. Reducing delay saves travel time (and therefore economic losses, as 'time is money'), reduces exhaust emissions and improves safety for pedestrians and cyclists by improving compliance with the signal indications. Reducing travel time and its variability saves delay, improves the efficiency of travel planning, improves the quality of public transport services (by improving their speed and reliability), and reduces costs for public transport and other fleet operators. Reducing queue length reduces noise and exhaust emissions, prevents the blocking of upstream crossings and reduces driver stress. Reducing queue length is closely related to (but not the same as) reducing delay, as all vehicles in queues are delayed, but not all delayed vehicles are in queues.

Of these four measures, the first two are the most important, and queue length is generally introduced as a constraint at locations where there is a risk of a queue blocking an upstream intersection (referred to as *blocking back*).

Both the delay and the number of stops for a particular stream may be calculated from the cumulative arrival and cumulative departure curves, as shown in Fig. 26.6. An extremely useful simplifying assumption is made here, namely that the arrivals and departures all occur at one location, the stop-line. This is equivalent to assuming that the queue occupies no physical space (or is 'vertical'). Vehicles are assumed to travel undelayed to the stop-line before joining a queue, if any. When the light turns red (only effective red is considered here), the cumulative arrivals at the stop-line build up as shown. When the light turns green, departures from the stop-line occur at the saturation flow rate until the cumulative departures curve reaches the cumulative arrivals curve. As the vehicles are assumed to travel to the stop-line undelayed, the delay experienced (at least before crossing the stop-line) is equal to the time spent in the vertical queue. The size of the vertical queue at any particular time is given by the vertical distance between the cumulative arrivals and cumulative departures curves at that time. The area between the two curves gives the total delay. The point where the two curves meet gives the time at which the vertical queue disappears (on the horizontal axis) and the number of stops associated with the vertical queue (on the vertical axis).

The representation of the queuing process portrayed in Fig. 26.6 is fundamental to the cyclic flow profile traffic model described in Chapter 27. The size of the vertical queue is, however, an underestimate of the size of the physical queue because some

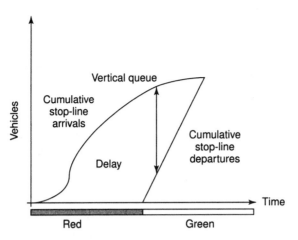

Fig. 26.6 Vertical queuing model

vehicles assumed to be still travelling undelayed to the stop-line will in reality have already joined the physical queue. Suppose that vehicles in the stream in question arrive at a constant rate of q veh/h with a density of k veh/km. The vehicles then join a physical queue with density k_j. During red, the vertical queue will grow at a rate of q veh/h whereas the physical queue will grow at a rate of $q' = q/(1 - k/k_j)$ veh/h. This implies a correction to the estimate of the number of stops. If r is the effective red time and s the saturation flow, then the number of stops associated with the vertical queue is $rsq/(s - q)$. Applying this correction would imply $rsq'/(s - q')$ stops, a somewhat larger number. However, the clearing wave (the front of the queue) travels backward faster than s, tending to reduce the total number of stops. The following two examples illustrate the calculation of the number of stops.

Example 26.10

Suppose that a traffic stream has a steady inflow of 500 veh/h (0.14 veh/s) at a speed of 50 km/h (13.9 m/s). This implies an average density of 0.01 veh/m (flow divided by speed) which implies an average spacing of about 100 m per vehicle (the inverse of the density). Suppose further that the effective red time is 30 s and the saturation flow is 1800 veh/h (0.5 veh/s). In the queue, each vehicle occupies on average 6 m including its headway. The average number of stops as estimated by the vertical queuing model is around 5.8. After the horizontal queue correction, the estimate is 6.2 stops. In practice the clearing wave (front of the queue) travels backward faster than one vehicle (about 6 m) every 2 s. As a rule of thumb, the nth vehicle in the queue begins to move after n seconds. This would imply 5.1 stops.

The total delay would be $0.5s^2r^2q/(s - q)^2$. If d is the average delay per vehicle, g the effective green time, c the cycle time (note that $c = r + g$), and if the queue clears every cycle, then:

$$d = \frac{0.5s^2(c-g)^2}{c(s-q)^2} \qquad (26.7)$$

This is the *uniform* component of delay in Webster's delay formula.[4]

Example 26.11

For Example 26.7, suppose additionally that the cycle time is 55 s, implying an effective green time of 25 s. The total delay is 121.5 veh/s and delay per vehicle is 15.8 s.

In reality, the arrival process will be to some extent random, and there will be some cycles when the queue does not clear. Those vehicles failing to get through will have to wait until the next green period. As the total queue length and the total delay are independent of the order in which the vehicles are served,[5] one can imagine a queuing process where the vehicles are served in reverse order. Any vehicle which is delayed in one cycle will not be served in the next unless there is some spare capacity after all the new arrivals have been served. Since the new arrivals will consume most, if not all, of the green interval, the residual vehicles will almost certainly wait all, or nearly all, of the next cycle time, *c*. Thus:

$$d = \frac{0.5s^2(c-g)^2}{c(s-q)} + \frac{Q}{q} \qquad (26.8)$$

where Q is the expected queue at the start of red.

There is an extensive literature on the evaluation of Q (see references 4 and 6). If arrivals and departures over the cycle were random with rates q and sg/c respectively then the expected queue excluding the vehicle being serviced would be:

$$Q = \frac{0.5\rho^2}{1-\rho} \qquad (26.9)$$

This provides the second term of Webster's delay formula:

$$d = \frac{0.5s^2(c-g)^2}{c(s-q)} + \frac{0.5\rho^2}{q(1-\rho)} \qquad (26.10)$$

Simulation results supplied Webster[4] with a third corrective term. It is Webster's three-term delay formula that is most commonly applied in practice.

In congested traffic conditions, the capacity of junctions is often temporarily exceeded. However, the above expression for delay tends to infinity as the degree of saturation ρ tends to 1, and cannot be used when the degree of saturation exceeds 1. This is because the expression applies to a *steady state* situation, and a degree of saturation greater than 1 is not sustainable indefinitely. Where capacity is temporarily exceeded, the *coordinate transformation method* has been proposed,[7] yielding the *sheared delay formula*. According to this formula, the expected delay tends to infinity as the degree of saturation tends to infinity. The rate with which infinite delay is approached is given by the queue growth implied over the interval considered by a unit increase in the degree of saturation.

26.7 Off-line signal plan generation

Having determined a cycle time to give sufficient capacity, the green time should be shared between the stages or streams. Conventionally, the first step is to identify the critical stream for each stage, namely the stream with the highest q/s ratio. Where the minimum cycle time has been chosen, the green time will be:

$$g_i = \frac{cq_i}{0.9s_i} \qquad (26.11)$$

where stream i is the critical stream of the stage. Where the cycle time is larger than the minimum value, the delay minimising green time for critical stream i is obtained approximately[8] by applying the *equisaturation* principle:

$$g_i = \frac{(c-1)q_i}{Ys_i} \qquad (26.12)$$

where as before $Y = \sum_i q_i/s_i$ (the summation is over the critical streams only).

Normally, the queue on the critical stream will be the last queue of the stage to discharge fully. As a rule of thumb, green time should be terminated when the queue on the critical stream has fully discharged, as beyond this time saturation flow ceases and the green time is less efficiently used. The service time for a vehicle lies somewhere between 1.6 s and 2.2 s, so a conservative estimate would be 2 s. Given the cycle time, the required green time for the critical stream may be approximated by:

$$g_i = (\text{service time})cq_i \qquad (26.13)$$

The following example illustrates the calculation of green times.

Example 26.12

Returning to Example 26.9, the stage durations should be 20 s and 27.2 s for stages 1 and 2 respectively when the cycle time is 58.2 s. When the cycle time is increased to 79.6 s, the stage durations should be 29.1 s and 39.5 s for stages 1 and 2 respectively.

Software is available for the off-line calculation of a traffic signal plan. A modern group-based optimisation method is implemented in the SIGSIGN program.[9] For a given sequence in which the streams are to receive green, the method generates optimal stages, stage durations, stage sequence and stage transition structures.

26.8 On-line microcontrol

26.8.1 Introduction

In principle, the performance of a traffic signal-controlled intersection can be significantly improved by responding to variations in traffic flow. Vehicle detectors offer the possibility of monitoring the state of traffic flow at the location of the loop (or other sensor). Modern microprocessor-based signal controllers may be programmed to respond to vehicle detector data, leading to what is referred to as *microcontrol*.

Microcontrol may be expressed in the form of a *flow diagram* showing precisely how and when the detected vehicle states (vehicle arrival, vehicle departure, headway, etc.) affect the start or end of green for a stage or a stream. A formal symbolism for such flow diagrams is often adopted,[3] where decision elements are denoted by hexagons and action elements by rectangles (see Fig. 26.7). A major feature of microcontrol in many

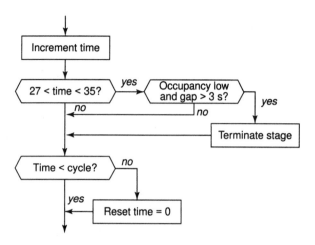

Fig. 26.7 Flow diagram for microcontrol

countries is the treatment of requests for priority from public transport (or indeed the emergency services)

Very often the freedom afforded microcontrol is limited by the need for coordination. One way of limiting the freedom of microcontrol is to impose a *frame plan*. This resembles the signal plan, as described in Section 26.4, but rather than giving the precise times when changes occur, it gives the earliest and latest times when changes can occur. The more common forms of intervention encountered in microcontrol are caused by *gapping out, occupancy, green request, queue detection* and *priority request*.

26.8.2 Gapping out

It was noted previously that when the queue on a stream has discharged, the efficiency of use of the green time by that stream falls significantly. As the saturation flow for a single lane lies around one vehicle every 2 s, it can be inferred that the saturation flow has ceased when headways in excess of say 3 s start to arise at the stop-line. However, by the time a headway of 3 s had been measured, 3 s will already have been wasted. It is desirable therefore to locate the vehicle sensor upstream at a distance determined by the approach speed. The minimum green time needs to be sufficient to clear all the vehicles between the sensor and the stop-line at the start of green. Gapping out is not suitable for an opposed turning stream. The following example illustrates the calculation of the minimum green time.

Example 26.13

If the approach speed is 50 km/h (or 13.9 m/s), the sensor should be at least 42 m from the stop-line in order to give adequate advance notice of a 3 s gap. If the average vehicle is 6 m long, there would be space for 7 average vehicles between the sensor and the stop-line, which at a rate of a vehicle every 2 s would require 14 s to clear. The minimum green time would normally be 14 s. However, if the traffic flow is light, the minimum green time could be variable and depend on the number of vehicles counted past the sensor since the end of green.

26.8.3 Occupancy

Larger vehicles can sometimes lead to premature gapping out, so often an *exponentially smoothed* occupancy measure is also used to identify the end of saturation flow (larger headways implies lower traffic density and therefore lower occupancy). Figure 26.7 shows an example of two-stage control where, within certain windows of opportunity defined by the frame plan, a large headway coupled with a low occupancy leads to the termination of the stage.

Example 26.14

Suppose occupancy is measured every 6 s. Consider the following series of measurements:

0.65   0.69   0.61   0.19   0.21   0.18   0.25   0.31   0.38   0.41

If $o_m(t)$ is the measured occupancy for period t, then the exponentially smoothed occupancy for that period, $o_s(t)$, is obtained from

$$o_s(t) = \lambda\, o_m(t) + (1 - \lambda)\, o_s(t - 1) \qquad (26.14)$$

where the value of λ lies between 0 and 1. If $o_s(0) = 0.6$ and $\lambda = 0.4$, then the following values for $o_s(t)$, $t = 1\ldots10$, are obtained:

0.62   0.65   0.63   0.46   0.36   0.29   0.27   0.29   0.32   0.36

The smoothing effect of the operation is apparent in the second series.

26.8.4 Green request

Sometimes a stage does not receive green unless requested by a detector. In this case, one sensor would be located near the stop-line to reduce the possibility of vehicles going undetected. There may also be one or more sensors upstream, perhaps also used for gapping out. In the case of a protected opposed turn where the vehicles queue in the centre of the intersection, there may be a sensor in the centre of the intersection located where the front of the queue would be, causing an *early cut-off* of the opposing flow.

26.8.5 Queue detection

Sometimes the time-occupied counter for a loop located upstream of the stop-line is used to detect when a queue has grown to such an extent as to risk blocking a lane entrance or upstream intersection. To recognise when a queue has reached the loop, a time-occupied threshold of 5 to 15 s is generally chosen. When a queue has been detected, there are two courses of action. If the queue is on a stream currently with green, the stage can be *extended* sufficiently to allow all the vehicles between the sensor and the stop-line to clear. If the queue is on another stream, a relevant stage can be *recalled* as soon as possible.

26.8.6 Priority request

Where green is requested for a public transport vehicle, whether by selective detection or by radio transmission, the point of initial request should lie sufficiently far upstream to allow enough time to be able to intervene, when necessary, to ensure that the delay to the vehicle is equal to or close to zero. When there are no stops on the approach to the intersection, the point of initial request should lie about 500 m from the stop-line. There may then be secondary request points on the approach to the intersection to confirm the progress of the vehicle.

Where there is a stop on the approach in the vicinity of the intersection, there must be either an allowance for the *stopped time* or a *request point* at the exit of the stop. As stopped time is variable, depending on the number of passengers boarding or alighting, the accuracy of the intervention will be reduced where the request point is before the stop. On the other hand, where the initial request is made at the exit to the stop, there may be insufficient time to intervene to give priority, sometimes causing delay to public transport.

Often there is a *cancellation point* just beyond the stop-line, where the priority request is terminated. In some cases, the request must also give information about the vehicle (for example, whether it is running ahead of or behind schedule) or its route (to determine what movement it will be making in the intersection). Where there is another intersection downstream at a distance of less than 500 m, the cancellation point may act as the initial request point.

There are basically two forms of intervention, *stage extension* (where the stage currently with green is extended to allow the passage of the bus or tram) and *stage recall* (where the current stage is ended as soon as the minimum green time will allow and the required stage is recalled).

26.9 On-line proprietary systems

The forms of microcontrol described previously are primarily concerned with the efficiency of usage of green by the stage currently with right of way. The costs imposed on streams currently without green are not taken into account. MOVA, developed by the Transport Research Laboratory[10] to overcome this limitation, has two modes of operation. When the intersection is operating at significantly below its capacity, MOVA switches to the next stage when the cost of so doing is outweighed by the cost of not switching, evaluated over the next cycle (in this context, cost will be a weighted combination of delay and stops). When the capacity of the intersection is approached, MOVA changes to a capacity maximising strategy, favouring stages with higher capacities over those with lower capacities. An interesting feature of MOVA is the use of simulation to predict the trajectories of vehicles over space and time after their detection at the upstream sensors and until they have cleared the stop-line.

Siemens have a number of pre-programmed modules, such as SAEWA,[11] which seeks to divide effective green time between stages so as to equate the predicted degrees of saturation, and VS-PLUS,[12] which makes stage change decisions on the basis of priority rules.

26.10 References

1. Lapierre, R., and Obermaier, A., Entwicklung der Lichtsignalsteuerung in städtischen Straßennetzen. In *88 Jahre Straßenverkehrstechnik in Deutschland*. Bonn: Kirschbaumverlag, 1988, 87–92.
2. Salter, R.J., *Highway traffic analysis and design*. London: MacMillan, 1989.
3. RiLSA, *Richtlinien für Lichtsignalanlagen*. Köln: Forschungsgesellschaft für Straßen- und Verkehrswesen, 1992.
4. Webster, F.V., *Traffic signal settings*. Road Research Technical Paper 39. London: HMSO, 1961.
5. Newell, G.F., *Theory of highway traffic signals*. Institute of Transportation Studies, Berkeley: University of California, 1989.
6. Wardrop, J.G., Some theoretical aspects of road traffic research. *Proceedings of the Institution of Civil Engineers*, 1952, **1**, 325–62.
7. Kimber, R.M., and Hollis, E.M., *Traffic queues and delays at road junctions*. Laboratory Report 909, Crowthorne, Berks: Transport Research Laboratory, 1979.
8. Webster, F.V., and Cobbe, B.M., *Traffic signals*. Road Research Technical Paper 56. London: HMSO, 1966.
9. Silcock, J.P., and Sang, A.P., SIGSIGN: A phase-based optimisation program for individual signal controlled intersections. *Traffic Engineering and Control*, 1989, **31** (5), 291–298.
10. Vincent, R.A., and Peirce, J.R., *MOVA. Traffic responsive, self-optimising control for isolated intersections*. Research Report LR170. Crowthorne, Berks: Transport Research Laboratory, 1988.
11. Böttger, R., SAEWA: Ein Grünzeit-Bemessungsverfahren nach Auslastungsgrad (Sättigungsgrad) für Siemens M-Geräte, *Grünlicht*, 1988, **25**, 4–11.
12. Albrecht, H., Backes, A., and Kaul, H., VS-PLUS: Ein neuer Weg zur Realisierung verkehrsabhängiger Steuerung, *Grünlicht*, 1993, **31**, 4–11.

CHAPTER 27

Signal control in networks

M.G.H. Bell

27.1 Off-line control

27.1.1 General principles

In urban networks, the distance between neighbouring intersections is frequently too small for platoons of traffic released by one intersection to disperse completely before arrival at the next intersection (about 1 km is required for complete platoon dispersion). Benefits can therefore be obtained by signal coordination, as was already recognised by the 1930s.[1] Through coordination, it is possible in certain circumstances to establish a *green wave*, whereby a platoon of traffic may pass through a sequence of intersections without stopping. Alternatively, the risk of excessive queuing between two intersections may be reduced by coordination. The other benefits of coordination include: the attraction of traffic by green waves to major roads away from minor roads in environmentally sensitive (typically residential) areas; the reduction of speed variation on major roads, thereby improving road safety; and the improvement of the comfort and speed of public transport services by reducing their need to stop at intersections.

A good way to represent the coordination of traffic signals along a path is through a time-distance diagram (see Fig. 27.1). If distance is represented by the horizontal axis and time by the vertical axis, vertical barriers may be located on the horizontal axis at the positions of the stop-lines. Gaps in these barriers correspond to effective green periods. A green wave is represented by a band passing through the gaps in the barriers at a speed v_p in the range $0.85v_{max} \leq v_p \leq v_{max}$, where v_{max} is the maximum permissible speed. The trajectories of buses or trams can be superimposed on the bands, allowing these to decelerate and accelerate at between 0.7 m/s^2 and 1.2 m/s^2 for trams and 1.0 m/s^2 and 1.5 m/s^2 for buses. Allowance must of course be made for the scheduled time spent at bus or tram stops.

There is already a difficulty in obtaining good coordination along one path in both directions. The problem is more severe for networks, as good coordination for one path generally comes at the expense of good coordination for other paths. There are two basic approaches. The first is *proactive*, and begins by identifying the paths to which one wishes to attract traffic. Such paths will generally be high-capacity arterials. Coordination is then used to reduce travel times on these paths at the expense of travel times on the other paths from which one wishes to displace traffic.

The second approach is *reactive*, whereby no particular path is prioritised. A coordination is sought that optimises some *performance index*, such as a weighted

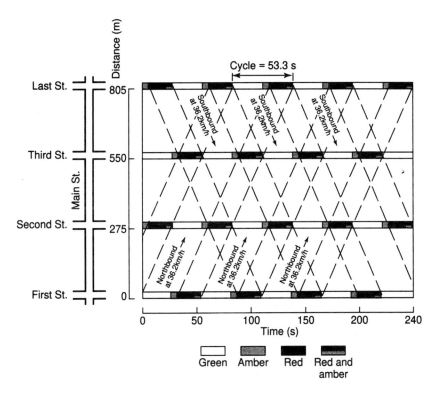

Fig. 27.1 Time-distance diagram

combination of delay and stops. Until recently, the latter approach has tended to dominate in the UK. However, increasing concern about the detrimental environmental effects of traffic is shifting the balance toward a more proactive approach.

27.1.2 Platoon dispersion

The case for the coordination of traffic signals is based largely (although not entirely) on the existence of *platoons* of traffic. Platoons are formed by traffic signals; a platoon is built while the light is red and is released while the light is green. As the platoon travels to the next intersection, differences in the speeds of the vehicles will cause the platoon to disperse. While it is possible that slow vehicles, combined with an absence of overtaking opportunities, may cause platoons to form between intersections, the tendency of platoons to disperse has been found to dominate. If platoons have substantially dispersed by the time they arrive at the next intersection, the case for coordination is significantly weakened.

A useful way to describe the flow of traffic passing a point (say a sensor) over the cycle is the *cyclic flow profile*. When the cycle is divided into small intervals of say 1 s, the cyclic flow profile is a table of flow rates for each interval. Based on the concept of the cyclic flow profile, the following recursive platoon dispersion formula[2]

$$q_s(i + t) = Fq_0(i) + (1 - F)q_s(i + t - 1) \tag{27.1}$$

is widely used to predict the arrival rate over time of vehicles at a point (generally taken to be the stop-line) given the departure rate over time at a point upstream; $q_s(i + t)$ is the stop-line arrival rate during interval $i + t$, $q_0(i)$ is the known departure rate during interval i at a point some distance upstream of the stop-line (the entrance to the link or the location of a sensor), t is the free flow travel time from the point at which q_0 is measured to the stop-line (sometimes taken as 0.8 times the average travel time), and F is the dispersion parameter. F is generally treated as a function of t, viz.:

$$F = \frac{1}{(1 + 0.5t)} \tag{27.2}$$

when the interval chosen for the cyclic flow profile is 1 s. Implicit in the platoon dispersion formula is the assumption that the vehicles travel to the stop-line undelayed (namely the 'vertical queue' assumption).

It has been demonstrated[3] that the recursive platoon dispersion formula is equivalent to an assumption that travel times follow a geometric distribution. Reference to Example 26.14, which related to the exponential smoothing of occupancy measurements, shows that the platoon dispersion formula in effect exponentially smooths the flows.

27.1.3 Cyclic flow profile traffic model

Cyclic flow profiles together with the recursive platoon dispersion formula define a traffic model that underlies the traffic simulation model SATURN,[4] the off-line traffic signal optimisation tool TRANSYT[2] and the SCOOT on-line traffic control system.[5] The cyclic flow profile at the entrance to a link (the *in-flow profile*) is dispersed along the link according to the platoon dispersion formula to give an *arrival profile* at the stop-line. Queues are assumed to occupy no physical space, so the dispersion occurs over the full length of the link. The signals are assumed to be either *effectively green* or *effectively red* (see Section 26.5 for the definition of these terms). An *accept profile* gives the capacity of the stop-line to service vehicles; this is zero while the light is red and equal to the saturation flow while the light is green. The *departure profile* describes the rate at which vehicles pass the stop-line. During green, the queue discharges at the saturation flow rate, after which the arriving vehicles pass unhindered. The departing vehicles are split according to pre-defined turning proportions and added to give the in-profiles for the downstream links. The area between the cumulative arrival and departure curves at the stop-line gives the total delay (as in Fig. 26.8), from which average delay per vehicle can be calculated. The steps of the cyclic flow profile traffic model are illustrated in Fig. 27.2.

The following example illustrates the operation of the cyclic flow profile traffic model.

Example 27.1

For the cyclic flow profile given in Table 27.1, and taking $t = 5$ s and $F = 0.29$, the recursive platoon dispersion formula yields the arrival profile also given in Table 27.1. The cycle is 60 s and the interval is 5 s. For intervals 1 to 6, the signal indication is red so the accept profile has value 0. For intervals 7 to 12, the signal indication is green, so the

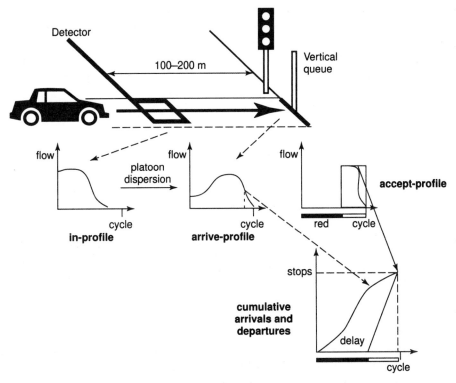

Fig. 27.2 The cyclic flow profile traffic model

Table 27.1 Example cyclic flow profiles

Light	Red						Green					
Interval (5 s)	1	2	3	4	5	6	7	8	9	10	11	12
In (veh/5 s)	1.30	1.40	1.20	0.90	0.50	0.20	0.30	0.10	0.20	0.10	0.10	0.20
Arrival (veh/5 s)	0.21	0.38	0.68	0.83	0.85	0.75	0.59	0.51	0.39	0.33	0.26	0.21
Accept (veh/5 s)	0	0	0	0	0	0	2.50	2.50	2.50	2.50	2.50	2.50
Depart (veh/5 s)	0	0	0	0	0	0	2.50	2.09	0.39	0.33	0.26	0.21

accept profile has a value given by the saturation flow, assumed here to be a vehicle every 2 s (or 2.5 vehicles per 5 s). In the first 6 intervals, 3.49 vehicles arrive on average. The saturation flow therefore lasts through interval 7 and into interval 8. By interval 9, the queue has discharged and the departures correspond to the arrivals.

Table 27.2 shows the corresponding cumulative arrivals, departures and approximate delays. An estimate of the number of vehicles queuing in each interval is the difference between the cumulative arrivals and cumulative departures up to and including that interval. As the intervals are 5 s, total delay in each interval is equal to the number of vehicles queuing multiplied by 5 s. The total delay is estimated as 63.05 vehicle-seconds.

Table 27.2 Example cumulative arrivals, departures and delay

Light	Red						Green					
Interval (5 s)	1	2	3	4	5	6	7	8	9	10	11	12
Cumulative arrivals (veh/5 s)	0.21	0.59	1.27	2.10	2.95	3.70	4.29	4.80	5.19	5.52	5.78	5.99
Cumulative departures (veh/5 s)	0	0	0	0	0	0	2.50	4.80	5.19	5.52	5.78	5.99
Delay (veh-s)	1.05	2.95	6.35	10.5	14.75	18.5	8.95	0	0	0	0	0

An early approach to network coordination was the Combination Method.[6] The method is based on a pre-calculated relationship (in the form of a table) between average delay per vehicle and the *offset* between two adjacent intersections, where the offset is the difference between the start of the effective green times at two adjacent intersections along a defined path. The next example illustrates the relationship between offset and delay.

Example 27.2

Returning to Example 27.1, a shift (or *offset*) of the start of green forward by 10 s (or two intervals) has the effect illustrated in Table 27.3. Note that the arrivals are accumulated since the start of red. The total delay is reduced to 38.5 vehicle-seconds.

Table 27.3 Example cumulative arrivals, departures and delay

Light	Red				Green						Red	
Time interval(5 s)	1	2	3	4	5	6	7	8	9	10	11	12
Cumulative arrivals (veh/5 s)	0.68	1.06	1.74	2.57	3.42	4.17	4.76	5.27	5.66	5.99	0.26	0.47
Cumulative departures (veh/5 s)	0	0	0	0	2.50	4.17	4.76	5.27	5.66	5.99	0	0
Delay (veh-s)	3.40	5.30	8.70	12.85	4.60	0	0	0	0	0	1.30	2.35

For a given arrival profile, the model just described is used to estimate the total delay per cycle for a given offset. The delays may then be combined in various ways, depending on whether the links are in parallel or in series. The following example illustrates the calculation of the optimum delay for two links in parallel.

Example 27.3

The total delays given in the first two rows of Table 27.4 refer to two links in parallel, one from intersection A to intersection B and the other from intersection B to intersection A.

There is a 50 s cycle time, so that an offset of 50 s is equivalent to no offset. The delays may be summed, remembering that an offset of A relative to B of 10 s is equivalent to an offset of B relative to A of -10 s (or 40 s, because of the cyclic nature of the process), that an offset of A relative to B of 20 s is equivalent of an offset of B relative to A of -20 s (or 30 s), etc. The relationship between delay and offset for the two links combined is shown in the third row of Table 27.4. The optimum offset occurs at an offset of 40 s.

The next example illustrates the calculation of optimum offset for two links in series.

Table 27.4 The relationship between offset and total delay (in vehicle-seconds)

Offset of A to B (s)	0	10	20	30	40
Link A to B (veh-s)	15	18	20	19	18
Link B to A (veh-s)	20	18	16	12	10
Combined (veh-s)	35	36	35	32	28

Example 27.4

The total delays given in the first two rows of Table 27.5 refer to two links in series, one from intersection A to intersection B and the other from intersection B to another intersection C. As before, the cycle time is 50 s. An offset of 0 s between intersections A and C may consist of an offset of 0 s between intersections A and B and 0 s between intersections B and C, or an offset of 10 s between intersections A and B and -10 s (or 40 s) between intersections B and C, etc. The delay per vehicle for all combinations of offsets giving an offset of 0 s between intersections A and C is given in the third row of Table 27.5. The optimum combination is an offset of 0 s between intersections A and B and therefore of 0 s between intersections B and C. When the same operation is performed for all possible offsets between intersections A and C, the best combination of offsets overall can be found.

Table 27.5 The relationship between offset and total delay (in vehicle-seconds)

Offset of A to B (s)	0	10	20	30	40
Link A to B (veh-s)	10	15	30	20	25
Link B to C (veh-s)	5	15	20	10	20
Link A to C (veh-s)	15	35	40	40	40

To find all the offsets of a network using the Combination Method, the links of a network are progressively combined in pairs (either in parallel or in series) until there is only one link left. The sequence in which the combinations are performed is arbitrary because the arrival profile used in the calculation of delay is assumed to be unaffected by the offsets elsewhere in the network. The TRANSYT simulation model[2] was developed in order to relax this assumption.

27.1.4 TRANSYT

The TRANSYT program uses the traffic model described in the previous section. For given signal settings, the cyclic flow profiles are repeatedly updated until an equilibrium is achieved, whereby the cyclic flow profiles in one cycle would be the same as those in the next. This is also the nature of the equilibrium sought in the SATURN traffic simulation model.[4] The optimisation of the offsets is achieved by 'hill climbing', whereby the effect of changing each offset by a fixed amount is evaluated on the basis of the cyclic flow profile traffic model, retaining only those changes that produce an improvement to the performance index (generally taken to be a linear combination of delay and stops). The offset changes are made in sequence, and both positive and negative changes are evaluated.

The division of green between stages at each intersection is done so as to equalise the *degree of saturation* (defined in Section 26.5). A common cycle time is chosen for the network so that the maximum degree of saturation at any intersection is 0.9.

The TRANSYT program has been remarkably successful for a number of reasons. The underlying cyclic flow profile traffic model is parsimonious, in the sense that only those aspects about the traffic process that are relevant for the assessment of performance are represented. The 'hill climbing' optimisation method used by TRANSYT appears to work well in practice, despite the non-convex nature of the optimisation problem (the performance index may have many local maxima with respect to the offsets). Moreover, the model can be extended in various ways, for example to allow for different user classes, such as public transport and pedestrians.

27.2 On-line control

27.2.1 General principles

It has long been appreciated that significant extra benefits at a network level may be obtained by allowing traffic signals to be responsive to variations in vehicle flows. Responsiveness may occur either at the intersection level or at the network level through signal coordination. The balance between local and network responses depends, among other things, on the level of traffic. Under light traffic conditions, for example at night, local responsiveness would be sufficient and coordination would result *de facto*. However, during a peak period, the benefits of an imposed coordination may be significant.

There is still an issue as to whether the control should be reactive or proactive. At an operational level, purely reactive control can lead to problems in congested situations. Decisions made reactively on the outskirts of a town, for example, may result in too much traffic being given access to a town centre. To avoid this problem, proactive *gating* strategies have been developed, whereby access to an area threatened with congestion is rationed so that congestion does not arise. In effect, this is a redistribution of queues to roads with sufficient capacity to contain them. Recent environmental concerns are giving added weight to proactive gating strategies.

Because of the lags involved in the traffic process itself (queues take some time to form and to dissipate), decisions affecting the control of traffic signals should be based on their predicted effects on performance. This is referred to as a form of *feed forward control*. If control is based only on the current situation, referred to as *feedback control*, there may be a significant loss of efficiency and instabilities may arise. Prediction requires a traffic model, such as the cyclic flow profile traffic model described earlier.

27.2.2 Plan selection

The first form of on-line control consisted of *plan selection*, whereby pre-calculated signal plans for the network are implemented according to the currently prevailing traffic conditions. There are a number of limitations to this approach: the number of plans required to suit all feasible network conditions could be very large; each time the plan is changed there are *transitional losses* because the pattern of traffic is disrupted; and decisions relating to when to switch plans are based on *current* traffic conditions, potentially leading to a loss of efficiency and instability.

27.2.3 SCOOT

An on-line implementation of the cyclic flow profile traffic model described earlier is to be found in the SCOOT (Split Cycle and Offset Optimisation Technique) system.[5] The control variables are the *splits* (the division of green time between the stages at each intersection), the offsets (defined earlier) and the cycle time (which is common to a region). The split at each intersection is changed to achieve an equal degree of saturation for each stage, the offsets (defined earlier) are changed to maximise the performance index, and the cycle time is changed to keep the maximum degree of saturation close to 0.9. The 'hill climbing' procedure is also adopted here, with the difference that only one step change is normally made to any split, offset or cycle time before implementation. The traffic signal settings are therefore moved *toward* the optimum suggested by the cyclic flow profile traffic model; the procedure therefore is not a true optimisation, such as will be discussed in the next section. The splits are changed most frequently and the cycle time is changed least frequently.

SCOOT cannot simply be regarded as an on-line version of TRANSYT. The 'hill climbing' is incomplete in SCOOT, as in normal operation only one step is taken before the new set of signal timings are implemented. Furthermore, the in-flow profiles are measured by inductive loop, while in TRANSYT they are calculated so as to achieve an equilibrium.

27.2.4 Discrete time, rolling horizon control

For good control, simulation experiments have shown that prediction for up to the next 2 minutes is helpful. For horizons longer than 2 minutes, the accumulation of uncertainties more than counteracts any potential benefits. When decisions regarding whether or not to change the signal indications are taken at regular intervals, say every 3 or 5 seconds, all feasible sequences of decisions may be represented as a *decision tree*. It is possible to associate a cost with each decision, where cost may be delay, a linear combination of delay and number of stops, or some other measure, perhaps of environmental damage. Figure 27.3 shows an example of such a tree plotted against cumulative cost. Beyond the horizon there is a terminal cost which represents in an approximate way costs beyond the horizon.

The best sequence of decisions is that which incurs least cumulative cost, including the terminal cost. The purpose of the terminal cost is to counteract a bias toward sequences of decisions that are low cost over the horizon but high cost thereafter. A number of algorithms are suitable for finding the optimal sequence of decisions.

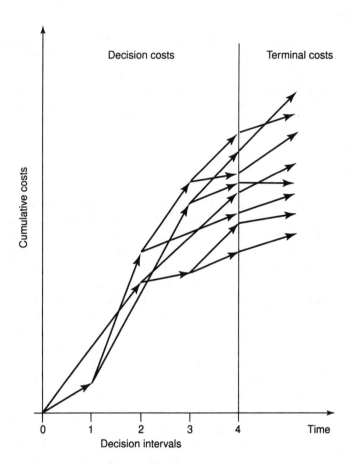

Fig. 27.3 Decision tree plotted against cumulative cost

Backward dynamic programming has been used in the OPAC system developed in the USA,[7] forward dynamic programming in the PRODYN system developed in France[8] and a branch-and-bound heuristic in the UTOPIA system developed in Italy.[9] It has also been pointed out[10] that a shortest path algorithm, such as Dijkstra's algorithm, offers an efficient way to determine the optimum sequence of decisions.

The cyclic flow profile traffic model is used here as well. The in-flow profile is measured by sensors located at the entrance of the link, as in the SCOOT system. As the queue is modelled as occupying no physical space, the vehicle sensors provide advanced information about arrivals at the stop-line. The extent of the forewarning is determined by the undelayed travel time from the sensor to the stop-line. For a sensor that is 42 m from the stop-line, the prediction would be for 3 s if the desired speed is 50 km/h. Predictions beyond 3 s would be based on either the pattern of arrivals in previous cycles or the predicted departures from upstream intersections (or both).

Once the optimum sequence of decisions has been determined, the first decision is implemented, the horizon is rolled forward and the whole process of optimisation is repeated. If decisions are taken every 3 seconds, the horizon is rolled forward every 3 seconds. When priority for public transport is required, the complexity of the optimisation is reduced. The requirement for green at a particular time for a particular stream reduces the number of feasible sequences of decisions. Priority for trams is an important feature of the UTOPIA system installed in Turin.[9]

The discrete time, rolling horizon approach is attractive for its flexibility and has therefore been the subject of much research and development activity.

27.2.5 Other proprietary systems

There are a number of other systems that have achieved varying degrees of international success. The SCATS system of Australia[11] is characterised by vehicle sensors located at the stop-line measuring the efficiency of use of the green time. The efficiency of use of green time determines the green split for each intersection, the cycle time for critical intersections and the offsets. The offsets are based on pre-determined plans that are selected on the basis of the current efficiency of use of green times.

Siemens offer a range of software modules which are combined to form an urban traffic control system. The hardware consists of microprocessor-based signal controllers linked by cable to a traffic computer, so the software modules may run either on the signal controllers or on the traffic computer, depending on the functions to be performed. The SAEWA module,[12] for example, runs on the local controller and endeavours to equalise the degree of saturation. The STAUKO module[13] modifies offsets locally to prevent a downstream queue from exceeding a certain size. The VERON module[14] runs centrally on the traffic computer and seeks to optimise offsets along certain routes. The traffic computer determines a framework plan which sets limits for the start and end of green. The logic in the local controller decides exactly when to start or end green for each signal group within the limits laid down in the framework plan. The framework plan may be more or less restrictive, depending on the degree of coordination required.

27.3 References

1. Lapierre, R., and Obermaier, A., Entwicklung der Lichtsignalsteuerung in städtischen Straßennetzen. In *88 Jahre Straßenverkehrstechnik in Deutschland*. Bonn: Kirschbaumverlag, 1988, 87–92.
2. Robertson, D.I., *TRANSYT – A traffic network study tool*. Laboratory Report LR 253. Crowthorne, Berks: Transport Research Laboratory, 1969.
3. Seddon, P.A., The prediction of platoon dispersion in the combination methods of linking traffic signals. *Transportation Research*, 1972, **6**, 125–130.
4. Van Vliet, D., SATURN: A modern assignment model. *Traffic Engineering and Control*, 1982, **23**, 578–81.
5. Hunt, P.B., Robertson, D.I., Bretherton, R.D., and Winton, R.J., *SCOOT – A traffic responsive method of coordinating signals*. Laboratory Report LR 1014. Crowthorne, Berks: Transport Research Laboratory, 1981.
6. Salter, R.J., *Highway traffic analysis and design*. London: Macmillan, 1989.
7. Gartner, N.H., OPAC: A demand responsive strategy for traffic signal control. *Transport Research Record 906*, 1983.

8. Henry, J.J., Farges, J.L., and Tuffal, J., The PRODYN real time traffic algorithm, *4th IFAC-IFIP-IFORS Conference on Control in Transportation Systems*. Baden Baden, September 1983, 307–311.

9. Mauro, V., and Di Taranto, C., UTOPIA. *6th IFAC-IFIP-IFORS Conference on Control, Computers and Communication in Transport*. Paris, September 1989.

10. Bell, M.G.H., A probabilistic approach to the optimisation of traffic signal settings in real time. In M. Koshi, (ed.), *Transportation and traffic theory*. New York: Elsevier, 1990, 619–632.

11. Lowrie, P.R., *SCATS – Sydney coordinated adaptive traffic system: A traffic responsive method of controlling urban traffic*. Sydney: NSW Roads and Traffic Authority, 1991.

12. Böttger, R., SAEWA: Ein Grünzeit-Bemessungsverfahren nach Auslastungsgrad (Sättigungsgrad) für Siemens M-Geräte, *Grünlicht*, 1988, **25,** 4–11.

13. Böttger, R., Koordinierung von signalanlagen in Stausituationen (STAUKO), *Straßenverkehrstechnik*, 1987, **3,** 85–90.

14. Böttger, R., On-line optimisation of the offset in signalised street networks. *Proceedings of the International Conference on Road Traffic Signalling*, London: IEE Conference Publication 207. 1982, 82–85.

CHAPTER 28

Driver information systems

M.G.H. Bell, P.W. Bonsall and C.A. O'Flaherty

This chapter covers a variety of techniques for communicating with drivers with the aim of assisting them in the driving task. These techniques range from conventional regulatory, warning and information signs, road markings and roadside post delineators, through variable message signs, to in-vehicle systems which are being promoted as part of a comprehensive intelligent transport system (ITS). Although traffic signals might also be considered to be part of these information systems, they are addressed separately (see Chapters 26 and 27) because of their importance.

28.1 Conventional traffic signs

A particular problem associated with the modern-day use of the motor vehicle is the difficulty which motorists from different countries, speaking different languages, have in understanding traffic signs on roads far away from home. As a result, nations have begun to move toward the international standardisation of traffic signs. Nonetheless, there is still no worldwide uniformity in road signing.

In large geographic regions such as the United States, where language differences are not perceived to be of major importance, the tendency has been to make use of signs which mainly rely upon word legends. In Europe, by contrast, emphasis is placed on the use of signs which communicate their message by ideographic representations rather than by inscriptions which may be incomprehensible to non-local motorists. Current practice for road signs in Britain is described in a series of publications[1] prepared by the Department of Transport. The British system is compatible with an international convention on signs that was agreed at Geneva in 1971 (and a protocol on road markings in 1973) under the auspices of the European Community.

Three major types of traffic sign can be identified according to their function: regulatory, warning and route information signs.

Regulatory signs

These signs carry instructions which must by law be obeyed. They can be divided into mandatory signs, which require the driver to take a positive action, e.g. 'Stop' or 'Turn left', and prohibitory signs which prohibit certain manoeuvres, e.g. 'No right turn' or 'No entry'.

Most British *mandatory signs* are circular with white or light-coloured symbols on a blue background. Important exceptions are the octagonal 'Stop' sign (used at non-signalised intersections with inadequate sight lines) and the inverted-triangular 'Give Way' sign (used on minor road intersections with major roads which are not controlled by traffic signals or 'Stop' signs) which have distinct shapes and colours, as well as capital letters, to make a more forcible impact on drivers.

With the exception of the waiting restriction and 'No entry' signs, all *prohibitory signs* are circular with a red or white centre and black or dark blue symbols.

Warning signs

These signs alert the motorist to a danger ahead, and usually result in a speed reduction or some other manoeuvre. They are used sparingly; over-frequent usage to warn of conditions that are obvious tends to bring them into disrepute and detract from their effectiveness.

Most warning signs are distinguished by an equilateral triangle with a red border encompassing a black symbol (usually a pictogram of the potential hazard) superimposed on a white background. Typically, they are used at the approaches to such features as intersections not indicated by advance direction signs, dangerous bends or hills, concealed or unguarded rail crossings, school and pedestrian crossings, converging traffic lanes, and a change from a dual to a single carriageway road.

Information signs

The most important of these signs are intended to assist drivers in getting to their destinations. These can be divided into three main groups: *advance direction signs* which are located in advance of intersections on high-speed roads and provide drivers with advice regarding the route choices that lie ahead; *direction signs* which are sited at intersections, i.e. where the turning manoeuvres take place, and repeat the information on advance direction signs; and *route confirmatory signs* which are used after intersections to confirm to drivers that they are on the correct route. Most signs are rectangular in shape, but simple direction signs (i.e. flag signs) usually have one end pointed. Information signs do not lose effectiveness by over-use.

Destinations used in direction signing are applied in accordance with hierarchical principles, with the directions becoming more specific further down the hierarchy. Thus, the first level guides drivers toward general destinations, e.g. 'The North'; these are termed 'super primary' destinations. Next, drivers are advised about locales that are major destinations, e.g. 'Leeds'; these are 'primary' destinations. The third level guides drivers to 'non-primary' destinations; these are towns on non-primary routes. Finally, drivers are guided toward 'local' destinations, e.g. suburbs, industrial estates, tourist attractions, available services.

28.1.1 Design and siting of information signs

The following discussion relates to some principles underlying the design and location of informatory signs (especially direction signs).

Lettering

A driver starts scanning the words on a sign as soon as they become legible, and this is determined by the size of the lettering. As this occurs the driver's line of sight is diverging from the straight-ahead position, reaching a maximum with the finding of the desired word; it is generally accepted that the reading should be completed before this divergence exceeds 10 degrees. There is no significant difference between the distances at which signs of equal area but with different types of lettering (i.e. good lower case, and upper case with or without serifs) can be read, and most drivers can read a 25.4 mm high lower-case letter such as 'x' at a distance of 15.25 m. Experimental work[2] has shown that the x-height, in millimetres, can be determined from the formula

$$x = \frac{1.47VT + 3.28S \text{ Cot } 10°}{1.97} \quad\quad (28.1)$$

where S = offset distance to the centre of the sign (m), V = 85th percentile approach speed (mile/h), and T = time required to read the sign(s). The 99.9th percentile reading times of the words (N = number of words) on stack-type and map-type advance direction signs are given by

Stack signs $\quad\quad\quad\quad\quad\quad\quad T = 2.135 + 0.333N \quad\quad\quad\quad\quad (28.2)$

Map signs $\quad\quad\quad\quad\quad\quad\quad \text{Ln}T = 1.043 + 0.054N \quad\quad\quad\quad (28.3)$

For advance direction signs N should preferably not be greater than 6 words. For flag signs at intersections the 99th percentile reading time is

$$T = 2.403 + 0.238N \quad\quad (28.4)$$

Layout

For a given sign area stack signs are legible at greater distances than diagrammatic signs because the letters in the stack layout are larger (see Fig. 28.1). Stack signs are best used at straightforward intersections, while diagrammatic signs are likely to produce fewer driver mistakes at complex junctions.[2]

Symbols

Symbolic signs are normally superior to alphabetic signs because they improve driver comprehension.[3] Also, they do not rely on any specific language. Bold symbols of simple design provide the best legibility distance for all age groups, but especially for older drivers.

Colour

To be easily read letters should be of light colour on a dark background, or dark on white background. It is cheaper to reflectorise letters rather than the whole sign; hence light letters on a dark background tend to be preferred where large signs are required; for instance, British practice is for directional signs on motorways to have white letters on a blue background. If the sign is small, say less than about 1.85 m², its readability can be increased by using dark letters on a light background, as is used on non-primary main roads.

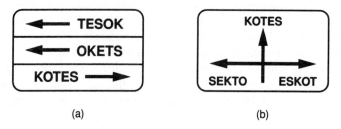

Fig. 28.1 Examples of sign layout: (a) stack and (b) diagrammatic

Reflective materials in common use on signs include reflector buttons, corner-cube prism sheeting, or spherical reflective sheeting. Where particularly conspicuous signs are desired, they may be illuminated by arranging lamps to shine directly on the message or, at greater expense, lighting signs indirectly from behind or using light-emitting diodes. This lighting should be as uniform as possible over the sign-face to avoid bright spots and shadow spots that cause part of the message to be hidden. An average sign luminance ranging from 25 to 150 cd/m^2 is generally adequate, except at locations with high background brightness.

Siting

With some obvious exceptions, such as signs detailing parking restrictions (usually parallel to the kerb), overhead signs, or flag direction signs which must point in the direction to be taken, most information signs are sited transverse to and at the side of the road. Typically they are set at an angle of 93–95 degrees from the approach-side line of travel; this minimises glare and reduces minor reflection without affecting distance legibility. A sign should be close to the edge of the carriageway to maximise reflectivity, but not so close that it is a hazard to traffic or likely to become splattered in wet weather; thus it should be at least 1.2 m from the carriageway edge of a high-speed road without shoulders and behind the shoulder of a road with hard shoulders. Ideally, also, the lower edge of a sign should be close to the eye-height of the driver (1.05 m); however, if splashing is likely it should be higher.

Advance direction signs should be sited sufficiently in advance of an intersection to enable the motorist to reduce speed sufficiently to make the appropriate manoeuvre safely following completion of the reading task.

Overhead signs are used in highly competitive driving environments, where it is not desirable for drivers to move their eyes horizontally away from the line of travel, e.g. over heavily trafficked multi-lane carriageways, or at complicated intersections (especially if they are close together), or where sight distances are restricted. To avoid being lost against the sky, overhead signs usually have white letters imposed on a dark background. They are more costly than roadside signs because of the large structures needed to support them.

Support posts

Most supports, including bracing and fixing clips (and the backs of signs) should be a nondescript colour so as to blend into the background. The supports should be strong enough to hold the sign permanently and resist vandalism and wind displacement. If supports in non-urban locations are so close to the carriageway edge that they could be involved in an accident they should be designed to yield or give way when hit by a motor vehicle.

28.2 Variable message signs

Variable message signs (VMS) differ from the conventional traffic sign in that they can be configured to show a range of different messages (including blank – no message) which vary according to current need. In their most basic form they may simply consist of a hinged panel that is manually switched to display, or cover up, the message. In their most advanced form they comprise a matrix of multi-coloured light points that can be controlled remotely or automatically to display a variety of messages (including text and symbols). The technology used in current VMS installations varies from mechanical systems based on roller blinds, flaps and rotating prisms, to electronic matrices based on batteries of light-emitting diodes or fibre-optics.

Variable message signs have three main uses: to issue instructions, to warn of dangers ahead, or to give advice regarding parking or routing. When not used for these purposes they may be used to provide general information (e.g. the time or temperature) or to give general advice (e.g. 'Drive Safely').

Instructions

Typically, these messages relate to lane usage or speed (e.g. 'use hard shoulder' or 'max speed 50') and, depending on the legislation, they may or may not have the force of mandatory instructions.

A recent innovation has been the use of VMS to communicate publicly with individual drivers. For example, a 'slow down' message may be triggered by automatic detection of a speeding vehicle. Similarly, VMS can be used to communicate with a driver who is sensed to be travelling too close to the vehicle in front.

Warnings

VMS may also be used to relay advice regarding local weather conditions (e.g. 'ice' or 'fog') or to incidents ahead (e.g. 'accident 5 km ahead' or 'lane closure ahead'). Weather warnings are often triggered automatically by an environmental sensor at the site; alternatively, these and other messages may be activated from a remote coordination centre or from an authorised vehicle travelling past the VMS installation.

Unfortunately, VMS warnings are often ignored by drivers. Over-use of these signs, and (the common) failure to turn them off promptly after the hazard has ceased, has led to their losing some degree of credibility.

Parking or route advice

This advice may be in the form of instructions or information in that it may indicate which direction, route or car park should be taken to a specific destination (e.g. 'for York follow A61', or 'turn left for Leeds', or 'for city centre use station car park'). If the VMS board is sufficiently large, a combined message may be given (e.g. 'roadworks on M1, use A1').

The potential for VMS to be used to manage the demand for car parking and road space is being increasingly realised. Research has indicated that compliance with direction advice depends crucially on the phrasing of the message, and that different categories of driver respond to VMS in different ways (e.g. parking guidance systems are often taken more seriously by visitors than by locals). A major theme of current experimentation with VMS is the production of semi- or fully-automatic systems which, having detected traffic congestion or an incident downstream of the sign, will select and display a message which will result in the most likely degree of rerouting.

28.3 Road markings

Technically, road markings include all lines, symbols, words or other devices that are applied to, set into, or attached to a carriageway of a sealed road. Their main functions are to guide vehicles into definite positions on the carriageway, and to supplement the regulations and warnings of traffic signs and signals.

28.3.1 Marking materials

Road markings come in a variety of materials including thermoplastic, paint, pre-formed tapes or sheets that are applied with a special adhesive, and road studs (also known as raised pavement markings).

Thermoplastic is a mixture of sand and pigment that is heated and then laid in a variety of thicknesses by hand, machine-drawn screed box, extrusion, or spray equipment. It is more easily attached to, and more durable on, bituminous surfacings than concrete ones. The road lines laid in Britain are mostly thermoplastic whereas on the European Continent and in the USA they are mainly paint.

Thermoplastic has the following advantages:

- it has a longer life compared with paint
- it fills the interstices of rough-textured roads, whereas paint soon wears from surface-dressing peaks and the interstices fill with dirt
- hot-laid thermoplastic will fuse with a bituminous surfacing, sometimes even when the road is cold or slightly damp
- it is proud of the carriageway, and this assists visibility on a wet night by facilitating drainage of the water film
- it contains sand and a binder, which ensures good skid-resistance as it erodes
- it sets instantly so that traffic can travel on it immediately after it is laid
- at hazardous locations that are away from buildings the thermoplastic can be profiled to provide a tactile and audible rumble warning to drivers when, say, an edge line is traversed by vehicle tyres.

The disadvantages of thermoplastic are:

- it has a greater initial cost than paint
- large-scale application is more difficult due to the large bulk of material that has to be melted down
- an excessive build-up of thickness from successive applications can be dangerous to motorcycles
- it may creep in very hot weather
- on dirty roads carrying light traffic it discolours more easily than paint.

Paints available for road marking can be classified by the type of base, i.e. alkyd, epoxy, rubber, vinyl, water-base and high polymer. The machines used to apply paint include air sprays (a mixture of air and paint is sprayed through a spray gun mounted between masking discs) and airless sprays (pressurised paint is sprayed through a tip, without masking discs, to produce a clean line). Paint can be laid more quickly than thermoplastic, but dries more slowly and must be protected from traffic until dry.

Thermoplastic or paint markings should incorporate tiny glass spheres or *ballotini* to improve their visibility at night under vehicle headlights. The ballotini beads are pre-mixed and/or 'dusted' onto the marking as it is being laid. Premixing, followed by dusting, is the best practice with thermoplastic; it ensures immediate reflectivity and, as the binder is worn by traffic, further beads are exposed to reflect the light from the headlights back to the driver. Dusting is more common with paint; however, reflectivity only exists as long as the beads remain in place.

The reflectivity of markings containing ballotini is reduced with the onset of rain and fog, i.e. as water films start to cover the beads and reflect the light from vehicle headlights away from the driver. When the beads are submerged, as more easily happens with paint, the lines become nearly invisible; however, the larger the beads the greater the amount of retroreflectivity that is retained in wet weather.

Tape markings with embedded glass beads are also available which are particularly useful at construction zones. Traffic can run on the tape immediately it is laid and, when the work is completed, the tape is easily removed.

Reflecting road studs, or *raised pavement markings,* such as cat's-eye and corner cube studs, can be used on their own, to replace other markings, or to supplement thermoplastic or paint markings. When used in conjunction with white lines to 'fix up' accident-prone, previously non-marked road locations they will generally result in reductions in both accident numbers and injury severity.[4]

The advantages of road studs over conventional line markings are:

- greater effectiveness in wet weather
- greater durability
- the vibration and audible noise created by vehicle tyres crossing road studs act as a secondary warning to the driver
- the use of differently coloured reflex lenses permits the imposition of directional control upon the motorist, e.g. to convey a 'wrong way' message.

Disadvantages include:

- their high initial cost
- their susceptibility to damage during snow-clearing
- they must be attached to a high-quality surfacing that does not require an early overlay or surface dressing.

The *cat's-eye reflecting road stud* (reflex lens type) was invented in Britain in the mid-1930s. It consists of a metal base that must be embedded in the road surface, and a separate rubber pad insert into each side of which (for two-way roads) or in one side only (for one-way roads) two longitudinal biconvex reflectors are fitted. As vehicle tyres pass over the rubber pad, its centre part is depressed so that the faces of the reflectors are automatically wiped clean by the front part of the pad. The life of the insert rubber pad varies according to the speed and density of traffic as well as the lateral placement of the stud on the carriageway.

Corner-cube road studs are easily bonded directly to the road surface with an epoxy-cement. They are composed of a plastic shell containing a reflector face (or two faces, if used for dual guidance purposes on a two-way road) made up of numerous reflective 'corner-cubes'; each of these reflectors consists of three sides of a cube, and a headlight ray is reflected through all three sides before returning to the eye of the motorist. A new corner-cube road stud (reflective area = 2000 mm^2) can return about 20 times more light than a cat's-eye stud (reflective area = 130 mm^2) at a distance of 100 m. Although tyre abrasion causes the face of the corner-cube reflector to become etched with fine scratches (which typically reduce the reflecting power to about 10 per cent of its original value in a year), it can still be brighter than a cat's-eye (typically reduced by 50 per cent after a year) at long distances. The life expectancy of a corner-cube detector on a motorway can be as low as six months and as high as eight years, depending on the severity of the traffic; a life in excess of 10 years is not uncommon on rural roads with low traffic.

In some countries *non-reflective ceramic road studs* are used for daytime visibility and/or to supplement corner-cube studs at night when wet (when the studs are intended to simulate less prominent painted lines). They are also used to mark spaces in car parks.

28.3.2 Longitudinal delineation

The various forms of longitudinal delineation have three main functions: they charac-terise the road, provide path guidance, and/or act as tracking references.

Characterisation implies that the delineation provides motorists with information about the nature of the road which leads them to expectations about the ease of the driving task; for instance, a yellow centre line is used in American road marking prac-tice to indicate that a road is two-way whereas a white centre line (or lane line) indicates a one-way two-lane road. By *path guidance* is meant the situation where a particular form (or colour) of delineation is continued through, say, one leg of a Y-junction to make it clear to the driver which route to follow; examples of this are the treatments at motor-way exits and entrances, and lane drops which involve merging or diverging. The *tracking reference* function refers to the situation whereby motorists use the centre and edge of the carriageway for directional guidance, and align their vehicles laterally relative to them; strong markings at these locations, especially in the face of lighting glare from oncoming vehicles and/or at curves at night, can greatly ease the driving task on high-speed roads that are unlit.

In Britain white materials are used for longitudinal markings intended for moving traffic, i.e. centre lines, lane lines and edge lines.

Centre lines have the obvious function of dividing carriageways that carry traffic in opposing directions. They are most usually denoted by broken single lines; these may be crossed at the discretion of the driver. The underlying design principle is for the dashed lines to be 'weak', i.e. short with long spaces in between, on safe sections of road, and for them to be strengthened, i.e. made longer within the same overall module, as locations are approached which require increasing caution; at dangerous locations, the dashed (strengthened) warning line becomes a solid continuous line which should not be crossed. Two contiguous parallel line markings are now used in most countries as a centre line at horizontal and vertical curves on two-way roads where overtaking is dangerous because of restricted sight lines or some other condition; if the nearside line is broken, drivers may legally cross the lines at their discretion; if the nearside line is continuous crossing is prohibited.

Lane lines are usually indicated by broken lines, with the marks being much shorter than the gaps; at locations requiring extra caution stronger markings may be used to alert the driver. Lane lines organise traffic into channels and can be used to increase the efficiency of carriageway usage at congested locations. They are normally used on rural two-way roads with three or more lanes, on one-way carriageways with two or more lanes, at approaches to major intersections and pedestrian crossings, and on congested streets where lines may be provided to maximise the number of lanes of traffic.

Edge lines are broken or continuous line markings which keep the drivers away from the edge of the carriageway and act as reference lines for drivers faced with oncoming bright headlights at night. Edge lines were initially developed to improve safety for drivers on unlit rural roads; now, however, they are also used on essentially all high-speed roads, in non-built-up areas, and at intersections. Continuous edge lines (which should not be crossed) are generally used:

- where the demarcation between the carriageway and verge is poor
- where headlight dazzle from oncoming vehicles is severe
- on roads subject to fog
- at changes to carriageway width
- on approaches to narrow bridges and bends that are indicated by warning signs.

Since a traffic lane in Britain must be at least 2.75 m wide, a road must be at least 5.8 m wide to carry both a centre line and edge lines.

28.3.3 Words and symbols

Road markings in the form of text and symbols are also used to communicate with the driver. *Directional information*, often abbreviated, is usually provided as text; for instance, at the entrance to a complex intersection drivers might be encouraged to 'get in lane' and major destinations might then be written in the appropriate lanes. Similarly, an instruction to 'slow down' might be written on the carriageway prior to a hazard. *Symbols* can be more effective than text instructions provided that their meaning is understood. Thus, arrows are widely used in most countries to provide instructions to merge or change lanes.

A major disadvantage associated with the use of text and symbol markings on the carriageway is that they cannot be easily read in heavy traffic conditions.

Various lines and other markings have particular statutory meanings which road users are required to know. Thus, kerb lines and markings have been adopted in many countries to indicate a variety of parking, waiting and loading restrictions, transverse lines on the carriageway are used to indicate stop lines, and continuous longitudinal lines on the carriageway cannot be crossed.

28.4 Guide posts

Guide posts with (usually) reflectorised delineators, placed on one or both sides of the road, can assist the night driver and reduce accidents by outlining the horizontal and vertical road alignment. Red and white reflectors are generally used on posts on opposite sides of the road: if the same colour is used on both sides the two lines can be indistin-guishable from a distance; if white reflectors are used on the near side there is a danger that they may be confused with opposing headlights.

For safety reasons posts are best made from (flexible) neoprene, polyvinyl or rubber, or breakaway wood. Typically, the reflectors are mounted on posts about 1 m above ground, and set back at least 1 m from the carriageway edge (for roads without shoul-ders) or behind the shoulders; if the reflectors are too close to the carriageway their effectiveness is reduced quickly by dust and dirt and they must be hand-cleaned very regularly. On straight sections of road the posts may be 150 m apart; the underlying rule is that at least two to three posts should be continually visible to the driver. At bends and prior to intersections, or in fog-prone areas, closer spacings are used.

Although not widely used in Britain guide posts are extensively used on the Continent and overseas to delineate unsurfaced roads and to supplement line markings on surfaced roads.

28.5 In-vehicle information systems

A number of different systems can be categorised under this heading. The main distinc-tion is between systems with *autonomous units*, which carry all their intelligence around with them, and those with *communicating units,* which receive information about the current state of the road system by radio or other means. This is a rapidly growing field and a number of new systems are currently under development in Japan, Europe and North America. (A useful review of the state of the art in 1993 is given in reference 5, but readers should consult the technical press for updates.)

28.5.1 Autonomous units

Systems with autonomous units use an onboard computer, equipped with information about the road network and a means of determining the vehicle's current location, to assist with the navigation task and help the driver to select good routes to desired desti-nations. Such devices cannot, however, know about current traffic conditions and so cannot provide optimal route advice on any given day. Autonomous units are likely to be of particular value to travellers who are not familiar with the road network and whose main aim is to reach their destination without getting lost (rather than trying to optimise their route). They can be expected to reduce the excess mileage and travel time associ-ated with drivers who make navigation errors and to increase safety by reducing the level of stress, frustration and confusion among such people.

The onboard computers keep track of their current position by reference to global positioning satellites under GPNS, e.g. GPS or GLONASS, or to terrestrial beacons such as radio transmitters. To supplement or replace this information they make use of inter-nally calculated trajectories based on 'dead reckoning' using known distances (from an odometer) combined with magnetic or gyroscopic compass bearing information. None of these positioning techniques is completely accurate on its own: for example, significant parts of a city are likely to be in shadow for satellites and dead reckoning methods suffer from accumulation of errors. Thus, some navigation systems use a combination of tech-nologies and/or map-matching software whereby information about road alignments in the vicinity is used to determine which road the vehicle is most likely to be on.

Once the vehicle's current location is established, the in-vehicle unit (IVU) can then either display this to the driver by superimposing it on a representation of the street map stored on the IVU's memory, or it may calculate the 'best' route to the specified destination (previously keyed in by the driver) and issue instructions accordingly, e.g. as a series of turn instructions given by text, voice synthesis or symbolic arrows. Road safety specialists generally prefer the turn instruction, particularly if it is given by voice, over a map showing current location because of the likelihood that the map display will distract the driver from the driving task.

The current generation of autonomous navigation units include CARIN from Philips, TRAVELPILOT from Bosch and the Etak system. Their favoured medium for storing the map data is CD-ROM because of its capacity, price and robustness.

28.5.2 Communicating systems

There are a number of different types of communicating units in use or being developed.

Information broadcast by radio

Conventional radio broadcasts of traffic bulletins have long been an important source of information to the road user and are likely to remain a key element in any communicating system. The quality of information broadcast in this way has improved with the establishment of traffic information centres to collate and relay the information. The AA Roadwatch Scheme, established in the UK in 1973, is a good example of what can be achieved when an organisation establishes links with highway authorities, public utilities, and the police.[6] The use of eye-in-the-sky helicopters to supplement ground-based methods of detecting congestion means that, in theory, the traffic bulletins can be up to date and detailed.

The problem with conventional radio broadcasts as a source of information for drivers is that they are relatively untargeted and intermittent. This means that the information is often of little relevance to a given listener and even when it is, it may be out of date. This reduces its value as a reliable source of traffic information and causes drivers to seek an alternative.

Information provided by teletext services prior to a journey can also be useful. However, teletext is not practical as a source of continuously updated information during the journey itself.

Coded broadcasts (RDS-TMC and radio paging)

The *Radio Data System* (RDS) transmits digital information alongside 'normal' radio programmes.[7] In Europe one of the RDS channels is reserved for traffic information (the Traffic Message Channel, TMC – hence RDS-TMC). Radios with an appropriate decoder recover these messages and have them delivered in the drivers' preferred language by text display or speech synthesis. The equipment can be set up to interrupt an existing programme whenever a message is received. By using a network of local transmitters rather than national ones, the information can also be filtered to include only that information which is likely to be relevant to drivers in the local transmitter's catchment.

The widespread adoption of a single transmission protocol such as ALERT C allows quite detailed messages to be understood by equipment produced by a range of different manufacturers. There is the prospect of this same protocol being used to allow transmission of information from roadside beacons, thus allowing very detailed targeting of information (somewhat akin to that achieved by variable message signs).

Radio paging is a quite different concept to RDS-TMC but it achieves a similar end – the transmission of digital information to specially adapted radio receivers. The TrafficMaster system in the UK uses this technology to provide its subscribers with a continuous stream of information indicating the location of slow-moving traffic anywhere on the national motorway network; this is supplemented with explanatory information (e.g. 'roadworks on M25 at junction 6') taken from the Roadwatch System. This information is superimposed on maps displayed on the TrafficMaster unit's screen. The source of TrafficMaster's information about the location of slow-moving traffic is a series of infra-red speed detectors at regular intervals along the motorway network.

Hybrid systems – navigators with communications links

It is, of course, technically possible to link the use of an onboard map database, such as a CD-ROM, with real-time information provided by radio, telephone or local beacons. An example of such a system is the Japanese AMTICS demonstrator which, in its 1992 trial in Osaka, used a dedicated radio broadcast frequency to receive real-time traffic information on an in-vehicle terminal similar to teletext. One page of this information shows a map display of the road system with the driver's position marked and all the roads colour coded to indicate their current reported speed. Another example of this type of hybrid system is Volvo's DYNAGUIDE system[8] which uses RDS-TMC information to generate a symbolic representation of current congestion conditions on a map display.

A third example, this one using the GSM digital mobile telephone system as its communication link, is the SOCRATES system[9] under development in a series of projects supported by the European Union. The system builds on the CARIN navigator and allows route recommendations to be based on information about current road conditions.

An alternative approach, which uses the massive information-carrying capacity of roadside infra-red or microwave beacons, is to dispense entirely with the onboard map and use the roadside beacons to transmit the route information. This is the approach adopted by Siemens' EUROSCOUT system;[10] it uses short-range infra-red beacons attached to traffic signal posts or other roadside objects. These beacons transmit to passing vehicles information about the best routes to all possible destination zones. Knowing its driver's intended destination, the in-vehicle unit then uses its autonomous dead reckoning and map-matching functions to keep track of its position and to trigger turn instructions to the driver at appropriate points in the network. These instructions can be given by a pictogram display and/or synthesised voice.

An interesting feature of the EUROSCOUT system is that the central storage of the map database allows rapid updating of the network and theoretically allows real-time manipulation by a central authority of the route recommendations being transmitted to drivers.

28.6 Issues in the provision of in-vehicle information and guidance

28.6.1 What are the benefits to the community?

The potential benefits to drivers associated with the provision of in-vehicle information and guidance may be sufficient to persuade governments and road authorities to permit such systems to exist, with the costs incurred being paid for by individual drivers. A full evaluation of the incidence of costs and benefits could, however, also persuade governments that these systems have hidden costs which mean they should not be allowed – or, conversely, that they could yield community benefits which warrant public subsidy being provided to encourage their use.

Potential disbenefits are that those systems may (a) be dangerous because they distract drivers, (b) encourage drivers to use links which are environmentally sensitive, and (c) actually increase congestion by concentrating traffic on 'advised' routes or by encouraging people to use private transport more than they might otherwise have done.

On the question of *safety*, it is argued that, even if the in-vehicle units offer some distraction, the net stress effect is less than it would be if drivers were trying to read maps or trying to read all road signs while driving. Also, a well designed interface relying on speech synthesis and control, rather than text/map and keypad, reduces distraction to a minimum. In relation to the use of *environmentally sensitive links*, it is likely that these could (by agreement with the road authority) be left out of the guidance network – but this raises the issue of the quality and credibility of the advice given by individual systems; for example, would a system which omitted all residential streets be competitive with one which included them? A technical solution is available to overcome the problem of *concentrating traffic on 'advised' routes*; a technique known as multi-routing could be applied to spread the traffic over a number of different routes if the population of guidance vehicles ever became significant. Also, as in the case of VMS-based advice, the advice message could be so phrased that not all drivers would choose to use the same alternative route. On the wider policy question of possible *increases in the amount of private travel* in response to improvements in journey time and comfort offered by information and advice systems, there remains considerable uncertainty.

Additional potential benefits to the community of in-vehicle guidance and information to drivers include more efficient use of the available network capacity, and reductions in congestion and its associated environmental effects. Modelling work has confirmed that these benefits should be forthcoming but as yet no field trials have been of sufficient scale to verify them empirically. It is also likely that the benefits would be particularly impressive if the guidance and information systems were to be coordinated with, say, road user charging and conventional traffic control signals.

28.6.2 On what criteria should route advice be based?

Most drivers choose routes to minimise travel time but a significant proportion of routes are also chosen with other criteria in mind: avoidance of queues or of particular types of road or hazardous manoeuvre, scenic quality, and ease of route finding are three well-known examples. It is also hypothesised that one reason why drivers prefer to receive traffic information rather than routing instructions is that they prefer to select their own routes rather than submit themselves to a discipline set by someone else. It is suggested therefore that the navigation systems which perform the route selection in-vehicle are likely to be more successful than those (like EUROSCOUT) which have the route chosen externally because the in-vehicle choice could, in theory, be tailored to the user's own requirements.

Even if it could be agreed that routes should be based on minimisation of journey time, the question remains as to whether this minimisation should be from the point of view of the driver (user-optimum routes) or the network as a whole (community-optimum routes). Modelling work has shown that the routing recommendations in these two cases could differ quite considerably – particularly if the drivers receiving advice form a significant proportion of the total traffic. One problem with the community-optimum approach, as with approaches based on networks that are designed to exclude environmentally sensitive links, is that the users are likely to conclude that they are not receiving the best possible advice and to decide not to use the system. If individual subscriptions are to be the main source of finance for in-vehicle systems, it is clearly vital that the users should believe they are receiving a valuable service and this consideration may therefore dissuade operators from any experimentation with community-optimum routing.

28.6.3 Sources of information

A key feature of many of the systems described above is a traffic information centre receiving data from a variety of sources. The major sources of data are likely to include:

- highway authorities for network changes and scheduled roadworks
- public utilities for utilities roadworks
- police for information about accidents and other incidents
- automatic network-state detectors, e.g. automatic traffic counters, TrafficMaster's infra-red detectors, and other specialist detectors
- probe vehicles which transmit speed profile data back to the control centre. (Note that this data source, which is used in the EUROSCOUT system, only becomes viable when the number of equipped vehicles becomes significant)
- historic data on high speeds that indicate the speeds likely to be experienced on a particular link at a particular time of day.

Taken together these sources offer the prospect of a high-quality real-time database on road network conditions which can service the needs of VMS and teletext, as well as radio broadcast, RDS-TMC and other in-vehicle information systems.

Another, quite different, source of data relates to the provision of 'yellow pages' information, i.e. information about services available to travellers and about potential attractions which might influence their route or destination. (A system which provides a driver with the quickest route to a theme park via a petrol station and a fast food outlet has already become a reality in US-based field trials.) The information to support such services would, most likely, be provided free by advertisers. Given the commercial forces at play it is possible that it is this kind of system that will most rapidly find its way into every car rather than a system sponsored by a road authority that is designed to help optimise network conditions.

28.7 References

1. Department of Transport, *Traffic Signs Manual*. London: HMSO (as amended from time to time).
2. Agg, H. J., *Direction sign overload*. Project Report 77. Crowthorne, Berks: Transport Research Laboratory, 1994.
3. Pline, J.L. (ed.), *Traffic Engineering Handbook*, 4th edition. Englewood Cliffs, NJ: Prentice-Hall, 1992.
4. Travers Morgan (NZ) Ltd, Accident counter-measures: Literature review, *Transit New Zealand Research Report No. 10*. Wellington, NZ, 1992.
5. Catling, I., *Advanced Technology for Road Transport*. Boston and London: Artech House, 1993.
6. Hofman, S., AA Roadwatch: A road traffic database for UK and Europe. *Proceedings of the 3rd International Conference on Vehicle Navigation and Information Systems*, 1992, 34–39.
7. Groot, M. Th. de, Rhine-Corridor: An RDS-TMC pilot for radio information. *Proceedings of the 3rd International Conference on Vehicle Navigation and Information Systems*, 1992, 8–13.
8. Hellaker, J., and Rosenquist, M., DYNAGUIDE: Presenting local traffic information using RDS/TMC. *Proceedings of the 3rd International Conference on Vehicle Navigation and Information Systems*, 1992, 20–25.
9. Catling, I., SOCRATES – What now? *Proceedings of the 3rd International Conference on Vehicle Navigation and Information Systems*, 1992, 448–54.
10. Sodeikat, H., Dynamic route guidance and traffic management operations in Germany. *Proceedings of the 3rd International Conference on Vehicle Navigation and Information Systems*, 1992, 1–7.

Index

Accident blackspots, 265–7
Accident investigations
  area action approach, 264
  mass action approach, 262
  planning process for, 265–9
  route action approach, 264
  single site approach, 262
Accident prevention, what it involves,
    261–2, *see also* Safety audits
Accident reduction
  area action, programmes for, 264
  importance of enforcement to, 314–16
  importance of engineering to, 316–18
  mass action approach to, 262
  planning process for, 264–9
  proven remedies leading to, 263
  route action approach to, 264
  single site approach to, 262
  what it involves, 262
Accident risk at at-grade pedestrian
    crossings, 478
Accidents, road
  and drinking, 311–12, 314–15
  and drugs, 312
  and elderly people, 304–5, 310–11
  and parking, 458
  and pedestrians, 305–6
  and public service vehicles, 307
  and seat belts, 313, 315
  and size of vehicle, 317
  and speeding, 313, 316
  and two-wheeled vehicles, 306–7, 315–16
  and young people, 304–5, 307, 310
  at intersections, 303
  British Government policy re, 304
  by age group, 304
  car, 304–5
  economic costs of, 88–9, 307–8
  factors contributing to, 308–9
  importance of enforcement upon, 314–16
  international comparisons of, 299–301
  on built-up roads, 302–3
  on motorways, 303–4
  on non-built up roads, 303
  proven remedies for reducing, 263
  role of education in preventing, 309–16
  statistics, 302, 304
  terminology re, 261, 299
  trends re, 300–7
Administration of roads and public
    transport in Great Britain, *see* Local
    authorities *and* Department of
    Transport
Aggregation of individuals' decisions to
    produce population estimates in
    transport modelling, 107
Air quality
  assessment, 36
  surveys, 249–50
AMTICS communicating system, 528
Annual average daily traffic (AADT)
  estimating peak hours flows from the,
    390–1
  factors used to convert short period
    counts to 24-hour, in Britain, 235
  relationships with annual hourly
    traffic (AAHT) in the USA, 289–90
Anti-glare screens, 346, *see also* Glare
Appraisal criteria considered in assessing
    transport improvement projects, 95–8
Appraisal of a public transport project,
    99–100
Appraisal of a road transport project
  budget constraints in relation to, 95
  criteria used in the, 95–8
  definition of, 80
  economic regeneration considerations in
    the, 94–5
  equity considerations in the, 93–4
  pricing policy considerations relating to
    the, 98–9
  safety considerations in the, 88–9

shadow pricing in relation to the, 84
valuing costs and benefits in the, 84–9
valuing environmental effects in the, 89–92
see also Economic assessments
ARCADY roundabout program, 404–5
Arterial roads, function of, 141
At-grade pedestrian crossings, *see* Pedestrian crossings, at-grade
Attitudinal surveys, 257–9
Auxiliary lanes at at-grade intersections, 361–2

Bar markings at roundabout intersections, 370
Bendiness of a road section, definition of, 321
Bicycles
development of, 6
suggested target for increased travel by, 52
trends in the number and usage of, 14
see also Cyclists *and* Cycling
Block rate grant, 28
Braking distance, definition of, 325
Bright patch, the, in road lighting, 441–2
British design-standard approach to road capacity
rural road design flows, 290–3
urban road design flows, 293–6
British Rail
changing role of, 22–3
finance for, 27
see also Transport administration in Great Britain
Build-outs used in traffic calming, 471
Burrell network assignment modelling, 123
Bus-based transport systems
conventional, 188–9
guided, 189–91
see also Demand-responsive bus services *and* Unconventional transport facilities
Bus bays, design and location of, 344, *see also* Bus stops
Bus control, methods of, 144

Bus lanes, 456–8
Bus priority, *see* Priority for buses
Bus stops, 189, 459, *see also* Laybys and busbays
Buses
applications in relation to transport policy objectives, 59–60, 65–6
decline in travel by, 14–15
express, 144
first, 5–6
forecasts of numbers of, and travel by, 297
strategies to encourage greater usage of, 144–7
see also Park-and-ride

Cable fences, 348
Camber, 344–5
Capacity, road
British design-standard approach, 290–8
economic, 281
environmental, 281
Highway Capacity Manual approach, 282–90
theoretical estimation of, 279
traffic, 281–2
Car park standards used for planning purposes, 409–12
Car parks, multi-storey
adjacent-parking ramped garage, 427–9
attendant-parking mechanical garage, 429–30
attendant-parking ramped, 429
clearway ramped garage, 426–7
design car concept used in the design of, 413–5
design considerations affecting patron-parking, 430–2
fee collection control and audit at, 432–4
lighting levels in, 431–2
locating, 412–13
long span and short span, 425
ramped garage, 426–9
types of, 425
underground, 425–6

Car parks, surface, off street
  design car concept used in the design of, 413–15
  entrances to, 415–17
  exits from, 421–2
  fee collection control methods and audit at, 416–17, *see also* Car parks, multi-storey
  locating, 412–13
  parking stalls and aisles in, 419–21
  pedestrian considerations at, 422–3
  traffic circulation within, 418
Car parking, *see* Parking
Car pooling, 64, 147–8
Car sharing, *see* Car pooling
Carriageway widths
  current practice in relation to, 340–2
  to ensure safe weaving, 390–2
Casualties, road, *see* Accidents, road
Category analysis models, 111
Central reservation, 342–3
Channel Tunnel, 16, 23
Channelization at at-grade intersections, types of, 361
Clawson approach, use of the, in relation to the environmental appraisal of a transport project, 90
Clearways, 459
Climbing lanes for roads, 334–5
Closed circuit television (CCTV)
  use of, in measuring journey speeds, 242
  use of, in measuring spot speeds, 240–1
Coaches, early, 4
COBA 9 program, 35, 89
Collector roads, function of, 141
Collision diagram used in accident analysis, 267–8
Commercial road vehicles
  accident trends, 301
  factors favouring the use of, for freight transport, 8–9, *see also* Freight Transport
  forecasts of numbers of and traffic by, 7, 297
  impact on rail freight, 16
  loading and unloading zones for, 459–60

  parking for, off-street, 423–5
  traffic by, 8–9
Community Impact Evaluation, 93
Comprehensive transport demand survey, 32–4
Computer-aided design (CAD)
  data input requirements for, at intersections, 402
  data input requirements for road alignment, 403
  outputs from intersection design programs, 404–5
  outputs from road alignment design packages, 404–8
  role of, 400–1
  what is, 401–2
Concentration, flow and speed relationships, 275–8
Concrete barriers, 347–8
Condition diagram used in accident analysis, 267
Condition surveys, road, 234
Congestion
  detrimental effects of, 16, 133, 183–4
  pricing, *see* Road pricing
  types of, 133
Congestion management, *see* Travel Demand Management
Contingent valuation, use of, in the environmental appraisal of a transport project, 90–1
Converted traffic, definition of, 31
Corrected moving observer method (Wardrop method) of measuring journey speed, travel time and delay, 244
Cost curve, in public transport service operations, 202–4
Crash-attenuation devices, 349
Crash barriers, 347
Cross classification models, 111
Crossroads, 366
Cross-section elements of a road
  anti-glare screens, 346
  camber, 344–5
  carriageway width, 340–2
  central reservation, 342–3

crash-attenuation devices, 349
laybys and busbays, 344
noise barriers, 349–53
safety fences, 346–8
shoulders, 343–4
side slopes, 345–6
Cross-sectional transport surveys, 230
Cultural heritage assessment, 36–7
Current traffic, definition of, 30
Curvature and centrifugal force in relation
   to the geometric design of roads,
   328–30
Curve widening, 332–3
Cycle routes
   design criteria for, 428
   design principles for, 176–7
   functions of, 481–2
   importance in relation to transport
    policy objectives, 60, 173
   objectives in relation to cyclists needs,
    176
Cyclic flow profile traffic model, 508–11
Cycling
   casualties, 480, *see also* Accidents,
    road
   usage in Great Britain, 480, *see also*
    Bicycles
Cyclists
   factors affecting use of bicycles by, 481
   who are they?, 480
Cyclist priority, *see* Priority for cyclists
Cyclist surveys, 247

Data collection of vehicle flows
   automatic methods of, 237–8
   manual methods of, 236–7
Delay surveys, 241–4
Delays at traffic signal-controlled
   intersections
   sheared delay formula, 500
   Websters delay formula, 499–500
Demand allocation: modelling the choice
   between alternatives
   all-or-nothing models, 123
   diversion curves, 123
   hierarchical logit model, 125–6
   logit model of individual choice, 124

market segmentation concept in relation
   to, 122
   stochastic all-or-nothing models, 123
Demand curve, in public transport
   operations, 202–4
Demand-responsive bus services, 145
Density of traffic, *see* Concentration, flow
   and speed relationships
Department of Transport
   administration of, 21–3
   finance sources for the, 27
Design car for parking, 413–5
Design flows, *see* British design-standard
   approach to road capacity
Design reference flows
   for roundabout intersections, 375–6
   for T-intersections, 368
Design speeds for urban and rural roads,
   320–4
Development plans, 26–7
Development traffic, definition of, 31
Dial-a-bus services, *see* Demand
   responsive bus services
Direction signing, *see* Traffic signs
Disabled people, *see* Elderly and disabled
   people
Disruption due to construction, assessment
   of, 37
District distributor roads, function of, 141–2
Docklands Light Railway, 23, 27
Drink-driving, *see* Accidents, road *and*
   Legislation
Driver information systems, importance in
   relation to transport policy objectives,
   67–8, *see also* In-vehicle information
   systems
Drivers' hours regulation in the road freight
   industry, 217
Drivers' licences, growth in, 16

Ecology and nature conservation,
   assessment of impact of construction
   upon, 37
Economic assessments, *see* Appraisal
   criteria *and* Appraisal of a road
   transport project *and* COBA 9
   program

Economic capacity of a road, 281
Economic efficiency and markets, 81–2
Economic regeneration considerations in
    the appraisal of transport improvement
    projects, 94–5
Edgelines, 525
Elasticity models, 121–2
Elderly and disabled people design for,
    at LRT stations, 193–4
    information for, 179
    ramp gradients for, 178
    steps for, 178
    street furniture for, 178–9
    walking surface quality for, 179
Emissions from motor vehicles
    effects upon global warming, 17–18
    impacts of, 17–18
    main, 17
    steps taken to reduce, 18
    suggested targets for, 52
Enforcement in relation to road safety, *see*
    Legislation
Enforcement of parking regulations, 463
Entrance ramp control on major roads, 146
Environment assessments, overview of,
    36–9, *see also* Leitch framework for
    evaluating environmental impacts of
    transport improvement projects *and*
    Valuing environmental effects in the
    appraisal of a transport project
Environmental capacity of a road, 281
Environmental impact surveys, 249–50
Equilibrium forces in transport modelling,
    106–7
Equisaturation principle used in off-line
    traffic signal plan generation, 501
Equity considerations in the appraisal of
    transport improvement projects, 93–4
Exponentially-smoothed forecast models,
    114
Express bus services, 144
EUROSCOUT communicating system,
    528
Eye heights used in the design of roads,
    326, *see also* Sight distance

Fatalities, *see* Accidents, road
Fee collection control and audit at off-
    street car parks, 416–7, 432–4
Finance for transport in Great Britain, 27
First year economic rate of return, in
    accident analysis, 269
Floating car method of measuring journey
    speed, travel time and delay, 243–4
Footway design and capacity, 475–6
Forecasts of traffic and vehicles in Great
    Britain (1995–2025), 297
Four-stage (sequential) travel demand
    model, 128–9
Freight transport
    by mode, 9
    by rail and water, potential for, 218–19
    governmental policy issues relating to,
        216–18
    impacts of road improvement upon,
        215–16
    provision for, in relation to transport
        policy objectives, 61
    trends in, 214–15
Freight transport policy issues, 216–18
Fuel taxes and transport policy objectives,
    70
Furness procedure in traffic modelling,
    118–20

Garages for off-street parking, *see* Car
    parks, multi-storey
Gateways for traffic calming, 472
Generalised cost, 86–8, 106, 207–8
Generated traffic, definition of, 31
Geometric design standards for roads, 323
Glare
    from road lighting, 443–4
    screens, anti-, 346
Global warming, 17–18
Grade-separated intersections
    diamond, 383
    five-bridge three-level roundabout, 384
    half cloverleaf, 384
    two-bridge two level roundabout, 384
    two-level dumbbell roundabout, 384
    see also Intersections with grade
        separation

Gradients, design of road, 333–4
Gravity models, 118–19
Growth factor models, 118–20
Guide posts, 525–6

Hall survey, 255
Headways, space and time, 274
Hedonic pricing, usage in environmental
    appraisal of transport projects, 90
High occupancy vehicle (HOV) priority,
    65
Highway Capacity Manual design
    approach
    level-of-service concept, 282–5,
    287–90
    level-of-service measures of
    effectiveness, 285–7
Highways Agency, 21–2
Home interview surveys, *see* Surveys,
    participatory
Horizontal alignment design for roads,
    327–33
Household activity travel simulator (HATS)
    survey method, 254–5
Households, trends in the average size of,
    10

In-vehicle information systems
    and accidents, 317
    autonomous units, 526–7
    communicating systems, 527–8
    issues in the provision of, 529–30
Induced traffic, definition of, 31
Inductive loops
    at traffic signals, 485–6
    usage of, in gathering data, 237
Information signs
    advance direction and direction, 518
    colours used in, 519
    hierarchical principles used in, 518
    layout of and lettering for, 518–19
    lighting of, 520
    route confirmatory, 518
    siting of, and support posts for, 520–1
    use of symbols in, 519
Input-output method of measuring speed
    and delay, 243

Integrated transport studies, 32–3
Interchanges
    all-directional, 388
    collector-distributor roads for, 386–8
    cyclic, 386
    definition of, 381
    trumpet, 386
Intersection design, at-grade
    and approach alignment, 361
    and parking, 361
    and sight distance, 362–3
    design vehicle for use in, 360–1
    overview of the process of, 357
    principles of, 357–60
    types of channelization used in, 361
Intersection spacing, 363–4, 389
Intersections at-grade
    auxiliary lanes at, 361–2
    basic forms of, 356
    lighting for, 442–3
    parking at, 361
    priority, 364–9
    priority for high occupancy vehicles
      (HOV) at, 456–8
    roundabout, 369–76
    sight-distances at, 362–3
    spacings of, 363–4
    traffic signal controlled, 377–81
    see also Intersection design, at-grade
Intersections with grade-separation
    definitions of, 381
    design considerations re weaving
      sections, merges and diverges,
      388–98
    examples of, 383–8
    overview of the design process for,
      388
    usage of, 382
Inventory surveys, 232–4

Journey speed, travel time and delay
    surveys, 241–4
Junctions, *see* Intersections

Kiss and ride parking, 168, *see also* Park-
    and-ride
Kontiv survey, 254

Land use and transport relationship, 181–3

Land use planning
environmental effects relating to, 38
mechanisms of, used to achieve transport objectives, 72–4, 138–40
statutory process, 26–7

Landscape, assessing the impact of construction upon the, 37

Laybys and busbays, 344–5, *see also* Bus stops

Legislation
effect of drink driving, 314–15
effect of motor cycle, 315–16
effect of seat belt, 315
enforcing speed limit, 316
see also Accidents, roads

Leitch framework for evaluating environmental impacts, 91–2, 97, *see also* Environmental assessments, overview of

Level of service
for pedestrian movement, 475
for vehicle movement, 282–90

License plate surveys
as used in participatory surveys, 257
at cordons and screenlines, 244–5
for journey speed measurement, 242
for parking, 246

Lifts in multi-storey car parks, 432

Light rail transport (LRT) systems, design considerations relating to, 192–6

Lighting
for information signs, 520
in multi-storey off-street car parks, 431–2
in surface car parks, 423

Lighting, road
column spacing and siting for, 447
discernment by dark silhouette in, 441
discernment by reverse silhouette in, 442–3
glare from, 443–4
installation terms used in, 437–40
lamps used in, 444
luminaire arrangements used in, 439, 445–6

luminaires used in, 444–5
mounting heights used in, 445
objectives of, 435
overhangs, bracket projections and set backs for, 446–7
photometric terms used in, 436–7

Line markings, longitudinal, 524–5

Local Authority administration in Great Britain, 24–5

Local distributor roads, function of, 141–2

Local roads and streets, function of, 141

Logit models of individual choice
hierarchical, 125–7
standard, 124

London, administration in, 24–5

London Area Transport Survey (LATS), 32–3

London Transport operations, 23

Marginal cost, definition of, 204

Matrix estimation models, 118–20

Maximum car method of measuring journey speed, travel time and delay, 244

Merges and diverges at intersection with grade separation, 392–8

Metropolitan railway (metro) systems, 196

Models, transport
classes of, 110–28
concepts underlying the development of, 106–8
current trends in, 129–30
desirable features of, 104–5
effects of constraints of time and money upon, 109
packages available, 128–9
range of available, 108–9
role of, in the planning process, 103–4
specification, calibration and validation of, 105, 108–9

Models, types of transport
averaging and smoothing-forecasts without a trend, 114
demand allocation: modelling the choice between alternatives, 122–7

elasticity, 121–2
matrix estimation, 116–20
regression analysis, 114–16
simple formulae, 110–11
simulation, 127–8
time series, 111–14
Monorail transport systems, 197–8
MOSS road alignment program, 406–8
Motor car
  accident trends associated with the,
    304–5
  beginning of the age of the, 6–8
  demographic changes associated with
    the, 10–11
  forecast numbers and usage, 7, 297
  households with regular use of a, 8–9
  impact on public transport of the, 14
  impact on the environment of the,
    16–18
  trends in the numbers and usage of the,
    8, 12, 14
  trip patterns by, 12
  trip rates by, 13
MOVA program for on-line microcontrol of
    traffic signals, 504
Moving average forecast models, 114
Moving observer methods of measuring
    speeds, travel times and delays,
    243–4

National Bus Company, privatisation of
    the, 25
Net Present Value of a transport
    improvement project, 95–7
NOAH road alignment program, 403
Noise barriers, 353
Noise from traffic
  and noise barriers, 345–53
  desirable limits of, 17, 52
  flow chart for estimating, 351
  Governmental actions to tackle, 17
  surveys of, 248–9
  thresholds and public attitude re, 54,
    248
Noise Insulation Regulations, 349–50
Non-operational parking, 410
Normal traffic growth, 31

Objectives underlying transport policy
  accessibility, 48
  checklist of practicability issues, 51
  conflicts, constraints and double-
    counting in, 50–1
  economic efficiency, 46–7
  economic regeneration, 48–9
  environmental protection, 47
  equity, 49
  finance, 49
  possible impact matrix, 50
  practicability, 49–50
  quantified, 45–6
  safety, 47
  statements of vision in relation to, 45
  suggested indicators for, 56
  sustainability, 48
  types of, 45–6
  see also Freight transport policy issues
Off-line traffic signal control in networks
  cyclic flow profile traffic model,
    508–11
  general principles for, 512–13
  platoon dispersion, 507–8
  time–distance diagram, 507
  TRANSYST program for, 512
Off-street car parks, *see* Car parks,
    surface, *and* Car parks, multi-storey,
    and Commercial road vehicles
On-line micro-control of traffic signals in
    networks
  discrete time, rolling horizontal control,
    513–15
  general principles, 512
  plan selection, 513
  SAEWA module, 515
  SCATS system of, 515
  SCOOT program, 513
  STAUKO module, 515
  terminology, 502–4
On-street parking, *see* Parking
One-way streets, 452–4
Omnibus, *see* Buses
Operational parking, 410
Origin-destination cordon and screenline
    surveys, 244–5
OSCADY traffic signals program, 405

Park-and-ride
   applications in relation to transport
     policy objectives, 60
   criteria for successful schemes, 167–9
   factors that influence usage of, 145
   interchanges, location of, 168, 413
   see also Kiss and ride parking
Parking
   beat survey, 246–7
   charges and transport policy objectives,
     70
   control, 64, 146, 154–5, 460–3
   discs, 462
   planning for town centre, 155–63
   policy, 154–5
Parking use surveys
   accumulation, 245–6
   concentration, 245–6
   duration, 246
Passenger transport
   Authorities in Great Britain, 24–5
   emergence of, 4–5
Passenger transport systems
   energy consumptions of, 184
   in rural areas, 186
   inter-urban travel by different, 185
   roles in urban areas of, 182–3, 201–2
Path trace survey method, 243
Pay and display parking control, 461–2
Pedestrian
   and cyclist facilities, 170
   areas, 60–1, 177
   crossings, at-grade, 476–8
   crossings, segregated, 478–9
   seating, 176
   streets, 142
   surveys, 247
   walking, speeds and distances, 157,
     175
   see also Elderly and disabled people
Pedestrian priority
   overall objectives of, 473
   some design considerations relating to,
     473–4
Pedestrians, cyclists and disabled people
   identifying current needs of, 171–2
   identifying priorities for, 173–4

   see also Elderly and disabled persons
Pedestrianisation in town centres, 479–80
Pelican crossings, 477–8
Perception-reaction time, 325
Photo-electric beams for traffic surveys,
   237–8
PICADY priority intersection program, 405
Piezo-electric cables used to collect data,
   237, 239
Planning Balance Sheet, 93
Pneumatic tubes used to collect data, 237,
   340
Policy, transport, *see* Transport policy
Population
   changes in town, 11
   of England and Wales (1800 and 1900),
     5
   of Great Britain, 10
Price elasticity of demand for public
   transport services, 204
Primary distributor roads, 141–2
Primary routes, 141
Priority for buses
   at intersections, 458
   bus lanes, 456–8
   bus roads/busways, 456
   measures for, 65, 144–5
Priority for cyclists
   cycle paths and routes, 481–2
   TOUCAN crossings, 481–2
Priority T-intersection
   capacity of a, 366–8
   delay at a, 368–9
   example of a layout of, 365
   types and usage of, 364, 366
Public Inquiries in Great Britain, 41
Public participation in road planning,
   39–41
Public transport
   appraisal, 99–100
   desired characteristics of a, system,
     186–7
   development of, 4–6
   examples of market segmentation in, 15
   features that attract people to, 209–10
   forecasts of numbers and traffic for bus,
     297

impact of the motor car upon, 14–16
importance of fare levels and structures
to achieve policy objectives, 71–2
target for passenger-kilometres via, 52
trends in usage of, 14–15
typical cost structures for, 206
user surveys, 247–8, 257
Public transport services
alternative objectives in relation to
subsidised, 205–7
appropriate modes for different
circumstances, 182–3, 201–2
market and planned approaches to the
provision of, 210–11
practical considerations in planning,
209–10, 212
rules for successful commercial, 202–5
socially optimal pricing and service
levels in, 207–9
Puffin crossings, 478

Questionnaire surveys, 256
Queue analysis method of measuring delay,
243
Queuing at traffic signals, 500–3

Radio-data systems (RDS), 527–8
Radio paging, 528
Rail transport
administration, 22–3
applications in relation to transport
policy objectives, 58–9
conventional, 196–7
factors favouring usage of, 143–4
initiation of, 5–6
market segmentation for, 15
privatisation of, 23
trends in usage of, 7, 9, 14, 16
Railheading, 165
Raised pavement markers, *see* Road
studs
Ramp metering, 146
Rationing road space, 63–4, 151
Reassigned traffic, 30
Redistributed traffic, 30
Regression analysis models, 114–17
Residents parking at the kerb, 464

Restriction of turning movements on
streets, 451–2
Reverse-flow traffic operation, 454–6
Rio Declaration, 18
Road hierarchy, 140–2
Road humps, 468–70
Road lighting, *see* Lighting, road
Road pricing, 70–1, 150–1
Road markings, 522–5
Road safety
campaign against speeding, effects
upon, 313
drinking and driving campaigns,
311–12
engineering and, 316–8
importance of enforcement to, 314–16
programmes for the elderly, 310–11
teaching, 310
see also Accidents *and* Accident
reduction *and* Safety audits
Road studs, 523–4
Roads
arterial, 141
development of the current system,
18–19
during the Dark and Middle Ages, 4
early, 2–3
motorway, the first in Britain, 18
ring and radial, 19
Roman, 3–4
Roadside interview surveys, 255–6
Roundabout intersections
capacities of, 374–6
delays at, 376
design features of, 372–3
types of, 370–1
usage of, 369–70
Rumble devices used in traffic calming,
472

SAEWA program for on-line micro-control
of traffic signals, 504
Safe overtaking sight distance, *see* Sight
distance
Safe stopping sight distance, *see* Sight
distance
Safety audits, 262, 353–4

Safety during traffic surveys, 256
Safety fences, 346–8
Sampling, location, in mass action accident reduction programmes, 267
Sampling strategy in transport studies, 225–8
Saturation flows at traffic signals, 379–81, 494–8
SATURN traffic simulation model, 512
SCOOT on-line traffic control system, 513
Scottish Bus Group, privatisation of, 25
Seat belts, *see* Accidents *and* Legislation
Sentry parking survey, 246
Severance
    definition of, 38
    estimating the impact of, 250
Shopping trends in, 11
Shoulders, 343–4
Side-slopes, 345–6
Siemens' software modules used for on-line traffic signal control, 515
Sight distance
    definitions of, 324–5
    design, for crest curves, 336–8
    design, for horizontal curves, 330
    design, for intersections, 362–3
    design, for motorist comfort, 338
    design, for sag curves, 338–9
    safe overtaking, requirements, 326
    safe stopping, requirements, 325–6
Simulation models, 127–8
SOCRATES communicating system, 528
Space-mean speed, 239–40, 275
Speed, factors affecting choice of, 314
Speed cushions, *see* Road humps
Speed limits, 313–4, 450–1
Speed surveys
    journey, 241–4
    spot, 239–41
Speed tables, *see* Road humps
Spot speeds, 239–41, 275
Stated Preference (SP) analysis, 258
Station spacing for LRT operation, 192–3
Stationery observer method of measuring journey speed, travel time and delay, 242–3
Stop-line surveys, 256

Structure plans, 26
Subsidised public transport services
    benefits of, for London Transport, 208
    objectives for, 205–6, 213
Superelevation design, 329–30
Surveys, observational
    cyclist, 171–2, 247
    environmental impact, 248–50
    inventory and condition, 232–4
    journey speed, travel time and delay, 241–4
    origin-destination cordon and screen-line, 244–5
    parking use and duration, 245–7
    pedestrian, 171–2, 247
    public transport usage, 247–8
    spot speed, 239–41
    vehicle flow, 234–8
    vehicle weight, 238–9
Surveys, participatory
    attitudinal, 257–9
    household activity travel simulator (HATS), 244–5
    household interview, 252–3
    Kontiv, 254
    license plate, 257
    public transport user, 257
    questionnaire, 256
    roadside interview, 255
    telephone, 254
    trip end, 255
Survey planning, design and conduct, 222–31

Telecommuting, 68, 149
Teleconferencing, 68, 149
Telephone surveys, 254
Teleshopping, 149
Tidal flow operation of roads, 454–6
Time-distance diagram, 507
Time-mean speed, 240–1, 275
Time series
    models, 111–14
    surveys, 230
TOUCAN cyclist crossings, 481–2
Track-based transport systems, 191–6
Traffic calming, 63, 465–72